FLUID MECHANICS

MERLE C. POTTER
and
JOHN F. FOSS
BOTH OF MICHIGAN STATE UNIVERSITY

GREAT LAKES PRESS, INC.
Box 483
Okemos, MI 48864

Copyright © 1982 by Merle C. Potter and John F. Foss
ISBN 0-9614760-8-7

Preface

Fluid mechanics is an established discipline composed of a rich blend of theoretical developments and experimental observations. It is also one of the essential subjects of the engineering sciences and the information it contains provides an important ingredient in the solution of many problems which are faced by the various engineering disciplines. The motivation for, and the structure of, this text are predicated upon these two observations.

Each element of this subject has been introduced so that the student's learning will be continuous from the elementary to the more advanced levels. This characteristic of the text is particularly evident in such chapters as Laminar Flows, Turbulent Flows, and Boundary Layer Flows. The application of fluid mechanics principles to the solution of engineering problems represents an equally important aspect of the text.

The structure of this text was stimulated and sustained by our belief that the particular blend of theoretical developments, descriptions of important phenomena, and the creation of examples and homework problems would justify the effort to produce it. In addition to the integrated whole of this writing, there are several contributions to the subject of fluid mechanics that make this text unique. The definition of a laminar flow as one which moves in laminae or layers has been replaced with a discussion of the characteristics of a laminar flow; a definition is not attempted since examples can be found to make an attempted definition, such as the one above, dysfunctional. (The viscous sublayer under a turbulent flow is an example.) A second benefit of this approach is realized when the motion of a turbulent flow is treated in a similar manner. Two unique discussions are found in the presentation of similarity. The discussion of Reynolds-number-independence for high Reynolds number flows should prove to be extremely useful in its application to many engineering problems; however, the replacement of "dimensional analysis" with dimensionless variables" may prove to be unfamiliar to the instructor. It should prove, though, to be more natural to the student since it follows directly from the Navier-

Stokes equations without reference to the Buckingham Pi-Theorem. Since there is no loss of technical capability and since this is a more streamlined approach, we trust that it will be a welcome addition. Another contribution is the complete development of the deformable control volume equations. The presentation parallels that of the nondeformable derivation and demands no additional mathematics. Another unique feature is the relatively thorough presentation of vorticity and circulation included in Chapter 5. This material, which could be omitted in an introductory course, is of considerable importance in the development of more sophisticated descriptions of fluid mechanics phenomena.

Our consistent objective is to establish the student's ability to execute a deductive problem solving approach to a sufficiently well posed problem. It is often true that the most difficult, if not the most important, part of the engineering task is to "define the problem"; maturity of judgment and experience with previous attempts often contribute substantially to this inductive process. It is the intent of this text to implicitly cooperate with the instructor who can articulate and develop this subjective capability within his students. However, once a problem has been identified, it is inefficient and potentially inaccurate to proceed inductively. It is better to proceed deductively from an established analytical structure. The control volume equations are valid representations of the fundamental laws; it is only necessary to select the appropriate equations and control volume to solve the problem. The motion of any Newtonian, isotropic fluid (and there are many) is "described" by the Navier-Stokes equations; one would be wise to start with these equations if they were of use in the problem solving task. It is our belief, and the basis for the structure of our text, that the most effective procedure to develop the student's capability to utilize these general formulations is to derive them from first principles and then to continually use, and hence reiterate, them in their application to various problems.

Consequently, the generalized control volume formulation is developed in Chapter 2 following introductory and elementary notions in Chapter 1. This general structure, which can be covered to include the deformable control volume, unsteady flow analyses and non-inertial coordinate systems, is used as the basis for the description of the conservation laws. Similarly, the Navier-Stokes equations are developed in Chapter 3 and utilized in: Chapter 4 to develop the basic elements of similitude, Chapter 5 to describe various types of flow phenomena, Chapter 6 to develop exact solutions for viscous flows, Chapter 7 to identify the phenomena of and equations for turbulent flows, Chapter 8 to develop the analytical descriptions of inviscid flows, Chapter 9 to rationalize the presence of a boundary layer from the nature of the governing equations in addition to the physical arguments for its existence, and Chapter 10 which presents the phenomenon of separation and other effects in external flows.

This text was developed for utilization in an undergraduate program in Engineering. A course in dynamics which uses vector algebra and a course in differential equations satisfactorally prepares the student for the fluid mechanics presentation herein. Since all topics are introduced without assuming prior work in fluid mechanics the text is appropriate for the introductory course in this subject. However, the constraints of time or appropriate depth of concentration may disallow a complete coverage of the more advanced topics which are included with the basic material. Consequently, such topics as the deformable control volume, non-inertial reference frames, vorticity and circulation, turbulence fundamentals and so forth may be omitted at the discretion of the instructor. At the minimum, these extended topics will provide reference material for the student (in a familiar notation and style) should he be faced with problems requiring such information after he has completed the basic course of instruction.

The basic needs in the area of fluid mechanics are the same for all fields of technology. Thus, every effort has been made to present the subject from a generalized point of view. The approach is to stress the fundamental relations in their most broad terms and then treat particular flow problems as special cases. For practical use the text may be divided into two parts. The first four chapters cover the basic elements of fluid mechanics while Chapters 5 through 11 represent their practical application to flow situations. Each subject is self-contained so that topics from the latter chapters may be chosen if desired on a selective basis.

Examples have been used liberally throughout the text. Some illustrate a simple notion, others are more complex to demonstrate the magnitude of a problem which is susceptible to deductive analysis. Included at the end of many of the examples are "extensions to the example." These extensions are intended to be part of the student's homework assignments. We have attempted to construct these in such a manner that their successful completion requires an understanding of the essential point of the example. The homework problems then provide a different physical situation requiring the application of basic theoretical concepts. Emphasis has been placed throughout on physical understanding to make the student aware of the wide variety of fluid flow phenomena.

There are many people who have helped in the preparation of this text to whom we express our appreciation. Drs. M. C. Smith, R. W. Zeren, M. C. Karnitz, and S. J. Kleis taught from and suggested improvements for the class notes while the text was being developed. The notes were typed by the Mechanical Engineering secretaries including Ms. Kathy

Winne, Ms. Barbara Coolen, Ms. Gloria Mannino, Ms. Jan Furtaw, Ms. Alberta Richburg, and Ms. Constance Geraci. We are also indebted to the numerous students who used the text while it was only available in the form of class notes and especially to those who shared their constructive criticisms of its content; Mr. Cha deserves special thanks in this regard.

MERLE C. POTTER
JOHN F. FOSS

East Lansing, Michigan
May, 1975

Contents

1 INTRODUCTION 3

- 1.1 Fluid Properties — 4
- 1.2 Dimensions and Units — 14
- 1.3 Mathematical Considerations — 16
- 1.4 Description of Fluid Motion — 25
- 1.5 Fluid Statics — 36
- 1.6 Bernoulli Equation — 49
- 1.7 Incompressible Gas Flow — 55

2 THE CONTROL-VOLUME FORM OF THE FUNDAMENTAL LAWS 70

- 2.1 Introduction — 70
- 2.2 The Control Volume — 72
- 2.3 Further Aspects of the Control Volume — 82
- 2.4 The Fundamental Laws — 87
- 2.5 Closure — 132

3 THE FUNDAMENTAL LAWS—DIFFERENTIAL FORM 166

- 3.1 Introduction — 166
- 3.2 The Differential Equations — 168
- 3.3 Velocity Gradients — 179
- 3.4 The Constitutive Equations — 185
- 3.5 The Governing Differential Equations — 188
- 3.6 Boundary Conditions — 194

4 SIMILITUDE 205

    4.1 Introduction 205
    4.2 Similitude 206
    4.3 Integral Quantities 213
    4.4 Parameters for More Complex Flows 214
    4.5 Compressible Flow 220
    4.6 Dimensionless Variables 223
    4.7 Closure 230

5 FLUID-FLOW PHENOMENA 237

    5.1 Introduction 237
    5.2 Phenomena Associated with Reynolds Number 238
    5.3 Boundary-Layer Flow, Inviscid Flow, and Separation 245
    5.4 Vorticity and Circulation 250
    5.5 Compressible-Flow Phenomena 263
    5.6 Cavitation 265

6 LAMINAR FLOWS 270

    6.1 Introduction 270
    6.2 Laminar Flow in a Pipe 274
    6.3 Laminar Flow Between Parallel Plates 282
    6.4 Laminar Flow Between Rotating Cylinders 286
    6.5 Sinusoidal Oscillations of a Flat Plate 289
    6.6 General Observations on Laminar Flows 293
    6.7 Laminar-Flow Instability 295

7 TURBULENCE 306

    7.1 Introduction 306
    7.2 The Describing Equations 307
    7.3 Approaches to the General Turbulence Problem 315
    7.4 Turbulent Flow in a Circular Pipe 331
    7.5 Closure 345

8 INVISCID FLOWS 352

    8.1 Introduction 352
    8.2 The Euler-s and Euler-n Equations 356

8.3	Equations of Potential Flow	360
8.4	Some Simple Plane Potential Flows	367
8.5	Superposition	375
8.6	Axisymmetric Flows	386
8.7	Rotational Inviscid Flows	390
8.8	Hele-Shaw Flow	392

9 BOUNDARY-LAYER FLOWS   400

9.1	Introduction	400
9.2	The Laminar-Boundary-Layer Equations	405
9.3	Laminar-Boundary-Layer Flow over a Flat Plate	411
9.4	Control-Volume Equations for the Boundary Layer	415
9.5	Turbulent-Boundary-Layer Flow over a Flat Plate	422
9.6	Power-Law Form of the Turbulent Boundary Layer	428
9.7	Additional Considerations Concerning the Turbulent Boundary Layer	434
9.8	Effects of a Pressure Gradient	436
9.9	Closure	438

10 EXTERNAL FLOWS   445

10.1	Introduction	445
10.2	Some Phenomena Associated with External Flows	446
10.3	Separation	450
10.4	Lift and Drag	454
10.5	Steady-Flow Drag Characteristics for Immersed Blunt Objects	460
10.6	Flow Characteristics for Airfoils	464
10.7	Effect of Compressibility on Drag	468
10.8	Effect of Cavitation on Drag	468
10.9	Added Mass	470

11 COMPRESSIBLE FLOWS   474

11.1	Introduction	474
11.2	Compressibility and the Speed of Sound	477
11.3	The Normal Shock Wave	481
11.4	Nozzle Flow	490
11.5	Steady, Uniform Compressible Flow in a Constant-Area Conduit	505

CONTENTS

11.6	The Oblique Shock Wave	518
11.7	Isentropic Turning by Compression and Expansion Waves	523
11.8	Closure	528

APPENDIX

A	Fluid Properties	539
B	Dimensions, Units, Conversions	542
C	Vector Relationships	545
D	Compressible-Flow Tables for Air	547

ANSWERS TO SELECTED EVEN-NUMBERED PROBLEMS   577

INDEX   581

FLUID MECHANICS

1

Introduction

The study of fluid mechanics plays a vital role in many fields of engineering. The aeronautical engineer is concerned with flow around bodies, to provide necessary lift and aerodynamic control with minimum drag penalty. The chemical engineer deals with fluid-flow phenomena as they relate to and influence the mass- and energy-transfer processes which underlie the chemical processing industry. Civil engineers are concerned with hydrodynamic transport of river sediment, with flood plain and erosion control, and with the hydraulic effects associated with irrigation, dams, piping systems, and sewage transport and treatment. Marine and ocean engineering and atmospheric meteorology also rely on fluid-mechanical principles. For the mechanical engineer, to design the turbine's stator and rotor so as to attain the proper flow pattern over these elements is a governing factor in the thermodynamic efficiency of the machine. Similarly, the solution of a heat-transfer problem, such as the cooling of an internal combustion engine, requires the understanding of the internal flow through engine core passages and the external flow over the radiator fin elements. To these examples should be added current developments in analysis of blood circulation, aerodynamic acoustics, and pollution control equipment. These extended contributions are made possible by the thorough grounding in fluid mechanics which has characterized modern engineering education. To provide such a grounding is a principal motive for this book.

A second objective is to provide this grounding in such a way that the student may develop and strengthen his capacity to carry out engineering analyses of problems as he might encounter them in professional practice. Meeting this objective is not a simple task for either author or student but it is essential that it be achieved. The introduction of the scientific method into engineering education has properly resulted in abandoning the once-common procedure of "training" the student to solve certain types of problems or to operate certain pieces of equipment. The engineering sciences have sought to provide the understanding necessary to allow the student to approach any problem on the basis of fundamental principles. But the scientific disciplines are concerned with the operational description of nature, in effect, the search for natural "truths"; and this emphasis may seem so far to overshadow the need for problem-solving relevance at the heart of engineering education as to rule that objective out completely. Yet such an outcome would be a distortion: properly understood, there need be no incompatibility between the two objectives. Take, for example, the use of the continuum assumption. In classical thermodynamics and in fluid mechanics, matter is considered to be spread continuously through space and endowed with various properties, such as pressure and entropy. This conjecture does not aim to ignore modern physics and its atoms and molecules; it is an expedient to generate a problem-solving capability.

This text will deal primarily with the development of general formulations which allow problems to be attacked by the deductive method, going from a general statement to a specific problem. The development of these general statements of the conservation of mass, momentum, and energy and of the numerous simplifications which follow from them provides the fundamental grounding; the examples and homework problems throughout the text are to provide the experiences from which the student may develop his ability to apply these principles to new situations.

1.1 FLUID PROPERTIES

1. Definition of Fluid

The substances referred to as fluids include both liquids and gases.*
We recall the definition of a liquid:

*Data on fluid properties are presented in the Appendix.

A state of matter in which the molecules are relatively free to change their positions with respect to each other but restricted by cohesive forces so as to maintain a relatively fixed volume.*

A gas is defined as follows:

A state of matter in which the molecules are practically unrestricted by cohesive forces. A gas has neither definite shape nor volume.†

These attributes differ from those of a solid, in which molecules tend to retain a definite fixed position. The fluids to be considered in this book are restricted to *those liquids and gases which move under the action of a shearing stress, no matter how small that shearing stress may be*. All gases fall naturally into this category. However, certain substances, such as plastics and catsup, can resist, without motion, small shearing stresses and in that condition appear to be solids. Then, at larger shearing stresses, the substances begin to flow and appear to be liquids. The explanation for such anomalies is provided by rheology.

We will further restrict our attention to *Newtonian fluids*, in which the shearing stress and the velocity field are linearly related; that is, if all the velocities in the flow are doubled then the shearing stresses are also doubled. This will be more obvious when the mathematical model for a Newtonian fluid is presented in Art. 3.4. Non-Newtonian fluids, which do not possess this linear character, are dealt with in the study of rheology.

To better understand the definition of a liquid which moves under the action of a shearing stress, consider a container filled with an unknown substance. If the container is set at an angle, as shown in Fig. 1.1, a shear force exists on each surface particle, owing to the body force due to gravity. If the substance were a liquid it would flow out of the container because of the action of these shearing forces, and would continue to flow until all shearing stresses vanished. The shear forces are zero when the surface is normal to the gravity vector, (Fig. 1.1b).

2. The Fluid as a Continuum

In fluid mechanics we postulate that the fluid is a *continuum*; that is, it is continuously distributed in the region of interest. Similarly, we regard all of the properties as continuous functions of position. As

---

Handbook of Chemistry and Physics, 40th ed., Chemical Rubber Publishing Co., p. 3104.
†*Ibid.*, p. 3095.

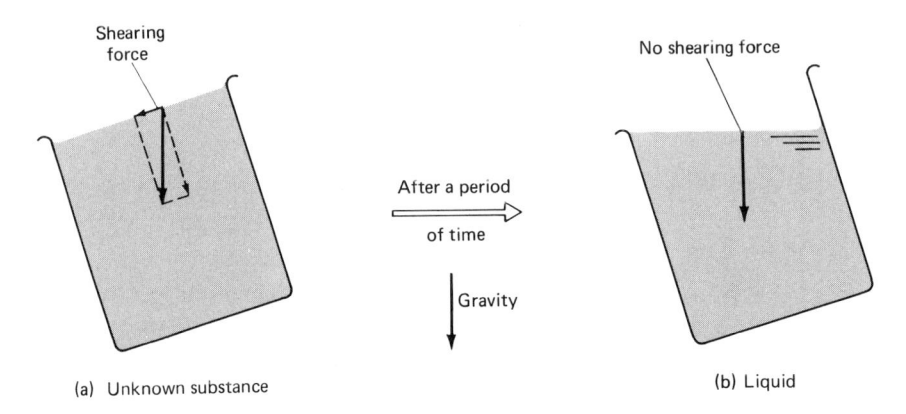

Fig. 1.1. Test for a liquid.

an example, consider the definition of density, ρ, given by

$$\rho = \lim_{\Delta V \to 0} \frac{\Delta M}{\Delta V} \tag{1.1}$$

where ΔM is the mass contained in ΔV.* Physically, ΔV cannot be allowed to shrink to zero since, if ΔV became extremely small, ΔM would vary discontinuously, depending on the number of molecules in ΔV. So the zero in the definition of ρ should be replaced by some quantity ϵ, small, but large enough to eliminate molecular effects. (Recall that there are about 2.7×10^{19} molecules in a cubic centimeter of air at standard conditions.) In some problems, such as those involving the reentry of a spacecraft into the upper atmosphere, the continuum assumption cannot be invoked because ϵ is too large to approximate zero. That is, the ϵ required such that ρ is well defined is an appreciable fraction of the spacecraft volume. However, the same elevation may allow ρ to be defined for meteorological studies involving large masses of air. A method often used to determine whether the continuum model is applicable is to compare the typical dimensions in the problem of interest to the mean free path of the molecules: If the dimensions in the problem are much (say, 1000 times) greater than the mean free path, the continuum model is acceptable. The mean free path for air at standard conditions is approximately 2.5×10^{-6} in., at 60 miles' elevation it is 4 in., and at 100 miles' it is 2000 in. At large elevations the subject of rarefied gas dynamics is introduced to deal

---

*The symbol V will stand for "volume." The vector quantity "velocity" will be represented by **V**, in heavy type, and its counterpart magnitude simply by V.

3. Thermodynamic Relationships

Thermodynamics enters the study of fluid mechanics for two rather general reasons. First, we are often concerned with matters such as the power, heat transfer, and kinetic energy associated with fluid flow, so that we must employ some form of the first law of thermodynamics. Secondly, when we describe the motion of a compressible gas we will find that thermodynamic as well as mechanical properties influence its behavior and hence must be included in our mathematical description. The thermodynamic considerations given here are meant to remind the student of what he has previously learned; alternatively, for the reader who has not had a formal exposure to the subject of thermodynamics, these notions may be viewed as self-contained facts (similar to definitions). There are available many excellent thermodynamics textbooks in which to pursue these concepts further.

A *thermodynamic system* is a fixed, identifiable quantity of mass that may interact with its *surroundings*, the remainder of the universe. A *property* is any attribute associated with the state of a system. Thermodynamic properties, such as temperature and pressure, determine the state of a system. Properties such as velocity and stress are *mechanical* properties. *Intensive* properties, such as temperature, are independent of the mass of the system; whereas *extensive* properties such as internal energy* $\tilde{U}$ depend upon the quantity of the mass which is described by the property. The term *specific* distinguishes the intensive property from the corresponding extensive property: for example, specific internal energy $\tilde{u} = \tilde{U}/M$.

Gibbs' phase rule indicates the number of independent intensive properties required to specify the state of a thermodynamic system in equilibrium; it may be written as

$$n_v = n_c - n_p + 2 \qquad (1.2)$$

where n_v is the number of independent variables, n_c is the number of components in the mixture, and n_p is the number of phases. From this it is seen that any two properties fix the state of a pure-substance

---

*In our usage, the wavy line (called a *tilde*) above an algebraic symbol merely distinguishes it from the same letter symbol without the tilde. For example, $\tilde{U}$ here means "internal energy," as noted. Later, the same letter without the tilde, U, will stand for the x-component of the velocity.

(single-component), single-phase system. Hence, any property may be expressed as a function of any other two properties for such a system.

The *perfect-gas law*

$$p = \rho RT \tag{1.3}$$

is a particularly simple relationship among three properties, pressure p, density ρ, and temperature T, which is valid for gases at sufficiently high temperatures and low pressures. R is the gas constant.*

Enthalpy is a property which finds considerable use in problems involving flowing fluids; it is defined as

$$h = \tilde{u} + p/\rho \tag{1.4}$$

and its introduction allows the *specific heat at constant pressure* to be defined as

$$c_p = \frac{\partial h}{\partial T}\bigg|_{\substack{\text{constant}\\\text{pressure}}} \tag{1.5}$$

Specific heat at constant volume is

$$c_v = \frac{\partial \tilde{u}}{\partial T}\bigg|_{\substack{\text{constant}\\\text{volume}}} \tag{1.6}$$

For a perfect gas, $\tilde{u} = \tilde{u}(T)$ and $h = h(T)$; hence $\tilde{u} = \tilde{u}_0 + c_v T$ and $h = h_0 + c_p T$, where $\tilde{u}_0$ and h_0 are arbitrary constants.

The ratio of specific heats will be designated by

$$\gamma = c_p/c_v \tag{1.7}$$

The foregoing relationships may be combined to relate c_p, c_v, and R:

$$c_p = c_v + R \tag{1.8}$$

The first and second laws of thermodynamics describe the interaction of a system with its surroundings. There are several possible constructions which may be followed in the presentation of the first law, with the primary differences expressed in terms of how "heat" and "energy" are defined. One reasonable procedure, followed here, is to define *heat*, Q, as "energy in transition across a system boundary associated only with temperature differences." From the continuum viewpoint of thermodynamics, internal energy may be considered to be

*For air, R is 1716 ft-lb$_f$/slug-°R or 53.3 ft-lb$_f$/lb$_m$-°R ("f" = force; "m" = mass).

a primitive concept; that is, it is not defined in terms of other quantities. Internal energy may be related to the energy states of the molecular constituents of the material, providing a bridge between thermodynamics and the more definite descriptions of quantum mechanics. Other forms of energy of importance in fluid mechanics are *kinetic energy* $MV^2/2$, and *potential energy* Mgz, where a reference frame from which to measure V and a reference elevation from which to measure z are implied. *Work* is a concept carried over directly from mechanics, where it is defined as $W = \int \mathbf{F} \cdot d\mathbf{s}$. The force $\mathbf{F}$ acts on the object (system), which moves along the path s; the value of the integral is the work of the surroundings on the object (system) to accomplish the motion. Thermodynamics generalizes the notion of work to be any interaction between a system and its surroundings whose *sole* effect could be the raising of a weight in a gravitational field. The most common example is the transmission of electrical energy across a system boundary; if passed through a (frictionless) motor, it could be used to raise a weight. The first law of thermodynamics is stated as

$$Q_{1-2} - W_{1-2} = E_2 - E_1 \qquad (1.9)$$

where "1" and "2" are states of the system before and after the work, respectively, Q_{1-2} is the net thermal-energy exchange with the surroundings, that is, heat transferred, and W_{1-2} is the net work necessary to alter the state of the system from 1 to 2. The energy E is the sum of all the various forms of energy; but we will restrict our attention to the kinetic, potential, and internal forms (omitting chemical, electric, magnetic, and so forth) such that

$$E = (V^2/2 + gz + \tilde{u})M \qquad (1.10)$$

where M is the mass in the system.

The second law requires the definition of a reversible process, that is, a process whose steps can be carried out in reverse in such fashion that the original state is regained and there is no alteration of the surroundings. Necessary conditions for a reversible process are that no frictional effects occur and that all heat effects occur under the action of infinitesimal temperature differences, which imply a transfer time approaching infinity for a finite amount of energy transferred. The notion of a reversible process is clearly motivated by theoretical, not practical, considerations, although some processes with negligible heat transfer do approximate the reversible condition. The second law of thermodynamics may be stated, verbally, in terms of heat engines and the necessity for such engines to reject energy to a lower-temperature reservoir.* For our purposes, a mathematical corollary,

$$\oint \frac{\delta Q}{T}\bigg|_{\text{reversible}} = 0 \qquad (1.11)$$

is of interest: the cyclic integral denotes a condition of passing through a series of states and returning the system to its initial condition; δQ is a small increment of heat (not a differential); and the heat transfer takes place reversibly. Any quantity whose cyclic integral is zero meets the necessary and sufficient conditions to be defined as a property; hence the extensive property *entropy* S is a consequence of the second law. Entropy is defined by

$$dS = \frac{\delta Q}{T}\bigg|_{\text{reversible}} \qquad (1.12)$$

Entropy as a specific property may be expressed as a function of any other two properties for a pure substance in a single phase, for example, $s = s(p, T)$. Entropy is a property significant for many thermodynamic considerations; it plays an important role in the fluid mechanics of compressible flows and acoustics. Given the property of entropy, we proceed to manipulate various equations to create useful relationships.

The definition of entropy (Eq. 1.12) and the first law in differential form for a reversible process may be consolidated to

$$\delta Q - \delta W = dE \qquad (1.13)$$

(where $\delta W|_{\text{reversible}} = p\, dV$) or

$$T\, ds = p\, d\left(\frac{1}{\rho}\right) + d\tilde{u} \qquad (1.14)$$

In Eq. 1.13, $dE = M\, d\tilde{u}$, omitting potential- and kinetic-energy changes, and $dV = M/\rho$ (M is the small quantity of mass contained in dV). The significant characteristic of Eq. 1.14 is that it is the first law of thermodynamics only for a reversible process *but* it is a valid relationship among properties for any process. Consequently, we can use Eq. 1.14 to define relationships among properties for any type of process. Making use of Eqs. 1.6 and 1.3, we can rewrite Eq. 1.14, for a

---

*The Kelvin-Planck statement of the second law may be put as follows: It is impossible for any device to operate in a cycle and produce work while exchanging heat only with bodies at a single fixed temperature.

perfect gas, as

$$T\,ds = c_v\,dT + p\,d\left(\frac{1}{\rho}\right) \quad (1.15)$$

or

$$ds = c_v\frac{dT}{T} - R\frac{d\rho}{\rho} \quad (1.16)$$

and, integrating from state 1 to state 2, we obtain (using Eq. 1.8)

$$s_2 - s_1 = c_v\,\ln\frac{T_2}{T_1} + R\,\ln\frac{\rho_1}{\rho_2}$$

$$= c_v\,\ln\left[\frac{T_2}{T_1}\left(\frac{\rho_1}{\rho_2}\right)^{\gamma-1}\right] \quad (1.17)$$

Alternate forms can be developed by making use of the perfect-gas relationship $p = \rho R T$:

$$s_2 - s_1 = c_v\,\ln\left[\frac{p_2}{p_1}\left(\frac{\rho_1}{\rho_2}\right)^{\gamma}\right] \quad (1.18)$$

and

$$s_2 - s_1 = c_v\,\ln\left[\left(\frac{T_2}{T_1}\right)^{\gamma}\left(\frac{p_1}{p_2}\right)^{\gamma-1}\right] \quad (1.19)$$

From these two expressions may be developed the isentropic-process relationships, which are of considerable importance in compressible flow. For a frictionless, adiabatic ($Q = 0$) process, $s_2 = s_1$, and

$$\frac{T}{\rho^{\gamma-1}} = \text{const} \quad (1.20)$$

$$\frac{p}{\rho^{\gamma}} = \text{const} \quad (1.21)$$

$$Tp^{\frac{1-\gamma}{\gamma}} = \text{const} \quad (1.22)$$

An isentropic process of considerable importance is the propagation of a small disturbance (acoustic) wave. The speed with which a wave traverses a medium is given by a, where

$$a = \sqrt{\left.\frac{\partial p}{\partial \rho}\right|_s} \qquad (1.23)$$

and by making use of the perfect-gas law and the isentropic relationship, the result

$$a = \sqrt{\gamma RT} \qquad (1.24)$$

can be easily obtained. (The derivation of Eq. 1.23 is given in Art. 11.2.)

4. Compressibility

Compressibility is a property which describes the material's resistance to a change in volume for a fixed mass. Its reciprocal, termed the *bulk modulus of elasticity*, is more often used:

$$B = \rho \frac{\partial p}{\partial \rho} \qquad (1.25)$$

The condition under which the process takes place—whether isothermal, isentropic, or otherwise—must of course be designated. For liquids, these distinctions are not particularly important. The bulk modulus for water is approximately 312,000 psi at standard conditions; note that this value indicates that water is approximately 100 times more compressible than steel.

5. Viscosity

A property which is quite important in describing the response of a fluid to an imposed shear stress is the *viscosity* μ. Viscosity may be thought of as the "internal stickiness" of a fluid. For a fluid flow the shearing stress is proportional to the rate of shearing strain, the constant of proportionality being the viscosity. We will be interested only in fluids that exhibit a linear relationship between the stress and the rate of strain. The equations which relate the shearing stresses to the rate of strain (the velocity field) are known as *constitutive equations*. Viscosity, stresses, rates of strain, and constitutive equations will be discussed in Chapter 3 when the equations which describe the velocity at each point in the flow are derived.

An important consequence of the property of viscosity is that, at the interface between a fluid and a solid, the velocity of the fluid is equal to that of the solid surface. This is termed the *no-slip* condition; it is considered in detail in Chapter 3.

It is sometimes more convenient to work with the *kinematic viscosity* ν, which is defined as

$$\nu = \mu/\rho \qquad (1.26)$$

6. Surface Tension

For liquids, another important property is *surface tension* σ. A free surface, separating a liquid from a gas, and an interface, separating two different liquids, involve many phenomena associated with surface tension, such as bubbles, capillary effects, and surface ripples. A free surface or an interface behaves like a membrane; its energy, in foot-pounds per unit area, is given by the surface tension in units of pounds/foot. The magnitude of the surface tension is strongly dependent upon the molecular state of the materials involved: for example, a detergent added to tap water can effect a large change in the surface-tension magnitude. For the interface, surface tension varies for the two liquids as well; for the free surface, it is dependent on the liquid but nearly independent of the gas. It is only slightly temperature-dependent. Additional details, along the lines of this continuum-mechanics description of a complicated molecular phenomenon, are given in Art. 1.6, on fluid statics.

7. Vapor Pressure

The last property to be reviewed is *vapor pressure*, the pressure at which a liquid is in equilibrium with its vapor; that is, if the vapor pressure is raised to 14 psi and the local pressure is 14 psi then the liquid and vapor will be in equilibrium with each other; we usually call this *boiling*. However, if a confined liquid is flowing, through a turbine, for example, and the local pressure is lowered to the vapor pressure, the result is *cavitation*, generally an undesirable phenomenon. Vapor pressure is quite sensitive to temperature, as we know from our experience with water.

Example 1.1

A hot-wire anemometer probe makes use of a fine tungsten wire; a diameter of 10^{-4} in. is not uncommon and allows rapid response as a result of a small thermal inertia. In analyzing the flow about the wire, it occurs to the engineer that the continuum approximation may not be valid. Determine the size of a particle containing 300 molecules and compare it to the diameter of the wire.

Solution. There are 2.7×10^{19} molecules per cubic centimeter; hence, the side of a cube containing 300 molecules is

$$\delta = [3 \times 10^2 / 2.7 \times 10^{19}]^{1/3} = 2.2 \times 10^{-6} \text{ cm}$$

and

$$d/\delta = 2.54 \times 10^{-4} / 2.2 \times 10^{-6} = 115$$

These conditions would appear to be quite adequate to consider the flow over the wire to be modeled as a continuum.

Extension 1.1.1. Particulate pollutants are often found which are approximately 10 microns in diameter (1 $\mu = 10^{-6}$ m). How many molecules of standard air are there in a particle which has a diameter of 0.01 times such a particle's diameter? *Ans.* 14,100

Example 1.2

Calculate the isentropic bulk modulus of elasticity for air at standard conditions and compare the value with that quoted for water.

Solution. For an isentropic process $p = \text{const}\, \rho^\gamma$. Hence

$$B = \rho \left.\frac{\partial p}{\partial \rho}\right|_s = \rho \, \text{const}\, \gamma \rho^{\gamma-1} = \rho \frac{p}{\rho^\gamma} \gamma \rho^{\gamma-1} = \gamma p$$

For air at standard conditions, this value is 20.6 psi, that is, a factor of 15,000 smaller than that for water.

Extension 1.2.1. Determine the isothermal bulk modulus of elasticity for air at standard conditions and explain how you would compress something isentropically and isothermally. *Ans.* 14.7 psi

1.2 DIMENSIONS AND UNITS

Four dimensions will primarily be used in this text: length (L), time (t), mass (M), and force (F). These four quantities are considered to be primary or basic in the sense that they represent four distinct physical effects. Because it is possible to independently select the unit of measure for each dimension and because the four quantities are related by Newton's second law, it is necessary to introduce a *dimensional constant* g_c in the following manner:

$$F = \frac{Ma}{g_c} \tag{1.27}$$

For the system of units most common in American engineering usage, the value of g_c is determined experimentally by observing that 1 pound-mass (lb_m) undergoes an acceleration of 32.2 ft/sec² when acted upon by 1 pound-force (lb_f). Therefore, g_c is in these units determined from Eq. 1.27 to be

$$g_c = 32.2 \frac{\text{ft}}{\text{sec}^2} \frac{lb_m}{lb_f} \qquad (1.28)$$

Another unit of mass, termed the *slug*, is defined as 32.2 lb_m. From Eq. 1.27, we then see that 1 slug accelerates at 1 ft/sec² when acted upon by 1 pound-force. Using M as slugs we can then determine

$$g_c = 1 \frac{\text{ft}}{\text{sec}^2} \frac{\text{slug}}{lb_f} \qquad (1.29)$$

Remember that g_c is a dimensional constant; it is not to be confused with the acceleration due to gravity (g).

An alternate set of definitions may be adopted. If three rather than four dimensions are accepted as basic, then we may define force in terms of mass, length, and time. In such a system the term "pound-force" is only shorthand notation for "slug-ft/sec²," and the defining relationship becomes

$$F = Ma \qquad (1.30)$$

where M is always in slugs. No dimensional constant is used.

In fluid mechanics, it is traditional to ignore the introduction of the pound as a mass unit and to circumvent the problem by always using slugs as the mass unit (since g_c is then of unit magnitude, it may be omitted from the equations). Such a policy is adopted for this text: g_c is not used in the equations, "pound" refers to a force unit, and the Newtonian relationship is written simply as $F = Ma$.

Example 1.3

Show that the quantity p/ρ has the same units as $V^2/2$ if density is expressed in slugs/ft³, p in psf, and V in fps.

Solution. The units are

$$\frac{p}{\rho} = \frac{\text{lb}}{\text{ft}^2} \bigg/ \frac{\text{slug}}{\text{ft}^3} = \frac{\text{lb}}{\text{ft}^2} \bigg/ \frac{\text{lb-sec}^2}{\text{ft}^4} = \frac{\text{ft}^2}{\text{sec}^2}$$

These units are obviously the same as the units on $V^2/2$.

Extension 1.3.1. Determine the magnitude of the sum $p/\rho + V^2/2$ when $V = 20$ fps, $p = 0.5$ psi, and $\rho = 0.076$ lb$_m$/ft^3.

Ans. 30,700 ft^2/sec^2

1.3 MATHEMATICAL CONSIDERATIONS

This article will review some of the mathematical relationships necessary in our study of fluid mechanics. Relationships will be presented without derivation or proof; consequently, it may prove worthwhile to review the material in a mathematics text. We will use these mathematical relations throughout without further comment. The course in fluid mechanics tends to utilize many of the physical phenomena presented in physics and calls on most of the mathematics required of the undergraduate. The analytical description of physical quantities such as velocity and pressure, which vary from point to point in a fluid flow, demands the use of relatively sophisticated mathematics. The use of vector calculus and partial differential equations is largely responsible for this added sophistication. A brief review of these subjects will be presented as a preparation for a mathematical description of the various laws governing fluid flow.

1. Vectors

Many of the quantities encountered in fluid mechanics are vector quantities; hence vector algebra and calculus are very useful. A vector will be represented by a boldface symbol (for example, **A**); a symbol in ordinary italic type stands for the magnitude of the vector (for example, A). We will define some vector operations, using a cartesian coordinate system.*

Recall that the *dot product*, where $\mathbf{A} = A_x\hat{i} + A_y\hat{j} + A_z\hat{k}$ ($\hat{i}, \hat{j},$ and $\hat{k}$ are the unit vectors in the x-, y-, and z-directions, respectively), is

$$\mathbf{A} \cdot \mathbf{B} = A_xB_x + A_yB_y + A_zB_z = AB\cos\theta \qquad (1.31)$$

in which θ is the angle between the two vectors and the x-, y-, and z-subscripts denote the components of the vectors in the x-, y-, and

---

*The term "cartesian coordinate system" will be used to denote a rectangular coordinate system. We will present expressions in cylindrical and spherical coordinates in the Appendix.

Art. 1.3 MATHEMATICAL CONSIDERATIONS

z-directions. The *cross product* is

$$\mathbf{A} \times \mathbf{B} = (A_y B_z - A_z B_y)\hat{i} + (A_z B_x - A_x B_z)\hat{j} + (A_x B_y - A_y B_x)\hat{k} \tag{1.32}$$

$$|\mathbf{A} \times \mathbf{B}| = AB \sin \theta$$

The *gradient* of a scalar function ϕ is denoted by $\nabla \phi$ and is given in cartesian coordinates as

$$\nabla \phi = \frac{\partial \phi}{\partial x}\hat{i} + \frac{\partial \phi}{\partial y}\hat{j} + \frac{\partial \phi}{\partial z}\hat{k} \tag{1.33}$$

where the gradient operator is

$$\nabla = \frac{\partial}{\partial x}\hat{i} + \frac{\partial}{\partial y}\hat{j} + \frac{\partial}{\partial z}\hat{k}$$

We recall that $\nabla \phi$ is normal to a constant-ϕ surface and that the rate at which ϕ changes in the n-direction is

$$\frac{\partial \phi}{\partial n} = \nabla \phi \cdot \hat{n} \tag{1.34}$$

where $\hat{n}$ is the unit vector in the n-direction. The gradient is useful in describing velocity fields, in expressing pressure variations, and in general manipulation of the vector equations.

The *divergence* of a vector function $\mathbf{A}$ is denoted by $\nabla \cdot \mathbf{A}$ and in cartesian coordinates is expressed as

$$\nabla \cdot \mathbf{A} = \frac{\partial A_x}{\partial x} + \frac{\partial A_y}{\partial y} + \frac{\partial A_z}{\partial z} \tag{1.35}$$

The divergence will be useful in expressing the conservation of mass in vector form.

The *curl* of a vector function $\mathbf{A}$ is denoted by $\nabla \times \mathbf{A}$ and, in cartesian coordinates, is expressed as

$$\nabla \times \mathbf{A} = \left(\frac{\partial A_z}{\partial y} - \frac{\partial A_y}{\partial z}\right)\hat{i} + \left(\frac{\partial A_x}{\partial z} - \frac{\partial A_z}{\partial x}\right)\hat{j} + \left(\frac{\partial A_y}{\partial x} - \frac{\partial A_x}{\partial y}\right)\hat{k} \tag{1.36}$$

The curl has basic application in determining rates of rotation of fluid particles.

The *Laplacian* of a scalar function ϕ is $\nabla \cdot \nabla \phi$ and is written as

$$\nabla^2 \phi = \frac{\partial^2 \phi}{\partial x^2} + \frac{\partial^2 \phi}{\partial y^2} + \frac{\partial^2 \phi}{\partial z^2} \tag{1.37}$$

Laplace's equation is $\nabla^2 \phi = 0$. It will be useful in solving problems involving incompressible fluids.

2. Functions of Several Variables

If ϕ is a function of x, y, z, then

$$d\phi = \frac{\partial \phi}{\partial x} dx + \frac{\partial \phi}{\partial y} dy + \frac{\partial \phi}{\partial z} dz \tag{1.38}$$

If we wish to transform coordinates from (x, y, z) to (ξ, η, ζ) we use

$$\begin{aligned}\frac{\partial \phi}{\partial x} &= \frac{\partial \phi}{\partial \xi}\frac{\partial \xi}{\partial x} + \frac{\partial \phi}{\partial \eta}\frac{\partial \eta}{\partial x} + \frac{\partial \phi}{\partial \zeta}\frac{\partial \zeta}{\partial x} \\ \frac{\partial \phi}{\partial y} &= \frac{\partial \phi}{\partial \xi}\frac{\partial \xi}{\partial y} + \frac{\partial \phi}{\partial \eta}\frac{\partial \eta}{\partial y} + \frac{\partial \phi}{\partial \zeta}\frac{\partial \zeta}{\partial y} \\ \frac{\partial \phi}{\partial z} &= \frac{\partial \phi}{\partial \xi}\frac{\partial \xi}{\partial z} + \frac{\partial \phi}{\partial \eta}\frac{\partial \eta}{\partial z} + \frac{\partial \phi}{\partial \zeta}\frac{\partial \zeta}{\partial z}\end{aligned} \tag{1.39}$$

In this context, it is important to recognize that a partial derivative is executed with the appropriate other variables held constant. For example, $\partial \phi / \partial \xi$ is evaluated at constant values of η and ζ whereas $\partial \phi / \partial x$ is evaluated at constant y and z. Higher-order derivatives follow from these relations; that is,

$$\frac{\partial^2 \phi}{\partial x^2} = \frac{\partial}{\partial x}\left(\frac{\partial \phi}{\partial x}\right)$$

and we simply substitute $\partial \phi / \partial x$ for ϕ in the first of Eqs. 1.39.

The Taylor series will also be of considerable use when we desire $\phi(x, y)$, knowing ϕ and its derivatives at a neighboring point. Taylor's series for two variables is

$$\begin{aligned}\phi(x + \Delta x, y + \Delta y) &= \phi(x, y) + \frac{\partial \phi}{\partial x}\Delta x + \frac{\partial \phi}{\partial y}\Delta y + \frac{\partial^2 \phi}{\partial x^2}\frac{(\Delta x)^2}{2} \\ &+ \frac{\partial^2 \phi}{\partial y^2}\frac{(\Delta y)^2}{2} + \frac{\partial^2 \phi}{\partial x \partial y}\Delta x\, \Delta y + \text{higher-order terms}\end{aligned} \tag{1.40}$$

where all the derivatives are evaluated at (x, y). If Δx and Δy are small, we may neglect the second-order-derivative terms and use

$$\phi(x + \Delta x, y + \Delta y) \cong \phi(x, y) + \frac{\partial \phi}{\partial x} \Delta x + \frac{\partial \phi}{\partial y} \Delta y \qquad (1.41)$$

3. Integral Transformation Theorems

Several integral theorems, theorems which enable us to transform surface integrals into volume integrals or line integrals into surface integrals, will now be presented, omitting proof in the interest of a quick review. These theorems will later be necessary for the basic laws governing the behavior of fluids. We introduce first the *theorem of Gauss*, which may take on any of the following special forms:

$$\int_V \nabla \phi \, dV = \int_S \hat{n} \phi \, dS$$

$$\int_V \nabla^2 \mathbf{A} \, dV = \int_S (\hat{n} \cdot \nabla) \mathbf{A} \, dS$$

$$\int_V \nabla \cdot \mathbf{A} \, dV = \int_S \hat{n} \cdot \mathbf{A} \, dS \qquad (1.42)$$

$$\int_V \nabla \times \mathbf{A} \, dV = \int_S \hat{n} \times \mathbf{A} \, dS$$

Here the symbol S represents a surface *completely surrounding* the volume V; $\hat{n}$ is an outward-pointing unit vector normal to the elemental area dS.

Another theorem which is often useful is *Stokes' theorem*, which may be stated as

$$\oint_L \mathbf{A} \cdot d\mathbf{l} = \int_S \hat{n} \cdot (\nabla \times \mathbf{A}) \, dS \qquad (1.43)$$

where S may be a three-dimensional surface and is bounded by the line L, $d\mathbf{l}$ is a directed line element of L, and $\hat{n}$ is normal to dS.

The circulation Γ of the vector $\mathbf{A}$ around the line L is defined as

$$\Gamma = \oint \mathbf{A} \cdot d\mathbf{l} \qquad (1.44)$$

It is of interest in determining the lift on an airfoil.

4. The Scalar Potential Function

If a vector **A** is given by the gradient of a scalar function ϕ, that is,

$$\mathbf{A} = \nabla\phi \tag{1.45}$$

then Stokes' theorem (Eq. 1.43) shows that the circulation is zero, or

$$\oint \mathbf{A} \cdot d\mathbf{l} = 0 \tag{1.46}$$

since $\nabla \times \nabla\phi = 0$. (The curl of a gradient is always zero.) The vector field **A** is called a *conservative vector field*, and the scalar function ϕ is a *scalar potential function* of the vector **A**. If the curl of **A** is not zero, then a scalar potential function does not exist. It follows from Eq. 1.46 that the line integral between any two points is independent of the path

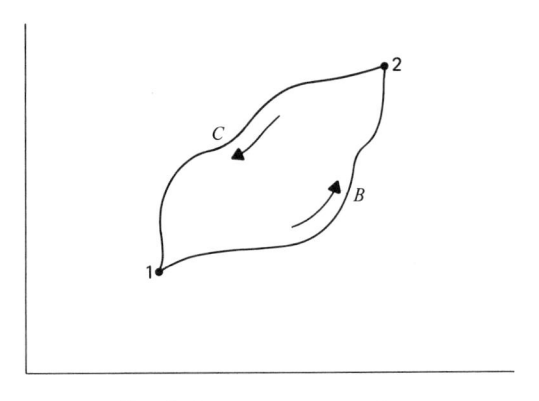

Fig. 1.2. Integration path.

between the two points, providing that the integrand is given by the gradient of a scalar field. To show this, consider the closed line integral with reference to Fig. 1.2:

Thus
$$\oint_L \mathbf{A} \cdot d\mathbf{a} = \int_1^2 \mathbf{A} \cdot d\mathbf{l} \underset{\text{along } B}{} + \int_2^1 \mathbf{A} \cdot d\mathbf{l} \underset{\text{along } C}{} = 0$$

$$\int_1^2 \mathbf{A} \cdot d\mathbf{l} \underset{\text{along } B}{} = -\int_2^1 \mathbf{A} \cdot d\mathbf{l} \underset{\text{along } C}{} = \int_1^2 \mathbf{A} \cdot d\mathbf{l} \underset{\text{along } C}{} \tag{1.47}$$

since, when the limits are exchanged, the negative sign is introduced. The path of integration is shown to be unimportant; only the end points are of interest.

Example 1.4

In Chapter 8 it will be shown that the fluid velocity at a point in some flows may be expressed as $\mathbf{V} = \nabla\phi$. Show that $\nabla\phi$ is normal to a constant-ϕ surface.

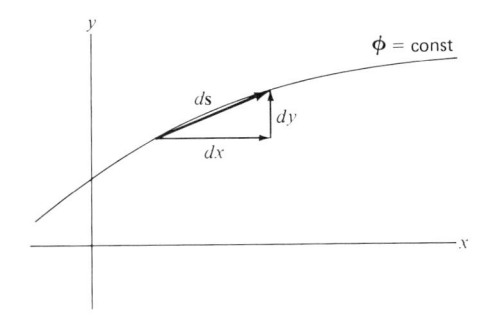

Fig. E1.4

Solution. Along a constant-ϕ surface $d\phi = 0$; hence

$$d\phi = \frac{\partial\phi}{\partial x}dx + \frac{\partial\phi}{\partial y}dy + \frac{\partial\phi}{\partial z}dz = 0$$

along this surface. Now

$$\nabla\phi = \frac{\partial\phi}{\partial x}\hat{i} + \frac{\partial\phi}{\partial y}\hat{j} + \frac{\partial\phi}{\partial z}\hat{k}$$

and an infinitesimal vector lying in the constant-ϕ surface (see Fig. E1.4) may be expressed as

$$d\mathbf{s} = dx\,\hat{i} + dy\,\hat{j} + dz\,\hat{k}$$

The dot product $\nabla\phi \cdot d\mathbf{s}$ is thus

$$\nabla\phi \cdot d\mathbf{s} = \frac{\partial\phi}{\partial x}dx + \frac{\partial\phi}{\partial y}dy + \frac{\partial\phi}{\partial z}dz$$

But this is zero, showing that $\nabla\phi \perp d\mathbf{s}$. Thus $\nabla\phi$ is normal to the constant-ϕ surface.

Example 1.5

A vector function is given by

$$\mathbf{A} = x^2\,\hat{i} + 2xy\hat{j} - z^2\hat{k}$$

Find the divergence and curl of $\mathbf{A}$ at $(1, -1, 2)$ and determine if it satisfies Laplace's equation. (If $\mathbf{A}$ is the fluid velocity at a point in the flow, then the divergence is related to the compressibility of a fluid element and the curl is related to the tendency of the element to rotate.)

Solution. The divergence is

$$\nabla \cdot \mathbf{A} = \frac{\partial A_x}{\partial x} + \frac{\partial A_y}{\partial y} + \frac{\partial A_z}{\partial z}$$

$$= 4x - 2z$$

$$= 0$$

The curl is

$$\nabla \times \mathbf{A} = \left(\frac{\partial A_z}{\partial y} - \frac{\partial A_y}{\partial z}\right)\hat{i} + \left(\frac{\partial A_x}{\partial z} - \frac{\partial A_z}{\partial x}\right)\hat{j} + \left(\frac{\partial A_y}{\partial x} - \frac{\partial A_x}{\partial y}\right)\hat{k}$$

$$= 2y\hat{k}$$

$$= -2\hat{k}$$

The Laplacian $\nabla^2 \mathbf{A}$ is

$$\nabla^2 \mathbf{A} = (\nabla \cdot \nabla)\mathbf{A} = \frac{\partial^2 \mathbf{A}}{\partial x^2} + \frac{\partial^2 \mathbf{A}}{\partial y^2} + \frac{\partial^2 \mathbf{A}}{\partial z^2} = 2\hat{i} - 2\hat{k}$$

and is not zero.

Example 1.6

We wish to express

$$\frac{\partial \phi}{\partial x} + \frac{\partial^2 \phi}{\partial x \, \partial y} + 4 \frac{\partial \phi}{\partial y}$$

in terms of ξ and η, where

$$\xi = x \quad \text{and} \quad \eta = y/x$$

This represents a transformation of coordinates from (x, y) to (ξ, η). (A transformation such as this will be used for a particular fluid flow in later chapters.)

Solution. From Eqs. 1.39,

$$\frac{\partial \phi}{\partial x} = \frac{\partial \phi}{\partial \xi}\frac{\partial \xi}{\partial x} + \frac{\partial \phi}{\partial \eta}\frac{\partial \eta}{\partial x}$$

$$= \frac{\partial \phi}{\partial \xi} - \frac{y}{x^2}\frac{\partial \phi}{\partial \eta}$$

$$= \frac{\partial \phi}{\partial \xi} - \frac{\eta}{\xi}\frac{\partial \phi}{\partial \eta}$$

where we have used $\eta/\xi = y/x^2$. We also have

$$\frac{\partial \phi}{\partial y} = \frac{\partial \phi}{\partial \xi} \overset{0}{\cancel{\frac{\partial \xi}{\partial y}}} + \frac{\partial \phi}{\partial \eta} \frac{\partial \eta}{\partial y}$$

$$= \frac{1}{x} \frac{\partial \phi}{\partial \eta}$$

$$= \frac{1}{\xi} \frac{\partial \phi}{\partial \eta}$$

Finally,

$$\frac{\partial^2 \phi}{\partial x \, \partial y} = \frac{\partial}{\partial x}\left(\frac{\partial \phi}{\partial y}\right) = \frac{\partial}{\partial \xi}\left(\frac{\partial \phi}{\partial y}\right)\frac{\partial \xi}{\partial x} + \frac{\partial}{\partial \eta}\left(\frac{\partial \phi}{\partial y}\right)\frac{\partial \eta}{\partial x}$$

$$= \frac{\partial}{\partial \xi}\left(\frac{1}{\xi}\frac{\partial \phi}{\partial \eta}\right) + \frac{\partial}{\partial \eta}\left(\frac{1}{\xi}\frac{\partial \phi}{\partial \eta}\right)\left(-\frac{y}{x^2}\right)$$

$$= \frac{1}{\xi}\frac{\partial^2 \phi}{\partial \xi \, \partial \eta} - \frac{1}{\xi^2}\frac{\partial \phi}{\partial \eta} - \frac{\eta}{\xi^2}\frac{\partial^2 \phi}{\partial \eta^2}$$

The expression is thus

$$\frac{\partial \phi}{\partial x} + \frac{\partial^2 \phi}{\partial x \, \partial y} + 4\frac{\partial \phi}{\partial y} = \frac{\partial \phi}{\partial \xi} + \frac{1}{\xi}\frac{\partial^2 \phi}{\partial \xi \, \partial \eta} + \left(\frac{4}{\xi} - \frac{\eta}{\xi} - \frac{1}{\xi^2}\right)\frac{\partial \phi}{\partial \eta} - \frac{\eta}{\xi^2}\frac{\partial^2 \phi}{\partial \eta^2}$$

Example 1.7

Determine the circulation of **A** around the triangle which has vertices at the origin (0, 4, 0) and (4, 0, 0). The vector function **A** is given by

$$\mathbf{A} = 4x^2\hat{i} + 2yz\hat{j} - (4y^2 + z^2)\hat{k}$$

Solution. The circulation is

$$\Gamma = \oint_L \mathbf{A} \cdot d\mathbf{l}$$

$$= \int_{\textcircled{1}} \mathbf{A} \cdot d\mathbf{l} + \int_{\textcircled{2}} \mathbf{A} \cdot d\mathbf{l} + \int_{\textcircled{3}} \mathbf{A} \cdot d\mathbf{l}$$

Referring to Fig. E1.7 and using $\mathbf{A} \cdot d\mathbf{l} = A_x \, dx + A_y \, dy + A_z \, dz$, we have

$$\Gamma = \int_{\textcircled{1}} A_x \, dx + \int_{\textcircled{2}} (A_x \, dx + A_y \, dy) + \int_{\textcircled{3}} A_y \, dy$$

$$= \int_0^4 4x^2 \, dx + \int_4^0 4x^2 \, dx + \int_0^0 2y\cancel{z} \, dy + \int_4^0 2y\cancel{z} \, dy$$

$$= 0$$

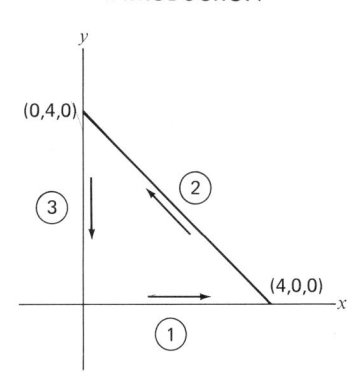

Fig. E1.7

Extension 1.7.1. Verify the result in the solution above by using the right-hand side of Eq. 1.43.

Example 1.8

Determine the scalar potential function of the vector function **A**:

$$\mathbf{A} = 2xy\hat{i} + x^2\hat{j} + 3z^2\hat{k}$$

Solution. From Eq. 1.45

$$\mathbf{A} = \nabla\phi = \frac{\partial \phi}{\partial x}\hat{i} + \frac{\partial \phi}{\partial y}\hat{j} + \frac{\partial \phi}{\partial z}\hat{k}$$

Hence

$$\frac{\partial \phi}{\partial x} = 2xy \qquad \frac{\partial \phi}{\partial y} = x^2 \qquad \frac{\partial \phi}{\partial z} = 3z^2$$

The first of these equations gives

$$\phi = x^2 y + f(y, z)$$

Substitute this in the second equation and

$$\frac{\partial \phi}{\partial y} = x^2 = x^2 + \frac{\partial f}{\partial y}$$

Thus $\partial f/\partial y = 0$ or $f = f(z)$. Substitute in the third equation and

$$\frac{\partial \phi}{\partial z} = 3z^2 = \frac{\partial f}{\partial z}$$

giving $f = z^3 + C$. We finally have

$$\phi = x^2 y + z^3 + C$$

1.4 DESCRIPTION OF FLUID MOTION

The curve-ball phenomenon is an application of fluid mechanics utilized by certain highly paid artisans of the national sport of baseball. It will also serve nicely as the basis for a discussion of the methods useful in describing the phenomena of fluid mechanics.

Consider a coordinate system affixed to the pitcher's mound; let $\mathbf{R}(t)$ be the displacement vector, which locates the ball at any time t. (This description of the motion of the ball is familiar from earlier courses.) $\mathbf{R}(t)$ defines the trajectory of the ball and consequently contains the information regarding the curved path of the ball which, if successfully executed, will mean that the final $\mathbf{R}$ position is the catcher's mitt. It is equally clear that the behavior of the ball is governed by

$$\mathbf{F} + \mathbf{W} = M \frac{d^2\mathbf{R}}{dt^2} \tag{1.48}$$

where $\mathbf{F}$ is the aerodynamic force (that is, the force of interaction between the ball and the surrounding air) which causes the horizontal curvature in the ball's path. A proper understanding of the curve-ball problem has logically led to the need to describe this aerodynamic force.

But how is this to be done? Consider, for example, the coordinate system used above to describe the trajectory of the ball, and consider a particle of air located along the path of the ball between the pitcher and catcher. Such a particle will execute a rather interesting motion: relatively stationary before the passage of the ball, in rapid motion as the ball passes, and subsequently in slower motion. This might be of interest to a micro-meteorologist, but it would be difficult to assemble and then apply such information to the curve-ball problem. Clearly, this description fails our engineering purposes because it focuses attention on an air particle and not explicitly on the air surrounding the ball. This fault is remedied for the curve ball (and indeed, by analogy, for the vast majority of fluid-dynamic analyses) by switching to a description in which the coordinate system originates at the center of the ball. The force $\mathbf{F}$ on the ball is equal to the appropriately integrated pressure and shear-stress distributions over the ball's surface. As we shall see in detail, the methods by which the pressure and shear-stress distributions are obtained from, or related to, the velocity of the surrounding fluid are an important part of fluid mechanics. Our present purpose is served by this example in that it demonstrates a principle of central importance in fluid mechanics: namely, whether for a turbine, a carburetor, or a rocket nozzle, our interest is almost invariably focused on a region in space (which may itself be moving, like the region surrounding the baseball), rather than on the history or behavior of individual particles.

1. Lagrangian and Eulerian Descriptions of Motion

In the *Lagrangian* description, individual particles are identified, and the thermodynamic properties (p, ρ, etc.) and flow properties (**R**, **V**, **a**, etc.) are functions of time only for the particle of interest. All the particles of the fluid are identified by locating them at some instant, say at time $t = 0$, and then by following each particle as time progresses. If the location of a particle at time $t = 0$ is (X_0, Y_0, Z_0) in a cartesian coordinate system, then the velocity and acceleration of the particle are simply

$$\mathbf{V} = \frac{d\mathbf{R}}{dt} \qquad \frac{d\mathbf{V}}{dt} = \frac{d^2\mathbf{R}}{dt^2} \tag{1.49}$$

where **R** is the displacement vector of the particle at time t, as shown in Fig. 1.3. Here $\mathbf{R} = \mathbf{R}(X_0, Y_0, Z_0, t)$ where, for a particular particle, X_0, Y_0, Z_0 are constants and "identify" the particle. If every particle is so located at some instant and followed for a period of time, a complete description of motion is then obtained during the period of interest.

In the introduction to this article, we pointed out that it was preferable to describe the behavior of the fluid in a given region in space (in that case, the flow field around the baseball) rather than the behavior of individual fluid particles. The former approach is identified with the *Eulerian* method of description. The thermodynamic and flow properties are, of course, associated only with particles of the fluid. The Eulerian method provides a formal basis for relating these properties to points in space. To do this, the velocity and pressure are described as $\mathbf{V}(x, y, z, t)$ and $p(x, y, z, t)$, with x, y, z, t the independent variables and **V** and p examples of dependent variables. Literally, this means that a particle occupying the point (x, y, z) at time t will have the velocity and pressure described by the functions **V** and p. Basically, there are two general benefits to be derived from the Eulerian description. Experimental data may be catalogued and analyzed with the Eulerian description; for example, a wind tunnel evaluation of a spinning sphere (curve ball) might involve velocity measurements at particular points upstream and downstream of the sphere.

Art. 1.4   DESCRIPTION OF FLUID MOTION   27

Also, the Eulerian method provides a rational framework in which an analytical solution may be sought. That is, differential equations involving functions such as $\mathbf{V}(x, y, z, t)$ and $p(x, y, z, t)$ may be formulated, boundary conditions established, and appropriate analytical or numerical techniques utilized for the solution.

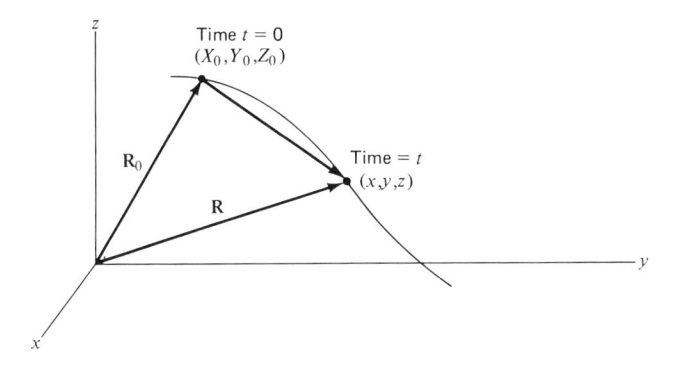

Fig. 1.3. Path of a particle.

2. Pathlines, Streaklines, Streamlines

Three lines will be defined which represent different ways to describe a fluid flow. A *pathline* is the curve that a particle traces as it moves over a time span in a flow field. The pathline would be most easily described in a Lagrangian frame of reference. It may also be thought of as a time-exposure photograph of a luminous particle.

A *streakline* is the curve composed of all particles which have originated at a specified fixed point. Photographically, a streakline would be the outcome of a snapshot of a flow field using visible markers. A streakline comprises all the particles which have passed a specified point (X_0, Y_0, Z_0) during a given time interval.

The pathline and the streakline are important because they can be observed by flow-visualization techniques in the laboratory and in nature. They are identical for *steady flow*, flow for which all fluid properties at a point are independent of time. This statement is reasonably obvious from a consideration of the definitions. Many flows which are locally unsteady in one region show steady behavior in other parts of the flow field. For example, the flow near the impeller of a centrifugal fan is clearly unsteady. However, if a duct is placed at the fan inlet then a steady flow can exist in the duct. To determine where the steady flow is established, a tracer could be injected, and the equivalence of the streaklines and pathlines would indicate a steady condition.

The third line to be considered is the *streamline*. A streamline is defined to be a curve in the flow field which has the property that the velocity vector of each fluid particle lying on the curve is tangent to the curve. There are an infinity of points and an infinity of streamlines in a flow field; a selected few are often shown (as in Fig. 1.4) at any one time for clarity. If $d\mathbf{R}$ is the infinitesimal displacement vector along a streamline, then

$$d\mathbf{R} \times \mathbf{V} = 0 \tag{1.50}$$

The streamline therefore has a clear mathematical meaning. Its relationship to identifiable particles in a flow field is most obvious for steady flow, for which a streamline, streakline, and pathline are identical (which can again be established from their definitions). If the flow is unsteady, the streamline patterns are constantly changing. All three lines are different for the general case.

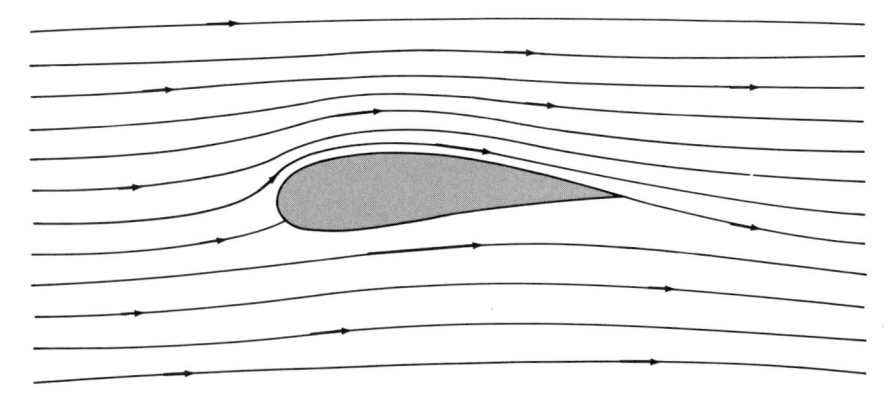

Fig. 1.4. Streamlines and velocity vectors in a flow field.

The streamline finds its primary use in mathematical description of fluid flow; the equations take on particularly interesting characteristics for some flows when expressed in streamline coordinates. An example of this is given in Art. 1.6. As noted, the streakline and pathline are important in flow-visualization work. When the flow is steady, the flow-visualization results may be used to infer the streamlines for purposes of analysis.

3. Acceleration and the Material Derivative

The concept of acceleration, the second derivative of the position vector with respect to time, is quite easily expressed in terms of the Lagrangian description:

$$\mathbf{a} = \frac{d^2\mathbf{R}}{dt^2} = \frac{d\mathbf{V}}{dt} \tag{1.51}$$

This expression, clearly, represents the acceleration of the particle whose radius vector is $\mathbf{R}(t)$ (see Fig. 1.5). Because this description

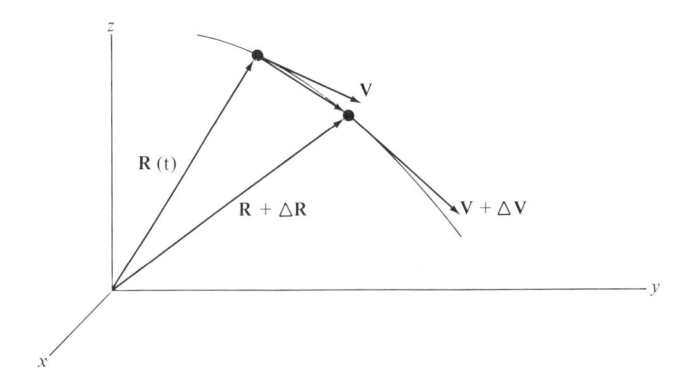

Fig. 1.5. Velocity of a particle at t and $t + \Delta t$.

relates to a single particle, the expression for $\mathbf{a}(t)$ is given directly by Eq. 1.51. However, it is precisely this attribute of dealing with a single particle that makes the Lagrangian description undesirable in fluid mechanics. As noted earlier, we are interested in the Eulerian description, the one which allows the conditions in a region of space to be described. Consequently, we proceed to formal development of the Eulerian description of acceleration of a particle. The time derivative of the particle velocity may be expressed as

$$\frac{d\mathbf{V}}{dt} = \lim_{\Delta t \to 0} \frac{\mathbf{V}(t + \Delta t) - \mathbf{V}(t)}{\Delta t} \tag{1.52}$$

Since the Eulerian description formally states that the particle velocity (or any other function) may be expressed as a function of position and time, $\mathbf{V} = \mathbf{V}(x, y, z, t)$, it is necessary only to determine where the particle, which passed through the point of interest at time t, is at time $t + \Delta t$. The location at time $t + \Delta t$ may be specified by considering a

Taylor series expansion of **V** about the point in space occupied by the particle at time t. (Note that only first-order terms are retained, in anticipation of letting $\Delta t \to 0$.) The first-order terms in the Taylor series expansion are

$$\mathbf{V}(x + \Delta x, y + \Delta y, z + \Delta z, t + \Delta t)$$

$$= \mathbf{V}(x, y, z, t) + \frac{\partial \mathbf{V}}{\partial x} \Delta x + \frac{\partial \mathbf{V}}{\partial y} \Delta y + \frac{\partial \mathbf{V}}{\partial z} \Delta z + \frac{\partial \mathbf{V}}{\partial t} \Delta t \quad (1.53)$$

This relationship allows us to approximate the velocity at a point $(x + \Delta x, y + \Delta y, z + \Delta z)$ at time $t + \Delta t$ if we know the velocity at (x, y, z) and its derivatives at time t. We wish to follow the particle so that Δx, Δy, and Δz are given by

$$\Delta x = u \, \Delta t \qquad \Delta y = v \, \Delta t \qquad \Delta z = w \, \Delta t \quad (1.54)$$

where the velocity vector **V** has components u, v, w. If we then form the derivative

$$\frac{D\mathbf{V}}{Dt} = \lim_{\Delta t \to 0} \frac{\mathbf{V}(x + \Delta x, y + \Delta y, z + \Delta z, t + \Delta t) - \mathbf{V}(x, y, z, t)}{\Delta t}$$

$$(1.55)$$

and use Eqs. 1.53 and 1.54, we obtain the acceleration of the fluid particle:

$$\mathbf{a} = \frac{D\mathbf{V}}{Dt} = u \frac{\partial \mathbf{V}}{\partial x} + v \frac{\partial \mathbf{V}}{\partial y} + w \frac{\partial \mathbf{V}}{\partial z} + \frac{\partial \mathbf{V}}{\partial t} \quad (1.56)$$

This is the acceleration of the fluid particle on which we fixed our attention when we used $\mathbf{V} \Delta t = \Delta \mathbf{R}$, expressed in scalar form by Eqs. 1.54. We will consistently use D/Dt whenever we wish to focus attention on a specific particle, or group of specific particles.

It should be noted that the technique of the Taylor expansion and of the subsequent limiting process is independent of whatever property of the fluid particle is under consideration. Consequently, the formulation above defines the *substantial*, *material*, or *total derivative* to be

$$\frac{D}{Dt} = u \frac{\partial}{\partial x} + v \frac{\partial}{\partial y} + w \frac{\partial}{\partial z} + \frac{\partial}{\partial t} \quad (1.57)$$

where the first three terms on the right side are called the *convective rate of change* and the last term the *local rate of change*. Steady flow occurs whenever the local rate of change is *identically* zero. The material derivative in Eq. 1.57 may operate on any property of the fluid —temperature T, density ρ, entropy s, pressure p, velocity $\mathbf{V}$.

Note that the left side of Eq. 1.57 is clearly Lagrangian in concept, since attention is fixed on a particle. The partial derivatives of the right-hand side are clearly Eulerian in nature. The u, v, w, in effect, form a bridge between the two in that the Lagrangian derivative must be evaluated at an instant in time and, at that instant, the u-, v-, and w-values of the particle are assigned to the point in space occupied by the particle.

The scalar components of the acceleration expressed in Eq. 1.56 are

$$\frac{Du}{Dt} = u\frac{\partial u}{\partial x} + v\frac{\partial u}{\partial y} + w\frac{\partial u}{\partial z} + \frac{\partial u}{\partial t}$$

$$\frac{Dv}{Dt} = u\frac{\partial v}{\partial x} + v\frac{\partial v}{\partial y} + w\frac{\partial v}{\partial z} + \frac{\partial v}{\partial t} \qquad (1.58)$$

$$\frac{Dw}{Dt} = u\frac{\partial w}{\partial x} + v\frac{\partial w}{\partial y} + w\frac{\partial w}{\partial z} + \frac{\partial w}{\partial t}$$

We can write the acceleration in vector form as

$$\frac{D\mathbf{V}}{Dt} = (\mathbf{V} \cdot \nabla)\mathbf{V} + \frac{\partial \mathbf{V}}{\partial t} \qquad (1.59)$$

where ∇ is the gradient operator, expressed in cartesian coordinates as

$$\nabla = \frac{\partial}{\partial x}\hat{i} + \frac{\partial}{\partial y}\hat{j} + \frac{\partial}{\partial z}\hat{k} \qquad (1.60)$$

By vector manipulation we can also show that the acceleration is

$$\frac{D\mathbf{V}}{Dt} = \nabla\left(\frac{V^2}{2}\right) + (\nabla \times \mathbf{V}) \times \mathbf{V} + \frac{\partial \mathbf{V}}{\partial t} \qquad (1.61)$$

which will be useful in the study of flows for which $\nabla \times \mathbf{V} = 0$.

4. Noninertial Reference Frames

The expression for the acceleration stated by Eq. 1.56 gives the acceleration relative to the reference frame in which the velocity field $\mathbf{V}(x, y, z, t)$ is measured. This may not be the absolute acceleration of the particle, which is required when, with Newton's second law, we relate the acceleration to the forces acting on the fluid. In a rotating reference frame, the unit vectors associated with the coordinate system are changing with respect to the inertial coordinate system. Because of this, the time rate of change of a vector **B** will be different in the two systems; the rates are related by*

$$\left.\frac{d\mathbf{B}}{dt}\right|_{XYZ} = \left.\frac{d\mathbf{B}}{dt}\right|_{xyz} + \boldsymbol{\Omega} \times \mathbf{B} \tag{1.62}$$

where $\boldsymbol{\Omega}$ is the rate at which the noninertial reference frame is rotating, as shown in Fig. 1.6. The acceleration in an inertial reference system may be obtained by twice differentiating the position vector $(\mathbf{S} + \mathbf{R})$ with respect to time, using Eq. 1.62. The term $d^2\mathbf{R}/dt^2$ is defined as the acceleration **a** and the term $d\mathbf{R}/dt$ is defined as the velocity **V** in the noninertial coordinate system. Consequently **V** and **a** are measured by an observer in the noninertial xyz-coordinate system. Carrying out the indicated operations, we find the absolute acceleration to be

$$\mathbf{A} = \mathbf{a} + \frac{d^2\mathbf{S}}{dt^2} + 2\boldsymbol{\Omega} \times \mathbf{V} + \boldsymbol{\Omega} \times (\boldsymbol{\Omega} \times \mathbf{R}) + \frac{d\boldsymbol{\Omega}}{dt} \times \mathbf{R} \tag{1.63}$$

where $\boldsymbol{\Omega}$ is the angular velocity of the xyz-reference frame and **V** and **a** are the velocity and acceleration, respectively, measured in the xyz-reference frame.

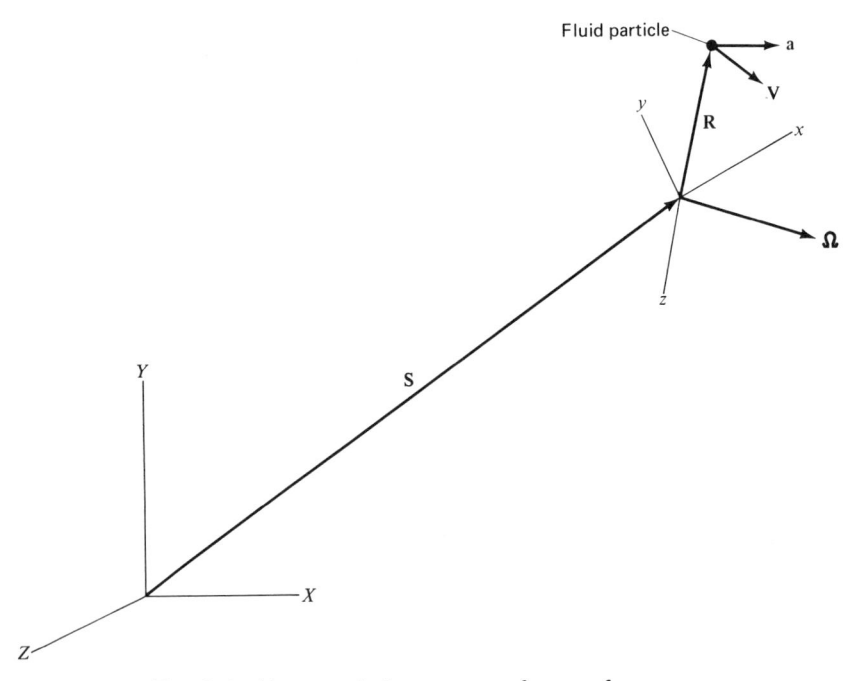

Fig. 1.6. Motion relative to two reference frames.

*See a text on dynamics for a development of this relationship.

Art. 1.4 DESCRIPTION OF FLUID MOTION

It is sometimes advisable to choose a rotating reference frame, especially when working problems involving fluid flows in rotating pieces of machinery such as turbines, dishwasher arms, and the like. In most engineering applications, however, we consider the acceleration referred to a coordinate system attached to the earth to be absolute. This is, of course, only an approximation which results from the fact that the last four terms on the right-hand side of Eq. 1.63 are much smaller than **a**, so that for reference frames attacted to the earth $\mathbf{A} \cong \mathbf{a}$. For problems in which **a** is small, such as the motion of the trade winds, it is necessary that these other terms be included in the expression for the absolute acceleration.

One should note that the acceleration measured in a reference frame for which $\Omega = 0$ and which is moving at constant velocity ($d^2\mathbf{S}/dt^2 = 0$) is identical to the acceleration measured in an absolute reference frame, and hence is itself an inertial reference frame. The reference frame moving at constant velocity is often used when interest is focused on moving objects, such as an aircraft where the reference frame is attached to the aircraft.

Example 1.9

A velocity field in an inertial reference frame is given by

$$\mathbf{V} = \left(10 + \frac{20x}{x^2 + y^2}\right)\hat{i} + \frac{20y}{x^2 + y^2}\hat{j}$$

Find the acceleration at (10, 10, 5).

Solution. The velocity field is independent of time; hence the acceleration is given by

$$\mathbf{a} = (\mathbf{V} \cdot \nabla)\mathbf{V}$$

and in component form is, from Eqs. 1.58,

$$a_x = \frac{du}{dt} = \left(10 + \frac{20x}{x^2 + y^2}\right)\left[\frac{20(x^2 + y^2) - 40x^2}{(x^2 + y^2)^2}\right] + \frac{20y}{x^2 + y^2}\left[\frac{-20x(2y)}{(x^2 + y^2)^2}\right]$$
$$= -0.1$$

$$a_y = \frac{dv}{dt} = \left(10 + \frac{20x}{x^2 + y^2}\right)\left[\frac{-20y(2x)}{(x^2 + y^2)^2}\right] + \frac{20y}{x^2 + y^2}\left[\frac{20(x^2 + y^2) - 40y^2}{(x^2 + y^2)^2}\right]$$
$$= -1.1$$

$$a_z = 0$$

Finally,

$$\mathbf{a} = -0.1\hat{i} - 1.1\hat{j}$$

Example 1.10

If a very thin plate is placed in a fast-moving air stream, as shown in Fig. E1.10A, a *boundary-layer* flow will develop on the plate. That is, a steady flow is obtained in which the retarding effect of the plate is felt only near the plate. The path of a particle which starts in the free stream and a typical velocity profile are shown on the sketch. Describe the acceleration of the particle from both a Lagrangian viewpoint and an Eulerian viewpoint.

Solution. In the free stream outside the boundary layer, the fluid feels no effect from the plate (no retarding force) and consequently feels no acceleration. Thus in the free stream:

Lagrangian: $\dfrac{Du}{Dt} = 0$ \qquad Eulerian: $\dfrac{\partial u}{\partial t} = 0, \quad u \neq 0$ but $\dfrac{\partial u}{\partial x} = 0$

As the fluid enters the boundary-layer region, the viscous shear effects of the plate have caused the fluid in the outer layers to be retarded. Since the force on the particle is in the negative direction, we have

Lagrangian: $\dfrac{Du}{Dt} < 0$ \qquad Eulerian: $\dfrac{\partial u}{\partial t} = 0, \quad u > 0, \left.\dfrac{\partial u}{\partial x}\right|_y < 0$

Because $\partial u/\partial x < 0$, the velocity u must decrease at a fixed y as x is increased. Hence the profiles appear as shown. The fluid velocity is zero at $y = 0$ because of the no-slip condition and u_∞ outside the boundary layer.

Note that this example emphasizes that the two descriptions give the acceleration *of a particle* at *some point* at *some instant of time*. They provide identical numerical values at the instant in time when the particle being followed by the Lagrangian description occupied the Eulerian point of interest.

Extension 1.10.1. Consider that the plate is heated at some distance from the leading edge. A thermal-boundary layer will be formed inside the velocity boundary layer, as shown in Fig. E1.10B. Examine the temperature change with respect to time of a particle following the path shown.

Example 1.11

As an application of Eq. 1.63, consider the description of the movement of atmospheric air resulting from a localized low-pressure region. For this problem, a coordinate system affixed to the sun may be considered inertial. (a) Identify the terms of Eq. 1.63 for this problem. (b) In which direction will the incoming air mass appear to be swirling to an observer on the ground? Does the answer depend upon the hemisphere under consideration?

Solution. Let the *xyz*-reference frame be fixed to the sun. Then **S** is the vector radius to the center of the earth; $d^2\mathbf{S}/dt^2$ accounts for the yearly rotation, which is unimportant in this problem. The angular velocity Ω expresses the earth's rotation about its own axis, 2π rad/day. **R** is the coordinate

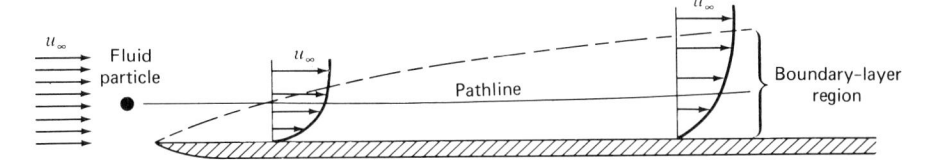

Fig. E1.10A. A boundary layer on a flat plate.

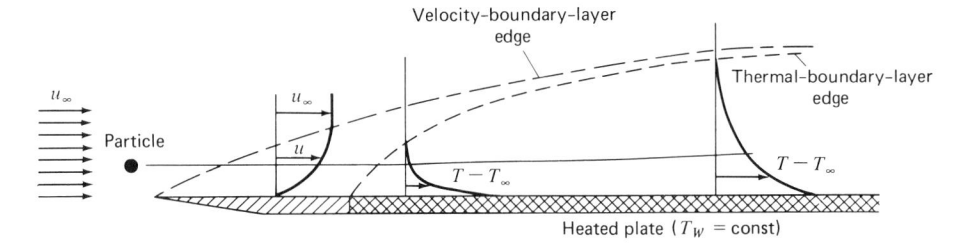

Fig. E1.10B. A thermal boundary layer.

(a) Earth   (b) Low-pressure spot

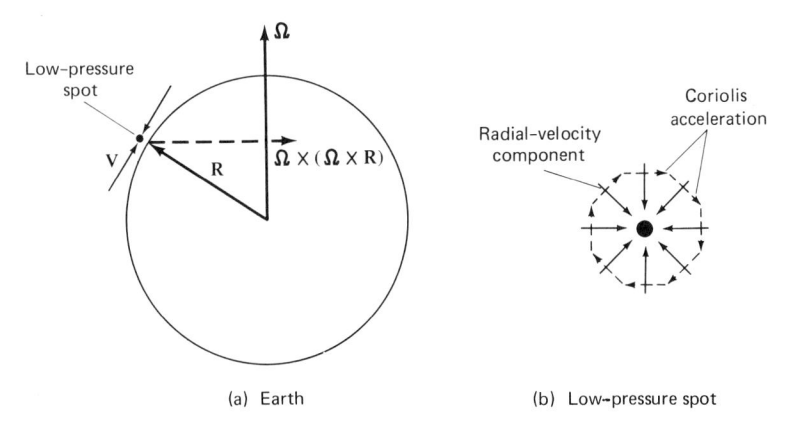

Fig. E1.11

location of an air mass corresponding to the low- pressure region, which means that it has a velocity **V** and acceleration **a** with respect to the earth's surface and directed toward the low-pressure region (see Fig. E1.11). The $d\Omega/dt$ term is zero and the $\Omega \times (\Omega \times R)$ term (the centripetal acceleration) acts on all particles toward the axis of rotation.

The term $2\Omega \times V$ (the coriolis acceleration) will be important in this atmospheric-flow problem. This vector lies in the plane of the earth's surface and is responsible for the swirling pattern seen in satellite pictures of the earth's clouds since the *negative* of this acceleration may be interpreted as a force effect on the incoming air. As the air flows in toward the low-pressure spot, the coriolis acceleration results in the rotational-velocity component. In the Northern Hemisphere this causes a counterclockwise swirl looking downward on the earth's surface. Is the direction reversed for the Southern Hemisphere? (Do not confuse this localized effect with the trade winds.)

1.5 FLUID STATICS

In this article we will consider a group of interesting problems in which the fluid does not move relative to a reference frame attached to the fluid container, a topic commonly referred to as fluid statics. Some of these problems deal with forces on submerged surfaces, manometry, and fluid in constantly accelerating containers. The common feature of all these problems is that the velocity field, relative to an appropriately chosen reference frame, is identically zero.

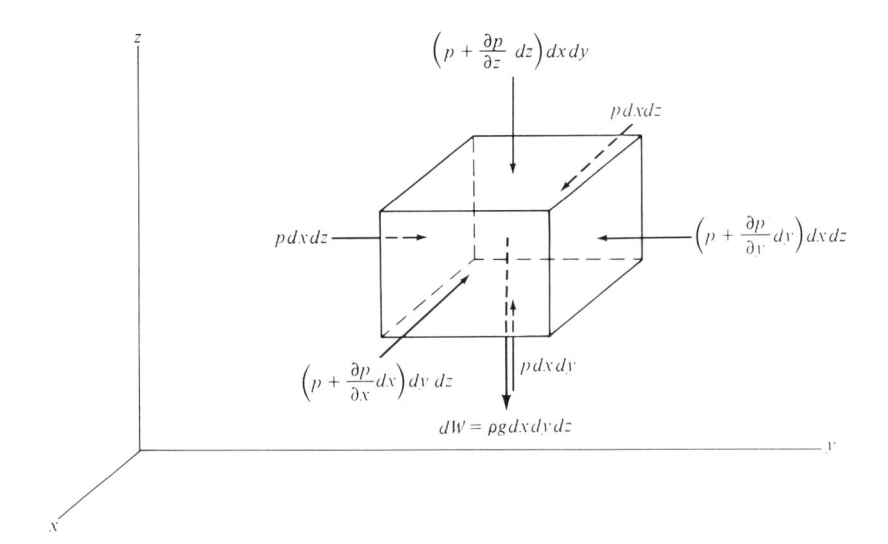

Fig. 1.7. Forces acting on an infinitesimal fluid element in static equilibrium.

Consider an infinitesimal fluid element, shown in Fig. 1.7. The resultant infinitesimal force acting on the element is

$$\Sigma d\mathbf{F} = -\frac{\partial p}{\partial x} dx\, dy\, dz\, \hat{i} - \frac{\partial p}{\partial y} dy\, dx\, dz\, \hat{j} - \frac{\partial p}{\partial z} dz\, dx\, dy\, \hat{k} - \rho g\, dx\, dy\, dz\, \hat{k}$$

(1.64)

where the z-direction is vertical. According to Newton's second law, the resultant forces are equal to the product of mass and acceleration. Considering only constant angular velocity or constant linear acceleration of the reference frame so that the fluid may be in static equilibrium with respect to the container, Eq. 1.63 reduces to

$$\mathbf{A} = \frac{d^2\mathbf{S}}{dt^2} + \mathbf{\Omega} \times (\mathbf{\Omega} \times \mathbf{R}) \tag{1.65}$$

Applying Newton's second law results in

$$-\left(\frac{\partial p}{\partial x}\hat{i} + \frac{\partial p}{\partial y}\hat{j} + \frac{\partial p}{\partial z}\hat{k}\right)dx\,dy\,dz - \rho g\,dx\,dy\,dz\,\hat{k}$$
$$= [\ddot{\mathbf{S}} + \mathbf{\Omega} \times (\mathbf{\Omega} \times \mathbf{R})]\rho\,dx\,dy\,dz \tag{1.66}$$

or

$$-\nabla p - \rho g\hat{k} = \rho[\ddot{\mathbf{S}} + \mathbf{\Omega} \times (\mathbf{\Omega} \times \mathbf{R})] \tag{1.67}$$

This is the equation that we will use for the various problems investigated in this article.

1. Manometry

Manometers are instruments which measure differences in fluid pressures by utilizing displacements of static liquid columns (see Fig. 1.8). We use Eq. 1.67 with both $\ddot{\mathbf{S}} = 0$ and $\mathbf{\Omega} = 0$. The scalar equations contained in vector Eq. 1.67 are, with the z-axis vertical,

$$\frac{\partial p}{\partial x} = 0 \qquad \frac{\partial p}{\partial y} = 0 \qquad \frac{\partial p}{\partial z} = -\rho g \tag{1.68}$$

For constant-density fluids, the solution of the above is

$$p_2 - p_1 = -\rho g(z_2 - z_1) \tag{1.69}$$

With this equation we can determine the differences in pressure between the two points of interest for a particular manometer.

Example 1.12

Determine the pressure p in the pipe if, in Fig. 1.8, $H = 4$ ft, $h = 2$ ft, $\rho = 1.5$ slugs/ft^3 for the oil, and $\rho_x = 10$ slugs/ft^3 for the manometer fluid.

Solution. A useful way to interpret Eq. 1.69 for manometer problems is to recognize that the pressure increases linearly with increasing depth in a

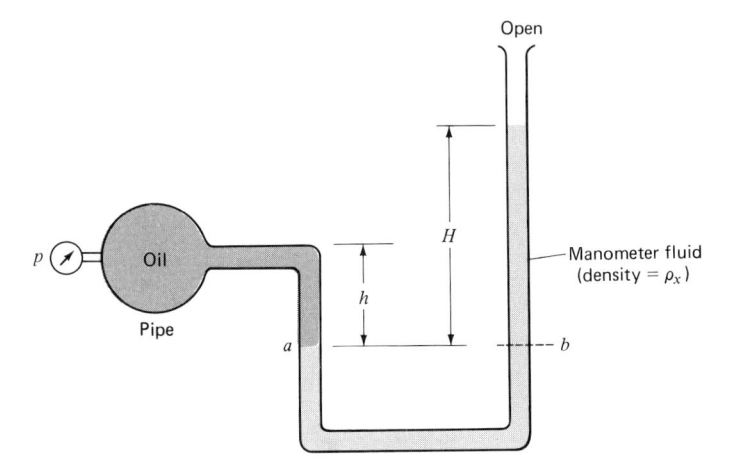

Fig. 1.8. An open manometer.

constant-density liquid and decreases with increasing height within the liquid. This may be used to formulate an algebraic statement of the pressures in a system such as that shown in Fig. 1.8. Starting at the location of the oil, we may write

$$p_{oil} + \rho_{oil} gh = p_a$$

since point a lies below the pipe elevation. The pressure at point b is the same as that at point a since they are at the same elevation in the same liquid. The algebraic expression of the pressure may be continued as

$$p_{oil} + \rho_{oil} gh = p_{atm} + \rho g H$$

from which it is easy to express the gage pressure in the pipe (using $p_{atm} = 0$ gage pressure) as

$$p_{oil} = \rho g H - \rho_{oil} gh$$

Quantitatively,

$$p_{oil} = 10 \times 32.2 \times 4 - 1.5 \times 32.2 \times 2 = 1190 \text{ lb/ft}^2$$

Extension 1.12.1. The quantities given in the example all remain unchanged except that the open end of the manometer is attached to a second pipe in which air is flowing at 50 psi. Determine the pressure in the oil pipe.

Ans. 8390 psf

Extension 1.12.2. A multiple-liquid manometer is sometimes used to obtain greater sensitivity when a single fluid is inappropriate. Such a situation can also occur when the pressure of a corrosive fluid is to be measured. Determine the pressure in the pipe of Fig. E1.12, using the indicated values for the multiple-fluid manometer.

Ans. 1050 psf

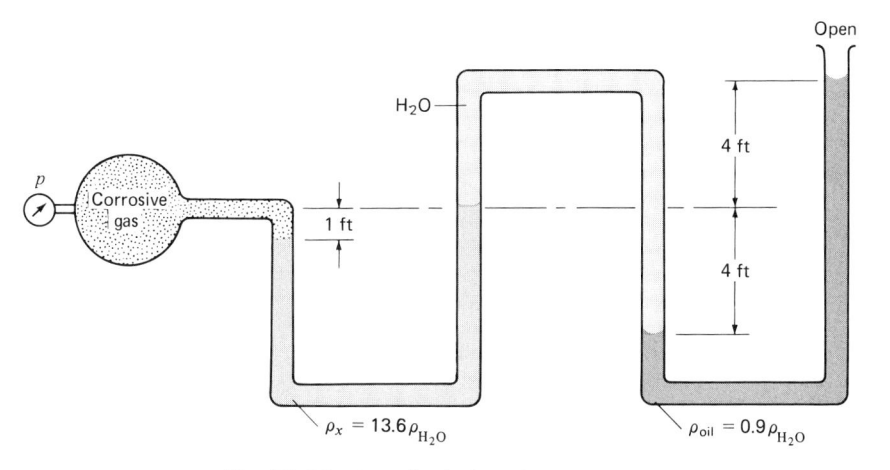

Fig. E1.12. A multiple-liquid manometer.

2. Forces on Submerged Surfaces

One of the most common application areas of fluid statics is that of computing forces on submerged surfaces. Forces on dams and gates and forces of buoyancy are of special interest.

Let us first determine the force on a plane submerged surface oriented at angle α as shown in Fig. 1.9. The total force, which acts normal to the area, is

$$F = \int_A p \, dA \tag{1.70}$$

By Eq. 1.69, we can express the pressure as

$$p = \rho g h \tag{1.71}$$

where h is measured down from the surface, assuming $p_{atm} = 0$. The force may then be stated as

$$F = \int \rho g h \, dA$$
$$= \rho g \sin \alpha \int l \, dA$$
$$= \rho g \sin \alpha \, l_c A$$
$$= \rho g h_c A \tag{1.72}$$

where h_c and l_c are the distances to the *centroid* of the area, measured from the free surface vertically and along the plane, respectively. Eq. 1.72 states that *the force on a plane surface is the pressure at the centroid multiplied by the area.*

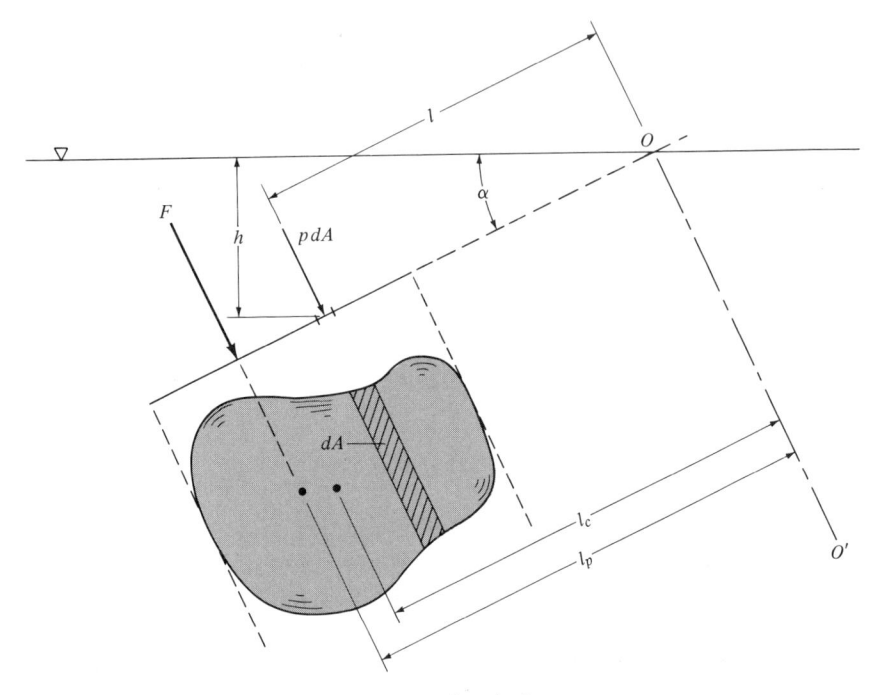

Fig. 1.9. An inclined plane area.

Let us now determine where the force acts. We do this by noting that the sum of the moments of all the infinitesimal forces acting on the area is equal to the moment of the resultant force F. The point at which the force acts is called the *center of pressure*, located vertically with h_p or along the plane with l_p. Equating moments gives

$$l_p F = \int l p \, dA$$
$$= \int l(\rho g l \sin \alpha) \, dA$$
$$= \rho g \sin \alpha \int l^2 \, dA$$
$$= \rho g I_O \sin \alpha \tag{1.73}$$

where I_O is the *moment of inertia* of the area about the axis passing through point O. Recalling that $I_O = I_c + l_c^2 A$, we may write Eq. 1.73 as

$$l_p = \frac{\rho g \sin \alpha \, (I_c + l_c^2 A)}{\rho g \sin \alpha \, l_c A}$$
$$= \frac{I_c}{l_c A} + l_c \tag{1.74}$$

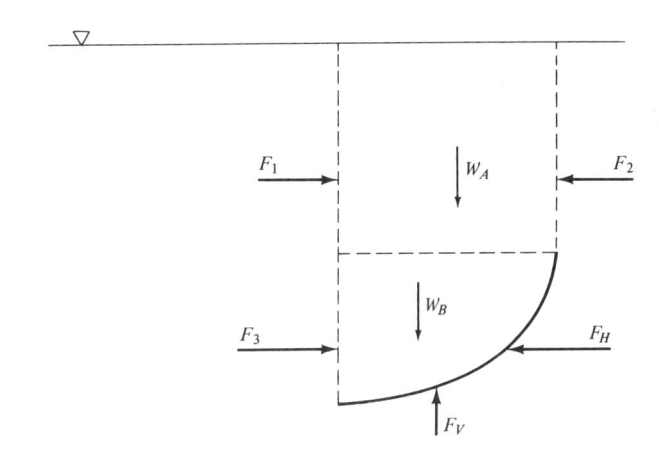

Fig. 1.10. Forces acting on a curved surface.

where I_c is the moment of inertia about the centroidal axis. We have now arrived at the magnitude of the force (Eq. 1.72) and its location (Eq. 1.74).

For a curved surface we resolve the force acting on it into components F_H and F_V. If we isolate the liquid contained above the surface, as shown in Fig. 1.10, we observe that

$$F_V = W_A + W_B = W \qquad (1.75)$$

where W is the weight of all the liquid above the curved surface, or of all the liquid that could be above the surface (see Example 1.13). The force component F_V acts through the centroid of the equivalent volume of the liquid contained above the surface, as can be shown by a summation of moments about the center of gravity. From Fig. 1.10 we see that $F_1 = F_2$, and thus

$$F_H = F_3 \qquad (1.76)$$

where F_3 is the force acting on the projection of the curved surface in the vertical plane. The force component F_H would act through the center of pressure of this projected area.

The equation for the pressure variation in a static fluid is linear; consequently, solutions to the equation can be superimposed. Such a technique is useful when the pressure above a liquid surface is different from the atmospheric value. For example, consider that the gas pressure above the liquid surface in Fig. 1.9 was 30 psig. The total force on the plate is given (as before) by

$$F = \int_A p\, dA \qquad (1.77)$$

However, the pressure p is affected by the boundary condition $p_{gas} = 30$ psig, not 0 psig as in Eq. 1.71. The pressure of a gas is spread uniformly over the containing vessel (the hydrostatic variations are negligible except in cases such as the atmosphere where the lengths are very large). Consequently the uniform-pressure solution may be added to the solution accounting for the liquid above the plane area. Then, F_T, the total force, is

$$F_T = F_{uniform} + F_{hydrostatic} = p_{gas} A + \rho g h_c A \qquad (1.78)$$

In order to evaluate the point of application of the total force, it is necessary to again sum the moments about an arbitrary reference point. For convenience, point O in Fig. 1.9 is selected. The uniform pressure clearly acts at the geometric centroid of the area; consequently, the length to the point of application l_T of the total force F_T is found from

$$l_T F_T = \left[\frac{I_c}{l_c A} + l_c \right] \rho g h_c A + l_c p_{gas} A \qquad (1.79)$$

Example 1.13

Determine the force P necessary to hold the 20-ft-wide gate in the position shown in Fig. E1.13A.

Solution. The pressure distribution on the bottom of the gate, which accounts for the force of the water on the gate, would be identical to the pressure distribution on the top of the gate if water were above the gate. Hence, we can determine F_H and F_V assuming water occupies the volume above the gate. Thus

$$F_V = \rho g \times \text{volume}$$
$$= 1.94 \times 32.2 \times 25\pi \times 20$$
$$= 98{,}100 \text{ lb}$$

and

$$F_H = \rho g \times h_c \times A$$
$$= 1.94 \times 32.2 \times 5 \times 200$$
$$= 62{,}400 \text{ lb}$$

The distance d_1 (Fig. E1.13B) is the distance to the centroid of the quarter circle. It is

$$d_1 = 4r/3\pi = 4.25 \text{ ft}$$

Art. 1.5   FLUID STATICS   43

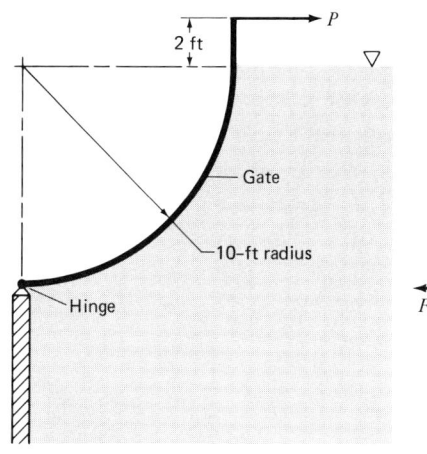

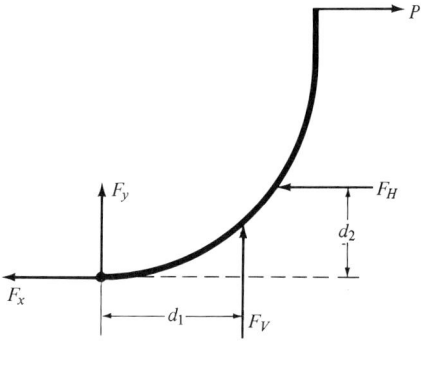

Fig. E1.13A. A water gate.   **Fig. E1.13B.** Free-body diagram.

The distance d_2 is

$$d_2 = 10 - l_p$$

$$= 10 - \frac{I_c}{l_c A} - l_c$$

$$= 10 - \frac{20}{12} - 5$$

$$= 3.33 \text{ ft}$$

Summing moments about the frictionless hinge gives

$$12P = d_1 \times F_V + d_2 \times F_H$$

$$= 4.25 \times 98{,}100 + 3.33 \times 62{,}400$$

so that

$$P = 52{,}100 \text{ lb}$$

Extension 1.13.1. Since the infinitesimal pressure force always acts normal to the surface element, the resultant force $\mathbf{F}_H + \mathbf{F}_V$ due to the pressure must act along a line which passes through the center of the circular arc. Using this fact, determine the force P *without calculating* F_V, d_1 or d_2.

Extension 1.13.2. Compute the force required to hold the gate closed if a uniform pressure of 5 psi is imposed on the liquid surface.   *Ans.*   172,000 lb

---

3. Fluid Statics in an Accelerating Reference Frame

In the two preceding sections of this article we have considered fluid at rest relative to an inertial reference frame. We turn now to phenomena in which the fluid accelerates relative to the inertial frame but remains at rest relative to a noninertial frame. One such problem involves liquid in a container rotating with constant angular velocity, and the other, liquid in a container accelerating with constant linear acceleration. We will limit the problems to a steady-state situation such that there is no motion relative to a reference frame attached to the container. The problems of sloshing and other unsteady phenomena will not be considered here.

For a tank accelerating as shown in Fig. 1.11, the fluid static equation (1.67) reduces to

$$\frac{\partial p}{\partial x} = -\rho a_x \qquad \frac{\partial p}{\partial y} = 0 \qquad \frac{\partial p}{\partial z} = -\rho g \qquad (1.80)$$

where a_x is the x-component of the acceleration $\ddot{\mathbf{S}}$. The solution of these equations is

$$p_2 - p_1 = -\rho a_x(x_2 - x_1) - \rho g(z_2 - z_1) \qquad (1.81)$$

If we choose points ① and ② on the surface of the liquid, as shown, then $p_2 = p_1$ and we have

$$\rho a_x(x_2 - x_1) + \rho g(z_2 - z_1) = 0 \qquad (1.82)$$

or

$$\frac{z_2 - z_1}{x_1 - x_2} = \frac{a_x}{g} = \tan \alpha \qquad (1.83)$$

where $\tan \alpha$ is the slope of the surface.

In solving problems we must often utilize conservation of mass; that is, if we know the volume of liquid in a container before acceleration, then in the accelerated state the same volume will occupy a different region of the container if the fluid is free to move. (See Example 1.14.)

Let us consider a rotating container containing liquid, as shown in Fig. 1.12. Choosing cylindrical coordinates, our equations become

$$\frac{\partial p}{\partial r} = \rho r \Omega^2 \qquad \frac{\partial p}{\partial \theta} = 0 \qquad \frac{\partial p}{\partial z} = -\rho g \qquad (1.84)$$

with the solution

$$p_2 - p_1 = \frac{\rho \Omega^2}{2}(r_2^2 - r_1^2) - \rho g(z_2 - z_1) \qquad (1.85)$$

Let us again pick two points on the surface, as shown. Then $p_2 = p_1$,

Art. 1.5   FLUID STATICS   45

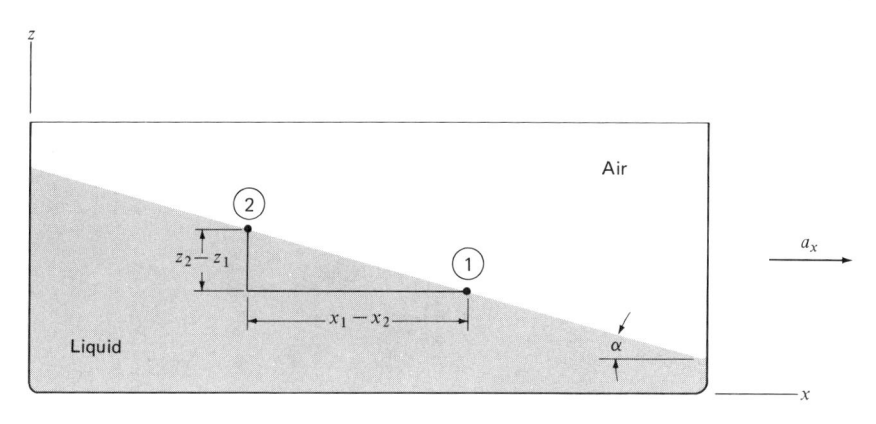

Fig. 1.11. A linearly accelerating tank.

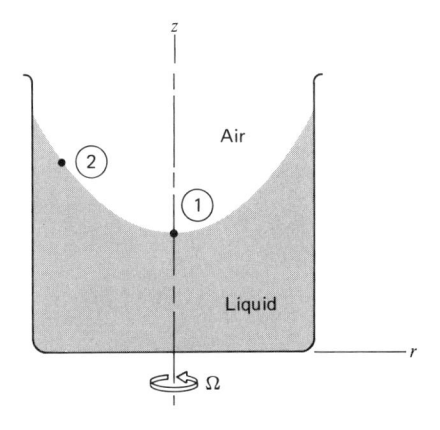

Fig. 1.12. A rotating container.

and with $r_1 = 0$,

$$\frac{\Omega^2 r_2^2}{2} = g(z_2 - z_1) \tag{1.86}$$

which is the equation of a parabola. The free surface, along which $p_2 = p_1$, is thus a paraboloid of revolution. Along with the conservation of mass, which may be necessary, Eq. 1.85 enables us to solve problems involving rotating fluids.

Example 1.14

The closed tank of Fig. E1.14 is accelerated at a constant rate a_x. Determine a_x necessary for the liquid surface to just meet the corner at A.

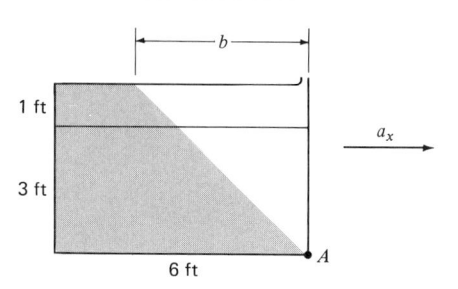

Fig. E1.14

Solution. The original volume of air in the tank is $6w$ ft^3, where w is the width. Since the liquid does not change volume the volume of air must remain at $6w$ ft^3. After accelerating the air volume will be triangular, with volume

or
$$\tfrac{1}{2}(4bw) = 6w$$

$$b = 3 \text{ ft}$$

Thus, using Eq. 1.81,

$$0 = \rho a_x(3) - \rho g(-4)$$

or

$$a_x = 42.9 \text{ ft/sec}^2$$

Extension 1.14.1. Determine the force acting on the bottom of the tank if $w = 4$ ft. Water is in the tank. *Ans.* 6000 lb.

Example 1.15

At what angular speed Ω will the liquid of Fig. E1.15 spill out?

Solution. Using Eq. 1.85 with points ① and ② as indicated, we have

$$0 = \rho \frac{\Omega^2}{2}(4) - \rho g(z_2 - z_1)$$

To determine $(z_2 - z_1)$, we reason that the initial volume of air is 4π ft^3 and that this must equal the volume of air in the paraboloid; hence (the volume of a paraboloid is $\pi R^2 H/2$)

$$4\pi = \tfrac{1}{2}[4\pi(z_2 - z_1)]$$

or

$$z_2 - z_1 = 2 \text{ ft}$$

Art. 1.5　　　　　　　　　　FLUID STATICS　　　　　　　　　　47

Finally,

$$\Omega^2 = g$$

or

$$\Omega = \sqrt{g} \text{ rad/sec}$$

Extension 1.15.1. Determine the value of Ω at which the water will just touch the bottom of the container.　　　　　　*Ans.*　8.02 rad/sec

Extension 1.15.2. At what rotational speed will 3/4 of the water spill out?
　　　　　　　　　　　　　　　　　　　　　　　　Ans.　13.1 rad/sec

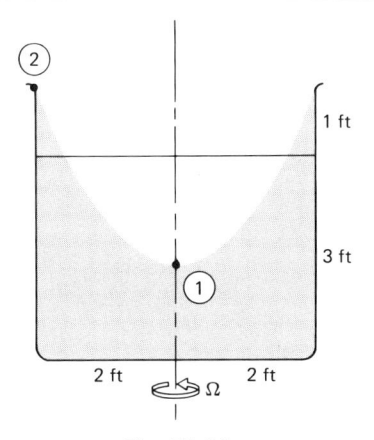

Fig. E1.15

4. Surface Tension

Our last example of fluid statics concerns phenomena in which surface tension is important. Three common cases involving surface tension are the droplet of a liquid, the soap bubble, and the capillary tube, all shown in Fig. 1.13.

Surface tension is a simple description of a rather complicated phenomenon. It is introduced to account for the behavior observed when two or three materials form a common boundary. The boundary materials in parts a and b of Fig. 1.13 are liquid-air; the materials in part c involve the liquid, the tube, and air. In these conditions, the *surface tension* σ is introduced to account for a discontinuous jump in

(a) Liquid droplet (b) Soap bubble (c) Capillary tube

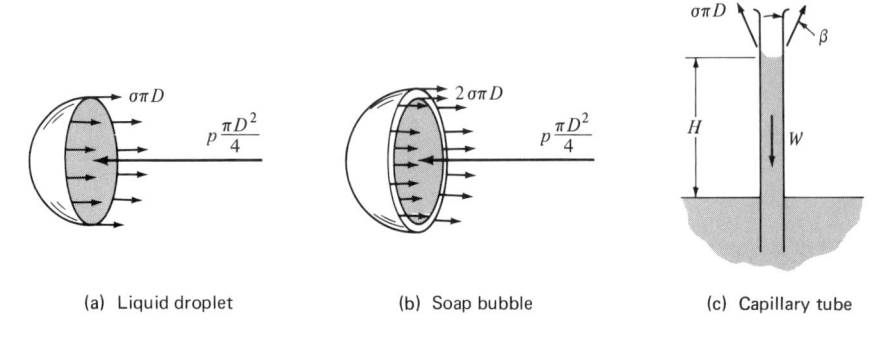

Fig. 1.13. Examples of surface-tension phenomena.

the pressure magnitude across the surface. This jump is supported by the surface-tension effect, as shown in the following examples. The "simple description" is to combine the unknown effects due to surface chemistry into the single coefficient σ.

Surface tension σ multiplied by a distance yields a force. From a force balance, the pressure inside a liquid droplet can be determined (Fig. 1.13a) to be

$$p = \frac{4\sigma}{D} \qquad (1.87)$$

where D is the diameter of the spherical droplet.

The pressure inside a soap bubble is

$$p = \frac{8\sigma}{D} \qquad (1.88)$$

since there is an inside and outside surface, both in contact with air.

A liquid tends to adhere to a solid at some *contact angle* β, as shown in the capillary tube in Fig. 1.13c. It adheres because of surface tension. If $\beta < 90°$ it "*wets*" the surface. The rise of liquid in a capillary tube is found by balancing the weight with the surface-tension force, yielding

$$H = \frac{4\sigma \cos \beta}{\rho g D} \qquad (1.89)$$

For a curved surface the general expression which relates the pressure difference across the surface to the surface tension is

$$\Delta p = \sigma \left(\frac{1}{r_1} + \frac{1}{r_2} \right) \qquad (1.90)$$

where r_1 and r_2 are the principal radii of curvature of the surface.

1.6 THE BERNOULLI EQUATION

In this article we will review an equation which you have undoubtedly used before, probably in physics. It is a simple relationship and has great utility; however, its potential for misuse is also great. We will indicate its limitations and in later chapters develop more general equations governing fluid flow and deduce this relationship as a special case. The justification for considering the Bernoulli equation at this point (perhaps prematurely, from another viewpoint) is its role in the solution of the control volume problems. The occurrence of flows which may be assumed steady and inviscid is common enough to make the Bernoulli equation often useful in describing many of the flow situations to be considered in Chapter 2.

Consider a fluid particle occupying an infinitesimal volume in a flow situation such that it is acted upon only by a gravitational body force and a surface force from the pressure of the surrounding fluid. We will neglect the viscous shearing stresses acting on the surface. The infinitesimal forces acting on the particle occupying the infinitesimal volume at an instant are shown in Fig. 1.14. The fluid particle is moving along a streamline. Newton's second law for the particle along the streamline gives

$$\rho \, ds \, dA \, \frac{DV}{Dt} = p \, dA - \left(p + \frac{\partial p}{\partial s} ds\right) dA - \rho g \, ds \, dA \, \cos \theta \quad (1.91)$$

where V is the particle velocity, tangent to the streamline, with s measured along the streamline.

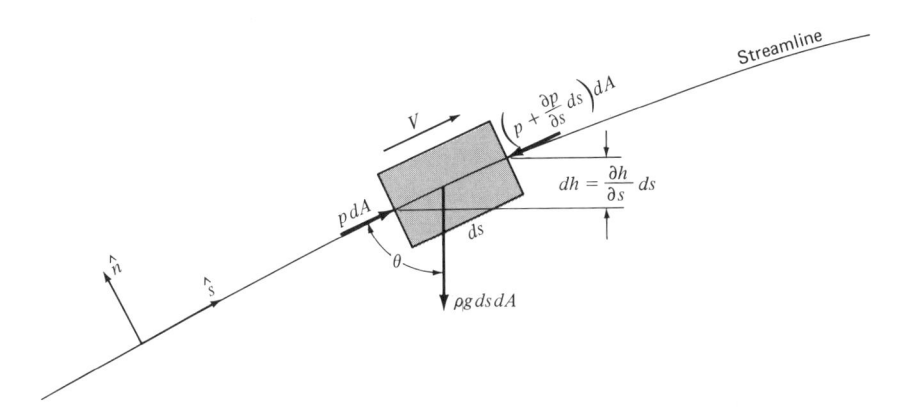

Fig. 1.14. Flow of a particle along a streamline. The particle occupies the infinitesimal fixed volume shown at the instant t.

If we divide both sides of Eq. 1.91 by $dA\,ds$, use $dh = (\partial h/\partial s)ds = ds\cos\theta$, and assume steady flow so that $DV/Dt = V\,\partial V/\partial s$, then there results*

$$\rho V \frac{\partial V}{\partial s} = -\frac{\partial p}{\partial s} - \rho g \frac{\partial h}{\partial s} \tag{1.92}$$

or, for *constant density*,

$$\rho \frac{\partial}{\partial s}\left(\frac{V^2}{2} + \frac{p}{\rho} + gh\right) = 0 \tag{1.93}$$

where h is measured in the vertical direction.
If we integrate *along the streamline* we obtain

$$\frac{V^2}{2} + \frac{p}{\rho} + gh = \text{const} \tag{1.94}$$

This result is *Bernoulli's equation*. We have assumed steady, incompressible, inviscid flow along a streamline in an inertial reference frame. The equation has application in flow around airfoils (low Mach number, so compressibility effects are unimportant), in a nostril, in a carburetor, and in many other situations. It will find particular application in measuring the velocity of a fluid with a *pitot tube*, a device shown in Fig. 1.15. If we apply Eq. 1.94 between a point ① in the flow and a point ② in the pitot tube where the fluid is not moving, we obtain

$$\frac{V_1^2}{2} + \frac{p_1}{\rho} = \frac{p_2}{\rho} \tag{1.95}$$

So, if we know pressure p_1 and measure pressure p_2 in the pitot tube, we can determine the velocity. Pressure p_2 is the *stagnation*, or *total*, *pressure*.

Also shown in Fig. 1.15 is a *piezometer tube*. The *static pressure* in a flowing fluid is that pressure which would be exerted upon a surface moving with the velocity of the fluid. Since the streamlines are parallel to the wall, the static pressure acts upon the opening (as well as the wall) and the pressure is indicated by the height of the liquid column.

*If we use

$$\nabla = \hat{s}\frac{\partial}{\partial s} + \hat{n}\frac{\partial}{\partial n}$$

and $\mathbf{V} = V\hat{s}$, then along the streamline the acceleration is

$$\frac{D\mathbf{V}}{Dt} = \frac{\partial \mathbf{V}}{\partial t} + (\mathbf{V}\cdot\nabla)\mathbf{V} = \frac{\partial(V\hat{s})}{\partial t} + V\frac{\partial(V\hat{s})}{\partial s} = \frac{\partial(V\hat{s})}{\partial t} + V\frac{\partial V}{\partial s}\hat{s} - \frac{V^2}{R}\hat{n}$$

Art. 1.6 THE BERNOULLI EQUATION 51

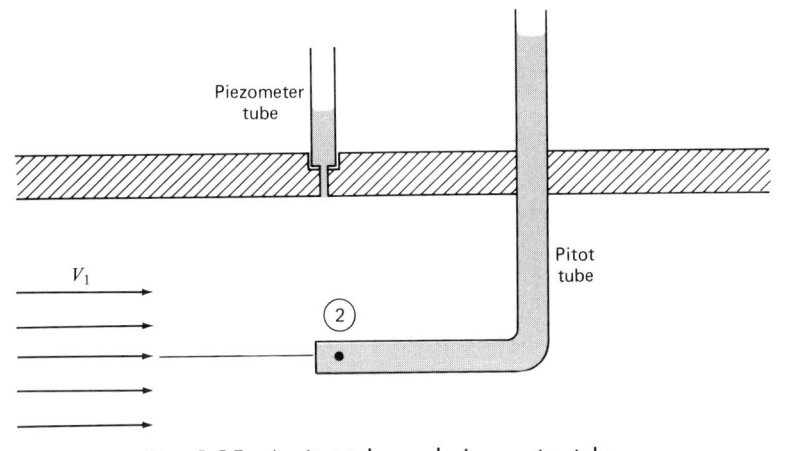

Fig. 1.15. A pitot tube and piezometer tube.

Example 1.16

An automobile is traveling at 75 mph. Approximate the maximum force acting on the front of a 6-in.-dia. headlamp. This value could be used as a conservative estimate for design purposes.

Solution. To a stationary observer, a moving car in stagnant air generates an unsteady flow. To simplify the situation, the car is assumed stationary with the air flowing by it, as is done in wind tunnels. This is a steady flow, for which Bernoulli's equation can be used. Actually, we are observing the flow from two different inertial reference frames; we choose the one that simplifies our calculations.

Assuming the car to be fixed and the air moving, Bernoulli's equation is written as

$$\frac{V_1^2}{2} + \frac{p_1}{\rho} + gh_1 = \frac{V_2^2}{2} + \frac{p_2}{\rho} + gh_2$$

where point ① is upstream in the undisturbed flow and point ② is on the headlamp. Assuming no change in elevation ($h_1 = h_2$), atmospheric pressure at point 1 ($p_1 = 0$), and the headlamp to be in a stagnation region ($V_2 = 0$), Bernoulli's equation simplifies to

$$\frac{V_1^2}{2} = \frac{p_2}{\rho}$$

The density of air at standard conditions is 0.00237 slug/ft³. Thus the pressure is (75 mph is equal to 110 fps)

$$p_2 = \frac{110^2}{2} \times 0.00237 = 14.3 \text{ lb/ft}^2$$

The total force is then

$$F = pA$$

$$= 14.3 \times \frac{\pi}{16} = 2.81 \text{ lb}$$

Note that the force will be somewhat less, since the pressure p_2 is only realized at the stagnation point.

Example 1.17

The Bernoulli equation has the same appearance as a particular form of the energy equation from thermodynamics. Compare the restrictions on the two equations; also, indicate the physical situations described by the two equations.

Solution. The operational differences between the two equations are noted below. It should first be recognized that the equations are different in principle since the Bernoulli equation is a special form of the momentum equation. Energy conservation and momentum conservation are independent physical laws. The restrictions on the Bernoulli equation are indicated below Eq. 1.94.

It is expected that you are familiar with the equation

$$\frac{p_1}{\rho} + \frac{V_1^2}{2} + gz_1 = \frac{p_2}{\rho} + \frac{V_2^2}{2} + gz_2$$

as an energy equation (from thermodynamics) for these special restrictions:

1. One-dimensional flow into and out of an "open system"
2. Steady flow
3. Incompressible flow
4. $\dot{Q} - \dot{W}_s = 0$
5. No "head loss," that is, no increase in the internal energy of the fluid caused by shear stresses in the fluid

Note that, although the restrictions of steady flow and incompressible flow are the same and that the restriction of an inviscid fluid (for the Bernoulli equation) is compatible with the restriction of no head loss (for the energy equation), the remaining restrictions make the two equations quite different in what they represent.

A major difference between the two equations is the physical region in a flow field for which they are appropriate. The energy equation is for a fixed region in space where fluid enters at one end and the same quantity leaves at the other end; the ends are of finite area. This means that the surface area is finite and a finite $\dot{Q}$ and $\dot{W}_s$ may cross the boundary. Even though it has been divided on both sides of the equation, a mass flux of magnitude $\rho A V$ exists at both ends of the region in space. Conversely, the Bernoulli equation expresses

Art. 1.6 THE BERNOULLI EQUATION 53

the constancy of the sum

$$\frac{p}{\rho} + \frac{V^2}{2} + gz$$

at different points along a streamline. There is no implication that the velocity is nonzero along this streamline; in fact, it is often useful to examine the stagnation streamline for a flow field. Examples of a stagnation streamline are shown in Fig. E1.17A. For no change in elevation, the stagnation pressure p_0 would be given by

$$p_0 = p_\infty + \rho \frac{V_\infty^2}{2}$$

where p_∞ and V_∞ are measured in the free stream away from the body.

Stagnation points

Airfoil at attack angle

Stagnation streamlines

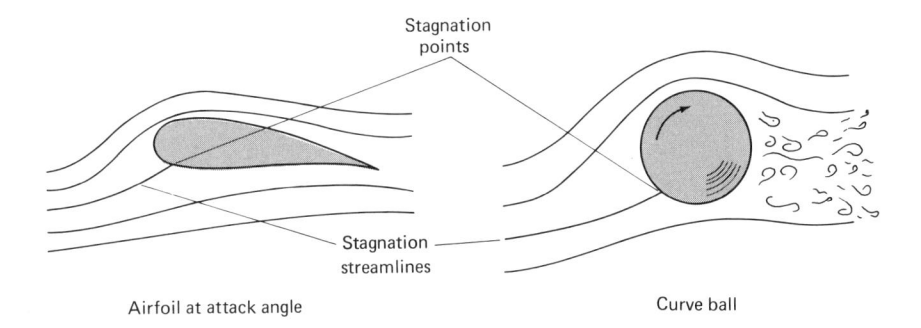

Curve ball

Fig. E1.17A

The essential difference between the two equations is emphasized by the requirement that (1) the Bernoulli equation was integrated along a streamline and the assumptions must therefore apply along the entire streamline, and that (2) the energy equation is evaluated only at the end states and is explicitly independent of the interior behavior of the flow. A somewhat unrealistic example of this difference is shown in Fig. E1.17B, where the work output just balances the heat transfer $\dot{Q}$ and losses are neglected. For the conditions indicated, the energy equation would result in the expression

$$\frac{V_1^2}{2} + \frac{p_1}{\rho} + gz_1 = \frac{V_2^2}{2} + \frac{p_2}{\rho} + gz_2$$

although the Bernoulli equation could not be used along a streamline from point ① to ② .

Extension 1.17.1. Explain why the Bernoulli equation, rather than the one-dimensional steady-flow energy equation, is the appropriate equation for analyzing the response of a pitot tube.

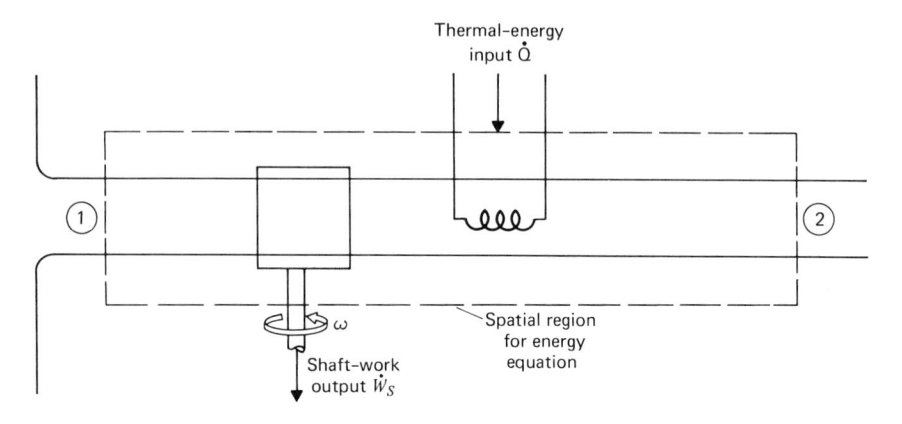

Fig. E1.17B

Extension 1.17.2. Assume that the Bernoulli equation can be used for the flow from point ① to ② and point ①' to ③ (Fig. E1.17C). Write the one-dimensional steady-flow energy equation for this problem and compare the two equations.

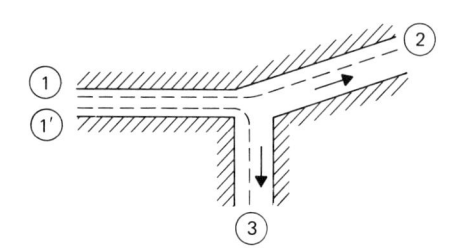

Fig. E1.17C

Example 1.18

A common technique for determining the magnitude of the velocity in a low-speed air flow is to measure the difference between the total pressure p_0 and the static pressure p_s and to calculate the velocity magnitude from this value. Find a simple form of the Bernoulli equation which allows the velocity magnitude to be determined if the pressure difference is measured in inches of water h and the temperature T and the barometric pressure H are known in °R and inches of mercury, respectively.

Solution. Bernoulli's equation takes the form of Eq. 1.95, written as

$$V = \left[\frac{2}{\rho} (p_0 - p_s) \right]^{1/2}$$

The density ρ can be calculated from the atmospheric conditions because the flow may be assumed incompressible (see Art. 1.8); thus,

$$\rho = \frac{p}{RT} = \frac{H}{12} \times (1.94 \times 32.2 \times 13.6) \times \frac{1}{1716T} \text{ slugs}$$

The conversion from h in inches of water to psf yields

$$p_0 - p_s = \frac{h}{12} \times 1.94 \times 32.2 \text{ psf}$$

Finally we have

$$V = \left[\frac{2 \times 1716}{13.6} \right]^{1/2} \sqrt{\frac{Th}{H}}$$

$$= 15.9 \sqrt{\frac{Th}{H}} \text{ fps}$$

For typical conditions ($T = 530°R$ and $H = 29.5$ in. Hg), there results

$$V = 67.4\sqrt{h} \text{ fps}$$

Note that a similar form would be obtained with a different constant if another gas were used.

1.7 INCOMPRESSIBLE GAS FLOW

The notion of incompressible flow of a liquid is easily visualized from our everyday experiences. We are also familiar with many physical cases in which a gas behaves in a compressible manner. But the notion of incompressible gas flow may not be equally clear. Nearly all examples of compressible behavior of a gas occur with respect to a closed volume such as an automobile tire or a vacuum-packed can of coffee. When the gas is free to move, the attempt to pressurize it will simply result in its motion from that region in space. Depending upon the restriction to the motion, it may or may not be incompressible. For example, when you "crack open" the coffee container and let the atmospheric air in or when you place your hand in the jet flow created by air leaking from the tire stem, you are observing an incompressible flow. The flow inside a hand-operated air pump provides an interesting example of a flow which is either compressible or incompressible, depending upon the rate at which the pump handle is depressed and the size of the outlet orifice with respect to the internal area of the cylinder.

A compressible flow can also result from thermal effects. The air which escapes from above the water in a heating tea kettle is an example of a condition in which the thermal-energy input increases the temperature and decreases the density, assuming that the air is free to move out of the region.

The physical problems which involve a confined or thermally caused compressible flow are rather easily identified. Compressible flows associated with the high-speed motion of a gas are not part of common experience and consequently not as easily identified. Such flows are, however, of considerable technological importance in terms of the motion of projectiles, high-speed aircraft, and high-velocity flows in turbines and compressors. High-speed compressibility effects are considered in detail in Chapter 11; a simple example will be discussed here to identify quantitatively the conditions under which such compressible behavior takes place or, alternately, when an incompressible gas flow may be expected. This quantitative evaluation will make use of the Bernoulli equation and the thermodynamic considerations of this chapter.

Consider a pressurized tank of fluid which is allowed to exhaust, through a nozzle, to the atmosphere as shown in Fig. 1.16. Let us determine the maximum pressure difference for which the flow can be considered incompressible for a perfect gas. A particle of fluid following the indicated streamline will be accelerated by the pressure difference ($p_0 - p_{atm}$). If the flow is incompressible, and if the motion is steady and essentially frictionless, then Bernoulli's equation gives

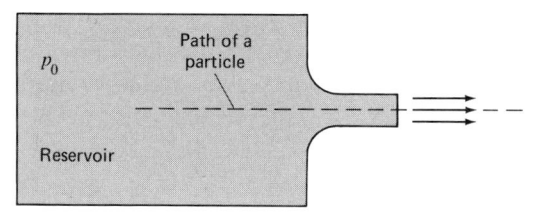

$$V_{exit} = \left[\frac{2}{\rho} (p_0 - p_{atm}) \right]^{1/2} \quad (1.96)$$

Fig. 1.16. An imcompressible flow of a gas from a reservoir.

That is, the exit velocity is controlled only by the pressure difference and is limited only by the pressure which can be created in the tank. If we consider a flow of a perfect gas, we would have to recognize that, as the pressure drops, the relationship $p = \rho RT$ indicates that some alteration in density, temperature, or both may be expected. If we stipulate that the fluid is moving so rapidly that little time is available for significant heat transfer (adiabatic) and that the approximation of a frictionless flow is still reasonable, then we would recognize the flow to be isentropic. Whether or not the flow is incompressible depends upon the *percentage of change* in the density as the particle travels along the dotted line; we can determine the pressure difference necessary to cause this small percentage of change in the density by the isentropic relationship of Eq. 1.21, p/ρ^γ = constant. In order to carry out this calculation, assume p_{atm} = 2120 psf, and the air in the tank to be at room temperature, 540°R. If we take a 3-percent change in ρ (that is, $\rho_{exit} = 0.97\rho_0$) to be the "limit" of our "incompressible flow" regime, we can calculate p_0. Recognizing that $p_{exit} = p_{atm}$, from

$$\frac{p_0}{\rho_0^\gamma} = \frac{p_{atm}}{\rho_{exit}^\gamma} \tag{1.97}$$

p_0 may be written as

$$p_0 = p_{atm} \left(\frac{\rho_0}{\rho_{exit}} \right)^\gamma \tag{1.98}$$

or, numerically

$$p_0 = 2120 \left(\frac{1}{0.97} \right)^{1.4} = 2212.4 \tag{1.99}$$

This corresponds to a gage pressure in the tank of

$$p_{gage} = 2212.4 - 2116.8 = 95.6 \text{ psf or } 0.664 \text{ psi} \tag{1.100}$$

Using Bernoulli's equation in the form of Eq. 1.96, the exit velocity associated with this slight pressure increase is

$$V_{exit} = [2(p_0 - p_{atm})/\rho]^{1/2}$$

$$= 280 \text{ fps} \quad \text{or} \quad 190 \text{ mph} \tag{1.101}$$

where we have determined ρ from $p_0 = \rho_0 R T_0$ to be 0.0023 slug/ft³.

A speed of 190 mph is one which most of us do not experience in our daily activities; hence we are not very familiar with compressible-flow effects. It is possible to express this fluid velocity in terms of the *Mach number* M, that is, the velocity of interest divided by the speed of sound (M = V/a). The speed of sound is found from Eq. 1.24 to be $a = \sqrt{\gamma RT} = \sqrt{1.4 \times 1716 \times 540} = 1140$ fps for air at 80°F. The Mach number is then

$$M = \frac{V}{a}$$

$$= \frac{280}{1140} = 0.25 \qquad (1.102)$$

This result is characteristic; in fact, incompressible-gas flow analyses are often used up to a Mach number of 0.3. Flow around automobiles, landing and take-off of commercial jet liners, flow in heating and air-conditioning ducts, and meteorological air flows are all examples of incompressible gas flows. We will discuss the importance of the Mach number and show its relationship to compressible-flow phenomena (M > 0.3) in Chapters 4 and 11.

PROBLEMS

1.1 An engineer is considering the design of a positive-displacement vacuum pump. As the region being evacuated reaches low pressure levels, the engineer realizes that the inlet flow to the pump may no longer be in the continuum regime. The inlet valve is 1 in. in diameter and opens a maximum distance of $\frac{1}{8}$ in. on the intake stroke. The mean free path for a gas is given by $\lambda = m/(\sqrt{2}\pi d^2 \rho)$ where m = mass of molecule = molecular weight/Avogadro's number. For air, $m = 4.8 \times 10^{-23}$ gram, and $d \simeq 3.7 \times 10^{-8}$ cm. If the air tank is at a condition of $T = 50°F$ and $p = 0.25$ psia, is the continuum assumption for the inlet to the pump still valid?

1.2 The momentum flux for one-dimensional flow is given by $\dot{m}V$, where $\dot{m}$ is mass rate of flow and V is velocity. (a) What relationship is there between the units of $\dot{m}V$ and those of force? (b) If the exit velocity is 20 fps and the rate is given as 5 lb_m/min, determine an appropriate momentum flux.

1.3 Work is defined as $\int \mathbf{F} \cdot d\mathbf{s}$. (a) What is the work done in one revolution by the string acting on a ball that is being whirled in a horizontal circle, assuming negligible drag? (b) What is the work done by the aerodynamic drag D on the ball in one revolution? ("Drag" is the force opposing the motion of the air on the ball.)

1.4 A scalar potential function ϕ is given by $\phi = 4x^2 + yz$. What is the associated conservative vector field? Is the curl of this vector field zero? Determine the circulation of this vector field around a unit circle in the xy-plane with center at the origin.

PROBLEMS

1.5 A vector field **A** is given by $\mathbf{A} = x^2\hat{i} + 2xy\hat{j} - z^2\hat{k}$. Is there a scalar potential function associated with this vector field? Determine $\nabla \cdot \mathbf{A}$ and $\nabla \times \mathbf{A}$, at $(2, -1, 4)$.

1.6 Determine the rate at which a function $\phi = x^2 + y^2$ changes along the line $x = 3y$ at the point $(6, 2)$.

1.7 Calculate the circulation of $\mathbf{A} = y^2\hat{i} + xy\hat{j} + z^2\hat{k}$ around a triangle with vertices at the origin, $(4, 4, 0)$ and $(0, 4, 0)$. Find the answer both by integrating around the triangle and by using Stokes' theorem.

1.8 (a) Find the scalar potential function associated with the vector function $\mathbf{A} = e^{2x}\hat{i} + 2yz\hat{j} + y^2\hat{k}$. (b) Determine the divergence of **A** at $(0, 2, 2)$.

1.9 By expanding in cartesian coordinates, show that (a) the divergence of the curl of a vector function $\nabla \cdot \nabla \times \mathbf{A} = 0$ and (b) the curl of the gradient of a scalar function $\nabla \times \nabla\phi = 0$.

1.10 The velocity vector field in a fluid flow is given by $\mathbf{V} = 2xy\hat{i} + t^2\hat{k}$. Determine $(\nabla \times \mathbf{V}) \times \mathbf{V}$ and $(\mathbf{V} \cdot \nabla)\mathbf{V}$ at the point $(1, -2, 0)$.

1.11 The temperature in a flow is measured at $120°F$ at the point $(2, 1)$. The vector gradient is determined to be $\nabla T = 10\hat{i} - 5\hat{j}$. Approximate the temperature at the point $(2.3, 0.5)$.

1.12 Pressure forces act on the infinitesimal elements shown. For the first element the pressure acting on the left and bottom faces is assumed to be equal to the pressure at the lower left corner. For the second element the pressure is $p(x, y)$ at the center of the element and the pressure on each face is assumed to be equal to the pressure acting at the middle of the face. For the third element the pressure is $p(x, y)$ at the lower left-hand corner and the pressure on each face is also assumed to be equal to the pressure acting at the middle of the face. The infinitesimal pressure forces shown on the last two elements are determined using Taylor's series with higher-order terms neglected. (a) Determine all the pressure forces on the top and right-hand surfaces. (b) Sum pressure forces in the x-direction on each element and show that identical results occur. (c) Determine which element is the simplest to work with.

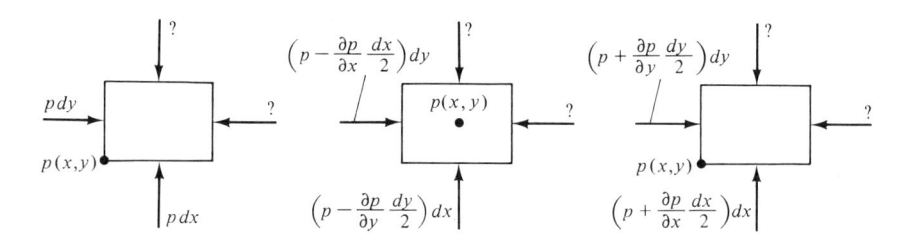

Prob. 1.12

1.13 Verify that $(\mathbf{V} \cdot \nabla)\mathbf{V} = \nabla(V^2/2) + (\nabla \times \mathbf{V}) \times \mathbf{V}$ by expanding both sides, using

$$\mathbf{V} = u\hat{i} + v\hat{j} + w\hat{k}$$

1.14 In cylindrical coordinates

$$\nabla = \hat{e}_r \frac{\partial}{\partial r} + \frac{\hat{e}_\theta}{r} \frac{\partial}{\partial \theta} + \hat{e}_z \frac{\partial}{\partial z}$$

and

$$\mathbf{V} = v_r \hat{e}_r + v_\theta \hat{e}_\theta + v_z \hat{e}_z$$

Determine $\nabla \cdot \mathbf{V}$ and compare with that of Table C.1.

1.15 Determine the acceleration vector field for a particular fluid flow if the velocity vector is given by the Eulerian description $\mathbf{V} = 2x\,\hat{i} - 2y\,\hat{j}$.

1.16 The density in a flow field is given by $\rho = 0.2(10 - y)e^{x/20}$, and the velocity field is $\mathbf{V} = 10(2y - y^2)\hat{i}$. Determine the rate of change of ρ at the point $(1, 1/2, 0)$. Also calculate the acceleration.

1.17 A body is inserted into an otherwise uniform stream of air and then withdrawn. Sketch the pathline of a fluid particle that had to move around the body, tracing the history of the particle from a time before the body was inserted until a time after the body was withdrawn. Also sketch a typical streakline of fluid particles and the streamline pattern at some instant after the body has been withdrawn for some time.

1.18 The temperature distribution is $T(y, t) = 10(4 - y^2)\sin\frac{\pi}{2}t$ in a flow in which $\mathbf{V} = 40y\,\hat{i}$. Find the rate at which the temperature of a fluid particle is changing at the point $(0, 1)$.

1.19 Water is flowing out of a large farm crop sprinkler which has 8-ft horizontal arms. Nozzles at the ends of the arms are at a 45° angle with respect to the ground and 90° to the arms. If the sprinkler is rotating at 25 rpm, what is the absolute acceleration of the water as it leaves the sprinkler? The velocity of the water relative to the nozzle is 50 fps.

1.20 The Mississippi River is about 1 mile wide. The surface of the river is normal to the "body-force" vector. At a latitude of 45°, determine how much higher one side is than the other. Which side is higher? The water travels at 5 fps.

1.21 An automobile is spinning out of control on an icy street. Its motion with respect to a fixed observer can be described as $\mathbf{S}$ to the center of gravity of the car plus $\mathbf{\Omega} = \Omega\hat{k}$ as the car spins about its center of gravity. (The center of gravity of the vehicle is above the transmission in the passenger compartment.) Working from your own sketch, describe what effective body forces this motion would result in for the gasoline droplets in the intake manifold. That is, if one were to express the acceleration of these droplets as $\mathbf{a}$ in Eq. 1.50, what body-force effects would have to be accounted for?

1.22 Determine the pressure (or vacuum) in the water pipe shown, in both psi and inches of water.

1.23 Determine the difference in pressure between the water pipe and the oil pipe shown.

PROBLEMS

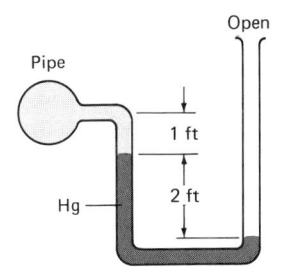

Prob. 1.22

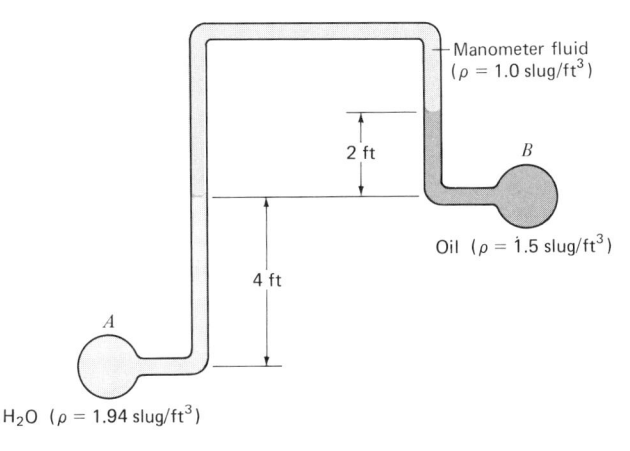

Prob. 1.23

1.24 The manometer shown was originally open, with the mercury levels 1 ft below the horizontal section. The top was then closed and the valve opened to allow an air pressure of 20 psi to be applied as shown. Determine the final reading H of the manometer.

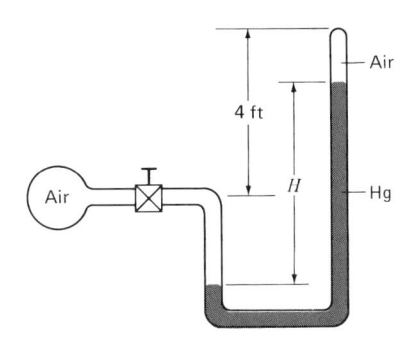

Prob. 1.24

1.25 For an assumed isothermal atmosphere, determine the pressure variation as a function of elevation. At what elevation would you expect a pressure of 2 psia? of 0.2 psia? $T = 60°F$.

1.26 Assume that the temperature varies linearly in the atmosphere, is 70°F at zero elevation, and $-50°F$ at 30,000 ft. Determine the pressure at 30,000 ft.

1.27 What force is necessary to hold shut the 5-ft-wide gate shown? Neglect the weight of the gate.

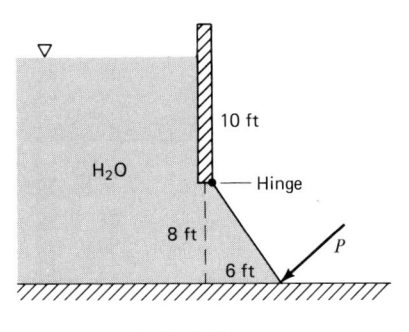

Prob. 1.27

1.28 The gate shown opens when the water reaches a certain height. Determine that height.

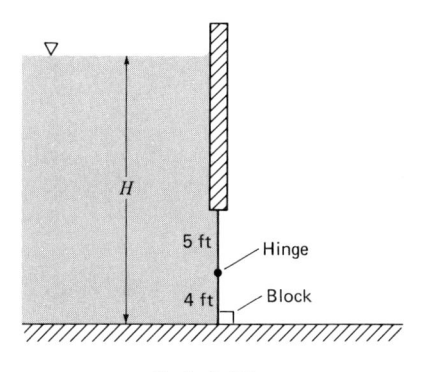

Prob. 1.28

1.29 The gate shown opens and closes automatically. At what depth of water H will it open? Neglect the weight of the gate.

1.30 Below Eq. 1.75 the statement is made that F_V acts through the centroid of the equivalent volume of liquid contained above the surface. Show that this is true by summing moments about an arbitrary reference axis. Does this result depend upon the liquid being of constant density?

1.31 The 800-lb, 10-ft-wide semicircular gate is to be opened as shown.

PROBLEMS

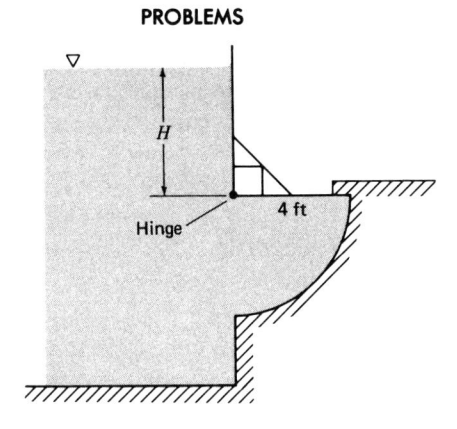

Prob. 1.29

What force P is needed if the center of gravity of the gate is 1 ft to the left of the hinge?

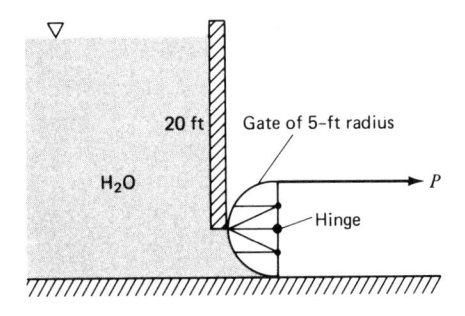

Prob. 1.31

1.32 The spring shown provides a force of 80,000 on the 5-ft-wide curved gate. At what height H will the gate just open?

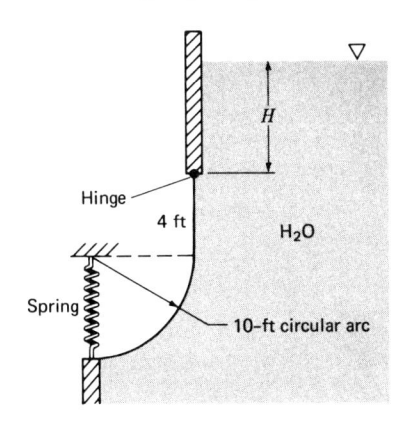

Prob. 1.32

1.33 A circular cylinder is used as a control device as shown. Assume the cylinder is held in place at the bottom contact. Determine the total reaction between the cylinder and the bottom contact surface. The cylinder weighs 5000 lb per foot of length.

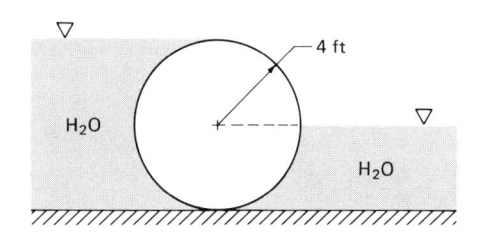

Prob. 1.33

1.34 Using the fact that the vertical force acting on a curved surface is equal to the weight of liquid contained above it (or that could be contained above it), show that the buoyancy force is equal to the weight of displaced liquid.

1.35 The tank is accelerated at a rate of 16.1 ft/sec². Determine the force acting on the rear end of the 8-ft-wide tank.

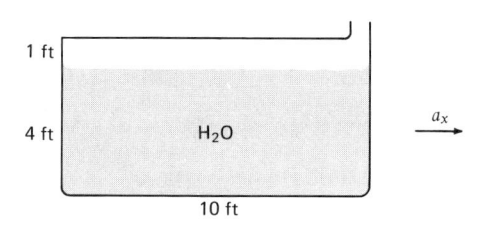

Prob. 1.35

1.36 A tank with dimensions identical to those of Prob. 1.35 is filled with water and initially pressurized to 10 psi at the top. It is then accelerated at 20 ft/sec² up a 30° incline. Find the maximum pressure in the tank.

1.37 The U-tube shown is rotated about its left leg at 400 rpm. Determine the pressure at point A. Now the tube is rotated about the right leg at the same rate. What is the pressure at point A now?

1.38 The 2-ft-dia. can shown is rotated at 100 rpm (a) Find the force acting on the bottom of the can. (b) What value of Ω would be necessary to just start to bare the bottom of the can?

1.39 The can shown is pressurized to 5 psi. Determine the value of Ω necessary to produce a pressure of 10 psi at point A.

PROBLEMS

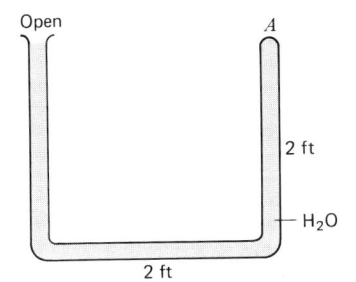

Prob. 1.37

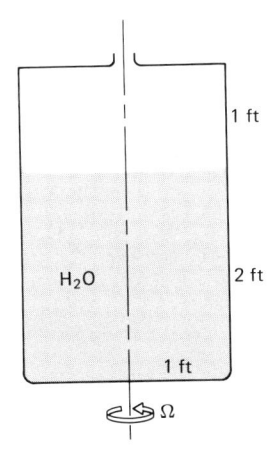

Prob. 1.38

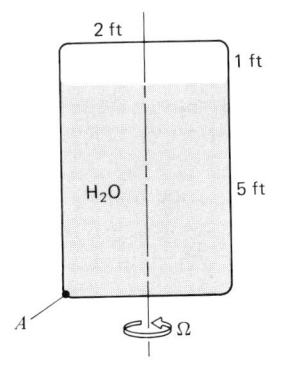

Prob. 1.39

1.40 Assume that water wets the surface of a capillary tube an angle of 30°. How high could water climb if the capillary tube is 10^{-5} in. in diameter? The water temperature is 50°F.

1.41 A low-volume fruit sprayer distributes droplets of highly concentrated spray in a jet of air from a large fan. What pressure would exist in the 40-micron-diameter droplets? (One micron equals one millionth of a meter.) Use $\sigma = 0.004$ lb/ft.

1.42 Two plates oriented as shown are placed in water. Determine the rise of water between the plates, that is, $h(x)$. $\beta = 0°$. Use $\sigma = 0.005$ lb/ft.

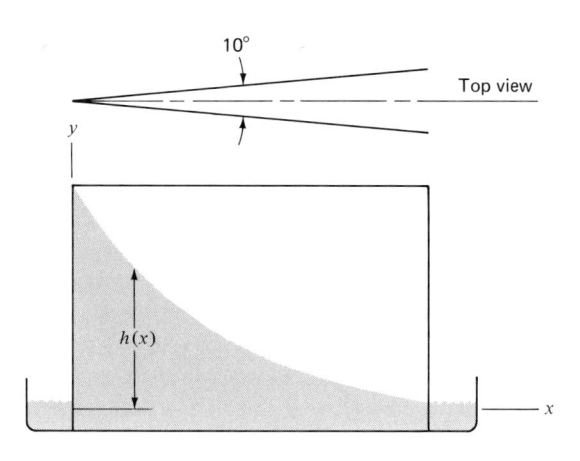

Prob. 1.42

1.43 A vacuum cleaner is rated to create a vacuum of 4 in. of water just inside the hose. What velocity of the air would you expect there? Use $\rho = 0.0024$ slug/ft^3.

1.44 Air is drawn from the atmosphere through a wind tunnel by a blower at the exit end. The velocity of the air in the rectangular test section is 100 mph. The top of the $\frac{1}{2}$-in-thick test section can be considered to be a simply loaded beam. From solid mechanics the maximum deflection of a simple beam due to a distributed load is $5wL^4/384EI$, where w is the load per foot of length. The material is steel, for which $E = 30 \times 10^6$ psi. Determine the maximum deflection of the 10-ft-wide top.

1.45 The flow from a large container is depicted in the figure. This phenomenon can be analyzed by the Bernoulli equation to obtain

$$V_e = \sqrt{2gh}$$

Develop this equation and carefully note *all* the restrictions which must be invoked to obtain this simple form.

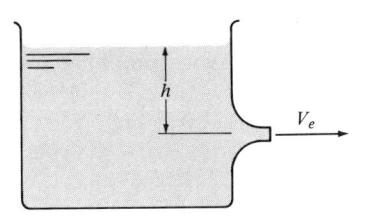

Prob. 1.45

1.46 In Prob. 1.45 a rather simple flow field was investigated and the result $V_e = \sqrt{2gh}$ was established. Is V_e an Eulerian or a Lagrangian quantity? What physical object has the velocity V_e? Consider a streamline extending from the free surface to the exit plane and consider a point along this streamline just before the exit nozzle and at the same elevation as the nozzle. At this point, is $\partial V/\partial t > 0$? < 0? $= 0$? At this point, is $a > 0$? < 0? $= 0$? Explain the difference between these two quantities.

1.47 A pitot tube is used to measure the air speed of a Piper Cub airplane. The instrument measures 6 in. of water. What is the speed of the Piper Cub?

1.48 Air flows in the pipe shown at 100 fps. (a) What H value will the manometer read if the pressure in the pipe is 20 psi? (b) If the open end is connected to the pipe, determine H. The density of mercury is $13.6\rho_{H_2O}$.

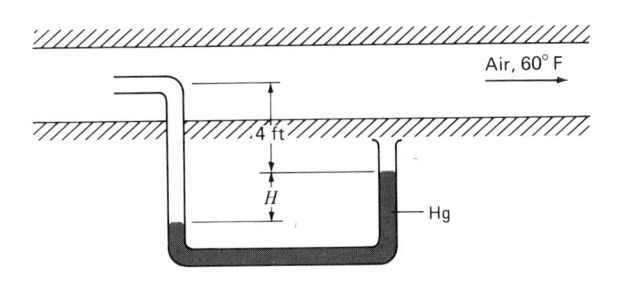

Prob. 1.48

1.49 In the preceding problem, let it be water that is flowing at 100 fps. Determine H for the manometer as shown and also calculate H if the open end of the manometer is connected to the pipe.

1.50 A burglar is siphoning water from a water tank on a steam tractor. (The burglar doesn't need the water but he loves to steal!) If the $\frac{1}{2}$-in-dia. hose terminates 6 ft below the water level how long will it take him to steal a gallon of water? Neglect viscous effects.

1.51 The muzzle velocity of a 30-caliber-rifle is of the order of 3000 fps. Can the flow around the projectile as it is traveling through the air be assumed to be incompressible?

1.52 The aerodynamic foil used on high-speed racing cars is designed by an engineer to give maximum force with minimum drag. The cars travel at speeds up to 200 mph in temperatures of 110°F. He assumes an incompressible air flow. Is this reasonable?

SELECTED REFERENCES

Three motion pictures in the NCFMF/EDC Series* are recommended as supplements to this chapter (numbers are catalog numbers): *Flow Visualization* (S. J. Kline, film principal; No. 21607), *Eulerian and Lagrangian Descriptions in Fluid Mechanics* (J. L. Lumley, film principal; No. 21621), and *Surface Tension in Fluid Mechanics* (L. M. Trefethen, film principal; No. 21609).

Such introductory concepts as *continuum, fluid*, and the like, are discussed in greater detail by G. K. Batchelor in *An Introduction to Fluid Dynamics*, Cambridge University Press, London, 1967. An even more comprehensive and fundamental discussion is given by M. J. Lighthill (see Part II of Section I) in the volume edited by L. Rosenhead, *Laminar Boundary Layers*, Oxford University Press, Oxford, 1963.

For a more detailed discussion of the thermodynamic considerations, see: J. E. Lay, *Thermodynamics*, Merrill Books, Columbus, Ohio, 1963; J. B. Jones and G. A. Hawkins, *Engineering Thermodynamics*, John Wiley & Sons, 1960; and G. J. Van Wylen, *Thermodynamics*, John Wiley & Sons, New York, 1959.

The description of fluid motion presented by Y. S. Yuan in his discussion of that subject parallels our own nicely. See Chapter 3 of *Foundations of Fluid Mechanics*, Prentice-Hall, Englewood Cliffs, N.J., 1967.

For a discussion and derivation of acceleration in a noninertial reference frame, see the following: G. W. Housner and D. E. Hudson, *Applied Mechanics and Dynamics*, Van Nostrand, New York, 1950; S. W. McCuskey, *Introduction to Advanced Dynamics*, Addison-Wesley Publishing Co., Reading, Mass., 1959; and R. L. Hoffman, *Dynamics*, Vol. 1, Addison-Wesley Publishing Co., Reading, Mass., 1962.

Fluid statics is covered adequately in nearly all reference books on fluid mechanics. For a presentation which emphasizes the generalized static condition, including noninertial effects, see A. G. Hansen, *Fluid Mechanics*, John Wiley & Sons, New York, 1967. For a discussion of capillary and surface-tension effects, see G. K. Batchelor, *An Introduction to Fluid Dynamics*, Cambridge University Press, London, 1967, and J. W. Daily and D. R. F. Harleman, *Fluid Dynamics*, Addison-Wesley Publishing Co., Reading, Mass., 1968. And for characteristic problems and solutions, see R. V. Giles, *Fluid Mechanics and Hydraulics*, 2d ed., Schaums Outline Series, McGraw-Hill Book

---

*Available from Encyclopaedia Britannica Educational Corporation, 425 N. Michigan Ave., Chicago, Ill. 60611. All catalog numbers refer to this series if not otherwise credited. A supplement to these films is provided by the reproduction (still pictures) and discussion of the experiments in *Illustrated Experiments in Fluid Mechanics*, MIT Press, Cambridge, Mass., 1972.

SELECTED REFERENCES

Co., New York, 1962, and W. F. Hughes and J. A. Brighton, *Fluid Dynamics* Schaums Outline Series, McGraw-Hill Book Co., New York, 1967.

The tables of thermodynamic properties included in the Appendix are meant to be characteristic and instructive, not exhaustive. For a more complete listing of certain thermodynamic (and gas dynamic) properties of the common gases (viscosities not included), see J. H. Keenan and J. Kayes, *Gas Tables*, John Wiley & Sons, New York, 1945; similarly, for steam: J. H. Keenen and F. G. Kayes, *Thermodynamic Properties of Steam*, John Wiley & Sons, New York, 1936. A good source of information on physical constants is to be found in the Chemical Rubber Co. (C.R.C.) Handbook Series. See, in particular, *Handbook of Applied Engineering Science*, edited by R. E. Bolz and G. L. Tuve; The Chemical Rubber Co. Press, Cleveland, 1970.

Most listings of engineering physical constants and thermodynamic data are presented in the English system of units. A document which provides a complete discussion of the basic dimensions and the conversions between systems of units (and one which the reader may wish to obtain for his personal library) is that by E. A. Mechtly, "The International System of Units." The Scientific and Technical Information Division of the National Aeronautics and Space Administration lists it as NASA-SP-7012 (1964).

2

The Control-Volume Form of the Fundamental Laws

2.1 INTRODUCTION

If the material of an undergraduate fluid mechanics course were subdivided into large topical areas, the control-volume form, that is, integral equation form, of the conservation laws* would be the most important; this material is covered in this detailed chapter and made lengthy by the many examples. Identifying this segment of the course as "most important" is clearly a subjective assessment; it deserves an explanation.

Essentially three paths may be foreseen for the student who completes this course: a career choice which does not require technical (or at least thermal-science) skills, engineering work with a bachelor's degree, or pursuit of graduate study. Because the control-volume formulation is essentially a problem-solving tool, this material will strengthen the analytical thought processes and skills which have allowed engineering graduates to do well in nontechnical capacities. For the student who will engage in technical work, the control-volume

---

*As you are well aware at this point in your studies, the word "law" does not imply the result of a legislative process, nor is it something we consider to be too sacred to question. It is a statement of an event which occurs uniformly in nature under the same conditions. The mathematical description of the event has been developed from universal scientific experience (i.e., experiments) and a law has never been violated within the stated constraints.

formulation of the fundamental laws will be the primary tool which he will use to solve engineering problems involving fluid mechanics. The student who pursues graduate studies in the thermal-science area will find that the dominant (if not exclusive) character of his course work will involve the differential equation version of these basic laws. Consequently, the material covering the differential equations, included in the later chapters, will be an important prerequisite to graduate study; however, this may be his only formal exposure to the control-volume statements of the fundamental laws.

Because of the considerable length of this chapter, a brief description of its organization and intent is in order. A formal mathematical expression is developed which allows the time rate of change of an extensive property for a given quantity of mass, a system, to be expressed in terms of quantities related to a specific region in space, a control volume. (The control volume may or may not be deformable.) Various forms of the general equations are established for an arbitrary extensive property such as momentum or energy. The specific equations for the fundamental laws of conservation of mass, Newton's second law of motion, and the first law of thermodynamics are next developed and specific examples are used to clarify the exact nature of each of the terms. Once sufficient exposure and experience is gained for the student to have a firm grasp of what each term in the various equations represents, the final portion of the chapter deals with the "artistic" part of the problem-solving process—the selection of an appropriate control volume and the appropriate equations. Since the equations are universally correct and since the earlier portions of the chapter will establish what the terms represent for a given control volume, the student's capability is established when he can choose a control volume which allows the desired answer to be obtained deductively from the general equations. This is quite similar to the choice the student had when choosing a free-body diagram in his mechanics courses.

This chapter is so organized that only nondeformable (fixed) control volumes need be considered. Fixed control volumes are encountered in the majority of engineering applications; and it is desirable to avoid the complexity of the "system to deformable control volume" transformation unless sufficient time is available for an adequate consideration of this more general approach. The deformable control volume, which allows the analysis of many interesting problems, is included in this chapter for current use as appropriate or for future reference in the event it is not covered explicitly in this course. There will be certain examples and problems involving moving control boundaries that are particularly suited to the deformable capability. These particular problems and examples will be starred. The starred sections, examples, and problems may be omitted with no penalty in the remaining chapters since the remaining chapters are not dependent on the deformable-control-volume capability.

2.2 THE CONTROL VOLUME

The fundamental laws governing the motion of a fluid are Lagrangian descriptions; for example, Newton's second law states that the vector summation of all the forces acting *on a particular mass* is equal to the rate at which the momentum of that mass is changing. Here the crucial point is that our basic laws apply to specified mass, or system, which is, in general, moving in space. This is a Lagrangian description of the motion of the mass particles. Since we are not fundamentally interested in a particular group of particles, we focus our attention on an Eulerian description where we define a region in space and observe the fluid flowing through it. This defined region in space is called a *control volume* and our objective will be to describe the rate at which an integral quantity associated with a system is changing as the flow passes into and out of the control volume. Fluid does not pass in and out of a system. Some typical control volumes are shown in Fig. 2.1; parts a and c represent fixed control volumes and part b represents a deformable control volume.

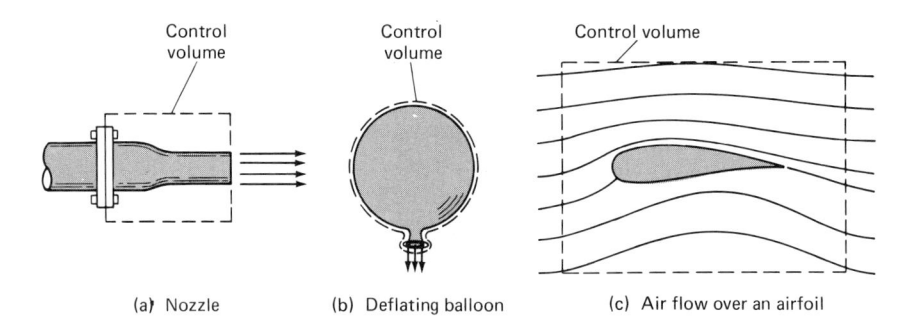

Fig. 2.1. Typical control volumes.

The relationship to be developed is one which will allow the rate of change of an *extensive property* for a system to be expressed in terms of a defined region in space. Mass, energy, and entropy are examples of extensive scalar properties; momentum and moment of momentum are examples of extensive vector properties. The word "system" as used here is defined as in thermodynamics: a fixed collection of mass particles.

As an example of these concepts consider an inflated balloon which is released at time $t = 0$, allowing the air to escape. The system is defined as the air within the balloon at time $t = 0$. The mass m of the system is an extensive property and the governing differential equation for m is $dm/dt = 0$. The center of gravity of the system has a velocity $\mathbf{V}$; the momentum of the system is $m\mathbf{V}$ and the governing equation for the momentum is $\mathbf{F} = d(m\mathbf{V})/dt$. It is clearly possible to describe this physical problem in these terms; it is also quite apparent that such a description will be of little value in calculating the behavior of the balloon. This is because the system, or Lagrangian, formulation focuses on the air; a more desirable description would allow us to focus attention on the balloon—or the region of space occupied by the balloon. It would be possible to make such a transformation from the system to a region in space for every new problem of interest; instead, a far more practical and useful scheme is proposed. First, we develop the following general relationship, which allows the time rate of change of a system's extensive property to be expressed in terms of a control volume, a defined region of space that, in general, is not fixed. We will then use this general relationship to solve particular problems.

Let N be the extensive property. It would be calculated by integrating its corresponding intensive property η over the volume of interest, that is,

$$N = \int_V \eta \rho \, d V \tag{2.1}$$

Thus, η is N per unit mass. The values of N for a system and a control volume, respectively, are given as

$$N_{\text{sys}} = \int_{\text{sys}} \eta \rho \, d V \qquad N_{\text{c.v.}} = \int_{\text{c.v.}} \eta \rho \, d V \tag{2.2}$$

The time rate of change of N_{sys} is

$$\frac{DN_{\text{sys}}}{Dt} = \frac{D}{Dt} \int_{\text{sys}} \eta \rho \, d V$$

$$= \lim_{\Delta t \to 0} \frac{N_{\text{sys}}(t + \Delta t) - N_{\text{sys}}(t)}{\Delta t}. \tag{2.3}$$

From Eq. 2.3, the generalized system-to-control-volume transformation will be derived for a fixed control volume (in Section 1 of this article, immediately following) and for a deformable control volume (in Section 2). (It is not necessary to study both sections; however, the starred examples and problems are dependent on the derivation in Section 2.)

1. The Nondeformable Control Volume

The nondeformable, or fixed, control volume is encountered in such engineering applications as flow over an aircraft, past a turbine blade, over a dam, through a pipe, around a ship, and out of a rocket nozzle. The defining terms are presented in Fig. 2.2, which represents a cross-sectional slice through the control volume.* The system and the control volume are coincident at time t. At time $t + \Delta t$ the control

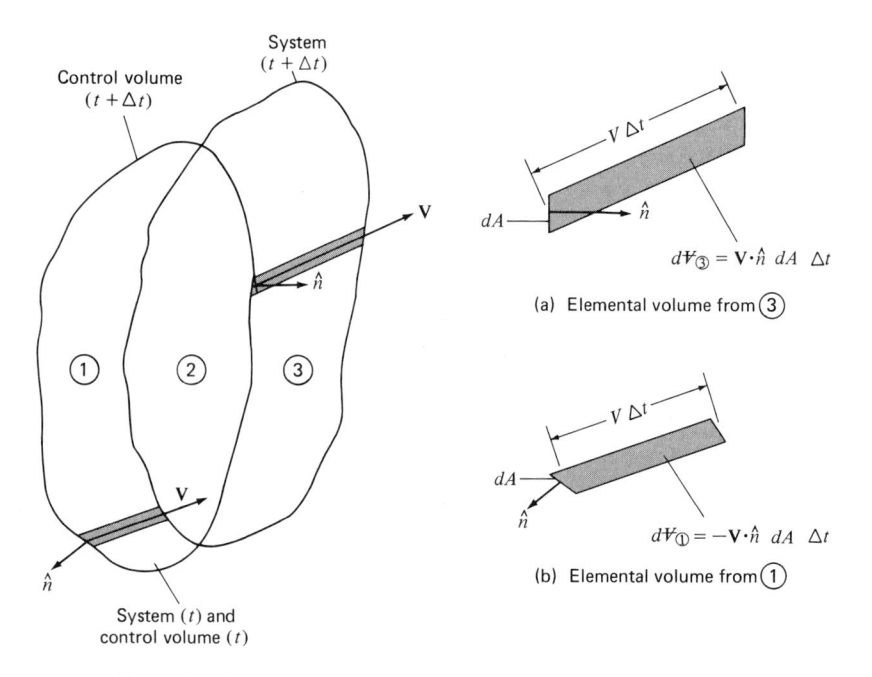

Fig. 2.2. The system and the nondeformable control volume.

*The potato-like shape is meant to suggest that the control volume may assume any form; in the application of the control-volume analysis you would place the control volume around the device or region of interest. For example, it is not difficult to imagine a control volume which would encompass the aircraft, turbine blade, dam, pipe, or nozzle mentioned above. The proper selection of the control volume is an important aspect of such problem solutions; this skill will be developed by practice.

Art. 2.2 THE CONTROL VOLUME

volume occupies its original region in space since it is fixed, but the system has moved to a new position, as indicated in the figure. The quantities in Eq. 2.3 may be expressed as

$$N_{\text{sys}}(t + \Delta t) = N_3(t + \Delta t) + N_2(t + \Delta t) \tag{2.4}$$

and

$$N_{\text{sys}}(t) = N_2(t) + N_1(t) \tag{2.5}$$

This allows us to write Eq. 2.3 as

$$\frac{DN_{\text{sys}}}{Dt} = \lim_{\Delta t \to 0} \frac{N_3(t + \Delta t) + N_2(t + \Delta t) - N_2(t) - N_1(t)}{\Delta t}$$

$$= \lim_{\Delta t \to 0} \frac{N_2(t + \Delta t) + N_1(t + \Delta t) - N_2(t) - N_1(t) + N_3(t + \Delta t) - N_1(t + \Delta t)}{\Delta t} \tag{2.6}$$

where, in the last expression, we have both added and subtracted $N_1(t + \Delta t)$ in the numerator to put it in the desired form.* Again referring to Fig. 2.2, we see that the expression may be written as

$$\frac{DN_{\text{sys}}}{Dt} = \lim_{\Delta t \to 0} \frac{N_{\text{c.v.}}(t + \Delta t) - N_{\text{c.v.}}(t)}{\Delta t} + \lim_{\Delta t \to 0} \frac{N_3(t + \Delta t) - N_1(t + \Delta t)}{\Delta t} \tag{2.7}$$

The first term on the right-hand side is by definition $dN_{\text{c.v.}}/dt$. Our equation now becomes

$$\frac{DN_{\text{sys}}}{Dt} = \frac{d}{dt} \int_{\text{c.v.}} \rho \eta \, dV + \lim_{\Delta t \to 0} \frac{N_3(t + \Delta t) - N_1(t + \Delta t)}{\Delta t} \tag{2.8}$$

The last term above may be written in integral form by recognizing that $N_3(t + \Delta t)$ is the total quantity of the extensive property N contained in region 3 at time $t + \Delta t$. Hence, $N_3(t + \Delta t)$ and likewise $N_1(t + \Delta t)$ are expressed as

$$N_3 = \int_{V_3} \eta \rho \, dV \quad \text{and} \quad N_1 = \int_{V_1} \eta \rho \, dV \tag{2.9}$$

---

*Note that the quantity $N_1(t + \Delta t)$ is not part of the system; rather, it refers to the value of the extensive property which is in the control volume but excluded from the system at time $t + \Delta t$.

each evaluated at time $t + \Delta t$. The incremental volume elements in each volume region may be written as shown in Fig. 2.2, so that

$$N_3(t + \Delta t) - N_1(t + \Delta t) = \int_{A_3} \eta\rho \, \mathbf{V} \cdot \hat{n} \, dA \, \Delta t - \int_{A_1} \eta\rho(-\mathbf{V} \cdot \hat{n}) \, dA \, \Delta t$$

$$= \int_{A_3 + A_1} \eta\rho \, \mathbf{V} \cdot \hat{n} \, dA \, \Delta t$$

$$= \int_{\text{c.s.}} \eta\rho \, \mathbf{V} \cdot \hat{n} \, dA \, \Delta t \quad (2.10)$$

where c.s. represents the *control surface*, which completely encloses the control volume. Substituting this expression in Eq. 2.8, the system-to-control-volume transformation is

$$\frac{DN_{\text{sys}}}{Dt} = \frac{d}{dt} \int_{\text{c.v.}} \rho\eta \, dV + \int_{\text{c.s.}} \eta\rho \, \mathbf{V} \cdot \hat{n} \, dA \quad (2.11)$$

Because the control volume (i.e., the limits on the volume integral) is not a function of time, the time derivative may be moved inside the integral,* so that Eq. 2.11 may be restated as

$$\frac{DN_{\text{sys}}}{Dt} = \int_{\text{c.v.}} \frac{\partial}{\partial t}(\eta\rho) \, dV + \int_{\text{c.s.}} \eta\rho \, \mathbf{V} \cdot \hat{n} \, dA \quad (2.12)$$

In either form—Eq. 2.11 or Eq. 2.12—this important equation is often referred to as the *Reynolds transport theorem*. We use D/Dt on the left since we are following a system, a particular collection of fluid particles, and d/dt on the right of Eq. 2.11 since we are simply differentiating a function of time.

We will refer to the first term on the right side of Eqs. 2.11 and 2.12 as the *time-rate-of-change term* and the second term as the *flux term*; both expressions refer to the extensive property N. Several characteristics of the equation are noteworthy. Only the normal component of the velocity contributes to the flux term. The flux term may reflect an unsteady-flow character in that ρ, η, and $\mathbf{V}$ may be time dependent. Note that only η evaluated at the control surface contributes to the flux term.

Many problems involve steady flows, so that $\partial(\eta\rho)/\partial t = 0$. The Reynolds transport theorem then takes the simplified form

$$\frac{DN_{\text{sys}}}{Dt} = \int_{\text{c.s.}} \rho\eta \, \mathbf{V} \cdot \hat{n} \, dA \quad (2.13)$$

*This follows directly from Leibniz' rule (from calculus).

Art. 2.2　　　　　　　　THE CONTROL VOLUME

Finally, it is of interest to indicate how the nondeformable-control-volume formulation relates to the deformable formulation. For the nondeformable control volume, the fluid crosses the control surface through the elemental area dA with the velocity $\mathbf{V}$ measured with respect to a coordinate system usually attached to the control volume. For the deformable control volume, the control surface may also be in motion with a velocity $\mathbf{V}_b$; the fluid crosses the control surface with a relative velocity $\mathbf{V}_r$ and the three velocities are related by the expression $\mathbf{V} = \mathbf{V}_b + \mathbf{V}_r$. Thus, the fixed-control-volume equations will result if we let $\mathbf{V}_b = 0$ and $\mathbf{V}_r = \mathbf{V}$ in the deformable-control-volume equations.

If only the fixed control volume is of interest, omit Section 2 of this article and Sections 1, 2, and 3 of Art. 2.3. The next section pertinent to the nondeformable control volume is Section 4 of Art. 2.3. The starred examples and problems are designed for the discussion of deformable control volumes.

Example 2.1

Consider the problem of an unsteady flow of fluid through a pipe contraction as shown in Fig. E2.1A. Identify regions ①, ②, and ③ after a short time Δt has elapsed. For an arbitrary extensive scalar property N, discuss the meaning of the control volume terms in Eqs. 2.11 and 2.12.

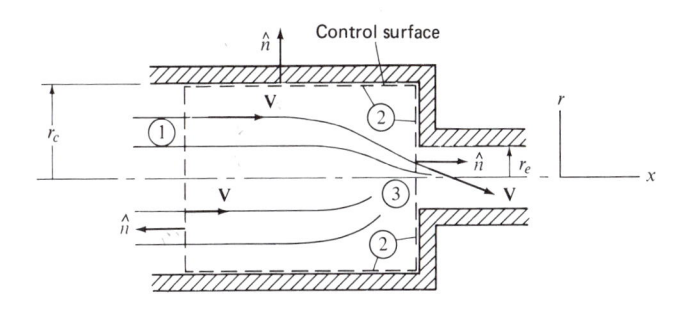

Fig. E2.1A

Solution. The volume designated as region ① of Fig. 2.2 contains the material inside the control volume which is not part of the system at time $t + \Delta t$. (The system is the collection of mass particles which occupy the control volume at time t.) The boundary between ① and ② as shown in Fig. E2.1B, is identified by the movement of non-system material into the control volume. Each segment of the boundary can be defined by the displacement of the original ($t = 0$) boundary as $\mathbf{V}\, \Delta t$. Similarly, the region designated ③ in Fig. 2.2 contains the material of the system which has left the control volume.

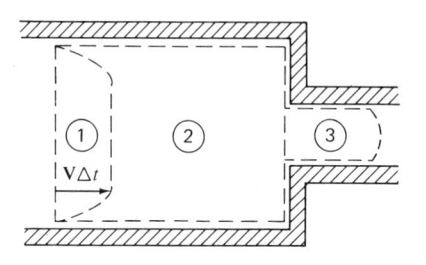

Fig. E2.1B

The time-rate-of-change term $\dfrac{d}{dt}\displaystyle\int_{c.v.} \eta\rho\, dV$ states that the values of η and ρ are to be evaluated at the location of each differential volume element and the product $\eta\rho\, dV$ is to be summed over the control volume. This provides $N(t)$ for the control volume. The time-rate-of-change term is then the derivative of $N(t)$.

The term $\displaystyle\int_{c.v.} \dfrac{\partial}{\partial t}(\eta\rho)dV$ in Eq. 2.12 implies that one should examine the time rate of change of the $(\eta\rho)$ product at all points inside the control volume at an instant. That is, at a particular point (x, y, z), $\eta(t)$ and $\rho(t)$ would provide the information from which we could find $\partial(\eta\rho)/\partial t$. With such values for all the points in the control volume, the above integral could be evaluated at time t.

The flux term is always zero for areas on which $\mathbf{V}\cdot\hat{n}$ is zero, that is, on areas where $\mathbf{V}$ is normal to $\hat{n}$. Some typical $\mathbf{V}$ and $\hat{n}$ vectors are shown in Fig. E2.1A. On area ② vector $\mathbf{V}$ is everywhere normal to the $\hat{n}$ vector (no flow occurs through the solid wall) so that the flux term is zero for area ②. On area ① the velocity vector $\mathbf{V}$ is opposite to the normal vector so that $\mathbf{V} = V_1\hat{i}$ and $\hat{n} = -\hat{i}$. The dot product on area ① then gives $\mathbf{V}\cdot\hat{n} = -V_1$ where V_1 is a function of r. Now consider the product $\mathbf{V}\cdot\hat{n}$ on area ③: The streamlines are not parallel to the exit pipe wall so that the dot product gives $\mathbf{V}\cdot\hat{n} = (V_x)_3$ where V_x is the component of the velocity vector in the x-direction. The flux term can now be written as

$$\int_{c.s.} \rho\eta\, \mathbf{V}\cdot\hat{n}\, dA = \int_1 \rho\eta\, \mathbf{V}\cdot\hat{n}\, dA + \int_2 \rho\eta\, \mathbf{V}\cdot\hat{n}\, dA + \int_3 \rho\eta\, \mathbf{V}\cdot\hat{n}\, dA$$

$$= -2\pi\int_0^{r_c} \rho_1\eta_1 V_1 r\, dr + 2\pi\int_0^{r_e} \eta_3\rho_3(V_x)_3 r\, dr$$

where the differential area element is written as $dA = 2\pi r\, dr$.

---

*2. The Deformable Control Volume

The deformable control volume differs from the fixed control volume in that the control boundary is allowed to move, as for a reciprocating piston, the lung during breathing, or an artery during the peristaltic pumping cycle of the heart. Figure 2.3 presents the defining terms for the derivation. It represents a cross-sectional slice through the control volume.* Note that the system occupies the control volume at time t. At time $t + \Delta t$, the system will have translated and deformed and the control volume will also have translated and deformed—but differently from the system (recall the balloon problem). Referring to Fig. 2.3, we may then write

$$N_{sys}(t + \Delta t) = N_5(t + \Delta t) + N_4(t + \Delta t) + N_3(t + \Delta t)$$

(2.14)

$$N_{sys}(t) = N_3(t) + N_2(t) + N_1(t)$$

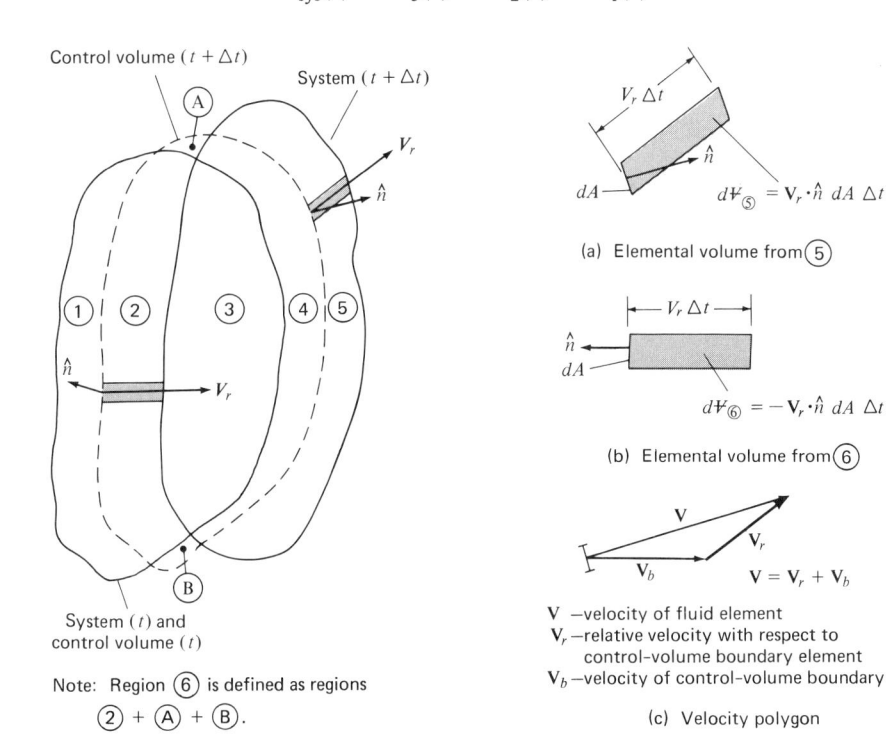

Fig. 2.3. The system and the deformable control volume.

*The irregular shape of the indicated control volume is meant to demonstrate that the applicability of the control volume is not restricted; in the application of the control-volume analysis you would place the control volume around the region or device of interest: for example, around the artery or the interior of the cylinder (excluding the piston) mentioned above. The proper selection of the control volume is an important aspect of such problem solutions; this skill will be developed by practice.

Substitute these into Eq. 2.3 and we have

$$\frac{DN_{sys}}{Dt} = \lim_{\Delta t \to 0} \frac{N_5(t + \Delta t) + N_4(t + \Delta t) + N_3(t + \Delta t) - N_3(t) - N_2(t) - N_1(t)}{\Delta t} \quad (2.15)$$

$$= \lim_{\Delta t \to 0} \left[\frac{N_4(t + \Delta t) + N_3(t + \Delta t) + N_6(t + \Delta t) - N_3(t) - N_2(t) - N_1(t)}{\Delta t} \right.$$

$$\left. + \frac{N_5(t + \Delta t) - N_6(t + \Delta t)}{\Delta t} \right]$$

You will note that we have added and subtracted $N_6(t + \Delta t)$ in the numerator to put it in the desired form. Referring again to Fig. 2.3, we see that Eq. 2.15 may be written as

$$\frac{DN_{sys}}{Dt} = \lim_{\Delta t \to 0} \frac{N_{c.v.}(t + \Delta t) - N_{c.v.}(t)}{\Delta t} + \lim_{\Delta t \to 0} \frac{N_5(t + \Delta t) - N_6(t + \Delta t)}{\Delta t}$$

$$= \frac{d}{dt} \int_{c.v.} \rho \eta \, dV + \lim_{\Delta t \to 0} \frac{N_5(t + \Delta t) - N_6(t + \Delta t)}{\Delta t} \quad (2.16)$$

The differential volume element in region ⑤ may be expressed as $dV_5 = \mathbf{V}_r \cdot \hat{n} \, dA \, \Delta t$ and in region ⑥ as $dV_6 = -\mathbf{V}_r \cdot \hat{n} \, dA \, \Delta t$, where $\hat{n}$ always points *out* of the control volume and $\mathbf{V}_r$ represents the velocity of the fluid relative to the control volume. Hence we may write

$$N_5(t + \Delta t) - N_6(t + \Delta t) = \int_{A_5} \rho \eta \, \mathbf{V}_r \cdot \hat{n} \, dA \, \Delta t + \int_{A_6} \rho \eta \, \mathbf{V}_r \cdot \hat{n} \, dA \, \Delta t$$

$$= \int_{c.s.} \rho \eta \, \mathbf{V}_r \cdot \hat{n} \, dA \, \Delta t \quad (2.17)$$

where c.s. represents the *control surface*, which completely encloses the control volume. Finally we arrive at the desired expression, which relates the rate of change in a quantity associated with a system to that associated with a volume in space, namely,

$$\frac{DN_{sys}}{Dt} = \frac{d}{dt} \int_{c.v.} \rho \eta \, dV + \int_{c.s.} \rho \eta \, \mathbf{V}_r \cdot \hat{n} \, dA \quad (2.18)$$

This is often referred to as the *Reynolds transport theorem*. We use D/Dt on the left since we are following a particular group of fluid particles, and d/dt on the right since we are only looking at a volume in space and *not* particular particles.

We will refer to the first term on the right-hand side as the *time-rate-of-change term* and the second as the *flux term*; both expressions refer to the extensive property N. Several characteristics of the equation are noteworthy. Only the normal component of the relative velocity contributes to the flux term. The flux term may reflect an unsteady-flow character in that $\mathbf{V}_r = \mathbf{V}_r(t)$ and a deformable control volume may lead to $\hat{n} = \hat{n}(t)$. Note that only η evaluated at the control surface contributes to the flux term.

The time-rate-of-change term may be nonzero for two distinct reasons: the integrand may be time-dependent, in which case $\partial(\eta\rho)/\partial t \neq 0$; or, the volume of the control volume $V_{\text{c.v.}}$ may not be constant; that is,

$$\frac{d}{dt} V_{\text{c.v.}} \neq 0$$

A deformable control volume is a necessary, but not sufficient condition for this latter inequality.

*Example 2.2

In order to help identify the physical meaning of the terms in Eq. 2.18, consider the problem of a pneumatic cylinder as shown in Fig. E2.2. For an arbitrary extensive property N, identify the meaning of the terms in Eq. 2.18 for the pneumatic cylinder and the control volume shown in these two cases: (a) zero leakage around the piston, and (b) leakage around the piston.

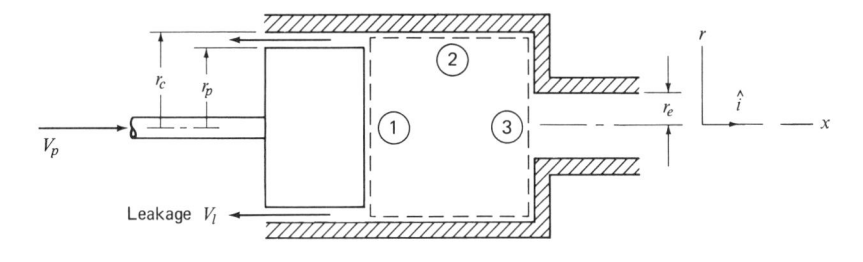

Fig. E2.2

Solution. The time-rate-of-change term and the flux term will be separately considered; each will be examined for the "instructions" provided by its mathematical nature. The time-rate-of-change term $\dfrac{d}{dt}\displaystyle\int_{\text{c.v.}} \eta\rho\, dV$ states that the values of η and ρ are to be evaluated at the location of each differential volume element and that the product of $\eta\rho\, dV$ is to be summed over the control volume at a given instant. This provides $N(t)$ for the control volume. The time-rate-of-change term is then the derivative of $N(t)$.

a. The flux term for the zero-leakage condition is

$$\int_{c.s.} \eta \rho \, \mathbf{V}_r \cdot \hat{n} \, dA = \int_{A_3} \eta_3 \rho_3 \, \mathbf{V}_3 \cdot \hat{n}_3 \, dA$$

since, over the remainder of the control surface, $\mathbf{V}_r \cdot \hat{n} = 0$. Because of the sharp-edged orifice at section ③, velocity $\mathbf{V}_3$ will not be perpendicular to or constant over area A_3. However, we may write $\mathbf{V}_3 \cdot \hat{n}_3 = (V_x)_3$ since $\hat{n}_3 = \hat{i}$. If η, V_x, and ρ are functions of r only, the flux integral may be written as

$$\int_{c.s.} \eta \rho \, \mathbf{V}_r \cdot \hat{n} \, dA = 2\pi \int_0^{r_e} (\eta \rho V_x)_3 r \, dr$$

b. If leakage occurs around the periphery of the piston, then such effects must be accounted for by the flux integral. Consequently, in addition to the foregoing, the flux term would include the term $\int_{A_1} \eta_1 \rho_1 \, \mathbf{V}_{r1} \cdot \hat{n}_1 \, dA$, where $\mathbf{V}_{r1}$ is the relative velocity at section ①. The unit vector $\hat{n}_1$ is given by $\hat{n}_1 = -\hat{i}$. Let the leakage velocity relative to the fixed $r - \theta - x$-coordinate system be $V_l(r)$; this represents fluid motion in the negative x-direction. The relative component of velocity which is responsible for the flux of N across the boundary is given by $\mathbf{V}_{r1} = -V_l \hat{i} + V_p \hat{i}$ since $\mathbf{V}_r = \mathbf{V} - \mathbf{V}_b$. Hence $\mathbf{V}_{r1} \cdot \hat{n} = V_l - V_p$. Again, for axisymmetric quantities,

$$\int_{c.s.} \eta \rho \, \mathbf{V}_r \cdot \hat{n} \, dA = 2\pi \int_0^{r_e} (\eta \rho V_x)_3 r \, dr + 2\pi \int_{r_p}^{r_c} (\eta \rho)_1 (V_l - V_p) r \, dr$$

This formulation recognizes that it is the velocity relative and normal to the control surface which accounts for the flux across the surface.

2.3 FURTHER ASPECTS OF THE CONTROL VOLUME

*1. Reformulation of the Time-Rate-of-Change Term

The time-rate-of-change term of the basic system-to-control-volume transformation may be reformulated to more easily account for the rate of deformation. Consider a generalized control volume at time t and let it deform, assuming the position shown in Fig. 2.4 at time $t + \Delta t$. We wish to reformulate the time-rate-of-change term $\dfrac{d}{dt} \int_{c.v.} \rho \eta \, d\mathcal{V}$. By definition of a derivative we may write

$$\frac{d}{dt} \int_{c.v.} \rho \eta \, d\mathcal{V} = \frac{dN_{c.v.}}{dt} = \lim_{\Delta t \to 0} \frac{N_{c.v.}(t + \Delta t) - N_{c.v.}(t)}{\Delta t} \quad (2.19)$$

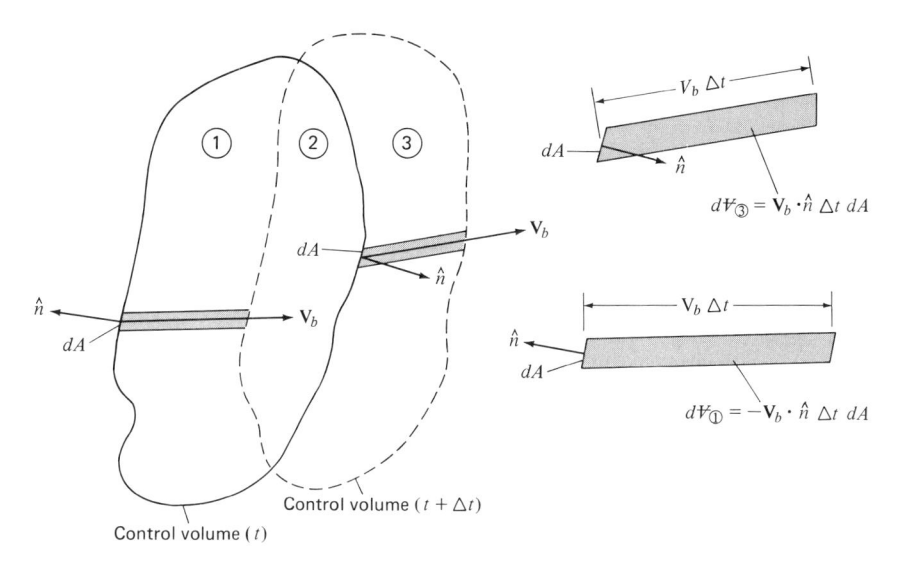

Fig. 2.4. The deformable control volume.

which, referring to Fig. 2.4, may be written

$$\frac{d}{dt}\int_{c.v.}\rho\eta\,dV = \lim_{\Delta t \to 0} \frac{N_3(t+\Delta t) + N_2(t+\Delta t) - N_2(t) - N_1(t)}{\Delta t} \tag{2.20}$$

or by rearranging,

$$\frac{d}{dt}\int_{c.v.}\rho\eta\,dV = \lim_{\Delta t \to 0} \frac{N_2(t+\Delta t) + N_1(t+\Delta t) - N_2(t) - N_1(t)}{\Delta t}$$

$$+ \lim_{\Delta t \to 0} \frac{N_3(t+\Delta t) - N_1(t+\Delta t)}{\Delta t} \tag{2.21}$$

The first term on the right accounts for the rate of change of N in the *fixed* volume which is occupied by the control volume at time t and thus is expressed as $\int_{c.v.} \frac{\partial}{\partial t}(\rho\eta)\,dV$.

The last term of Eq. 2.21 may be expressed, by reference to Fig. 2.4, as

$$\lim_{\Delta t \to 0} \frac{N_3(t+\Delta t) - N_1(t+\Delta t)}{\Delta t}$$

$$= \lim_{\Delta t \to 0} \frac{\int_{A_3}\eta\rho\,\mathbf{V}_b\cdot\hat{n}\,dA\,\Delta t + \int_{A_1}\eta\rho\,\mathbf{V}_b\cdot\hat{n}\,dA\,\Delta t}{\Delta t}$$

$$= \int_{c.s.}\eta\rho\,\mathbf{V}_b\cdot\hat{n}\,dA \tag{2.22}$$

There finally results

$$\frac{d}{dt}\int_{c.v.} \rho\eta \, d\mathcal{V} = \int_{c.v.} \frac{\partial}{\partial t}(\rho\eta) \, d\mathcal{V} + \int_{c.s.} \rho\eta \, \mathbf{V}_b \cdot \hat{n} \, dA \qquad (2.23)$$

For a nondeformable control volume, $\mathbf{V}_b$ is everywhere zero and the control volume is fixed in space.

We may now write the Reynolds transport theorem as

$$\frac{DN_{sys}}{Dt} = \int_{c.v.} \frac{\partial}{\partial t}(\rho\eta) d\mathcal{V} + \int_{c.s.} \rho\eta \, \mathbf{V}_b \cdot \hat{n} \, dA + \int_{c.s.} \rho\eta \, \mathbf{V}_r \cdot \hat{n} \, dA \qquad (2.24)$$

where $\mathbf{V}_b$ is the velocity of the control surface and $\mathbf{V}_r$ is the velocity of the fluid with respect to the control surface.

The vector sum of the integrands of the last two terms on the right-hand side of Eq. 2.24 leads to another form of the equation,

$$\frac{DN_{sys}}{Dt} = \int_{c.v.} \frac{\partial}{\partial t}(\eta\rho) \, d\mathcal{V} + \int_{c.s.} \rho\eta \, \mathbf{V} \cdot \hat{n} \, dA \qquad (2.25)$$

where $\mathbf{V} = \mathbf{V}_b + \mathbf{V}_r$ is the total velocity of the fluid with respect to the chosen reference frame.

Equations 2.18, 2.24, and 2.25 are identical on the left side; they differ in terms of the "instructions" provided by the nature of the terms on the right-hand side. These three forms are developed to enable the analyst to choose the most convenient formulation for a given problem.

*2. Nondeformable Control Volumes

Many applications of the control volume involve only fixed control volumes, for which $\mathbf{V}_b$, the velocity of the control surface, is everywhere zero (see parts a and c of Fig. 2.1). For fixed control volumes the Reynolds transport theorem (Eq. 2.24) becomes

$$\frac{DN_{sys}}{Dt} = \int_{c.v.} \frac{\partial}{\partial t}(\rho\eta) \, d\mathcal{V} + \int_{c.s.} \rho\eta \, \mathbf{V} \cdot \hat{n} \, dA \qquad (2.26)$$

where $\mathbf{V} = \mathbf{V}_r$ since $\mathbf{V}_b$ is everywhere zero. $\mathbf{V}$ is the velocity of the fluid at the control surface.

*3. Steady Flow

The condition of steady flow requires that no quantity may vary with time at a particular point; that is, the partial derivative with respect to time of any quantity is zero. For steady flow, the system-to-control-volume transformation is

$$\frac{DN_{sys}}{Dt} = \int_{c.s.} \rho\eta \, \mathbf{V}_b \cdot \hat{n} \, dA + \int_{c.s.} \rho\eta \, \mathbf{V}_r \cdot \hat{n} \, dA \tag{2.27}$$

Many problems encountered in fluid mechanics involve steady flow and nondeformable control volumes ($\mathbf{V}_b = 0$ and $\mathbf{V}_r = \mathbf{V}$). The Reynolds transport theorem is then

$$\frac{DN_{sys}}{Dt} = \int_{c.s.} \rho\eta \, \mathbf{V} \cdot \hat{n} \, dA \tag{2.28}$$

4. Uniform Flow and the Average of a Quantity

The assumption of a *uniform flow* is often made, that is, that the flow quantities are constant across the area where the fluid crosses the control surface. Assume, for a steady uniform flow, that the fluid enters a nondeformable control volume normal to area A_1 and leaves normal to A_2 (see Fig. 2.5a). Then we would have

$$\frac{DN_{sys}}{Dt} = \eta_2 \rho_2 V_2 A_2 - \eta_1 \rho_1 V_1 A_1 \tag{2.29}$$

where the negative sign is introduced because $\mathbf{V}_1 \cdot \hat{n}_1 = -V_1$. (Remember that the normal vector $\hat{n}$ always points *out* of the control volume.) Hence, the assumption of uniform flow allows the Reynolds transport theorem to be simplified from an integral to an algebraic

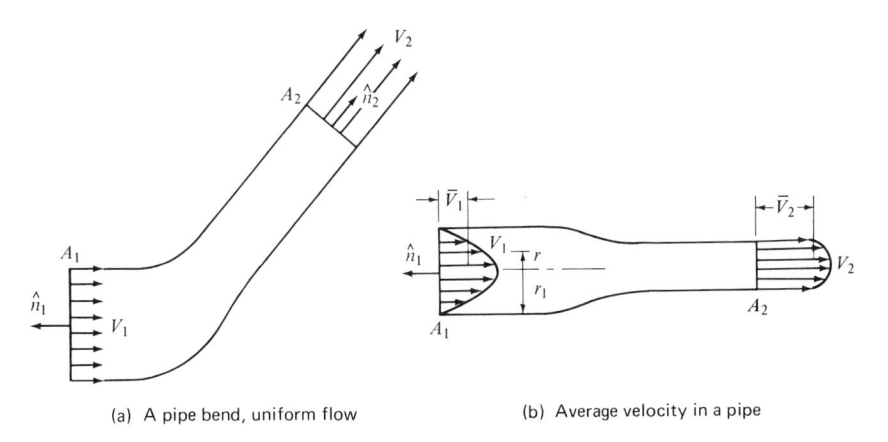

(a) A pipe bend, uniform flow   (b) Average velocity in a pipe

Fig. 2.5. Example of uniform flow and average velocity.

formulation. Because of this, uniform flow is often assumed; the assumption is justified if the resulting inaccuracy is within the acceptable tolerance of the engineering motivation for the problem.

We often use the average value of quantities when dealing with nonuniform flows, the average value being designated by a bar over the symbol for the quantity. See Fig. 2.5b for an example of average velocity. The average of a function over an area is given by

$$\bar{f} = \frac{1}{A} \int_A f \, dA \tag{2.30}$$

Applying this to a steady-flow, nondeformable situation, the Reynolds transport theorem becomes

$$\frac{DN_{sys}}{Dt} = (\overline{\eta \rho V})_2 A_2 - (\overline{\eta \rho V})_1 A_1 \tag{2.31}$$

It should be pointed out that $\overline{\eta \rho V} \neq \bar{\eta} \, \bar{\rho} \, \bar{V}$.

*Example 2.3

Given the physical problem of Example 2.2, reevaluate the time-rate-of-change term and the flux term, using Eqs. 2.24 and 2.25 instead of Eq. 2.18.

Solution. The flux term of Eq. 2.24 is identical with that of Eq. 2.18, so it remains unchanged. The instructions given by the first two terms of Eq. 2.24 will now be carried out. The term $\int_{c.v.} \frac{\partial}{\partial t} (\rho \eta) \, dV$ implies that one should examine the time rate of change of the $(\eta \rho)$ product at all the points inside the control volume at an instant. That is, at a particular point (x, y, z), $\eta(t)$ and $\rho(t)$ would provide the information from which we could find $\partial (\eta \rho)/\partial t$. With such values for all the points in the control volume the integral above could be evaluated at time t. An important feature is that only those points inside the control volume are considered. Consequently, in Example 2.2 a physical point near surface ① at time t will be excluded from the integral after the piston face passes the point at a later time $t + \Delta t$.

The operations described above are quite different from those described in Example 2.2. The results of the operations are also different. The difference is expressed by the term $\int_{c.s.} \eta \rho \, \mathbf{V}_b \cdot \hat{n} \, dA$. This is nonzero only at the piston face, where

$$\hat{n} = -\hat{i}$$

$$\mathbf{V}_b = V_p \hat{i}$$

$$\eta \rho = (\eta \rho)_1$$

For an axisymmetric condition the integral may be written as

$$\int_{c.s.} \eta\rho \, \mathbf{V}_b \cdot \hat{n} \, dA = -2\pi \int_0^{r_p} (\eta\rho)_1 V_p r \, dr$$

It is important to note that this term, like the flux term, is evaluated at the control surface only. So we see that

$$\frac{d}{dt} \int_{c.v.} \rho\eta \, dV = \int_{c.v.} \frac{\partial}{\partial t}(\rho\eta) \, dV - 2\pi \int_0^{r_p} (\eta\rho)_1 V_p r \, dr$$

For Eq. 2.25, the time-derivative term is the same as that for Eq. 2.24. The term involving the area integral cannot properly be called a net flux with respect to the control volume; it is a flux of N from the region of space occupied by the control volume at the instant at which the term is evaluated. The reference frame for the evaluation of the velocity may be located on the piston or on the cylinder. For convenience, we will use a coordinate system attached to the cylinder. The area integral term can then be expressed as the sum of three terms:

$$\int_{c.s.} \eta\rho \, \mathbf{V} \cdot \hat{n} \, dA = 2\pi \int_0^{r_e} (\eta\rho V_x)_3 r \, dr + 2\pi \int_{r_p}^{r_c} (\eta\rho)_1 V_l r \, dr - 2\pi \int_0^{r_p} (\eta\rho)_1 V_p r \, dr$$

Note that one of these is identical with a part of the flux term in Example 2.2.

Extension 2.3.1. The piston in this example is suddenly fixed. The higher pressure in the cylinder causes the compressible fluid to continue to flow out the tube and to leak past the piston. Describe the terms in Eqs. 2.18, 2.24, and 2.25 for this condition.

2.4 THE FUNDAMENTAL LAWS

We will now utilize our generalized system-to-control-volume transformation in the application of the basic laws governing fluid flow: the conservation of mass, of momentum, and of energy. Two of these laws —the conservation of mass and of energy—are expressed as scalar equations, and the other is expressed as a vector equation. In our development of the Reynolds transport theorem we used the symbol N to represent an extensive property and η to represent its corresponding intensive property. This property may be a scalar quantity such as energy or a vector quantity such as momentum. If it were a vector quantity we would use $\mathbf{N}$ and $\boldsymbol{\eta}$ to represent the generalized extensive and intensive properties.

In this article the fundamental laws governing fluid flow will be expressed in terms of the control-volume formulation. A variety of examples will illustrate the various manifestations of the control-volume form of the basic laws. Starred sections (2, 4, 6, and 8) will discuss the special features of the deformable-control-volume expressions of the basic laws; these sections extend the nondeformable-control-volume considerations presented in Sections 1, 3, 5, and 7.

1. Conservation of Mass

We have defined "system" as an identifiable and fixed quantity of matter. Since matter can neither be created nor destroyed (in the absence of fusion or fission processes) and since the quantity "mass" is a measure of the material present, it is clear that the mass M of a system is constant. Therefore, the conservation of mass for a system is expressed as

$$\frac{DM}{Dt} = \frac{D}{Dt} \int_{\text{sys}} \rho \, d\mathcal{V} = 0 \qquad (2.32)$$

Since the mass of the system is an extensive property, we can immediately transform Eq. 2.32 into the control-volume formulation by the use of the Reynolds transport theorem. Note that M per unit mass, and therefore η, is unity. For a nondeformable control volume we have

$$\frac{D}{Dt} \int_{\text{sys}} \rho \, d\mathcal{V} = \frac{d}{dt} \int_{\text{c.v.}} \rho \, d\mathcal{V} + \int_{\text{c.s.}} \rho \, \mathbf{V} \cdot \hat{n} \, dA = 0 \qquad (2.33)$$

which is the *integral form of the conservation of mass*. The special cases of constant density or steady flow occur quite often in engineering practice. For a steady flow and a nondeformable control volume the conservation of mass is

$$\int_{\text{c.s.}} \rho \, \mathbf{V} \cdot \hat{n} \, dA = 0 \qquad (2.34)$$

For a fixed control volume with constant density

$$\int_{\text{c.s.}} \mathbf{V} \cdot \hat{n} \, dA = 0 \qquad (2.35)$$

We have not assumed steady flow in Eq. 2.35.

For Eq. 2.34, note that the density at the entrance and exit locations to the control volume enters into the problem description; the changing density values inside the control volume do not enter the problem description. An easily visualized example would be the compressible flow of a gas through a nozzle, shown in Fig. 2.6a.

Art. 2.4 THE FUNDAMENTAL LAWS 89

The nature of Eq. 2.35 can be demonstrated by a finite tank of water exhausting through a converging exit tube (Fig. 2.6b). This flow is clearly time-dependent, but the equation (and the physical example) shows that only the flux term is needed to satisfy the instantaneous conservation-of-mass equation.

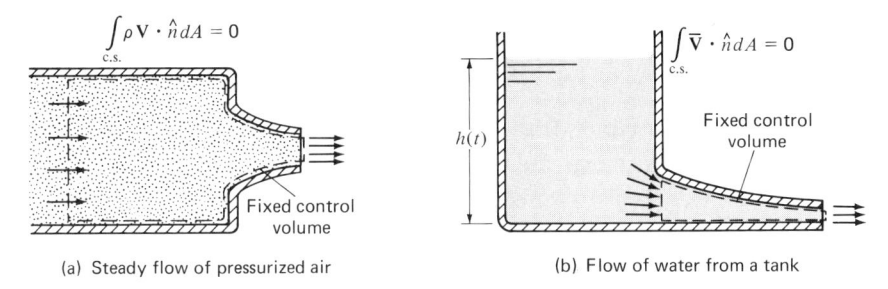

(a) Steady flow of pressurized air   (b) Flow of water from a tank

Fig. 2.6. Examples for the simplified conservation-of-mass equations.

Before working some examples illustrating the conservation of mass we should point out two flux quantities closely related to the subject. The *mass flux* $\dot{m}$, the mass rate of flow,* across an area is

$$\dot{m} = \int_A \rho \, \mathbf{V} \cdot \hat{n} \, dA \qquad (2.36)$$

and the *flow rate* q, the volume rate of flow,† is

$$q = \int_A \mathbf{V} \cdot \hat{n} \, dA \qquad (2.37)$$

The flow rate is usually used in incompressible flows and the mass flux in compressible flows. The mass flux is also expressed as

$$\dot{m} = \rho A \overline{V} \qquad (2.38)$$

where we have assumed ρ to be uniformly distributed over the cross-sectional area A, and $\overline{V}$ is the average velocity. The flow rate may similarly be expressed as

$$q = A\overline{V} \qquad (2.39)$$

*2. Conservation of Mass—Deformable Control Volume

The deformable-control-volume expression of the conservation of mass requires that the velocity in the flux term be recognized as that

---
*For mass flux, we use slugs/sec as the unit.
†For flow rate, we use ft³/sec (usually designated cfs) as the unit.

relative to the control surface and that the appropriate time-rate-of-change term be used. The two possible forms (see Eqs. 2.18 and 2.24) are

$$0 = \frac{d}{dt}\int_{c.v.}\rho\, d\mathcal{V} + \int_{c.s.}\rho\, \mathbf{V}_r\cdot\hat{n}\, dA \qquad (2.40)$$

and

$$0 = \int_{c.v.}\frac{\partial\rho}{\partial t}\, d\mathcal{V} + \int_{c.s.}\rho\, \mathbf{V}_b\cdot\hat{n}\, dA + \int_{c.s.}\rho\, \mathbf{V}_r\cdot\hat{n}\, dA \qquad (2.41)$$

The preference for which form of the equation to use is based upon the nature of the known and the desired information. For example, if the material behaves as if it were incompressible, then $\partial\rho/\partial t = 0$, and Eq. 2.41 would probably be the more convenient form. If the volume of the control volume is of interest, Eq. 2.40 would be used. If the thermodynamic data of the pressure and temperature time histories of the gas in a control volume are known, Eq. 2.41 would probably be most convenient to relate this information to $\mathbf{V}_b$ and $\mathbf{V}_r$.

Example 2.4

In many important applications the flow is steady and the control volume is fixed. Express the conservation of mass if the fluid enters normal to the inlet area A_1 and exits normal to the exit area A_2.

Solution. The continuity equation for steady flow and a fixed control volume is

$$0 = \int_{c.s.}\rho\, \mathbf{V}\cdot\hat{n}\, dA$$

For flow crossing the control surface at A_1 and A_2, this becomes

$$\int_{c.s.}\rho\, \mathbf{V}\cdot\hat{n}\, dA = \int_{A_1}\rho\, \mathbf{V}\cdot\hat{n}\, dA + \int_{A_2}\rho\, \mathbf{V}\cdot\hat{n}\, dA = 0$$

or, for the velocity vector normal to the areas,

$$\int_{A_2}\rho_2 V_2\, dA - \int_{A_1}\rho_1 V_1\, dA = 0$$

since $\mathbf{V}_1\cdot\hat{n}_1 = -V_1$. Assume ρ_1 and ρ_2 are constant over their respective areas; then

$$\rho_2\int_{A_2}V_2\, dA = \rho_1\int_{A_1}V_1\, dA$$

If V_1 and V_2 are constant over A_1 and A_2, respectively, then

$$\rho_2 A_2 V_2 = \rho_1 A_1 V_1$$

If V_1 and V_2 are functions of the areas, then

$$\rho_2 A_2 \overline{V}_2 = \rho_1 A_1 \overline{V}_1$$

where $\overline{V}_2$ and $\overline{V}_1$ are the average velocities over the areas.

Extension 2.4.1. Assume V_1 is a parabolic profile in a pipe, so that $V_1 = V_{max}(1 - r^2/r_1^2)$ where r_1 is the radius of the pipe. (See fig. 2.5b. on p. 85.) Sketch the velocity profile. Also, determine $\overline{V}_1$ if $V_{max} = 20$ fps.

Ans. 10 fps

Extension 2.4.2. In Example 2.4 an incompressible fluid exits through two areas, A_2 and A_3. Assuming uniform profiles, write the simplified continuity equation.

Example 2.5

Air enters the device of Fig. E2.5 at a temperature, pressure, and velocity of 1000°R, 500 psia, and 500 fps, respectively. It leaves at a temperature of 560°R and 14.7 psia. Determine the exit velocity.

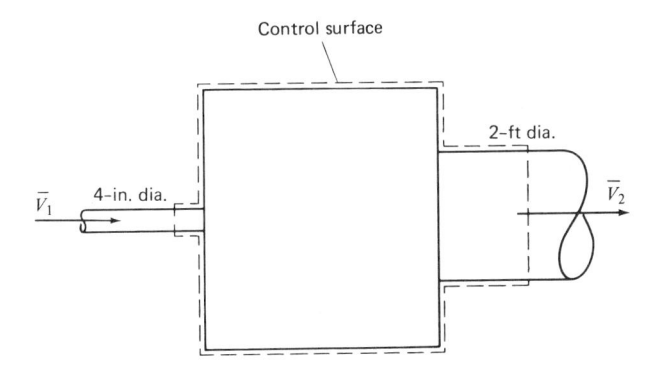

Fig. E2.5

Solution. The flow is assumed to be steady from the information given, and consequently, the control-volume form of the conservation of mass is

$$0 = \int_{c.s.} \rho \mathbf{V} \cdot \hat{n} \, dA$$

Note that, even though the density changes with respect to space, it is constant with respect to time; hence, the total mass $\int_{c.v.} \rho \, dV$ inside the control volume is constant. The density is constant over A_1 and over A_2 even though $\rho_1 \neq \rho_2$. The continuity equation can thus be written as

$$0 = \rho_2 \int_{A_2} V_2 \, dA - \rho_1 \int_{A_1} V_1 \, dA$$

where $\mathbf{V}_1 \cdot \hat{n} = -V_1$. Continuity thus becomes

$$\rho_2 \overline{V}_2 A_2 = \rho_1 \overline{V}_1 A_1$$

We find ρ_1 and ρ_2 from the equation of state. We have

$$\rho_1 = \frac{p_1}{RT_1} = \frac{500 \times 144}{1716 \times 1000} = 0.042 \text{ slugs/ft}^3$$

$$\rho_2 = \frac{p_2}{RT_2} = \frac{14.7 \times 144}{1716 \times 560} = 0.0022 \text{ slugs/ft}^3$$

Finally,

$$\overline{V}_2 = \frac{\rho_1 A_1 \overline{V}_1}{\rho_2 A_2} = \frac{0.042 \times \frac{4\pi}{144} \times 500}{0.0022 \times \pi} = 265 \text{ fps}$$

Extension 2.5.1. Suppose that a valve is suddenly turned in the device, reducing the exit velocity to 150 fps. If the volume of the device is 20 ft³, determine the rate of change of the average density in the device. Assume the inlet and exit pressures and temperatures remain unchanged.

Ans. 0.0399 slug/ft³/sec

★Example 2.6

Apply the conservation of mass to the problem of an inflating balloon.

Solution. Clearly, the physical problem involves an influx of mass to the balloon (represented by the last integral in Eq. 2.33), balanced by a change of mass inside the balloon. The integral over the volume in Eq. 2.33 represents the total mass inside the control volume (balloon). The air will stop flowing into the balloon when the pressure in the supply pipe equals the pressure in the balloon, as shown. The mass in the balloon changes as the volume changes and as the density changes, the density being dependent on the pressure and temperature of the air in the balloon. Equation 2.33 asserts that the slope of the curve of Fig. E2.6 is related to the entrance density, velocity, and area.

The flux term is nonzero only across the inlet, where it is reasonable to assert that the density is uniform over the inlet area; that is, the flux term may be expressed as

$$\int_{c.s.} \rho \, \mathbf{V} \cdot \hat{n} \, dA = -\rho_i \overline{V}_i A_i \tag{E2.1}$$

The full equation is then

$$\frac{d}{dt} \int_{c.v.} \rho \, dV - \rho_i A_i \overline{V}_i = 0 \tag{E2.2}$$

It is also reasonable to consider the density to be uniform over the interior of the balloon at any time t. On this assumption, and approximating the balloon

Art. 2.4   THE FUNDAMENTAL LAWS

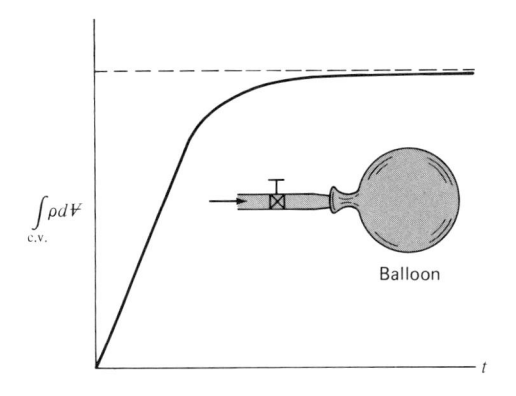

Fig. E2.6

as a sphere of radius R, there results

$$\frac{d}{dt}\int_{c.v.} \rho\, dV = \frac{d}{dt}\left(\rho\,\frac{4}{3}\pi R^3\right)$$

$$= \frac{4}{3}\pi R^3\frac{d\rho}{dt} + 4\pi\rho R^2\frac{dR}{dt} \qquad (E2.3)$$

and the conservation of mass equation becomes

$$\frac{4}{3}\pi R^2\left(R\frac{d\rho}{dt} + 3\rho\frac{dR}{dt}\right) = \rho_i A_i \overline{V}_i \qquad (E2.4)$$

This is the desired result. The following questions are posed to allow you to gain a firmer comprehension of the problem by exploring several variations. From these extended considerations, you should gain an appreciation for the nature of the terms in the control-volume form of the conservation of mass.

Extension 2.6.1. Consider the case where a rigid spherical tank is substituted for the balloon. Indicate the appropriate form of Eq. E2.4 for this case.

Extension 2.6.2. As the pressure increases in the balloon, the temperature is maintained directly proportional to the pressure. Assuming air to be an ideal gas ($p = \rho RT$), indicate the effect on Eq. E2.4.

Extension 2.6.3. For the special case of a rigid tank and $\rho_i = \rho_{\text{tank}}$, show that the density ratio at any time t is

$$\ln(\rho/\rho_1) = \frac{A_i}{V_{\text{tank}}}\int_{t_1}^{t}\overline{V}_i\, dt$$

Extension 2.6.4. How would Eq. E2.4 be changed if the balloon had a small leakage hole that allowed air to escape as it was being filled?

3. Linear Momentum

The product of the velocity **V** of a system's center of mass and its mass M is the linear momentum of the system. The conservation-of-momentum principle states that "the momentum of a system is constant if it is completely isolated from interaction with its surroundings." The observation that "the time rate of change of the linear momentum of a system is equal to the net force acting on the system" is often referred to as Newton's second law of motion. This equivalence is valid only if the time rate of change of momentum is evaluated in an inertial coordinate system. The equation expressing this equivalence is

$$\Sigma \mathbf{F} = \frac{D}{Dt} \int_{\text{sys}} \rho \mathbf{V} \, d\mathcal{V} \tag{2.42}$$

Since momentum is an extensive property, the force-momentum relationship may be expressed in control-volume form, using the Reynolds transport theorem. For the transformation to the fixed-control-volume form, note that momentum per unit mass is the vector velocity **V**, so that

$$\Sigma \mathbf{F} = \frac{D}{Dt} \int_{\text{sys}} \rho \mathbf{V} \, d\mathcal{V} = \frac{d}{dt} \int_{\text{c.v.}} \rho \mathbf{V} \, d\mathcal{V} + \int_{\text{c.s.}} \rho \mathbf{V} (\mathbf{V} \cdot \hat{n}) \, dA \tag{2.43}$$

which is the general form of the nondeformable *integral momentum equation*. This is a vector equation; the x-component form is written as

$$\Sigma F_x = \frac{d}{dt} \int_{\text{c.v.}} \rho V_x \, d\mathcal{V} + \int_{\text{c.s.}} \rho V_x (\mathbf{V} \cdot \hat{n}) \, dA \tag{2.44}$$

For a steady-flow, nondeformable control volume, with the velocity measured in an inertial reference frame, the momentum equation becomes

$$\Sigma \mathbf{F} = \int_{\text{c.s.}} \rho \mathbf{V} (\mathbf{V} \cdot \hat{n}) \, dA \tag{2.45}$$

In certain problems, considerable analytical simplification is gained by considering an accelerating coordinate system. Consider, for example, the simplification gained in describing atmospheric observations from a reference frame rotating with the earth instead of one fixed to the sun as discussed in Sec. 4 of Art. 1.4. However, the relationship between the net force and the time rate of change of momentum is valid only in terms of an inertial coordinate system. Therefore, it is necessary to reconsider the statement of the momentum equation if we wish to express the velocity **V** in a noninertial reference frame.

The relationship between the acceleration **A** in an inertial reference frame and the acceleration **a** in a noninertial reference frame is given by Eq. 1.63. It is

$$\mathbf{A} = \mathbf{a} + \frac{d^2\mathbf{S}}{dt^2} + 2\mathbf{\Omega} \times \mathbf{V} + \mathbf{\Omega} \times (\mathbf{\Omega} \times \mathbf{R}) + \frac{d\mathbf{\Omega}}{dt} \times \mathbf{R} \quad (2.46)$$

where **R**, **S**, **V**, and $\mathbf{\Omega}$ are defined in Fig. 1.6. The time rate of change of the momentum of the system in the noninertial reference frame is

$$\left.\frac{D\mathbf{N}_{sys}}{Dt}\right|_{xyz} = \frac{D}{Dt}\int_{sys}\rho\mathbf{V}\,dV \quad (2.47)$$

Hence, to generalize Eq. 2.42, it is necessary to form the expression

$$\Sigma\mathbf{F} = \int_{sys}\left[\frac{d^2\mathbf{S}}{dt^2} + 2\mathbf{\Omega} \times \mathbf{V} + \mathbf{\Omega} \times (\mathbf{\Omega} \times \mathbf{R}) + \frac{d\mathbf{\Omega}}{dt} \times \mathbf{R}\right]\rho\,dV$$
$$+ \frac{D}{Dt}\int_{sys}\rho\mathbf{V}\,dV \quad (2.48)$$

or, equivalently,

$$\Sigma\mathbf{F} - \mathbf{F}_I = \frac{D}{Dt}\int_{sys}\rho\mathbf{V}\,dV$$
$$= \frac{d}{dt}\int_{c.v.}\rho\mathbf{V}\,dV + \int_{c.s.}\rho\mathbf{V}(\mathbf{V}\cdot\hat{n})\,dA \quad (2.49)$$

where $\mathbf{F}_I$ is the "inertial body force" acting on the control volume and **V** is the velocity measured in the noninertial coordinate system. The designation "inertial body force" is an acknowledgment that an observer in a noninertial coordinate system observes the presence of these accelerations as apparent body forces.

It is convenient to distinguish between the various types of forces which may act on the control volume. In general, the net force may consist of surface forces $\mathbf{F}_s$ and body forces $\mathbf{F}_B$. The surface forces may result from pressure and shear-stress distributions acting on the fluid and from normal and shear stresses acting on mechanical members which extend through the control-volume surface. Body forces include the force effects of gravitational and other fields, such as magnetic or electric fields. The latter are important in magnetohydrodynamic and electrohydrodynamic flows.

Before considering some examples which examine the terms in Eq. 2.49 in detail, we will discuss two important flow features encountered when considering the momentum equation. The first involves the contribution of the pressure to the surface forces F_s. The absolute pressure can be written as

$$p_{abs} = p_{gage} + p_{atmospheric} \qquad (2.50)$$

The contribution of the pressure to the total surface force is

$$\int_{c.s.} p_{abs} \hat{n}\, dA = \int_{c.s.} p_{gage} \hat{n}\, dA + \int_{c.s.} p_{atm} \hat{n}\, dA \overset{0}{\nearrow} \qquad (2.51)$$

The gage pressure may vary over the area, but the atmospheric pressure is constant over the whole control surface; thus p_{atm} may be removed from the integral, leaving $\int_{c.s.} \hat{n}\, dA$, which is zero. Hence we may use either the absolute pressure or the gage pressure in determining the pressure force on the control surface.

A second and often-encountered feature is that any flow of fluid which has parallel streamlines experiences negligible variation in pressure normal to the streamlines. That is, if the streamlines are parallel, then there is no acceleration of the fluid particles normal to the streamlines and hence no pressure variation. This means that a jet with parallel streamlines issuing into an atmosphere will experience zero gage pressure across the jet and hence the contribution to the total surface force will be zero. For the fluid flowing from the pressurized tank shown in Fig. 2.7 the surface forces would be as shown.

The force summation in the horizontal x-direction is (note that the x-component body force is zero)

$$\Sigma F_x = p_1 A_1 - F_B \qquad (2.52)$$

where p_1 is gage pressure and F_B is the force acting on the exposed face of the bolts and the flange face. This would, of course, be equal to the rate of momentum change given by an evaluation of the terms in Eq. 2.45.

*4. Linear Momentum—Deformable Control Volume

The major features of the linear momentum equation for the deformable control volume are expressed in the time-rate-of-change term; the

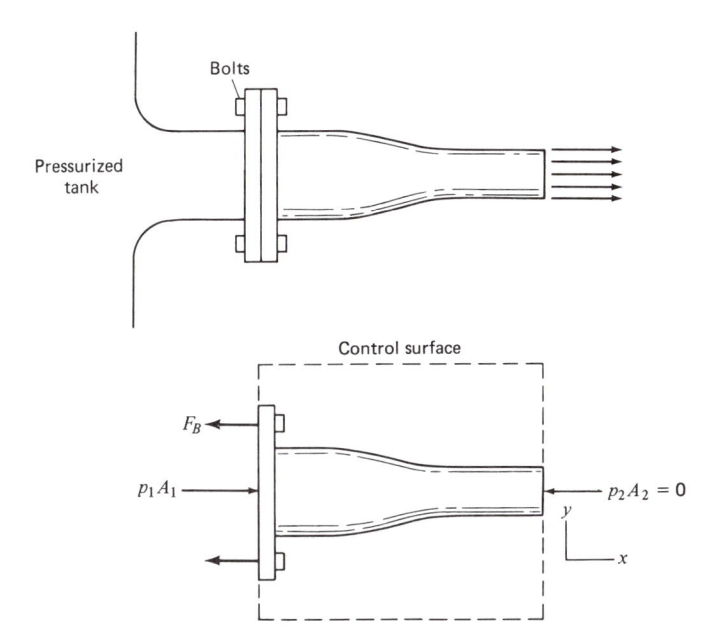

Fig. 2.7. An example illustrating pressure forces.

force terms are not affected except in the details of their evaluation, and the momentum flux term is altered by the inclusion of the relative velocity. The alternate forms of the deformable expression are

$$\Sigma \mathbf{F} - \mathbf{F}_I = \frac{d}{dt} \int_{c.v.} \rho \mathbf{V} \, d\mathcal{V} + \int_{c.s.} \rho \mathbf{V} \, \mathbf{V}_r \cdot \hat{n} \, dA \qquad (2.53)$$

and

$$\Sigma \mathbf{F} - \mathbf{F}_I = \int_{c.v.} \frac{\partial}{\partial t} (\rho \mathbf{V}) \, d\mathcal{V} + \int_{c.s.} \rho \mathbf{V} \, \mathbf{V}_b \cdot \hat{n} \, dA + \int_{c.s.} \rho \mathbf{V} \, \mathbf{V}_r \cdot \hat{n} \, dA \qquad (2.54)$$

As in the nondeformable case, if the velocity $\mathbf{V}$ is measured in an inertial reference frame the inertial body force $\mathbf{F}_I$ is zero. As in the case of the continuity equation, the nature of the known and the desired information will suggest whether Eq. 2.53 or Eq. 2.54 is the more useful.

Example 2.7

Many applications of the momentum equation involve a steady flow, a nondeformable control volume, and uniform quantities over the inlet and exit areas. Express the momentum equation for these conditions for flow in a pipe if the fluid enters normal to the inlet area A_1 and exits normal to the exit area A_2, as shown in Fig. E2.7.

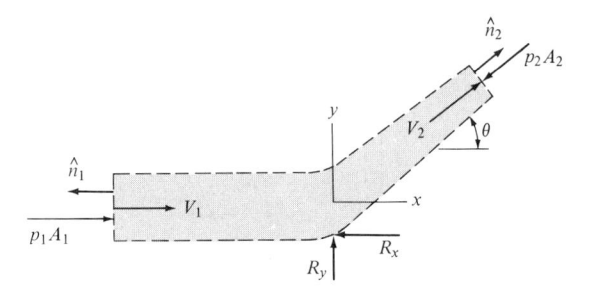

Fig. E2.7

Solution. The forces are as shown on the control volume. The control volume includes only the water which has been removed from the pipe. The vertical direction is z; hence there are no body forces required in the description of this problem. The reactions R_x and R_y are the components of the reaction of the bend acting on the fluid. That is R_x and R_y represent the net effect of the pressure and shear-stress distributions on the surface between the fluid and the pipe. The force of the fluid acting on the bend would, of course, act in the opposite direction. For steady flow and fixed control volume, the momentum equation is

$$\Sigma \mathbf{F} = \int_{c.s.} \rho \mathbf{V} \, (\mathbf{V} \cdot \hat{n}) \, dA$$

For the control volume shown in Fig. E2.7 the x-component equation is

$$\Sigma F_x = \int_{c.s.} \rho V_x \, \mathbf{V}_r \cdot \hat{n} \, dA = \int_{A_2} \rho V_{2x} V_2 \, dA + \int_{A_1} \rho V_{1x} (-V_1) \, dA$$

This may be written as

$$p_1 A_1 - p_2 A_2 \cos \theta - R_x = \rho_2 V_{2x} V_2 A_2 - \rho_1 V_{1x} V_1 A_1$$

or

$$p_1 A_1 - p_2 A_2 \cos \theta - R_x = \rho_2 V_2^2 A_2 \cos \theta - \rho_1 V_1^2 A_1$$

$$= \dot{m}(V_2 \cos \theta - V_1)$$

where

$$\dot{m} = \rho_1 A_1 V_1 = \rho_2 A_2 V_2$$

The y-component equation is

$$R_y - p_2 A_2 \sin \theta = \rho_2 V_2^2 A_2 \sin \theta$$

$$= \dot{m} V_2 \sin \theta$$

In vector form, the momentum equation would be

$$\Sigma \mathbf{F} = \rho_2 \mathbf{V}_2 V_2 A_2 - \rho_1 \mathbf{V}_1 V_1 A_1 = \dot{m}(\mathbf{V}_2 - \mathbf{V}_1)$$

Extension 2.7.1. Determine the force components acting on a horizontal, 45° bend in a pipe transporting 10 cfs of water. The diameter does not change and the pressure in the 6-in.-dia. pipe is 75 psi.   *Ans.* 909 lb. and 2200 lb

Extension 2.7.2. The 6-in.-dia. pipe splits into two exiting horizontal branches so that the total cross-sectional area of the two exiting branches equals the entering area. This allows the velocity and pressure in all sections to be equal. The exit areas are equal squares, the entering velocity and pressure are 100 fps and 50 psi, respectively, and one branch exits at −90° to the incoming pipe, with the other at 45°. Determine the components of the resultant force acting on the water.   *Ans.* 3380 lb and −764 lb

Example 2.8

Derive an expression for the drag acting on a width w of a two-dimensional airfoil. For the condition shown in Fig. E2.8, the flow will be assumed symmetric and two-dimensional. The velocity profile $u(y)$ at a downstream position is known.

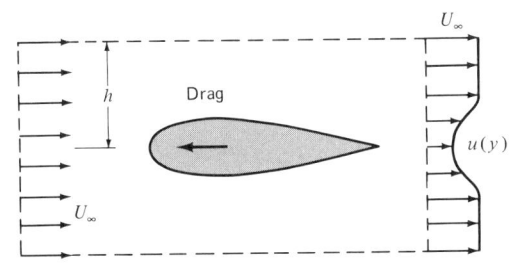

Fig. E2.8

Solution. The x-component force, the drag, is determined from the nondeformable, inertial, steady-flow form of the momentum equation (Eq. 2.45):

$$\Sigma F_x = \int_{c.s.} \rho V_x \, \mathbf{V} \cdot \hat{n} \, dA$$

The control surfaces are chosen to be at large distances up- and downstream from the airfoil so that the pressure is atmospheric. The control surface cuts through the airfoil at two z-locations, resulting in a width w. The force exposed at each surface is drag/2 acting in the x-direction. Hence, the force term for the indicated control volume is $(-D)$, and the full equation can be written as

$$-D = \int_{\text{front}} \rho V_x \, \mathbf{V} \cdot \hat{n} \, dA + \int_{\text{rear}} \rho V_x \, \mathbf{V} \cdot \hat{n} \, dA + \int_{\substack{\text{top and} \\ \text{bottom}}} \rho V_x \, \mathbf{V} \cdot \hat{n} \, dA$$

$$= 2\int_0^h \rho U_\infty(-U_\infty) w \, dy + 2\int_0^h \rho u^2(y) w \, dy + 2U_\infty \dot{m}_{\text{top}}$$

where $\dot{m}_{top}$ is the mass flux through the top surface (and through the bottom surface), and $(V_x)_{top} \cong U_\infty$. More air is flowing in through the front area than flows out through the rear area; $2\dot{m}_{top}$ quantitatively represents this difference. It is found from the continuity equation (2.34), that

$$0 = \int_0^h \rho(-U_\infty)w\, dy + \int_0^h \rho u(y) w\, dy + \dot{m}_{top}$$

or

$$\dot{m}_{top} = \int_0^h \rho(U_\infty - u)w\, dy$$

Substituting this in the momentum equation gives the desired expression for the drag:

$$D = 2\rho w \int_0^h u(U_\infty - u)\, dy$$

The value of h is large enough so that the velocity is U_∞ across the top surface. It can be increased to infinity since there is no contribution to the integral for $y > h$. Thus, the expression is often written as

$$D = 2\rho w \int_0^\infty u(U_\infty - u)\, dy$$

This expression would have to be modified for a real airfoil of finite length since we have neglected the component of flow along the airfoil and off the end. This component is usually small, so the expression developed is often used to approximate the drag for finite airfoils.

Extension 2.8.1. A crude approximation to the downstream symmetrical velocity profile in the wake of an airfoil is the linear profile $u(y) = 20|y| + 300$ fps. Determine the drag on the 50-ft-wide airfoil if $U_\infty = 500$ fps and $\rho = 0.0024$ slugs/ft³. *Ans.* 88,000 lb

Extension 2.8.2. A mistake is made in the calculation of the drag on an airfoil in that the mass flux through the top and bottom of the control volume of the example is ignored. Determine the percentage of error this would lead to in Extension 2.8.1. *Ans.* 136% high

Example 2.9

A plane jet of water 2 in. thick strikes a fixed flat plate oriented normal to the flow. For a flow rate of 10 cfs per ft of width, calculate the force needed to hold the plate in place. The vertical coordinate is z.

Solution. The control volume and forces are as shown in Fig. E2.9A. The velocity at section 1 is assumed uniform, of magnitude

$$V_1 = \frac{10}{2/12} = 60 \text{ fps}$$

Art. 2.4   THE FUNDAMENTAL LAWS   101

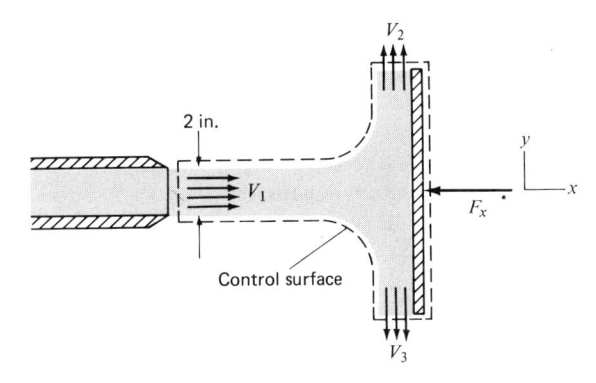

Fig. E2.9A

Bernoulli's equation gives the velocities at sections ② and ③ as $V_2 = 60$ fps and $V_3 = 60$ fps since the pressure is atmospheric on the surface. Using the momentum equation for this steady-flow, nondeformable control volume, we have

$$\Sigma F_x = \int_{c.s.} \rho V_x (\mathbf{V} \cdot \hat{n}) \, dA$$

or

$$-R_x = \int_{A_2} \rho \cancel{V_{2x}}^0 V_2 \, dA + \int_{A_3} \rho \cancel{V_{3x}}^0 V_3 \, dA + \int_{A_1} \rho V_{1x}(-V_1) \, dA = -\rho V_1^2 A_1$$

Hence

$$R_x = 1.94 \times 60^2 \times \frac{2}{12} = 1164 \text{ lb/ft of width}$$

In these calculations we have assumed the plate is large enough so that the streamlines are all parallel in the y-direction at exit areas ② and ③.

Extension 2.9.1. If a center body were placed in the nozzle in Fig. E.2.9A, a velocity profile at the exit of the shape shown in Fig. E2.9B could be achieved. For the same mass rate of flow from the nozzle, compute the net force on the plate.   *Ans.*   1552 lb

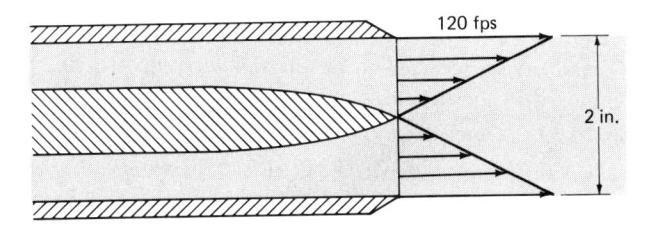

Fig. E2.9B

Example 2.10

The plate of Example 2.9 is moving to the right at 40 fps. Calculate the force on the plate and the velocity at the upper edge of the plate.

Solution. The plate is moving at constant speed, so the flow will be steady in a coordinate system attached to the plate. This reference frame is an inertial reference frame, so the momentum equation is

$$-R_x = \int_{A_1} \rho V_{1x}(-V_1)\, dA$$

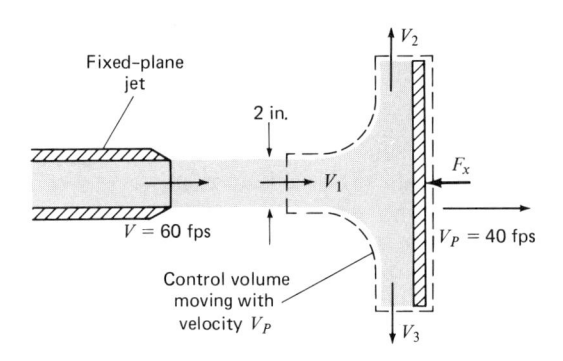

Fig. E2.10

The control volume is fixed in this moving reference frame (Fig. E2.10). The water crosses section ① with a velocity of 20 fps; hence, the momentum equation above gives

$$R_x = 1.94 \times 20^2 \times 2/12 = 129.3 \text{ lb/ft of width}$$

The velocities at sections ② and ③ in the moving-coordinate system would also be 20 fps, as predicted by Bernoulli's equation referred to the moving reference frame. To an observer on the ground the velocity at section ② would be

$$\mathbf{V}_2 = 40\hat{i} + 20\hat{j}$$

Note that we must always be careful to limit the use of Bernoulli's equation to the case of steady flow, referred to an inertial reference frame; these restrictions were imposed in the development in Art. 1.7.

Extension 2.10.1. Determine the magnitude of the force if the plate is moving to the *left* with a velocity of 40 fps. Also determine the angle with which the velocity leaves the plate as observed from the ground.

Ans. 3230 lb, 21.8°

Example 2.11

High-speed air jets strike the blades on the rotor of a turbine, are deflected by the blades, and exhaust to the surroundings, which may include several additional stages of rotor and stator blades. Determine the force component R_x of the air acting on the blades for a given turbine-blade speed V_B (Fig. E2.11A). Turbine blades are designed so that the fluid enters and leaves the blades tangentially as observed relative to the blade.

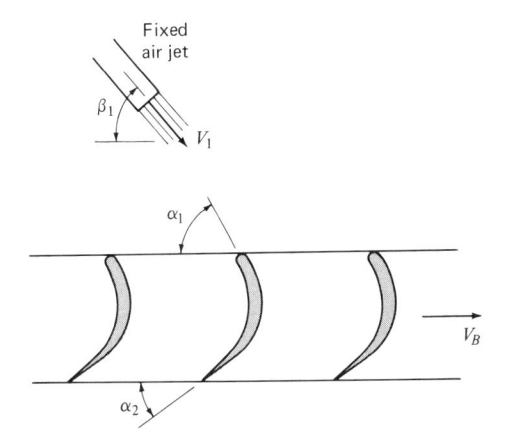

Fig. E2.11A

Solution. The flow will be assumed to be steady. Actually, a blade would enter the jet, be acted on by the jet, and then leave the influence of the jet, a sequence that may be viewed as a transient phenomenon for a single jet but a periodic phenomenon for the sequence of jets seen by the blade. However, we analyze the interaction of the moving blade and the fixed jet by assuming a time-averaged jet stream deflection, as shown in Fig. E2.11B. From the momentum equation (2.45) the time-averaged force would be

$$-R_x = \rho V_2 A_2 (-V_2 \cos \beta_2) - \rho V_1 A_1 (V_1 \cos \beta_1)$$

or

$$R_x = \dot{m}(V_2 \cos \beta_2 + V_1 \cos \beta_1)$$

The quantities V_1 and β_1 are known; we must determine V_2 and β_2. If the jet is oriented at angle β_1, so that the air stream enters tangentially, then

$$\mathbf{V}_{r1} = \mathbf{V}_1 - \mathbf{V}_B$$

where $\mathbf{V}_{r1}$ is the relative velocity entering, as observed from the blade. This is

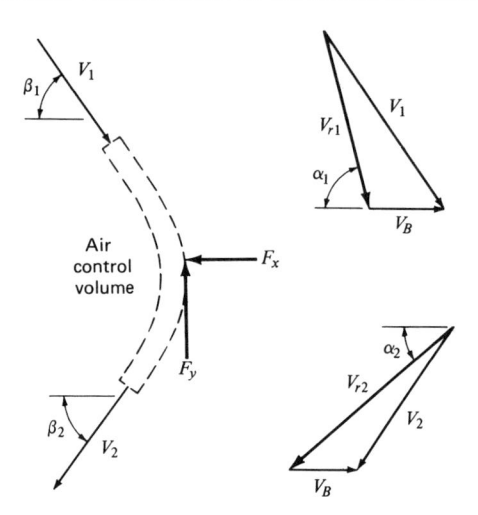

Fig. E2.11B

shown diagramatically in Fig. E2.11B. From this diagram we see that

$$V_{r1} \sin \alpha_1 = V_1 \sin \beta_1$$

For a particular blade, this relationship allows us to determine V_{r1}. This diagram also allows us to choose the proper β_1 for a given α_1, V_B, and V_1. Now, as the air moves over the blade this relative speed changes magnitude only as friction acts on the air. The friction is very small; it is often assumed that the relative fluid speed remains essentially unchanged as it leaves the blade. Hence we use

$$V_{r1} = V_{r2}$$

Knowing V_B and the blade angle α_2, it is now possible to determine V_2 and β_2 by using the velocity diagram at the exit, which is a graphical representation of the vector relationship

$$\mathbf{V}_2 = \mathbf{V}_{r2} + \mathbf{V}_B$$

The two component equations are

$$V_2 \sin \beta_2 = V_{r2} \sin \alpha_2$$

and

$$-V_2 \cos \beta_2 = V_B - V_{r2} \cos \alpha_2$$

A simultaneous solution yields V_2 and β_2. The momentum equation gives the desired force R_x.

Art. 2.4 THE FUNDAMENTAL LAWS 105

Extension 2.11.1. For a jet speed of 300 fps and a mass flux of 2 slugs/sec issuing from jets striking the blades of the turbine rotor of this example, determine the proper β_1 and the force R_x exerted on the blades if $\alpha_1 = 45°$, $\alpha_2 = 60°$, and $V_B = 100$ fps. *Ans.* 31.3°, 532 lb.

Extension 2.11.2. Repeat Extension 2.11.1, but assume that friction acts on the air moving over the blades so that $V_{r2} = 0.96 V_{r1}$. *Ans.* 31.3°, 524 lb.

Extension 2.11.3. A row of stator blades is to be oriented between the rotor blades of Extension 2.11.1 and a second row of rotor blades with identical angles and speed. Determine α_1 and α_2 of the stator blades. Assume no decrease in velocity as the air passes over the stator blades. *Ans.* 87°, 23.5°

Example 2.12 ─────────────────────────────────

In one-dimensional flows, it is often desirable to refer to average quantities because of the simplifications that result. The average velocity in one-dimensional flows such as pipe or channel flows is used in expressions for the losses, the flow rate, and the like. The momentum flux is related to the average velocity by the expression $\int \rho V^2 \, dA = \beta \rho A \overline{V}^2 = \beta \dot{m} \overline{V}$, where β, the *momentum flux correction factor*, depends on the velocity profile. For a parabolic velocity distribution in a circular pipe, calculate β.

Solution. A parabolic velocity distribution in a pipe is given by

$$u(r) = \frac{u_{max}}{r_0^2}(r_0^2 - r^2)$$

where u_{max} is the maximum velocity on the centerline ($r = 0$). The average velocity is then

$$\overline{V} = \frac{1}{A}\int_A u \, dA = \frac{1}{\pi r_0^2} \int_0^{r_0} \frac{u_{max}}{r_0^2}(r_0^2 - r^2) 2\pi r \, dr$$

$$= \frac{u_{max}}{2}$$

The momentum flux is

$$\beta \rho \overline{V}^2 A = \int_A \rho u^2 \, dA = \int_0^{r_0} \rho \frac{u_{max}^2}{r_0^4}(r_0^2 - r^2)^2 2\pi r \, dr$$

$$= \rho \frac{u_{max}^2}{r_0^4}\left(\frac{r_0^6}{6}\right) 2\pi$$

Using $u_{max} = 2\overline{V}$ and $A = \pi r_0^2$, we have

$$\beta = \frac{4}{3}$$

So we see that if the velocity varies over the area, the momentum flux is not $\dot{m}\bar{V}$ but must be multiplied by the factor β. If the velocity is constant over the area then $\beta = 1$, as is obvious.

Extension 2.12.1. Find β for a parabolic distribution between parallel plates. *Ans.* 1.2

Extension 2.12.2. Show that $\beta \geq 1$ by considering $u = \bar{V} + \Delta u$ where $\bar{V}$ is the average velocity.

*Example 2.13

The device shown in Figure E2.13, of cross-sectional area A, rests on a frictionless level track.* It is initially filled with water to the indicated height h_0. At time $t = 0$, the quick release valve allows the liquid to escape. Using an appropriate form of the momentum equation, determine an equation for the motion of the device in terms of the motion of the liquid for $h(t) > 0$.

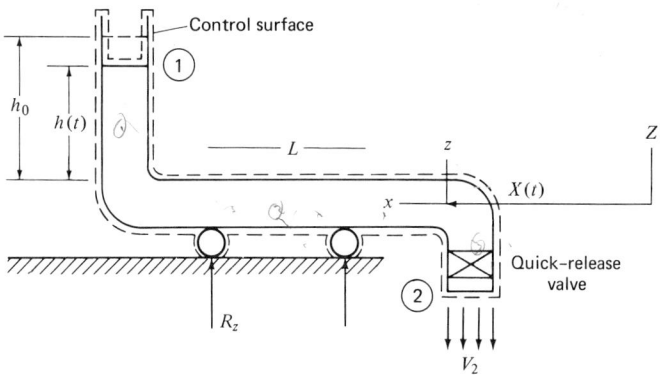

Fig. E2.13

Solution. Since the device can move only in the x-direction, only the x-equation will be considered. The deformable control volume is selected to extend over the outer surface of the device, across the top of the water surface, and between the rollers and the frictionless surface. For this control volume, there is no force acting in the x-direction. The momentum equation is then

$$-\int_{c.v.} \rho \frac{d^2X}{dt^2} dV = \frac{d}{dt} \int_{c.v.} \rho V_x \, dV + \int_{A_2} \rho V_{2x} \mathbf{V}_r \cdot \hat{n} \, dA \overset{0}{}$$

Note that V_x at the entrance and exit is zero. The time-rate-of-change term (see Eq. 2.23) may be written as

$$\frac{d}{dt} \int_{c.v.} \rho V_x \, dV = \int_{c.v.} \frac{\partial}{\partial t} (\rho V_x) \, dV + \int_{A_1} \rho V_{1x} \mathbf{V}_b \cdot \hat{n} \, dA \overset{0}{}$$

---
*A 16-mm motion picture of the movement of this device is available from the authors.

By restricting attention to the case of $h \geq 0$, V_x is zero on the deformable water surface. V_x is nonzero in the horizontal section of length L. In this region V_x may be related to the $h(t)$ for the water surface if the assumption of uniform flow is made. (The area is everywhere constant.) For uniform flow in the horizontal section, this velocity is

$$V_x = -\frac{dh}{dt}$$

With this assumption of uniform flow, we have (neglecting corner effects)

$$\int_{\text{c.v.}} \frac{\partial}{\partial t}(\rho V_x) \, d\mathcal{V} = -\rho A L \left(-\frac{d^2h}{dt^2}\right)$$

where the first negative sign accounts for the velocity being in the negative x-direction. The noninertial term may be written using $X(t)$ and the instantaneous mass of the complete control volume to obtain

$$\int_{\text{c.v.}} \rho \frac{d^2X}{dt^2} \, d\mathcal{V} = [M_f + M_{\text{device}} + \rho A h(t)] \frac{d^2X}{dt^2}$$

where M_f is the mass of the water below $h = 0$. The desired equation relating the position of the device to the height of the liquid is

$$-(M_f + M_{\text{device}} + \rho A h) \frac{d^2X}{dt^2} = \rho A L \frac{d^2h}{dt^2}$$

This equation cannot be solved, since both $X(t)$ and $h(t)$ are unknowns. The energy equation, to be developed later, must be used to find an additional relationship between the two variables. See Example 2.21 for this development.

Extension 2.13.1. When the valve is released, which way will the device move?

Extension 2.13.2. Experimental data indicate that the liquid surface falls at a constant rate during a large fraction of the period of its descent to $h = 0$. What may be said about $X(t)$ in this time period?

Extension 2.13.3. Experimental data also indicate that after h passes through zero, the device stops, then accelerates toward its original position. Explain this observation by reference to the general equation.

Extension 2.13.4. If a bucket is added on the tube to catch the efflux of water, how is the basic equation modified?

5. Moment of Momentum

For physical situations involving rotating devices or flows, it is often necessary to deal with the moment-of-momentum equation, which results directly from the linear momentum relationship. For an elemental particle of mass dM, that relationship is

$$d\mathbf{F} = \frac{d\mathbf{V}}{dt} dM \qquad (2.55)$$

Hence

$$\mathbf{R} \times d\mathbf{F} = \mathbf{R} \times \frac{d\mathbf{V}}{dt} dM$$

$$= \frac{d}{dt}(\mathbf{R} \times \mathbf{V}) dM - \frac{d\mathbf{R}}{dt} \times \mathbf{V} dM \qquad (2.56)$$

where $d\mathbf{R}/dt \times \mathbf{V} = 0$ since $d\mathbf{R}/dt = \mathbf{V}$. In fact, we may write

$$\mathbf{R} \times d\mathbf{F} = \frac{d}{dt}(\mathbf{R} \times \mathbf{V} dM) \qquad (2.57)$$

since $d(dM)/dt = 0$ for a particular particle. Summing over all the particles in the system, we have

$$\sum \mathbf{M} - \mathbf{M}_I = \frac{D}{Dt} \int_{\text{sys}} \mathbf{R} \times \mathbf{V} \rho \, dV \qquad (2.58)$$

where $\mathbf{M}_I$ is the moment of the inertial forces, and $\sum \mathbf{M}$ is the sum of the moments of all forces, about the origin of $\mathbf{R}$.

With the use of the Reynolds transport theorem, this becomes

$$\sum \mathbf{M} - \int_{\text{c.v.}} \mathbf{R} \times \left[\frac{d^2\mathbf{S}}{dt^2} + 2\boldsymbol{\Omega} \times \mathbf{V} + \boldsymbol{\Omega} \times (\boldsymbol{\Omega} \times \mathbf{R}) + \frac{d\boldsymbol{\Omega}}{dt} \times \mathbf{R} \right] \rho \, dV$$

$$\qquad (2.59)$$

$$= \frac{d}{dt} \int_{\text{c.v.}} \mathbf{R} \times \mathbf{V} \rho \, dV + \int_{\text{c.s.}} \mathbf{R} \times \mathbf{V} \rho (\mathbf{V} \cdot \hat{n}) \, dA$$

the *integral moment-of-momentum equation* for a nondeformable control volume.

*6. Moment of Momentum—Deformable Control Volume

For the deformable control volume, we again recognize that the velocity vector in the quantity $(\mathbf{V} \cdot \hat{n})$ in Eq. 2.59 must be the relative velocity $\mathbf{V}_r$, since only the relative velocity accounts for fluid flowing through the control surface. The integral moment-of-momentum equation is then

$$\sum \mathbf{M} - \mathbf{M}_I = \frac{d}{dt} \int_{\text{c.v.}} \rho \mathbf{R} \times \mathbf{V} \, dV + \int_{\text{c.s.}} \mathbf{R} \times \mathbf{V} \rho (\mathbf{V}_r \cdot \hat{n}) \, dA \qquad (2.60)$$

Art. 2.4 THE FUNDAMENTAL LAWS 109

In order to move the time derivative inside the volume integral we must account for the variable limits that exist in deformable control volumes; this introduces the velocity V_b and Eq. 2.60 may be written in the alternate form

$$\sum \mathbf{M} - \mathbf{M}_I = \int_{c.v.} \frac{\partial}{\partial t}(\rho \mathbf{R} \times \mathbf{V}) \, dV + \int_{c.s.} \rho \mathbf{R} \times \mathbf{V} (\mathbf{V}_b \cdot \hat{n}) \, dA$$
$$+ \int_{c.s.} \rho \mathbf{R} \times \mathbf{V} (\mathbf{V}_r \cdot \hat{n}) \, dA \quad (2.61)$$

Problems involving the deformable control volume requiring the moment-of-momentum equation are rare, and none will be included in this text. The foregoing equations have been presented for completeness.

Example 2.14

An interesting feature of a rotating arm which is driven by a jet (e.g., a dishwasher spray arm or a lawn sprinkler) is that a natural upper limit exists for the rotational speed. This is rather different from the case of linear momentum, where the only upper bound on the velocity of a propelled craft is imposed by drag or (in the case of space travel) relativistic effects.

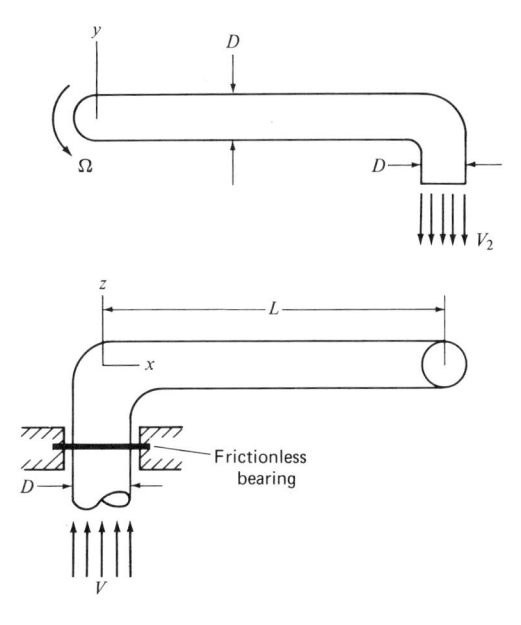

Fig. E2.14A

110 CONTROL-VOLUME FORM OF FUNDAMENTAL LAWS Ch. 2

Consider an arm of constant cross-sectional area with a shape as shown in Fig. E2.14A. Liquid enters the center of the shaft at the support and flows outward through the indicated nozzle. For a condition of zero frictional moment and with rotation in the horizontal plane only, determine $\Omega(t)$ for the arm. Assume that a steady flow is caused to flow through the arm and the arm starts from rest at time $t = 0$.

Solution. a. *Stationary Reference Frame.* Assume the reference frame to be stationary. In this stationary reference frame, assuming zero frictional moment, the moment-of-momentum equation (2.59) becomes

$$0 = \frac{d}{dt} \int_{c.v.} \mathbf{R} \times \mathbf{V}_I \, \rho \, d\mathcal{V} + \int_{c.s.} \rho \mathbf{R} \times \mathbf{V}_I \, (\mathbf{V} \cdot \hat{n}) \, dA$$

where the subscript on $\mathbf{V}_I$ reminds us that the velocity is measured in an inertial reference frame. The control volume is as shown in Fig. E2.14B. A cylindrical coordinate system is employed and the axial (z) components are independent of the θ location of the arm. The contribution of the short nozzle to the volume integral will be neglected; also, the area integral over section ①, the inlet area, has no contribution to the efflux integral since there is no entering angular momentum. The velocity $\mathbf{V}_I$ is the velocity as observed in the stationary reference frame; it includes the $\mathbf{\Omega} \times \mathbf{R}$ contribution. After expressing all vectors in the position shown (this choice does not affect the solution), the foregoing equation is written as

$$0 = \frac{d}{dt} \int_0^L R\hat{i} \times [V\hat{i} + (\Omega\hat{k} \times R\hat{i})]\rho A \, dR + \rho V_2 A_2 [L\hat{i} \times (-V_2\hat{j} + L\Omega\hat{j})]$$

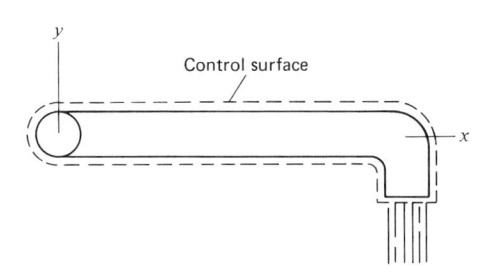

Fig. E2.14B

where $V_2 = V = q/A$. By carrying out the indicated cross-products we see that only the z-components (coefficients of $\hat{k}$) are nonzero; the z-component angular momentum equation gives

$$0 = \rho A \frac{L^3}{3} \frac{d\Omega}{dt} + \rho A V(-LV + L^2\Omega)$$

Art. 2.4 THE FUNDAMENTAL LAWS

In more conventional form,

$$\frac{d\Omega}{dt} + \frac{3V\Omega}{L} = \frac{3V^2}{L^2}$$

This is a first-order, ordinary differential equation with initial condition $\Omega(0) = 0$. The general solution is found by adding the particular solution to the homogeneous solution to give

$$\Omega(t) = Ce^{-\frac{3V}{L}t} + \frac{V}{L}$$

Using the initial condition the final solution is

$$\Omega(t) = \frac{V}{L}\left(1 - e^{-\frac{3V}{L}t}\right)$$

Clearly, this angular velocity has the upper bound V/L. This is not due to frictional effects since we have neglected frictional torque.

b. *Rotating Reference Frame.* The reference frame will be attached to the rotating arm (Fig. E2.14C). This is a noninertial frame so $\mathbf{M}_I$ must be retained. The angular momentum equation is then

$$-\int_{c.v.} \mathbf{R} \times \left[2\boldsymbol{\Omega} \times \mathbf{V} + \boldsymbol{\Omega} \times (\boldsymbol{\Omega} \times \mathbf{R}) + \frac{d\boldsymbol{\Omega}}{dt} \times \mathbf{R}\right]\rho \, dV$$

$$= \frac{d}{dt}\int_{c.v.} \mathbf{R} \times \mathbf{V}\rho \, dV + \int_{c.s.} \mathbf{R} \times \mathbf{V}\rho(\mathbf{V}\cdot\hat{n}) \, dA$$

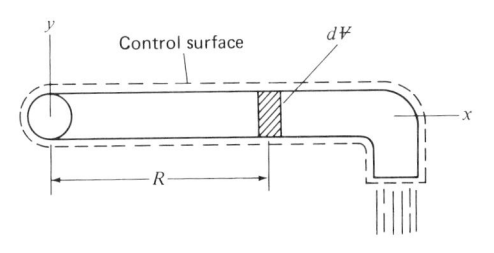

Fig. E2.14C

In this equation $\mathbf{V}$ is the velocity in the rotating reference frame and does not include $\boldsymbol{\Omega} \times \mathbf{R}$. The quantity $\mathbf{R} \times [\boldsymbol{\Omega} \times (\boldsymbol{\Omega} \times \mathbf{R})]$ is zero, the quantity $\mathbf{R} \times \mathbf{V}$ is zero since in the arm $\mathbf{V} = V\hat{i}$ and $\mathbf{R} = R\hat{i}$, and at the entrance there is no contribution to the control surface integral. The equation becomes

$$-\int_0^L R\hat{i} \times \left[2\Omega\hat{k} \times V\hat{i} + \frac{d\Omega}{dt}\hat{k} \times R\hat{i}\right]\rho A \, dR = L\hat{i} \times (-V_2\hat{j})\rho V_2 A_2$$

where $V_2 = V$ = constant. Only the z-components are nonzero and this component equation may be written as

$$-\Omega V \rho L^2 A - \frac{d\Omega}{dt} \rho A \frac{L^3}{3} = -L\rho V^2 A$$

or

$$\frac{d\Omega}{dt} + \frac{3V}{L}\Omega = \frac{3V^2}{L^2}$$

This equation is, of course, the same as in part (a) and has the identical solution.

Extension 2.14.1. Determine the time necessary for the arm to reach 99 percent of its steady-state speed if the flow rate is 2 cfs, $D = 2$ in., and $L = 2$ ft. *Ans.* 0.0333 sec.

Extension 2.14.2. In the example the rotating arm was of constant cross-sectional area. Assume the arm tapers from an area of $2A$ at $R = 0$ to an area of A at $R = L$. Determine $\Omega(t)$. Does the upper bound on $\Omega(t)$ change? Let the exit velocity be V.

Extension 2.14.3. In the example the nozzle exit area was the same as the arm area. Assume the nozzle suddenly tapers to an exit area $A_2 = A/4$. Determine $\Omega(t)$ and the upper bound on $\Omega(t)$. Let the exit velocity be V.

Extension 2.14.4 If the frictional torque M_f were constant, indicate its effect on the governing equation. Repeat for $M_f = C\Omega$.

Extension 2.14.5. It was assumed for this example that V is constant, that is, V is independent of Ω. This may not be true unless the flow is driven by a positive-displacement pump. For a constant-pressure reservoir the velocity of the water is not constant but would be dependent upon the "back pressure" effects caused by Ω. That is, $V = V(\Omega)$. Determine the effect of this on the governing equation. Is the resulting equation linear or nonlinear?

7. Energy Equation

The principle of conservation of energy is similar to that for momentum, namely, if a system is isolated from its surroundings its energy must remain constant; however, when a system is free to interact with its surroundings the first law of thermodynamics states that

$$\dot{Q} - \dot{W} = \frac{D}{Dt} \int_{\text{sys}} \rho e \, d\mathcal{V} \tag{2.62}$$

where the standard conventions of "rate of heat flow $\dot{Q}$ into the system is positive" and "rate of work $\dot{W}$ done by the system is positive" are implied. The specific energy is designated by e. The common thermodynamic definitions for heat and work will be used:

Work is an interaction between a system and its surroundings. Work is done by a system on its surroundings if the sole external effect of the interaction could be the lifting of a body. The magnitude of work is the product of the weight of the body lifted and the distance it could be lifted if the lifting of the body were the sole external effect of the interaction.*

Heat is energy in transition across a system boundary associated only with temperature effects.[†]

The dimensions of work are length times force; those of heat are the same. Both quantities are path functions, that is, dependent upon the process.

We will consider three forms of energy. For a particle of mass dm, they are given as kinetic ($\frac{1}{2} V_{I'}^2 \, dm$), potential ($gz \, dm$), and internal ($\tilde{u} \, dm$), where $V_{I'}$ is the velocity referred to an inertial reference frame[‡] and z is referred to a fixed datum. These forms are sufficient for many gineering problems—in any case, for all the problems to be considered in this text. Important exceptions are those problems involving chemical reactions and potential energy associated with magnetic or electric fields (we will only consider gravitational-field effects). With these restrictions, the nondeformable-control-volume form of the energy equation may be written (for $\eta = e = V_{I'}^2/2 + gz + \tilde{u}$) as

$$\dot{Q} - \dot{W} = \frac{d}{dt} \int_{\text{c.v.}} \left(\frac{V_{I'}^2}{2} + gz + \tilde{u} \right) \rho \, d\mathcal{V} + \int_{\text{c.s.}} \left(\frac{V_{I'}^2}{2} + gz + \tilde{u} \right) \rho \, \mathbf{V} \cdot \hat{n} \, dA$$

(2.63)

The terms $\dot{Q}$ and $\dot{W}$ which appear in the energy equation refer respectively to the heat-transfer rate and the work rate between the system and the surroundings. We assert that, for the instant at which we invoke the Reynolds transport theorem (i.e., as $\Delta t \to 0$), the symbols $\dot{Q}$ and $\dot{W}$ may be interpreted as interchanges between the control volume and the surroundings—if we properly account for all effects. The heat-transfer term is unambiguous; the work-rate term must be dealt with in an exacting and cautious manner. In fact, the manipula-

─────────

*J. B. Jones and G. A. Hawkins, *Engineering Thermodynamics*, John Wiley & Sons, New York, 1960, p. 22.

[†]See Chapter 1 for a discussion of the thermodynamic considerations of energy and heat.

[‡]It is convenient to measure the velocity in a stationary reference frame in order to avoid an infinite kinetic energy condition. For example, if the kinetic energy of the air entering a jet engine in constant-speed flight were measured with respect to the engine, the balance of the atmosphere would constitute an unbounded amount of kinetic energy. I' denotes the stationary reference frame, a special case of an inertial reference frame.

tions and special considerations required are sufficiently complex to make a brief introduction appropriate.

We wish to develop equations appropriate to the many types of flow fields which can be analyzed by use of the energy equation. Some of these are so common and important that various authors have developed their energy equations to deal with them but with insufficient generality to deal with other situations. For example, the form of the energy equation required to calculate the power input to a pump which provides a specified flow rate at a given pressure rise is much simpler than the one we shall derive. Conversely, our equation will be general enough so that the student will be able to extend its use to other appropriate situations.

To provide an equation which is more useful but no less general than Eq. 2.63 we will explicitly define certain components of the work-rate term, for utility, and leave others with rather broad definitions, for generality.

Consider a resultant force $\mathbf{F}$ acting on a system. If the force moves the system through a distance $\Delta \mathbf{r}$ the work done on the system is given by $\mathbf{F} \cdot \Delta \mathbf{r}$. The net force acting on the system is related to the motion of the system by Newton's second law,

$$\mathbf{F} = M \frac{d\mathbf{V}}{dt} \tag{2.64}$$

where M is the mass of the system. Integrating over the path of the motion results in an expression for the net work done on the system:

$$W = \int_1^2 \mathbf{F} \cdot d\mathbf{r} = M \int_1^2 \frac{D\mathbf{V}}{Dt} \cdot d\mathbf{r} = M \int_1^2 \mathbf{V} \cdot d\mathbf{V}$$
$$= M \frac{V^2}{2} \bigg|_1^2 = M(V_2^2 - V_1^2)/2 \tag{2.65}$$

where we have used $d\mathbf{r}/dt = \mathbf{V}$ and $\mathbf{V} \cdot d\mathbf{V} = d(V^2/2)$. Note that the net work results in a change in kinetic energy of the system. This development also indicates that this correspondence between work and kinetic energy requires that the velocity and displacement vectors must be referred to an inertial reference frame since Eq 2.64 is restricted in this manner.* The even more restrictive condition established earlier

---
*As an example of this restriction, consider yourself holding an object in an accelerating automobile. Even though your hand has not moved with respect to the car (the noninertial reference frame) you have done work on the object since the force of your hand on the object, required to accelerate it, has moved through a distance.

for kinetic energy will also be employed for the work-rate term, since our purpose in establishing a description of the work rate is for use in the energy equation, where the kinetic energy must also be described. However, the first law is not restricted to inertial reference frames. Therefore, the student should be aware of the motivation for this restriction and not consider it to be inherent in the energy equation itself.

The rate of doing work, called power, is given by $\lim_{\Delta t \to 0} \mathbf{F} \cdot \Delta \mathbf{r}/\Delta t$, or $\mathbf{F} \cdot \mathbf{V}_{I'}$, where the subscript I' reminds us that the velocity is referred to a stationary reference frame. The infinitesimal force $d\mathbf{F}$ acting on a differential area dA may be expressed as $d\mathbf{F} = \tau \, dA$, where τ is the stress vector resulting from the action of the surroundings on the system. The rate at which the system does work on the surroundings is

$$\dot{W} = -\int_A \tau \cdot \mathbf{V}_{I'} \, dA \tag{2.66}$$

It is useful to distinguish between the work-rate effects associated with the movement of the fluid and the motion of the control volume; this is done by noting that the velocity $\mathbf{V}_{I'}$ can be written as

$$\mathbf{V}_{I'} = \mathbf{V} + \dot{\mathbf{S}} + \mathbf{\Omega} \times \mathbf{R} \tag{2.67}$$

where $\dot{\mathbf{S}}$ and $\mathbf{\Omega}$ are the velocity and angular velocity, respectively, of the reference frame attached to the control volume referred to an inertial reference frame and $\mathbf{V}$ is the velocity in the reference frame attached to the control volume. The general expression for the work rate is then

$$\dot{W} = -\int_A \tau \cdot \mathbf{V} \, dA - \dot{\mathbf{S}} \cdot \int_A \tau \, dA - \int_A \tau \cdot \mathbf{\Omega} \times \mathbf{R} \, dA \tag{2.68}$$

Note that $\dot{\mathbf{S}}$ has been removed from the integrand since it is the velocity of the coordinate system attached to the control volume and thus does not vary with respect to the area A. The integral $\int \tau \, dA$ is the net force acting on a system or control volume. An example of such a net force would be the lift plus the drag on an aircraft; $\dot{\mathbf{S}}$ would be the velocity of the aircraft referred to the ground.

It is also useful to distinguish between the contribution of the shearing stress and that of the normal stress to the first integral in Eq. 2.68. Consider the system shown in Fig. 2.8, the stress vector may be decomposed into a normal part and a shearing part as

$$\tau = \tau_N + \tau_S \tag{2.69}$$

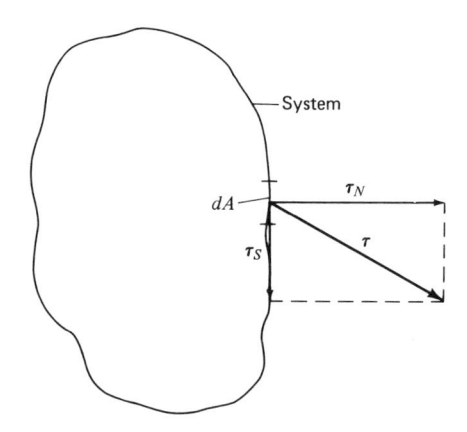

Fig. 2.8. Stress vector acting on elemental area.

where τ_N is the normal stress and τ_S is the shearing stress. The work is then

$$\dot{W} = -\int_{c.s.} \tau_S \cdot \mathbf{V}\, dA - \int_{c.s.} \tau_N \cdot \mathbf{V}\, dA - \mathbf{F} \cdot \dot{\mathbf{S}} - \int_{c.s.} \mathbf{\Omega} \times \mathbf{R} \cdot \tau\, dA \quad (2.70)$$

where $\mathbf{F} = \int \tau\, dA$. Note that the integration is over the surface surrounding the system or the control volume, since both are identical at the instant considered. The symbols τ_S and τ_N refer to stress effects at the boundary of the control volume. These stresses may exist in the fluid, in which the normal stress is assumed* to be the negative of the pressure, $\tau_N = -p$, or in the solid medium which occupies the boundary location. For an example of a solid occupying the control surface, consider a shaft connecting a motor to the impeller of a centrifugal fan (see Fig. 2.9) and consider control surface ①, which passes through the shaft. In the first term in Eq. 2.70, let τ_S be the shear stress at a radius r of the exposed shaft ($\tau_S = \tau_S \hat{\theta}$) and let the velocity be expressed as $\mathbf{V} = r\Omega \hat{\theta}$. Then,

$$-\int \tau_S \cdot \mathbf{V}\, dA = -\int \tau_S r\Omega\, dA$$

$$= -T\Omega \quad (2.71)$$

where the torque T on the shaft is given by $\int \tau_S r\, dA$. Note that this formulation properly indicates that the motor is doing work on the pump and the shaft-work term is negative. If the fluid device were a turbine instead of a pump, the shaft would transmit power from the control volume. This can be seen from the nature of the stress term: that is, $\tau_S = -\tau_S \hat{\theta}$ and $\mathbf{V} = r\Omega \hat{\theta}$. Hence the shaft work is positive.

*Using the pressure as the normal part of the stress vector in a fluid is an approximation that is generally applicable. See Eqs. 3.43 for the exact expression.

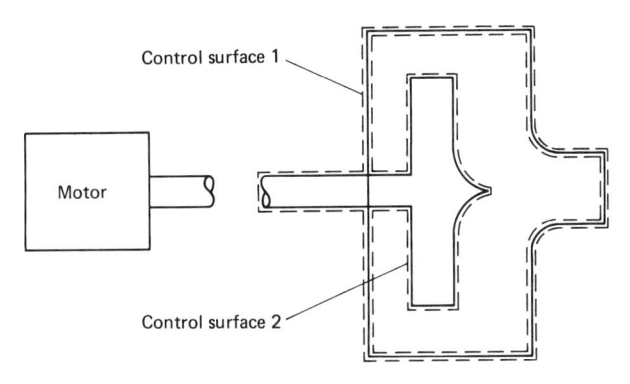

Fig. 2.9. Control volumes for work-rate-term evaluation.

The magnitudes of the integral terms on the right-hand side of Eq. 2.70 are not the same for control surface ① and control surface ②. The magnitude of the input power to the pump is, of course, independent of the control surface. The pressure integral term of Eq. 2.70, $\tau_N = -p$, respresents the dominant physical effect in the work rate of the impeller on the fluid for surface ②, as one can appreciate by considering the details of the impeller-fluid interaction.

The second term on the right-hand side of Eq. 2.70 can also represent the work-rate effect at a solid boundary exposed by the control surface. Consider Fig. 2.10. The work rate for the indicated control volume is transmitted through the piston rod. For the intake, compression, and exhaust strokes it is necessary to put work into the control volume. Work is extracted from the control volume after the chemical energy is released in the combustion process; the integral $\int \tau_N \cdot \mathbf{V} \, dA$ leads to this conclusion, if we use the appropriate sign for the normal stress, the velocity $\mathbf{V}$, and the outward-drawn normal $\hat{n}$.

In a fluid the normal stress vector is replaced by $\tau_N = -p\hat{n}$. Equation 2.70 is then written as

$$\dot{W} = -\int_{c.s.} \tau_S \cdot \mathbf{V} \, dA + \int_{c.s.} p \, \mathbf{V} \cdot \hat{n} \, dA + \dot{W}_{I'} \qquad (2.72)$$

where $\dot{W}_{I'}$ replaced the work-rate terms associated with the moving coordinate system. Specifically,*

$$\dot{W}_{I'} = -\dot{\mathbf{S}} \cdot \mathbf{F} - \int_{c.s.} \mathbf{\Omega} \times \mathbf{r} \cdot \boldsymbol{\tau} \, dA \qquad (2.73)$$

The pressure integral term is used to characterize an effect called "flow work." This effect occurs when a fluid crosses a boundary; in order to

---

*Note that $\dot{W}_{I'}$ may be nonzero for a control volume translating at constant velocity even though a constant-velocity reference frame is an inertial reference frame.

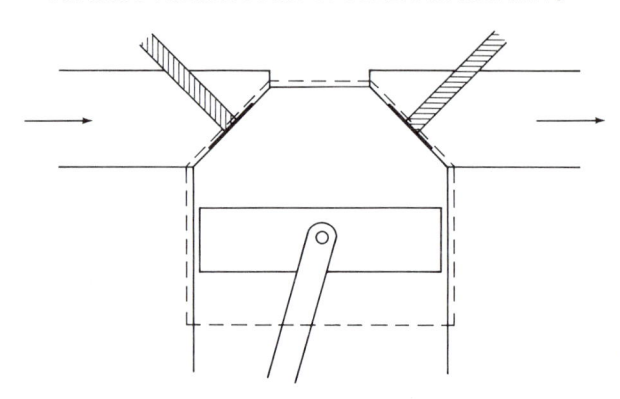

Fig. 2.10. Piston-cylinder control volume.

move fluid from or into a region the fluid must be moved against a resisting force, measured as the pressure times the exposed area of the fluid moved. For the control volume of Fig. 2.10, the fluid that enters the control volume at the inlet valve and the fluid that leaves at the exhaust port must overcome the resisting force due to the pressure inside the cylinder and that in the exhaust manifold, respectively. It is important to recognize that the flow-work effect is associated with the magnitude of the pressure at the control surface and not a pressure difference. The net flow work is given by the expression $\int_{c.s.} p \mathbf{V} \cdot \hat{n} \, dA$.

The shear-stress-related portion of the work rate is $\int \tau_S \cdot \mathbf{V} \, dA$ where the shear stress is evaluated in the control-surface plane. For the usual fluids of engineering interest, τ_S is assumed zero if the velocity field is assumed uniform; also the dot product is zero if the flow is perpendicular to the control surface. The integral is also zero if the velocity is zero on the control surface. Consequently, this term is rarely of importance. An example of the $\dot{W}_{shear}$ term would be an object moving through a fluid where the object would be excluded from the control volume, e.g. a paddle drawn through a liquid. At the deforming control surface the contribution to the work rate due to the shearing stresses would be $-\int \tau_s \cdot \mathbf{V} \, dA = -$ shear force $\times$ object velocity. When the control surface cuts a shaft (see Eq. 2.71), $\int \tau_N \cdot \mathbf{V} \, dA + \int \tau_S \cdot \mathbf{V} \, dA$ will be called shaft work $\dot{W}_S$.

The general work-rate term of the energy equation (Eq. 2.63) can now be written in terms of the particular quantities which have been identified. Specifically,

$$\dot{W} = \dot{W}_S + \dot{W}_{shear} + \int_{c.s.} p \mathbf{V} \cdot \hat{n} \, dA + \dot{W}_{I'} \qquad (2.74)$$

Art. 2.4   THE FUNDAMENTAL LAWS   119

where $\dot{W}_S$ = work rate associated with the motion of the shearing or normal stresses in a rotating or translating shaft or its equivalent (e.g., electrical power)

$\dot{W}_{shear}$ = work rate associated with the motion of the shearing stress caused by the fluid motion at the control surface

$\int p\mathbf{V}\cdot\hat{n}\,dA$ = flow work associated with the motion of the fluid as it crosses the control surface

$\dot{W}_{I'}$ = the work rate associated with the motion of the control volume with respect to an inertial coordinate system

Utilizing Eq. 2.74 for the work-rate term, the most general form of the energy equation to be considered for the nondeformable control volume is

$$\dot{Q} - \dot{W}_S - \dot{W}_{shear} - \dot{W}_{I'} = \frac{d}{dt}\int \rho\left(\frac{V_I^2}{2} + gz + \tilde{u}\right)dV$$
$$+ \int_{c.s.}\left(\frac{V_I^2}{2} + gz + \tilde{u} + \frac{p}{\rho}\right)\rho\,\mathbf{V}\cdot\hat{n}\,dA \quad (2.75)$$

Many problems in fluid mechanics involve undesired and unavoidable transfer of energy from the "mechanical" to the "thermal"* terms of the energy equation. Since frictional effects are responsible for this undesired transfer, the process is irreversible and always results in a decreased magnitude of the combined potential-energy, kinetic-energy, and flow-work terms. Because of the historical precedent of the hydraulic designation of these terms as "heads" (e.g., "velocity head," $V^2/2g$) the term *head loss* (h_L) is commonly used to designate the magnitude of this *dissipation effect*. Formally, the energy equation may be rewritten as

$$-\dot{W}_S - \dot{W}_{shear} - \dot{W}_{I'} = \frac{d}{dt}\int_{c.v.}\left(\frac{V_I^2}{2} + gz\right)\rho\,dV$$
$$+ \int_{c.s.}\left(\frac{V_I^2}{2} + gz + \frac{p}{\rho}\right)\rho\,\mathbf{V}\cdot\hat{n}\,dA \quad (2.76)$$
$$+ \left[\frac{d}{dt}\int_{c.v.}\tilde{u}\rho\,dV + \int_{c.s.}\tilde{u}\rho\,\mathbf{V}\cdot\hat{n}\,dA - \dot{Q}\right]$$

---

*It is helpful to identify potential and kinetic energy and flow work as mechanical terms and heat transfer and internal energy as thermal terms. In incompressible-flow problems the mechanical terms usually represent useful energy and the thermal terms unusable energy. Hence, it is usually undesirable to convert mechanical energy to thermal energy.

The term in brackets contains the losses; that is, the heat transfer and the internal energy change associated with frictional or dissipative effects represent the losses. Specifically,

$$\text{Losses} = \left[\frac{d}{dt} \int_{c.v.} \tilde{u}\rho \, dV + \int_{c.s.} \tilde{u}\rho \, \mathbf{V} \cdot \hat{n} \, dA - \dot{Q} \right]_{\text{dissipative}} \quad (2.77)$$

An example of a nondissipative contribution to the bracketed term of the energy equation (Eq. 2.76) is the compressible flow of a gas from a pressurized tank. If the gas expands without frictional effects the bracketed term will have a negative value, since the temperature at the exit will be less than that in the tank. If frictional effects are present, the net value will remain negative but will be of smaller magnitude.

The head loss would be written as

$$h_L = \frac{1}{\dot{m}g} \left[\frac{d}{dt} \int_{c.v.} \tilde{u}\rho \, dV + \int_{c.s.} \tilde{u}\rho \, \mathbf{V} \cdot \hat{n} \, dA - \dot{Q} \right]_{\text{dissipative}} \quad (2.78)$$

where the quantity $\dot{m}g$ is used to give the appropriate dimensions.

The losses are often expressed in terms of a *loss coefficient K* as

$$\text{Losses} = \dot{m} K \frac{V^2}{2} \quad (2.79)$$

where the velocity V is a characteristic velocity in the region where the losses occur. The consideration of such a loss term is somewhat subtle; it is dealt with at length in Chapter 8. We introduce the loss coefficient here as a matter of convenience so that we may use it for special cases before the more complete discussion.

The head loss, in terms of the loss coefficient, is

$$h_L = K \frac{V^2}{2g} \quad (2.80)$$

For those problems encountered in engineering applications which involve steady flow through nondeformable control volumes and for which the frame of reference is stationary with respect to the ground, the integral energy equation reduces to the form

$$\dot{Q} - \dot{W}_S = \int_{c.s.} \left(\frac{V^2}{2} + \frac{p}{\rho} + gz + \tilde{u} \right) \rho \, \mathbf{V} \cdot \hat{n} \, dA \quad (2.81)$$

or, using the loss coefficient,

$$-\dot{W}_S = \int_{\text{c.s.}} \left(\frac{V^2}{2} + \frac{p}{\rho} + gz\right)\rho\, \mathbf{V}\cdot\hat{n}\, dA + \dot{m}K\frac{V^2}{2} \quad (2.82)$$

The restrictions noted above are applicable to many of the problems encountered by the practicing engineer; hence, it is worthwhile noting this special form of the equation. But by means of the more complex form, Eq. 2.75, he can solve the more difficult problems which may arise.

The magnitude of the head loss depends on the geometric and flow conditions; in later chapters, we will consider the factors which affect the head loss. One note of caution is in order at this point. Since there are 778 ft-lb per Btu, a dissipative process can destroy a considerable amount of "mechanical energy" with the result of only a slight increase in the temperature level of the fluid. In order to provide a quantitative assessment of this observation, consider the case of water in a reservoir. If the reservoir surface level were 778 ft above a receiver pond, then each pound of water that passed through an ideal turbine would produce 778 ft-lb of work if the water exhausts at atmospheric pressure and negligible velocity. If, on the other hand, the water were merely allowed to drain into the lower receiver pond in such a manner that the initial 778 ft-lb of energy were simply dissipated, the final water temperature would be only 1 Fahrenheit degree above ambient (assuming $\dot{Q} = 0$) since $c_p = 1$ Btu/lb$_m$-°F. Therefore, in terms of our intuitive insight into physical problems, we must guard against considering slight temperature changes as indicating only negligible energy changes.

*8. The Deformable-Control-Volume Energy Equation

The possibility of a deformable control surface introduces several modifications into the energy equation developed in the previous section. These are considered in this section; however, the numerous considerations that apply equally to fixed- and deformable-control-volume forms will not be duplicated. Using the Reynolds transport theorem the deformable energy equation is written as

$$\dot{Q} - \dot{W} = \frac{d}{dt}\int_{\text{c.v.}}\left(\frac{V_{I'}^2}{2} + gz + \tilde{u}\right)\rho\, d\mathcal{V} + \int_{\text{c.s.}}\left(\frac{V_{I'}^2}{2} + gz + \tilde{u}\right)\rho\, \mathbf{V}_r\cdot\hat{n}\, dA \quad (2.83)$$

The inertial velocity in the expression for the work rate (Eq. 2.66) is given by $\mathbf{V}_I = \mathbf{V}_r + \mathbf{V}_b + \dot{\mathbf{S}} + \mathbf{\Omega} \times \mathbf{R}$. With this introduced the work-rate equation (2.74) becomes

$$\dot{W} = \dot{W}_S + \dot{W}_{shear} + \int_{c.s.} p\, \mathbf{V}_b \cdot \hat{n}\, dA + \int_{c.s.} p\, \mathbf{V}_r \cdot \hat{n}\, dA + \dot{W}_I \quad (2.84)$$

An easily visualized example of the $\int_{c.s.} p\, \mathbf{V}_b \cdot \hat{n}\, dA$ term would be the work rate associated with the pressure acting on the piston face in Fig. 2.10 if the control-volume boundary were placed on the piston face. An example which demonstrates the $-\int_{c.s.} \mathbf{\Omega} \times \mathbf{R} \cdot \boldsymbol{\tau}\, dA$ contribution to the $\dot{W}_I$ work-rate term would be a container of angularly accelerating water on a rotating turntable while the surface is deforming. Using R and H as the radius and height of the water, respectively, the work-rate term $-\int_{c.s.} \mathbf{\Omega} \times \mathbf{R} \cdot \boldsymbol{\tau}\, dA$ due to the shear stress on the side walls would be $-2\pi R^2 \Omega \int_0^H \tau_s\, dh$, where $dA = 2\pi R\, dh$.

The energy equation (2.83) can finally be put in the form

$$\dot{Q} - \dot{W}_S - \dot{W}_{shear} - \dot{W}_{I'}$$
$$= \int_{c.v.} \frac{\partial}{\partial t}\left[\rho\left(\frac{V_{I'}^2}{2} + gz + \tilde{u}\right)\right] dV + \int_{c.s.} \left[\frac{V_{I'}^2}{2} + \frac{p}{\rho} + gz + \tilde{u}\right] \rho\, \mathbf{V}_b \cdot \hat{n}\, dA$$
$$+ \int_{c.s.} \left[\frac{V_{I'}^2}{2} + \frac{p}{\rho} + gz + \tilde{u}\right] \rho\, \mathbf{V}_r \cdot \hat{n}\, dA \quad (2.85)$$

This is the most general form of the energy equation to be presented in this book.

Example 2.15

For situations involving fixed control volumes with only one inlet and one exit, such as pipes or turbines, the energy equation takes on a simple form. For a steady incompressible flow, assuming uniform flow quantities at the inlet area A_1 and exit area A_2, show Eq. 2.75 in its simplified form.

Solution. For the fixed control volume, Eq. 2.75 is used. It is, for steady flow,

$$\dot{Q} - \dot{W}_S = \int_{c.s.} \left[\frac{V^2}{2} + gz + \tilde{u} + \frac{p}{\rho}\right] \rho\, \mathbf{V} \cdot \hat{n}\, dA$$

Art. 2.4   THE FUNDAMENTAL LAWS   123

The shearing stresses on the control surface do no work and $\dot{W}_{\tau'} = 0$. For velocity normal to the inlet area A_1 and exit area A_2, the equation may be written as

$$\dot{Q} - \dot{W}_S = \left[\frac{V_2^2}{2} + gz_2 + \tilde{u}_2 + \frac{p_2}{\rho}\right]\rho V_2 A_2 - \left[\frac{V_1^2}{2} + gz_1 + \tilde{u}_1 + \frac{p_1}{\rho}\right]\rho V_1 A_1$$

Continuity considerations show that

$$\rho A_1 V_1 = \rho A_2 V_2 = \dot{m}$$

so that

$$-\frac{\dot{W}_S}{\dot{m}} = \frac{V_2^2}{2} + gz_2 + \frac{p_2}{\rho} - \frac{V_1^2}{2} - gz_1 - \frac{p_1}{\rho} + \text{losses}$$

where the losses are

$$\text{losses} = \tilde{u}_2 - \tilde{u}_1 - \frac{\dot{Q}}{\dot{m}}$$

If there are no losses or shaft work transferred then

$$\frac{V_2^2}{2} + gz_2 + \frac{p_2}{\rho} = \frac{V_1^2}{2} + gz_1 + \frac{p_1}{\rho}$$

which looks very much like Bernoulli's equation; however, this is not along a streamline but between sections ① and ②, which are the only two locations where a mass flux occurs.

Extension 2.15.1. Water flows through a 4-in.-dia. pipe into a 10-hp pump and out a 2-in.-dia. pipe. Determine the decrease in pressure, assuming 70-percent pump efficiency. Neglect all additional losses. The flow rate is 2 cfs.
Ans. 5720 psf

Extension 2.15.2. Write the energy equation for a flow field with inlet ① and outlets ② and ③. Do not make the assumption of incompressibility, but adopt the other assumptions in the example above. Neglect elevation changes; introduce enthalpy $h = p/\rho + \tilde{u}$ into the equation.

Example 2.16

A jet aircraft is moving at a velocity V_∞. Use the energy equation to relate the fuel consumption $\dot{m}_f$ to the flow situation. The energy released by the fuel is q_f Btu/slug.

Solution. a. Choose a fixed control volume as shown in Fig. E2.16A, with boundaries at large distances from the aircraft and moving with the aircraft. The pressure will be atmospheric over the entire control surface and will not contribute to the energy equation. No shear acts on the surface, the work rate is zero, and elevation changes are not important. Accounting for the energy released in the combustion process could be accomplished by noting the time

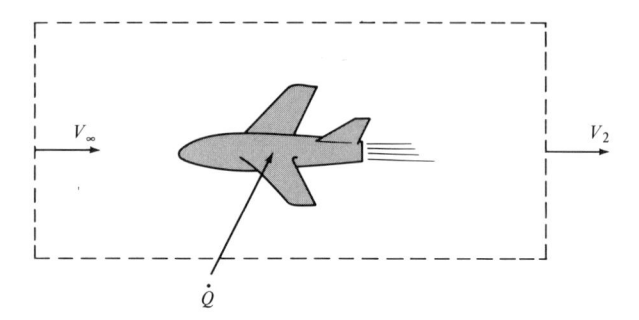

Fig. E2.16A

rate of change of chemical energy within the control volume; however, we have not considered this form of energy in our analysis. An equivalent technique, motivated by the physical processes occurring in the combustion chamber, is to consider the fuel exterior to the control volume and to consider the energy released by the combustion and transferred to the air to be a source of thermal energy flux $\dot{Q}$. For the given problem, $\dot{Q} = \dot{m}_f q_f$. The factor 778 ft-lb/Btu will be introduced to maintain the usual units for the various terms in the energy equation.

The energy equation (2.75) reduces to

$$\dot{Q} = \int_{c.s.} \left(\frac{V_I^2}{2} + \tilde{u} \right) \rho \, \mathbf{V} \cdot \hat{n} \, dA$$

or

$$778 q_f \dot{m}_f = \int_{A_2} \frac{(V_2 - V_\infty)^2}{2} \rho V_2 \, dA + \int_{A_2} \tilde{u} \rho V_2 \, dA - \tilde{u}_\infty \rho V_\infty A$$

Note that a reference frame attached to the control volume moving with the aircraft is an inertial reference frame and hence the velocities in the above are "inertial." The expression above, relating the fuel consumption to the flow properties, is not very useful since it requires a knowledge of the velocity profile and internal energy (temperature) distribution downstream from the aircraft; however, it does show that the energy from the fuel is converted to kinetic and internal energy of the air.

b. Now choose a control volume that just surrounds the aircraft itself, as in Fig. E2.16B; this differs from the previous control volume in that pressure and shear on the control volume surface now give rise to a force which includes the lift and the drag. Using Eq. 2.73, the energy equation (2.75) now becomes

$$\dot{Q} + \mathbf{F} \cdot \dot{\mathbf{S}} = \int_{c.s.} \left(\frac{V_I^2}{2} + \frac{p}{\rho} + \tilde{u} \right) \rho \, \mathbf{V} \cdot \hat{n} \, dA$$

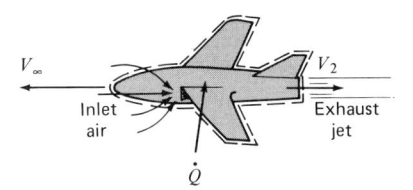

Fig. E2.16B

The lift force does no work, and thus $\mathbf{F} \cdot \dot{\mathbf{S}} = -DV_\infty$, where D is the drag force due to the pressure and shearing stresses acting on the aircraft. Air crosses the control surfaces at the inlet and exit to the jet engine. Assuming uniform profiles, the energy equation becomes

$$778 q_f \dot{m}_f - V_\infty D = \frac{(V_2 - V_\infty)^2}{2} \rho_2 V_2 A_2 + p_2 V_2 A_2 + \tilde{u}_2 \rho_2 V_2 A_2$$
$$- \frac{(V_1 - V_\infty)^2}{2} \rho_1 V_1 A_1 - p_1 V_1 A_1 - \tilde{u}_1 \rho_1 V_1 A_1$$

Continuity shows that

$$\dot{m}_f + \rho_1 V_1 A_1 = \rho_2 V_2 A_2$$

The fuel mass flux $\dot{m}_f$ is small compared to the air mass flux $\dot{m} = \rho_1 A_1 V_1$. Neglecting $\dot{m}_f$ compared to $\dot{m}$ the energy equation is

$$778 q_f \dot{m}_f - V_\infty D = \dot{m} \left[\frac{(V_2 - V_\infty)^2}{2} + \frac{p_2}{\rho_2} + \tilde{u}_2 - \frac{(V_1 - V_\infty)^2}{2} - \frac{p_1}{\rho_1} - \tilde{u}_1 \right]$$

or

$$778 q_f \dot{m}_f / \dot{m} = \frac{V_2^2 - V_1^2}{2} - (V_2 - V_1) V_\infty + h_2 - h_1 + \frac{V_\infty D}{\dot{m}}$$

The quantities at the engine exit are much easier to measure or determine; hence, the last version would be preferred. It shows that the energy from the fuel increases the kinetic energy of the exhaust gas, increases the exhaust gas temperature ($h = c_p T$), and overcomes the drag.

Extension 2.16.1. Consider an automobile moving at 60 mph with a drag force of 300 lb. The car's gas mileage is 12 mpg, gasoline weighs 58 lb/ft³, and the overall efficiency of the engine is 15 percent. Determine the energy released per slug of fuel. *Ans.* 675,000 Btu/slug

Example 2.17

Water passes through a turbine as it flows from a reservoir to the river below. The reservoir has no water inputs during the period of interest so that the surface slowly decreases in elevation. For a flow rate of 1000 cfs, an outlet area of 50 ft², losses to the outlet of $KV^2/2$ with $K = 10.0$, (K accounts for the losses in the pipe only), and a turbine-generator efficiency of 20 percent, determine the expected horsepower output. Work this problem using a nondeformable control volume.

Solution. A nondeformable control volume is chosen with the control surface just *below* the surface of the reservoir (Fig. E2.17A). There will be a very small velocity V_1 over this inlet to the control volume. The energy equation for this essentially steady flow is

$$-\dot{W}_S = \int_{A_1} \left(\frac{V^2}{2} + \frac{p}{\rho} + gz \right) \rho \, \mathbf{V} \cdot \hat{n} \, dA + \int_{A_2} \left(\frac{V^2}{2} + \frac{p}{\rho} + gz \right) \rho \, \mathbf{V} \cdot \hat{n} \, dA + \text{losses}$$

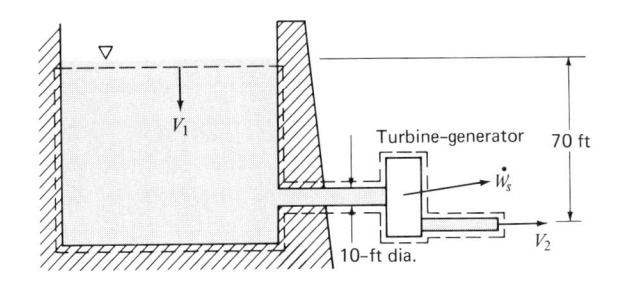

Fig. E2.17A

where A_1 is the reservoir surface area and A_2 is the outlet area. The datum is chosen so that $z_2 = 0$, the pressure is atmospheric at both the inlet and exit, and the velocity at the inlet is very small so that $V_1^2 \approx 0$. The energy equation then becomes

$$-\dot{W}_S = gz_1(-\rho V_1 A_1) + \frac{V_2^2}{2}(\rho V_2 A_2) + K\frac{V^2}{2}\dot{m}$$

Continuity yields $\rho V_1 A_1 = \rho V_2 A_2 = \dot{m}$, which is given as 1000ρ slugs/sec. The loss coefficient is based on the velocity in the 10-ft-dia. interior pipe, which is $V = 1000/100\pi/4 = 12.7$ fps. Hence,

$$-\frac{\dot{W}_S}{\dot{m}} = -gz_1 + \frac{V_2^2}{2} + K\frac{V^2}{2}$$

$$= -32.2 \times 70 + \frac{20^2}{2} + 10\frac{12.7^2}{2}$$

where $V_2 = 1000/50 = 20$ fps. Finally, the work extracted from the water by the turbine is

$$\dot{W}_S = \left(32.2 \times 70 - \frac{20^2}{2} - 10\frac{12.7^2}{2} \right) 1000 \times 1.94 = 2.42 \times 10^6 \text{ ft-lb/sec}$$

Art. 2.4 THE FUNDAMENTAL LAWS

The energy supplied by the generator is only 20 percent of this. In horsepower it is

$$\text{Energy output} = 2.42 \times 10^6 \times \frac{0.2}{550} = 880 \text{ hp}$$

Extension 2.17.1. Consider the reservoir to be continuously supplied by water (Fig. E2.17B). Assume V_1 to be uniformly distributed. For this steady-state condition, show that the flux term

$$\int_{A_1} \left(\frac{V^2}{2} + \frac{p}{\rho} + gz \right) \rho \, \mathbf{V} \cdot \hat{n} \, dA$$

can be simplified to the identical result as in the example, namely, $gz_1(-\rho A_1 V_1)$.

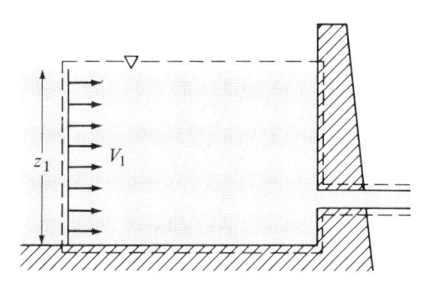

Fig. E2.17B

Example 2.18

A fan which is designed to deliver a volume rate of air has the characteristic curve* shown in Fig. E2.18A and is placed in a flow system as shown. If the blower is attached to a duct with a variable-area nozzle exhausting to atmospheric pressure, determine the pressure at the blower exit as a function of the diameter of the nozzle.

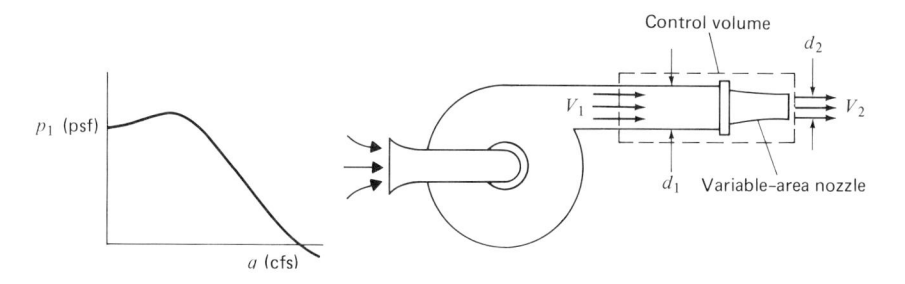

Fig. E2.18A

*The fan characteristic curve is experimentally determined. Fan test procedures have been established by the various professional societies and such standardized data are usually available from the manufacturer.

Solution. For a control volume which extends from the blower exit to the nozzle exit, the energy equation for one-dimensional steady flow with negligible head loss and elevation change is

$$0 = \int_{c.s.} \left(\frac{V^2}{2} + \frac{p}{\rho} \right) \rho \mathbf{V} \cdot \hat{n} \, dA$$

$$= \left(\frac{V_2^2 - V_1^2}{2} + \frac{\cancel{p_2}^{\,0} - p_1}{\rho} \right) \dot{m}$$

For one-dimensional flow, continuity gives $V_2 A_2 = V_1 A_1$ (air is incompressible for low speeds), or

$$V_1 = V_2 \left(\frac{d_2}{d_1} \right)^2$$

Since p_2 is zero gage pressure, the energy equation, with continuity substituted in, becomes

$$p_1 = \rho \frac{V_2^2}{2} \left(1 - \frac{d_2^4}{d_1^4} \right)$$

Physically, this may be interpreted as stating that the pressure is dependent upon the velocity; one equation is thus available for the two unknowns, V_2 and p_1. The volume flow rate is expressed in terms of the velocity by

$$q = A_2 V_2 = \frac{\pi d_2^2}{4} V_2$$

Hence, the pressure relationship can be written as

$$p_1 = \frac{8\rho q^2}{\pi^2 d_2^4} \left(1 - \frac{d_2^4}{d_1^4} \right)$$

For given d_2 and d_1 values, this equation is in the form of a parabola. A second equation for the same unknowns is provided by the fan characteristic curve. The two curves, when plotted on the same graph (Fig. E2.18B), intersect at the

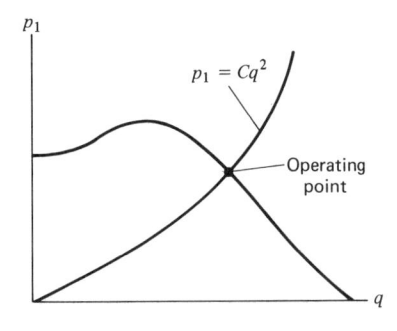

Fig. E2.18B

Art. 2.4   THE FUNDAMENTAL LAWS   129

simultaneous solution for the two equations; that is, they intersect at the operating point for the system. The solution would give p_1 for given d_2 and d_1.

Extension 2.18.1. Examine the limiting cases: $d_2 = d_1$; and $d_2 = 0$. Is the general formulation still valid? Explain.

Extension 2.18.2. If loss effects were included between section ①and section ②(as losses = $\dot{m}K V^2/2$), how would this change the form of the equation? For a given d_2 and d_1, describe the effect of K on the operating point.

Extension 2.18.3. Consider the common experience of placing your finger over the end of a garden hose. What will the p_1-vs.-q "characteristic curve" for the hose be like? Does the rest of the derivation for the example problem apply? What happens to q as the exit area is constricted? What happens to V_2?

*Example 2.19 ─────────────────────────────

Water is pumped from a reservoir to a device as shown in Fig. E2.19. It is desired that 2 ft^3/sec of water be provided to the device at 5 psi by the 2-in.-dia. pipe. Assume the losses to be $KV^2/2$, with $K = 5.0$. Determine the required horsepower input if the pump is 68-percent efficient.

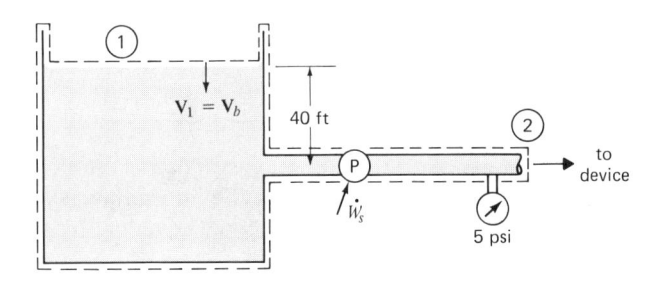

Fig. E2.19

Solution. An appropriate form* of the energy equation for this steady flow (see Eq. 2.85) is

$$\dot{Q} - \dot{W}_S = \int_{A_1} \left(\frac{V^2}{2} + gz + \tilde{u} + \frac{p}{\rho} \right) \rho \, \mathbf{V}_b \cdot \hat{n} \, dA + \int_{A_2} \left(\frac{V^2}{2} + gz + \tilde{u} + \frac{p}{\rho} \right) \rho \, \mathbf{V}_r \cdot \hat{n} \, dA$$

*The condition that the reservoir is quite large leads to two simplifications in this problem since the elevation of the free surface changes quite slowly. The time rate of change of the specific energy in the control volume is negligible $\left[\frac{\partial}{\partial t} \left(\frac{V^2}{2} + \tilde{u} + gz \right) \approx 0 \right]$ and the kinetic energy at the free surface is similarly negligible ($V_1^2/2 \approx 0$).

Section ① is on the deformable surface of the reservoir where $V_r = 0$. Using $p_1 = 0$, $z_1 = 40$ ft, and $z_2 = 0$, we have

$$-\dot{W}_S = 40g(-\rho V_b A_1) + \left(\frac{V_2^2}{2} + \frac{p_2}{\rho}\right)\rho V_2 A_2 + K\frac{V_2^2}{2}\rho A_2 V_2$$

where the internal energy change and heat transfer have been replaced by the losses, $\dot{m}KV_2^2/2$. The mass flow rate $\dot{m}$ is included to give the correct units on the losses. From continuity considerations,

$$\rho V_b A_1 = \rho V_2 A_2 = \dot{m}$$

Also

$$V_2 = \frac{2}{A_2} = \frac{2}{\pi/144} = 91.7 \text{ fps}$$

The energy equation is thus

$$-\dot{W}_S = -40g(2\rho) + \left(\frac{91.7^2}{2} + \frac{5 \times 144}{1.94}\right)2\rho + 5.0\frac{91.7^2}{2}2\rho = 94,300 \frac{\text{ft-lb}}{\text{sec}}$$

This is the work put into the fluid; hence, it is negative. The horsepower input to the pump may be calculated in terms of the pump efficiency as

$$\text{Pump hp} = \frac{94,300}{550} \times \frac{1}{0.68} = 252 \text{ Hp}$$

*Example 2.20

In Example 2.13, an expression was developed, relating the position $X(t)$ of the device shown there to the height $h(t)$ of its liquid column. The momentum and continuity equations were used to develop the equation

$$-\frac{d^2X}{dt^2}(M_f + M_{\text{device}} + \rho Ah) = \rho AL\frac{d^2h}{dt^2} \qquad h > 0$$

From an analytical viewpoint, this equation is indeterminate; that is, there is only one equation for the two unknowns, h and X. The physical phenomenon must also satisfy energy considerations, and the energy equation will thus provide a second equation for the time-dependent X and h values.

The heat-transfer and internal-energy effects will be accounted for by the loss term; the shaft work is zero, and we will neglect the air-drag contribution to the work rate. The energy equation (2.85) for this problem is then

$$0 = \int_{\text{c.v.}} \frac{\partial}{\partial t}\left[\rho\left(\frac{V_I^2}{2} + gz\right)\right]dV + \int_{\text{c.s.}}\left[\frac{V_I^2}{2} + gz + \frac{p}{\rho}\right]\rho \mathbf{V}_b \cdot \hat{n}\, dA$$

$$+ \int_{\text{c.s.}}\left[\frac{V_I^2}{2} + gz + \frac{p}{\rho}\right]\rho \mathbf{V}_r \cdot \hat{n}\, dA + \text{losses}$$

where we have written the velocity as V_I to indicate that it is referred to a stationary reference frame. The velocity in the expression can be written as

$$\mathbf{V}_{I'} = \mathbf{V} + \frac{dX}{dt}\hat{i}$$

where $\mathbf{V}$ is the velocity measured in the moving xyz-reference frame. Then

$$V_{I'}^2 = \mathbf{V}_{I'} \cdot \mathbf{V}_{I'} = V^2 + 2\frac{dX}{dt}\mathbf{V}\cdot\hat{i} + \left(\frac{dX}{dt}\right)^2$$

The kinetic-energy terms may then be evaluated for the problem as

$$\int_{c.v.} \frac{\partial}{\partial t}\left(\frac{V_I^2}{2}\right)\rho\, dV + \int_{c.s.} \rho\frac{V_I^2}{2}\mathbf{V}_b\cdot\hat{n}\, dA + \underbrace{\int_{c.s.} \rho\frac{V_I^2}{2}\mathbf{V}_r\cdot\hat{n}\, dA}_{\text{Cancel}}$$

$$= \int_{c.v.} \frac{\partial}{\partial t}\left[\frac{V^2}{2} + \mathbf{V}\cdot\hat{i}\,\frac{dX}{dt} + \tfrac{1}{2}\left(\frac{dX}{dt}\right)^2\right]\rho\, dV$$

From continuity, $\mathbf{V}_b\cdot\hat{n} = -\mathbf{V}_r\cdot\hat{n}$ and consequently the last two terms on the left-hand side cancel. The velocity magnitude V is given by $V = -(dh/dt)$ and $\mathbf{V}\cdot\hat{i}\,(dX/dt)$ may be assumed to be nonzero only in the horizontal section of length L. Using $\dot{h} = dh/dt$, $\ddot{h} = d^2h/dt^2$, etc., for simplicity, the integrand becomes

$$\frac{\partial}{\partial t}\left[\frac{V^2}{2} - V\dot{X} + \frac{\dot{X}^2}{2}\right] = \dot{h}\ddot{h} + \dot{h}\ddot{X} + \dot{X}\ddot{h} + \dot{X}\ddot{X}$$

Remembering that $V\,(= -\dot{h})$ is nonzero only for the water, we may write

$$\int_{c.v.} \frac{\partial}{\partial t}\left(\frac{V_I^2}{2}\right)\rho\, dV$$

$$= \dot{h}\ddot{h}[M_f + \rho Ah] + \rho AL[\dot{h}\ddot{X} + \dot{X}\ddot{h}] + \dot{X}\ddot{X}[M_f + \rho Ah + M_{\text{device}}]$$

where M_f = mass of water below $h = 0$. The potential energy terms may be most easily accounted for by choosing the datum such that $z_{\text{exit}} = 0$, and writing (see Eq. 2.23)

$$\int_{c.v.} \frac{\partial}{\partial t}(\rho gz)\, dV + \int_{c.s.} \rho gz\,\mathbf{V}_b\cdot\hat{n}\, dA = \frac{d}{dt}\int_{c.v.} \rho gz\, dV = \frac{d}{dt}\left[\rho g\frac{h}{2}hA\right]$$

The term $\int_{c.s.} \rho gz\,\mathbf{V}_r\cdot\hat{n}\, dA$ at the exit is zero since $z_{\text{exit}} = 0$. The pressure is zero at both the exit and the water surface. The energy equation then becomes

$$\dot{h}\ddot{h}[M_f + \rho A h] + \rho A L[\dot{h}\ddot{X} + \dot{X}\ddot{h}] + \dot{X}\ddot{X}[M_f + \rho A h + M_{\text{device}}] + \frac{3}{2}\rho g h \dot{h} + \text{losses} = 0$$

To develop a complete governing equation, the losses would have to be accounted for. It is often valid to set the losses $= \dot{m}KV^2/2 = (\rho A V)KV^2/2$. In this form K accounts for the loss-producing geometric effects, either a "long" conduit or regions of internal separated flow. In any event, the formulation of the present problem is complete since K is assumed known.

Extension 2.20.1. For the equations developed in this example, describe the equations in terms of their order, whether they are linear or nonlinear, and whether ordinary or partial.

Extension 2.20.2. How many initial conditions are required by these equations? What are they?

Extension 2.20.3. How would the energy equation change if a bucket were added to catch the efflux?

Extension 2.20.4. With $h(t) = \int_0^t \dot{h}\, dt$, the two equations can be expressed as two first-order equations involving η and ζ if we let

$$\eta = \dot{h} \quad \dot{\eta} = \ddot{h} \qquad \zeta = \dot{X} \quad \dot{\zeta} = \ddot{X}$$

Because of this, the equations for $\dot{\eta}$ and $\dot{\zeta}$ may be considered to be algebraic equations for an instant where the coefficients involving η and ζ are assumed constants. Then, using the initial conditions as starting conditions, the "algebraic" equations can be solved numerically for $\dot{\eta}$ and $\dot{\zeta}$. The values of η and ζ for the next time $t + \Delta t$ are approximated, for small Δt, as

$$\eta(t + \Delta t) = \eta(t) + \dot{\eta}(t)\Delta t \quad \text{and} \quad \zeta(t + \Delta t) = \zeta(t) + \dot{\zeta}(t)\Delta t$$

and the solution can "march forward" in time.

Without executing the details of the foregoing technique, follow its reasoning such that you could obtain a numerical $h(t)$ and $X(t)$ solution.

2.5 CLOSURE

This chapter has dealt with the basic equations in their most general forms, including the deformable control volume, noninertial reference frames, and unsteady-flow processes. The example problems have been selected to clarify the meaning of each term in the equations and to indicate how a term-by-term analysis may be utilized to evaluate the complete equation. The equations, and an understanding of how to evaluate each term for a given problem, constitute the essential tools to be used in the analysis of engineering problems. The equations are valid for any control volume; whether or not the control-volume equations may be used to determine the desired values depends upon the nature of the problem and the choice of the control volume. It is, of course, impossible to provide absolute guidelines to be used for the solution of all problems; however, it is possible to provide guidance for the general problem-solving task.

Fluid mechanics problems involve the determination of velocity or pressure within a flow field, or of the net force, torque, or power associated with physical devices; these are the quantities accounted for by the control-volume equations. Therefore, the selection of the proper control volume or volumes will either provide the desired answer or indicate what data are required to obtain the desired information. (It should be noted that many of the integrals require a knowledge of the velocity or pressure distributions and that this information must be obtained from either experimental data or the differential equations. The differential equations are the subject of much of the remainder of this book.)

The selection of the control volume is related to, and is often dependent upon, the information desired. For example, suppose that the velocity at a given location is to be determined from the energy or continuity equation; in such a case the control surface may pass through the wall of a conduit without additional complication. However, such a control surface will expose surface forces caused by the stresses in the conduit, stresses which must be accounted for in the momentum equation even if they are unknown and difficult to evaluate or estimate. Choice of a different control volume configuration might exclude the unknown stresses from the analysis and hence lead to a tractable momentum equation. Therefore, the control volume is selected with at least a recognition of which equation is to be utilized.

The "rules" for the selection of a control volume are rather simple: the control volume should explicitly include the desired information (e.g., if the exit velocity is desired then the control surface must include the exit plane); the control volume should involve other data which are known or which may be assumed or independently computed. But to recognize which equation should be used to obtain the desired information is a considerably more subtle process, and therefore a more difficult one for which to establish guidelines. The remainder of this

TABLE

Various Forms of

Continuity	Momentum

DEFORMABLE

$$0 = \frac{d}{dt}\int_{c.v.}\rho\, dV + \int_{c.s.}\rho\, \mathbf{V}_r\cdot\hat{n}\, dA \qquad \sum\mathbf{F} = \frac{d}{dt}\int_{c.v.}\rho\mathbf{V}\, dV + \int_{c.s.}\rho\mathbf{V}\,(\mathbf{V}_r\cdot\hat{n})\, dA$$

$$0 = \int_{c.v.}\frac{\partial\rho}{\partial t}\, dV + \int_{c.s.}\rho\, \mathbf{V}_b\cdot\hat{n}\, dA \qquad \text{or}$$

$$+\int_{c.s.}\rho\, \mathbf{V}_r\cdot\hat{n}\, dA \qquad \sum\mathbf{F} = \int_{c.v.}\frac{\partial}{\partial t}(\rho\mathbf{V})\, dV + \int_{c.s.}\rho\mathbf{V}\,(\mathbf{V}_b\cdot\hat{n})\, dA$$

$$+\int_{c.s.}\rho\mathbf{V}\,(\mathbf{V}_r\cdot\hat{n})\, dA$$

NONDEFORMABLE

$$0 = \frac{d}{dt}\int_{c.v.}\rho\, dV + \int_{c.s.}\rho\, \mathbf{V}\cdot\hat{n}\, dA \qquad \sum\mathbf{F} = \frac{d}{dt}\int_{c.v.}\rho\mathbf{V}\, dV + \int_{c.s.}\rho\mathbf{V}\,(\mathbf{V}\cdot\hat{n})\, dA$$

or $\qquad$ or

$$0 = \int_{c.s.}\frac{\partial\rho}{\partial t}\, dV + \int_{c.s.}\rho\, \mathbf{V}\cdot\hat{n}\, dA \qquad \sum\mathbf{F} = \int_{c.v.}\frac{\partial}{\partial t}(\rho\mathbf{V})\, dV + \int_{c.s.}\rho\mathbf{V}\,(\mathbf{V}\cdot\hat{n})\, dA$$

Steady

$$0 = \int_{c.s.}\rho\, \mathbf{V}\cdot\hat{n}\, dA \qquad \sum\mathbf{F} = \int_{c.s.}\rho\mathbf{V}\,(\mathbf{V}\cdot\hat{n})\, dA$$

Steady Uni

$$\rho_1 A_1 V_1 = \rho_2 A_2 V_2 = \dot{m} \qquad \sum\mathbf{F} = \rho_2 A_2 V_2 \mathbf{V}_2 - \rho_1 A_1 V_1 \mathbf{V}_1$$

$$= \dot{m}(\mathbf{V}_2 - \mathbf{V}_1)$$

Steady Non

$$\sum F_x = \dot{m}\,(\beta_2 \bar{V}_{2x} - \beta_1 \bar{V}_{1x})$$

$$\rho_1 A_1 \bar{V}_1 = \rho_2 A_2 \bar{V}_2$$

$$\sum F_y = \dot{m}\,(\beta_2 \bar{V}_{2y} - \beta_1 \bar{V}_{1y})$$

$$\sum \dot{W} = \dot{W}_S + \dot{W}_{I'} + \dot{W}_{\text{shear}}$$

β = momentum correction factor, defined by $\beta A \bar{V}^2 = \int_A V^2\, dA$.

α = kinetic-energy correction factor, defined by $\alpha A \bar{V}^3 = \int_A V^3\, dA$.

2.1

Fundamental Laws

Energy

CONTROL VOLUME

$$\dot{Q} - \sum \dot{W} = \frac{d}{dt}\int_{c.v.} \left(\frac{V_I^2}{2} + gz + \tilde{u}\right)\rho\, d\mathcal{V} + \int_{c.s.} \left(\frac{V_I^2}{2} + gz + \tilde{u} + \frac{p}{\rho}\right) \mathbf{V}_r \cdot \hat{n}\, \rho\, dA + \int_{c.s.} p\, \mathbf{V}_b \cdot \hat{n}\, dA$$

or

$$\dot{Q} - \sum \dot{W} = \int_{c.v.} \frac{\partial}{\partial t}\left[\rho\left(\frac{V_I^2}{2} + gz + \tilde{u}\right)\right] d\mathcal{V} + \int_{c.s.} \left[\frac{V_I^2}{2} + gz + \tilde{u} + \frac{p}{\rho}\right]\rho\, \mathbf{V}_b \cdot \hat{n}\, dA$$

$$+ \int_{c.s.} \left[\frac{V_I^2}{2} + gz + \tilde{u} + \frac{p}{\rho}\right]\rho\, \mathbf{V}_r \cdot \hat{n}\, dA$$

CONTROL VOLUME

$$\dot{Q} - \sum \dot{W} = \frac{d}{dt}\int_{c.v.} \rho\left(\frac{V_I^2}{2} + gz + \tilde{u}\right) d\mathcal{V} + \int_{c.s.} \left[\frac{V_I^2}{2} + gz + \tilde{u} + \frac{p}{\rho}\right]\rho\, \mathbf{V} \cdot \hat{n}\, dA$$

$$\dot{Q} - \sum \dot{W} = \int_{c.v.} \frac{\partial}{\partial t}\left[\rho\left(\frac{V_I^2}{2} + gz + \tilde{u}\right)\right] d\mathcal{V} + \int_{c.s.} \left[\frac{V_I^2}{2} + gz + \tilde{u} + \frac{p}{\rho}\right]\rho\, \mathbf{V} \cdot \hat{n}\, dA$$

Flow

$$\dot{Q} - \sum \dot{W} = \int_{c.s.} \left[\frac{V^2}{2} + gz + \tilde{u} + \frac{p}{\rho}\right]\rho\, \mathbf{V} \cdot \hat{n}\, dA$$

or

$$-\sum W = \int_{c.s.} \left[\frac{V^2}{2} + \frac{p}{\rho} + gz\right]\rho\, \mathbf{V} \cdot \hat{n}\, dA + \text{losses}$$

form Flow

$$\frac{\dot{Q} - \dot{W}_S}{\dot{m}} = \left(\frac{V_2^2}{2} + \frac{p_2}{\rho_2} + gz_2 + \tilde{u}_2\right) - \left(\frac{V_1^2}{2} + \frac{p_1}{\rho_1} + gz_1 + \tilde{u}_1\right)$$

or

$$-\frac{\dot{W}_S}{\dot{m}} = \left(\frac{V_2^2}{2} + \frac{p_2}{\rho_2} + gz_2\right) - \left(\frac{V_1^2}{2} + \frac{p_1}{\rho_1} + gz_1\right) + \text{losses}$$

uniform Flow

$$-\frac{\dot{W}_S}{\dot{m}} = \left(\alpha_2 \frac{\overline{V}_2^2}{2} + gz_2 + \frac{p_2}{\rho_2}\right) - \left(\alpha_1 \frac{\overline{V}_1^2}{2} + gz_1 + \frac{p_1}{\rho_1}\right) + \text{losses}$$

$\mathbf{V}_r$ = velocity of fluid relative to control surface boundary.
$\mathbf{V}_b$ = velocity of control surface boundary.
$\mathbf{V}$ = velocity of fluid in coordinate system attached to control surface.
$\overline{V}$ = average velocity, defined as $\overline{V} = \frac{1}{A}\int_A V\, dA$.

article and the examples that follow should assist the reader's recognition of what information is to be expected from each equation. However, it should be emphasized that actual experience in using the equations is the most effective method of gaining such an appreciation. Table 2.1 (immediately preceding), arranged both by deformable and nondeformable control volumes and, further, for a variety of types of flow, may assist the reader in choosing the appropriate form of equation.

The continuity equation governs the interrelationship of the scalar terms, density, velocity magnitude, cross-sectional area, and rate of volume change, for a control volume. As such, it can relate the conditions between stations but cannot, for example, solve for the velocity from a given area. Since the equation deals with the first power of the velocity, it is common to use the average velocity, which is a useful term if the density is constant over the area and if the flow is not separated, so that the geometric area is the effective area over which the mass flux occurs.

The momentum equation or the moment-of-momentum equation deals with vector quantities, and therefore three independent component equations, to determine forces and moments on physical objects associated with the flow field. The uniform, steady-flow form of the momentum equation for a conduit flow is

$$\sum \mathbf{F} = \dot{m}(\mathbf{V}_2 - \mathbf{V}_1) \tag{2.86}$$

This shows that the force is related to both the velocity change and the mass rate of flow. The primary usage of this equation is the calculation of forces. The uniform-flow assumption must be used with caution, since

$$\dot{m}\overline{V} = \rho A \overline{V}^2 \leqslant \int_A \rho V^2 \, dA \tag{2.87}$$

where the inequality exists if the flow is not uniform. Recall that β is a coefficient which has been introduced to make the foregoing an equality; specifically,

$$\beta \dot{m}\overline{V} = \int_A \rho V^2 \, dA \tag{2.88}$$

The energy equation may be used, for example, to compute the velocity from known potential head or pressure data or to determine the shaft-work rate associated with a given flow. The steady-, uniform-

Art. 2.5   CLOSURE   137

flow form for flow in a conduit,

$$-\dot{W}_s = \dot{m}\left[\frac{V_2^2 - V_1^2}{2} + \frac{p_2 - p_1}{\rho} + g(z_2 - z_1)\right] + \text{losses} \quad (2.89)$$

may be used for these general calculations when the appropriate restrictions are satisfied. The nonlinear effects also require that the uniform-flow approximation be used with caution, since

$$\dot{m}\overline{V}^2 \leq \int_A \rho V^3 \, dA \quad (2.90)$$

The *kinetic energy correction factor* α is defined to make the above an equality; specifically,

$$\alpha \dot{m}\overline{V}^2 = \int_A \rho V^3 \, dA \quad (2.91)$$

Example 2.21

Water flows through the nozzle of Fig. E2.21, exiting to the atmosphere. A manometer is attached as shown. Determine the force necessary to hold the nozzle on a hose.

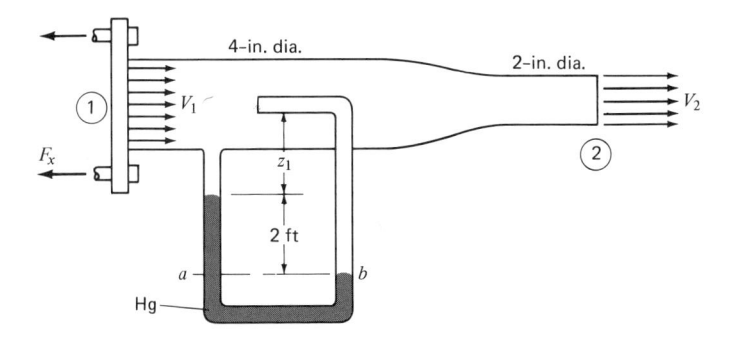

Fig. E2.21

Solution. The requirement to determine a force suggests that the momentum equation should be considered. Furthermore, since the force to secure the nozzle onto the hose is desired one of the control surfaces should pass through this plane. The other surfaces to close the control volume* are selected to lie outside of the nozzle. This excludes exposing any other forces in a solid material. The downstream surface is placed at the exit plane of the nozzle where the pressure and velocity are assumed known. Let F_x represent the unknown force. The momentum equation is

$$-F_x + p_1 A_1 = \int_{c.s.} \rho V_x \mathbf{V} \cdot \hat{n} \, dA$$

$$= \rho V_2^2 A_2 - \rho V_1^2 A_1$$

Using the continuity equation applied to the same control volume, $A_1 V_1 = A_2 V_2$ or $V_2 = 4 V_1$; hence, the momentum equation can be written as

$$F_x = p_1 A_1 - 3\rho A_1 V_1^2$$

The manometer system allows the velocity to be related to the pressure. We know that $p_a = p_b$ in the stagnant manometer fluid. Hence we may write

$$p_1 + \rho g z_1 + \rho_{Hg} g \times 2 = \frac{V_1^2}{2} \rho + p_1 + \rho g z_1 + \rho g \times 2$$

where $\rho = \rho_{H_2O}$. Thus the manometer reading shows us that, using $\rho_{Hg} = 13.6 \, \rho_{H_2O}$,

$$12.6 \times 2g = \frac{V_1^2}{2}$$

or

$$V_1 = 40.3 \text{ fps}$$

The energy equation which also relates the velocity and pressure in this steady flow, assuming no losses, is

$$0 = \int_{c.s.} \left(\frac{V^2}{2} + \frac{p}{\rho} \right) \rho \mathbf{V} \cdot \hat{n} \, dA$$

or

$$\frac{V_2^2}{2} + \cancelto{0}{\frac{p_2}{\rho}} = \frac{V_1^2}{2} + \frac{p_1}{\rho}$$

Continuity allows us to write $V_2 = 4V_1$, giving the energy equation as

$$\frac{15 V_1^2}{2} = \frac{p_1}{1.94} \quad \text{or} \quad p_1 = 23{,}600 \text{ psf}$$

* Note that an equally suitable control volume would be one which is interior to the nozzle; however, this would not allow the force F_x to be included in the momentum equation since the force $\int p \, dA$ on the nozzle wall would be the unknown force.

Art. 2.5   CLOSURE

Finally, combining these results, F_x is calculated to be

$$F_x = p_1 A_1 - 3\rho A_1 V_1^2$$

$$= (23{,}600 - 3 \times 1.94 \times 40.3^2)\,\frac{16\pi}{4 \times 144} = 1240 \text{ lb}$$

In this problem we have had to use the continuity, energy, and momentum equations, in addition to interpreting the reading on the manometer. This is not at all uncommon in many practical applications where forces are desired.

Extension 2.21.1. If the pitot tube part of the manometer were placed at section ②and if it read 2 ft of mercury, calculate the force, assuming no losses.
Ans. 1240 lb

Extension 2.21.2. If air were flowing with $\rho = 0.0024$ slug/ft^3, what would V_1 be in Example 2.21? Consider the air to be incompressible. If the local velocity exceeds $0.3a$, where the speed of sound $a = \sqrt{\gamma RT}$, then compressibility is important. Is it important here? Assume $T_{\text{exit}} = 60°$F and use $\gamma_{\text{air}} = 1.4$.
Ans. 1190 fps

---

*Example 2.22

An automotive shock absorber provides the viscous damping effect in a vehicle's suspension system. To properly perform its function in the suspension system, it must maintain the correct force-vs.-velocity relationship. (The ability to design such a device to achieve the desired relationship is not a common aspect of many engineers' professional activity and it is certainly not a handbook exercise. It is a good example of the advanced capability that we wish to develop in the reader.) We will consider the analysis of a simplified shock absorber and demonstrate the application of the rather sophisticated aspects of the deformable control volume to this common, but complicated, device. We will attempt to relate the force F on the piston to its velocity V_p. A schematic of such a device is shown in Fig. E2.22.

Solution. A coordinate system fixed to the frame of the shock absorber and with x aligned in the direction of the piston travel will be used for this problem. The selection of the control volume is based upon isolating the desired information and including known or calculable quantities. The deformable control volume shown in the sketch is selected in anticipation of meeting these requirements.

The general control-volume equation for the conservation of mass is

$$0 = \int_{\text{c.v.}} \cancel{\frac{\partial \rho}{\partial t}}^{0} dV + \int_{\text{c.s.}} \rho \mathbf{V}_r \cdot \hat{n}\, dA + \int_{\text{c.s.}} \rho \mathbf{V}_b \cdot \hat{n}\, dA$$

Let V_p be the velocity of the piston and V_{r1} and V_{r2} be the velocities relative to

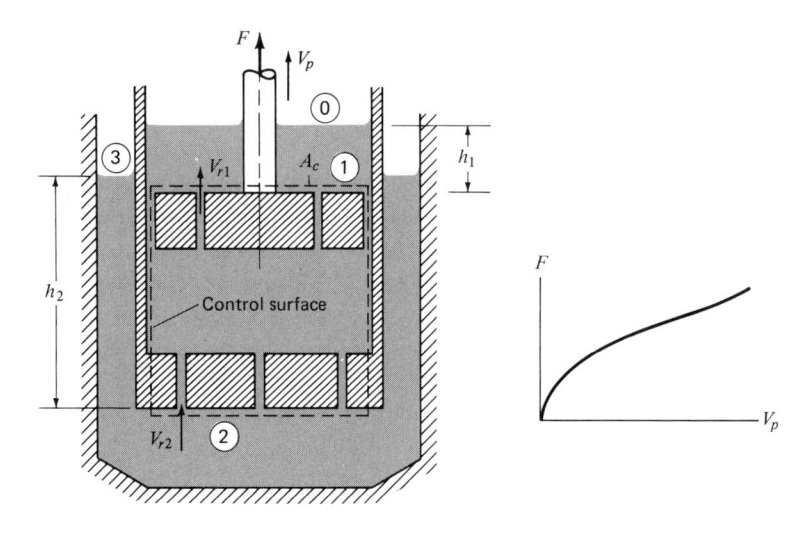

Fig. E2.22

the piston and to the shock absorber case, respectively. Then,

$$0 = V_p A_c + \sum_{i=1}^{N} [(V_{r1})_i A_i] - \sum_{j=1}^{M} [(V_{r2})_j A_j]$$

The notational aspects of this equation are important; they will allow us to do an accurate accounting of the various effects occurring in this problem. The velocities are considered to be x-component velocities; the sign of the velocity is dictated by its direction in our coordinate system. V_p is positive when the piston moves upward; hence, $\mathbf{V}_b \cdot \hat{n} = V_p$. The relative velocities are also referred to the coordinate system; hence, $\mathbf{V}_{r1} \cdot \hat{n} = V_{r1}$ and $\mathbf{V}_{r2} \cdot \hat{n} = -V_{r2}$. Whether or not the velocities themselves are negative or positive depends, of course, on the piston motion. For example, as the piston moves upward, $V_p > 0$, it can be anticipated that $V_{r1} < 0$ and $V_{r2} > 0$. The summations account for the presence of multiple holes (possibly of different sizes) and of a leakage path between the piston and the case.

The force on the piston rod is present in the work-rate term of the energy equation. It is also anticipated that the "losses" incurred as the fluid passes through the orifices are important physical effects in the operation of the device. Therefore, the energy equation will be important in the description of this problem and, based on Eq. 2.85, it can be expressed as

$$\mathbf{F} \cdot \mathbf{V}_p = \int_{c.v.} \frac{\partial}{\partial t} \left[\frac{V_I^2}{2} + gz \right] \rho \, dV + \int_{c.s.} \rho \left[\frac{V_I^2}{2} + \frac{p}{\rho} + gz \right] \mathbf{V}_r \cdot \hat{n} \, dA$$
$$+ \int_{c.s.} \rho \left[\frac{V_I^2}{2} + \frac{p}{\rho} + gz \right] \mathbf{V}_b \cdot \hat{n} \, dA + \text{losses}$$

Note that $\mathbf{F} \cdot \mathbf{V}_p$ is the force of the rod on the piston (surroundings on the system) times the associated velocity. Hence it is equal to the negative of the rate of work of the system on the surroundings. Since $\mathbf{F}$ and $\mathbf{V}_p$ are always in the same direction, $\mathbf{F} \cdot \mathbf{V}_p = FV_p$. We can simplify this expression by noting that the potential-energy effects are expected to be negligible; that is, three or six inches of oil creates pressure differences which are negligible with respect to those associated with damping a car's motion. The kinetic energy in the chamber (the region between the casing and the piston) may also be expected to be small; however, that at the exit locations may be significant. The rate of change of the kinetic energy of the piston is a possibly significant term; it will be included.

The pressures at surfaces ① and ② are influenced by the familiar gravitational hydrostatic pressure variation; they are also affected by the acceleration of the fluid between these surfaces and the free surface. These regions of fluid tend to move as a single mass (such a description is quite adequate to calculate the pressure). The inviscid equation for unsteady motion was developed in Chapter 1; for the distance s along a streamline, it is

$$\frac{\partial V}{\partial t} + V \frac{\partial V}{\partial s} = -\frac{1}{\rho} \frac{\partial p}{\partial s} - g \frac{\partial h}{\partial s}$$

Since the velocity V is essentially constant everywhere in the mass fluid, $\partial V/\partial s \approx 0$. Consequently, integrating from ① to ⓪ and from ② to ③ respectively yields

$$p_1 = \rho \left(g + \frac{\partial V_0}{\partial t} \right) h_1$$

$$p_2 = \rho \left(g + \frac{\partial V_3}{\partial t} \right) h_2$$

These are the pressures which appear in the flux integrals and p_1 also enters the problem in the $\int_{\text{c.s.}} p\, \mathbf{V}_b \cdot \hat{n}\, dA$ term.

The velocity V_0 may be related to V_{r1} and V_p by a control volume which contains the fluid in the region between the piston face and the free surface. Specifically, continuity gives

$$0 = (V_0 - V_p)(A_c - A_r) + \sum_{i=1}^{N} [(V_{r1})_i A_i]$$

Similarly, a control surface extended from surface ⓪ to surface ③ may be used to relate the two velocities for this third deformable control volume (but without flux effects):

$$V_0(A_c - A_r) + V_3 A_3 = 0$$

The viscous loss effect enters the problem description directly in the energy equation. This loss effect would result in a greater pressure in the control-volume interior, which is related to a greater force in the piston rod. In order to describe the losses, we will employ the loss-coefficient expression for each of the orifices:

$$\text{Loss per orifice} = \dot{m}_i K_i \frac{V_i^2}{2}$$

Bringing together all of the earlier considerations and using the kinetic-energy correction factor α (see Eq. 2.91), which allows the average velocity to be used in the equation, yields

$$FV_p = V_p M_p \frac{dV_p}{dt} - \sum_{j=1}^{M} \left[(\alpha_j + K_j) \frac{(V_{r2}^2)_j}{2} \rho (V_{r2})_j A_j \right]$$

$$+ \sum_{i=1}^{N} \left[K_i \frac{(V_{r1}^2)_i}{2} \rho (V_{r1})_i A_i \right]$$

$$+ \sum_{i=1}^{N} \left\{ \alpha_i [V_p + (V_{r1})_i]^2 \rho (V_{r1})_i A_i \right\} + p_1 V_p (A_c - A_p)$$

$$+ p_1 \sum_{i=1}^{N} [(V_{r1})_i A_i] - p_2 \sum_{j=1}^{M} [(V_{r2})_j A_j]$$

An important, but rather subtle, aspect of this equation is that α and K are functions of V_{r1} and V_{r2}. That is, the velocity profile, from which α is evaluated, and, to a lesser extent, the loss coefficients will be influenced by the direction of the flow and by the magnitude of the velocity. These would have to be determined by separate testing or predictive procedures. Since the pressures p_1 and p_2 are known in terms of V_0 and V_3, which can be related to V_{r1}, V_{r2}, and V_p, these latter three represent three independent variables for the (desired) dependent variable F. However, only two equations are available for the three unknowns. An additional computing strategy and different control volumes may be used to relate the two velocities V_{r1} and V_{r2}. During a compression stroke, the flow exits from the chamber; for the rebound stroke, the liquid enters the chamber from the reservoirs. In either condition, the common pressure p_c can be used to relate the V_{r1} and V_{r2} values. Consider a control volume extending over the exit plane of the hole and into the chamber (for a compression stroke) or into the fluid between the piston or casing and the free surface (for a rebound stroke).

For such a control volume, the energy equation may be written as

$$0 = \int_{\text{c.s.}} \rho \left(\frac{V_l^2}{2} + \frac{p}{\rho} + gz \right) \mathbf{V}_r \cdot \hat{n} \, dA + \text{losses}$$

The potential-energy effects are negligible. The kinetic-energy correction factor can be employed to allow the equation to be written in terms of the average velocity and the losses can be expressed in terms of the average velocity.

For a compression stroke, the energy equation for the flow through each orifice (for which the subscripts i and j are implied for the two following equations, respectively) may be written as

for the piston: $$0 = \alpha \frac{(\mathbf{V}_p + \mathbf{V}_{r1})\cdot(\mathbf{V}_p + \mathbf{V}_{r1})}{2} + \frac{p_1 - p_c}{\rho} + K\frac{V_{r1}^2}{2}$$

for the casing: $$0 = \alpha \frac{V_{r2}^2}{2} + \frac{p_2 - p_c}{\rho} + K\frac{V_{r2}^2}{2}$$

If a rebound stroke occurs, then the separate equations are

for the piston: $$0 = \alpha \frac{(\mathbf{V}_p + \mathbf{V}_{r1})\cdot(\mathbf{V}_p + \mathbf{V}_{r1})}{2} + \frac{p_c - p_1}{\rho} + K\frac{V_{r1}^2}{2}$$

for the casing: $$0 = \alpha \frac{V_{r2}^2}{2} + \frac{p_c - p_2}{\rho} + K\frac{V_{r2}^2}{2}$$

Note that the inertial velocity $(\mathbf{V}_{r1} + \mathbf{V}_p)$ is used for the kinetic-energy term when the control volume moves with the piston.

These equations, the conservation of mass, and the energy equation applied to the deformable control volume considered initially provide four equations for the five variables, F, V_p, V_{r1} and V_{r2}, and p_c. Consequently, F can be expressed in terms of V_p as desired.

The complexity of these equations suggests that a digital computer solution be employed to examine the influence of V_p on F. The spring-loaded orifice valves that would exist in an automotive shock absorber would be modeled as a pressure-differential-dependent loss coefficient.

Extension 2.22.1. If the control surface were placed at the chamber side of the piston face, would the desired quantity F enter the problem directly? If not, what would it be equal to in terms of the quantities appearing in the analysis?

Extension 2.22.2. If the liquid above the piston moves downward with an acceleration of 32.2 ft/sec², what is the pressure on surface ①? Would you expect a liquid acceleration of this magnitude at 50 mph on a rough road?

Extension 2.22.3. Identify the quantities that can be controlled by a designer to influence the F-vs.-V_p response of the shock absorber.

PROBLEMS

2.1 Make a list of the various extensive quantities encountered in fluid mechanics that may be treated using the Reynolds transport theorem. Also list the various intensive quantities encountered in fluid mechanics.

2.2 Consider an application where the control volume is identical to the system at all times. This might be a confined gas in a piston-cylinder arrangement. What would the Reynolds transport theorem reduce to?

*2.3 A certain control volume is defined to be the interior surface plus the exit area of a deflating balloon. Identify the regions referred to in the system-to-control-volume transformation equation (2.14). Also, indicate where the various terms in Eq. 2.24 are zero or nonzero.

*2.4 Two pistons move as shown. Consider the control volume between the pistons to be divided into three parts, V_1, V_2, and V_3, with the areas separating V_3 from V_1 and V_2 located at fixed distances l_1 and l_2, respectively. Show that

$$\frac{d}{dt}\int_{c.v.} \eta\rho\, dV = \int_{c.v.} \frac{\partial}{\partial t}(\eta\rho)\, dV + \eta_2\rho_2 A_2 V_2 - \eta_1\rho_1 A_1 V_1$$

by applying Leibniz' rule to $\dfrac{d}{dt}\int_{c.v.} \eta\rho\, dV$ after expressing the integral in appropriate form for the three volumes which make up the control volume. Assume all quantities to be functions of x and t.

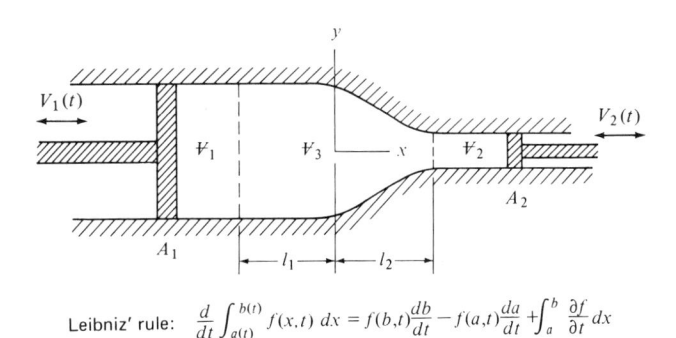

Leibniz' rule: $\dfrac{d}{dt}\int_{a(t)}^{b(t)} f(x,t)\, dx = f(b,t)\dfrac{db}{dt} - f(a,t)\dfrac{da}{dt} + \int_a^b \dfrac{\partial f}{\partial t}\, dx$

Prob. 2.4

2.5. Water flows in the 2-in.-dia. pipe and out radially between the two circular discs. Determine the average velocity at a radius of 2 ft.

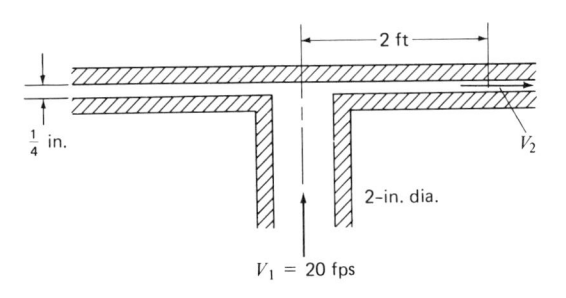

$V_1 = 20$ fps

Prob. 2.5

2.6 Water flows in a 2-ft-high channel with $V = 50(4y - y^2)$ fps, where y is measured from the bottom of the 20-ft-wide channel. The water is stratified so that $\rho = 0.25(10 - y)$ slug/ft^3. Determine $\dot{m}$. Also, show that $\dot{m} \neq \bar{\rho}\bar{V}A$.

2.7 Water flows into the tank shown with a velocity of 20 fps. A piston is being withdrawn at a rate of 5 fps. Determine the exit velocity V_2.

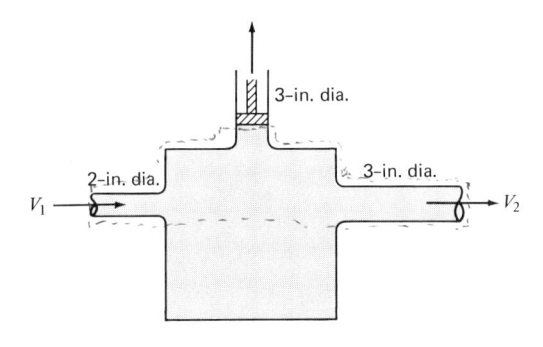

Prob. 2.7

2.8 The setup shown provides a liquid flow through the tissues for a physiological experiment. Taking readings $h_1(t)$ and $h_2(t)$, develop an expression for the storage rate of the liquid in the tissue. Identify your choice of control volume very carefully.

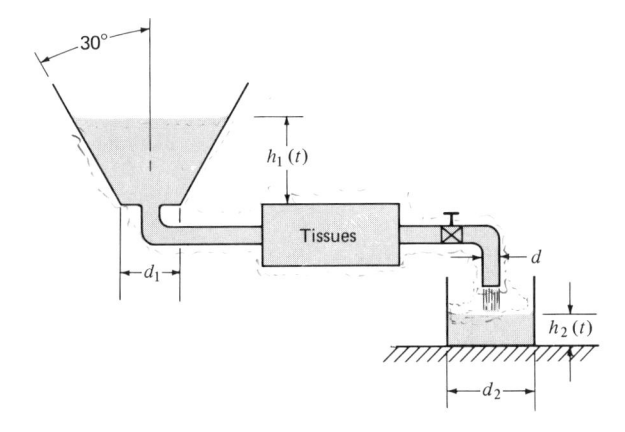

Prob. 2.8

2.9 The jet pump shown operates by inducing a secondary flow by the action of the flow in the 2-in.-dia. pipe. Across the section at the exit from the small pipe the velocity in the annulus is 10 fps; in the small pipe it is given by $V = 90[1 - (r/R)^2]$. Determine the average velocity at the jet pump exit.

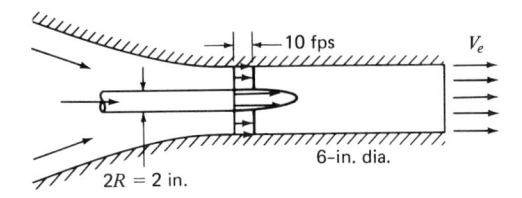

Prob. 2.9

2.10 The velocity leaves tangent to the blades of the centrifugal fan shown. Find an expression for the exit velocity V_2 in terms of the uniform inlet velocity V_1. Assume constant density.

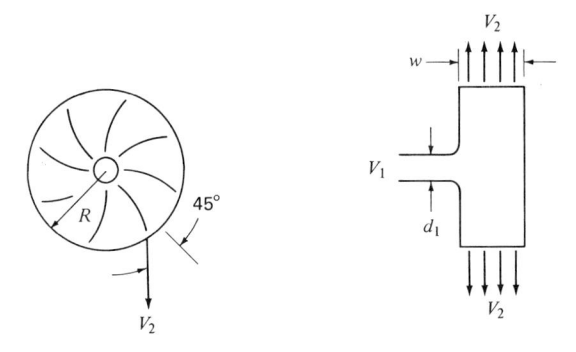

Prob. 2.10

*__2.11__ A plunger in a veterinarian's hypodermic needle is plunged at 2 fps. Determine the exit velocity if (a) no leakage past the plunger occurs and (b) the velocity of the fluid leaking past the plunger (relative to a stationary reference frame) is approximately equal to V_2. The diameter of the plunger is 0.997 in.

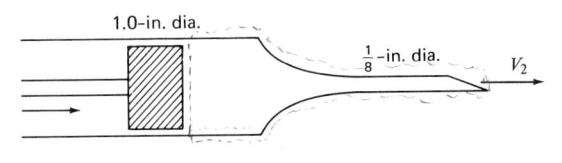

Prob. 2.11

2.12 Fluid flows over a flat plate with a boundary layer (a layer near the boundary where the velocity differs from U_0) formed in which the velocity is given by $u = U_0[(2y/\delta) - (y^2/\delta^2)]$. A streamline enters the boundary layer 2 in. above the plate, as shown. How far from the plate will the streamline be at the location where $\delta = 4$ in.?

PROBLEMS

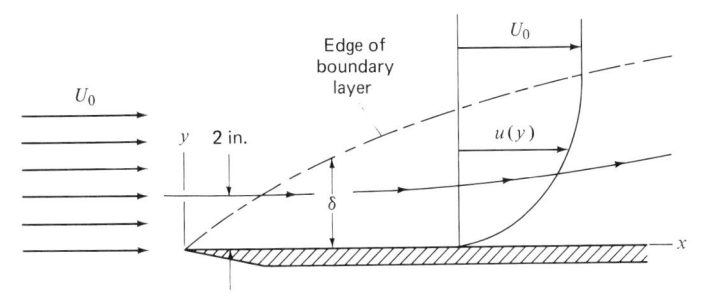

Prob. 2.12

2.13 Air is escaping from a tire through a $\frac{1}{4}$-in.-dia. valve core at a velocity of 100 fps. Estimate the rate at which the density in the tire is changing if the pressure in the tire is 25 psi and an isothermal process is assumed to the exit. The exit pressure and the tire temperature are at standard atmospheric conditions. The volume of the tire remains essentially constant at 10 ft^3.

2.14 A 4-in.-dia. stream of water strikes the conical deflector as shown. Determine the flow rate from the stationary jet if a 10-lb force is measured on (a) a stationary deflector and (b) a deflector moving to the right at 20 fps.

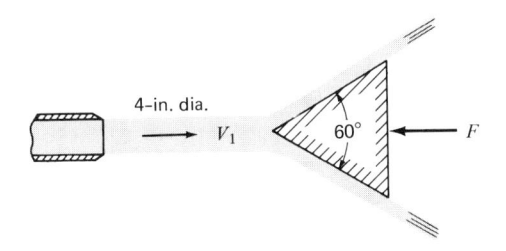

Prob. 2.14

2.15 A bend of 90° occurs in a piping system. The diameter changes from a 4-in.-dia. pipe to a 2-in.-dia. pipe. The pressure is 80 psi in the larger pipe. If the flow rate is 2 cfs of water, determine the force of the water on the bend. Neglect losses so that Bernoulli's equation is applicable.

2.16 Water flows in a wide channel and suddenly undergoes a "hydraulic jump" to a new height h_2, as shown (next page). Assuming no dragging force on the bottom of the channel, determine h_2. The pressure on the open surface is zero gage pressure.

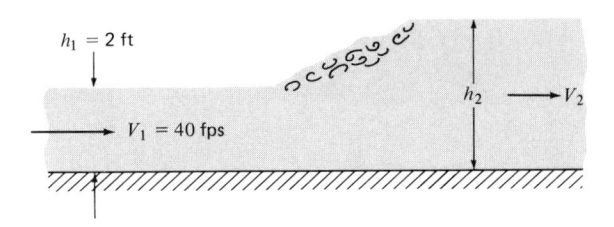

[figure: Prob. 2.16 — $h_1 = 2$ ft, $V_1 = 40$ fps, h_2, V_2]

Prob. 2.16

2.17 Spot welding requires a large and concentrated discharge of electrical energy; copper electrodes are used for this purpose. However, the current heats the copper, and it deforms plastically under the action of the stresses associated with being pressed against the piece. Consequently it is desirable to cool the electrode tip to reduce its temperature and hence its deformation. Since the tips need be replaced relatively often, even with the cooling, it would involve minimum "lost time" if they could be changed by hand. That is, if a tapered shaft and a frictional fit would hold the tip in position then the operator would not be required to use a wrench or other tool for its replacement. Determine the force required to hold the tip in place for the operating conditions indicated in the sketch. Neglect losses.

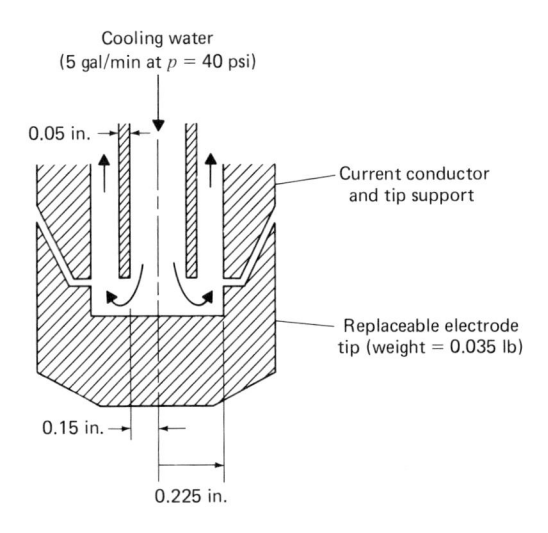

[figure: Prob. 2.17 — Cooling water (5 gal/min at $p = 40$ psi), 0.05 in., Current conductor and tip support, Replaceable electrode tip (weight = 0.035 lb), 0.15 in., 0.225 in.]

Prob. 2.17

2.18 A rectangular jet of air is so directed at a rigid flat plate that part of the jet is interrupted as shown. Bernoulli's equation implies that the velocity in all three streams is equal at positions where the streamlines are all parallel (the pressure is then atmospheric). Derive an expression for the force F on the plate if one-third of the mass flux $\dot{m}_1$ is diverted down the plate.

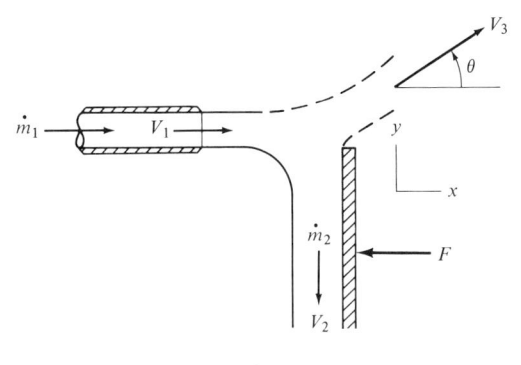

Prob. 2.18

2.19 The two-dimensional rectangular water jet strikes a flat plate as shown. Calculate the reactive force R. What percentage of the total mass flux is diverted down the plate?

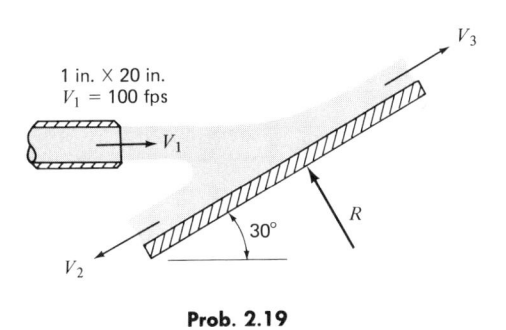

Prob. 2.19

2.20 The plate in Prob. 2.19 is attached to a 5000-lb object that moves on frictionless wheels in the direction of the jet. How fast will it be moving after 10 sec? Gravity acts normal to the jet.

2.21 A series of blades attached to a 2-ft-radius wheel interact with four water jets to produce a torque about the axis of the wheel. A typical blade-jet orientation is shown (next page). The speed of the water relative to the blade remains constant as the water moves over the blade (see Example 2.11). Determine the maximum power available from the wheel.

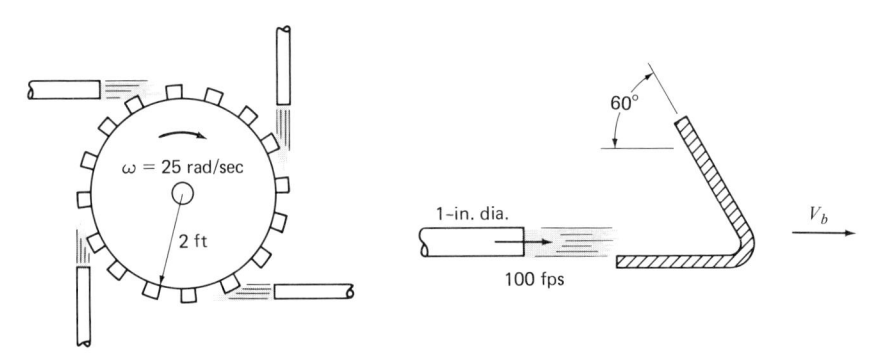

Prob. 2.21

2.22 A stage of blades on a turbine rotor is designed so that the blade on the inlet side is parallel to the direction of the blade motion. (In Example 2.11, $\alpha_1 = 0$.) The exit blade angle is such that no component of velocity exists in the direction of the blade motion. Determine the maximum possible horsepower output if the blade speed is 40 fps, and the water exits from a 2-in.-dia. nozzle at 100 fps. Also, determine the exit blade angle.

2.23 A large vehicle is initially decelerated by dipping a blade into a trough of water. Assume the water speed relative to the blade does not change as the water moves over the blade. The 100-ton vehicle is moving at 120 mph when the 2-ft-wide blade is suddenly dipped 4 in. below the surface of the water. Calculate the initial deceleration and the velocity $V(t)$ of the vehicle. Also determine the time and distance it would take to decelerate to 30 mph.

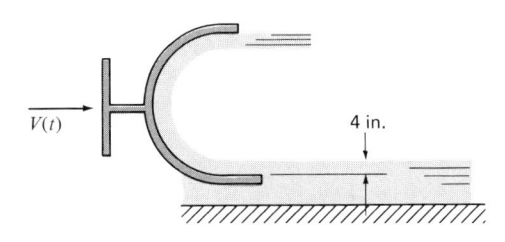

Prob. 2.23

2.24 Flat, rectangular formica sheets are to be loaded into a slightly larger box; there is a clearance of 1/4 in. between the walls of the box and a 4-ft² sheet. If the box is deep enough so that a terminal velocity of the sheet is obtained, the problem can be satisfactorily modeled without a deformable control volume. Determine the terminal velocity of the 20-lb sheet. (*Hint*: A control volume with one surface on the upper surface of the sheet would involve a flow with negligible losses.)

PROBLEMS

2.25 A fire-fighting nozzle is attached to a 3-in. flexible hose in which the pressure is 100 psi. The nozzle is adjusted down to 1 in. in diameter. If you were the fire chief on the job, how many men would you have holding the nozzle? How high a building do you think you could reach from the ground with such a device?

2.26 A manometer is attached to the nozzle as shown. Determine the force on the nozzle; $\rho_{Hg} = 13.6\rho_{H_2O}$. Assume uniform velocity profiles. The losses through the nozzle are assumed negligible.

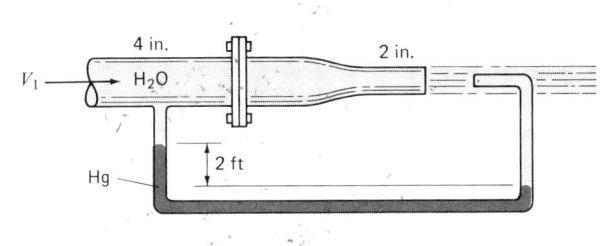

Prob. 2.26

2.27 The flow over the side of a cylindrical surface in a falling liquid jet is a complex flow field; observe such a flow by placing (1) a drinking glass or soft drink bottle and (2) a smaller cylinder (like a finger) in a falling water column. Observe the flow pattern. Using an appropriate control volume, develop an expression for the force component F perpendicular to the jet. Suppose this is a balloon in an air jet and the balloon is not restrained with a force F; would the balloon tend to remain in the air jet? Why?

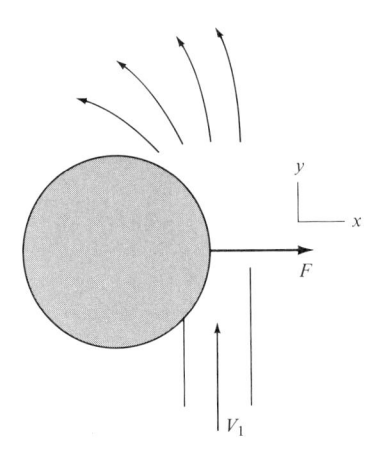

Prob. 2.27

2.28 Air flows over a flat plate, developing a boundary layer as shown in Prob. 2.12. Determine the total dragging force on the 10 ft wide plate between the leading edge and a position where $\delta = 4$ in. The free-stream velocity is 400 fps and the density of air is $\rho = 0.00238$ slug/ft^3. The pressure is everywhere atmospheric. Do not neglect the momentum of the air which leaves the top if a rectangular control volume is chosen.

2.29 As the air in a wind tunnel moves over the airfoil a defect in velocity is noted at some distance downstream. The wingspan is 100 ft wide and $\rho_{air} = 0.00238$ slug/ft^3. From the data given, estimate the minimum horsepower needed for the aircraft to cruise at 500 fps.

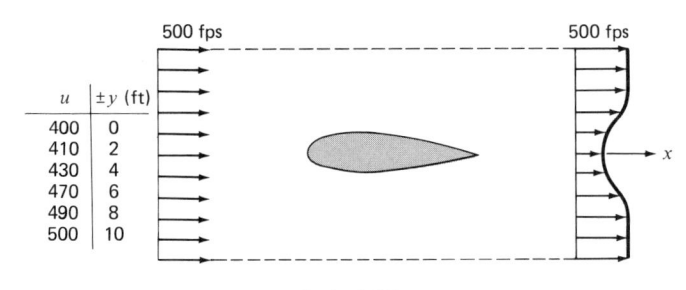

u	$\pm y$ (ft)
400	0
410	2
430	4
470	6
490	8
500	10

Prob. 2.29

2.30 A cart is moved by the action of a high-speed jet as shown. If the cart moves on frictionless wheels, determine the initial acceleration of the cart (the mass of the water in the cart is negligible). The mass of the cart is 10 slugs and the cart starts from rest. What equation governs the cart velocity V_c if the water collected in the cart is not neglected? A numerical solution using the computer could be developed.

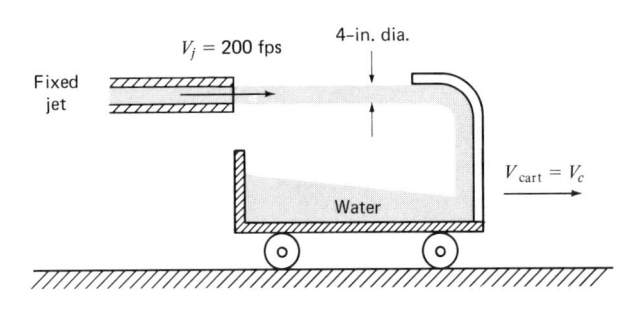

Prob. 2.30

2.31 Water flows into a $\frac{1}{8}$-in.-dia. pipe, and through the action of a viscous layer a parabolic velocity profile is eventually generated. In the inviscid core

PROBLEMS

Bernoulli's equation is applicable. (a) Determine the pressure drop from the entrance to the position where the flow is just parabolic. (b) The shearing stress at the wall is $\mu|\partial V/\partial r|_{wall}$; determine the pressure drop over 100 ft of the developed parabolic flow if $\mu = 10^{-5}$ lb-sec/ft².

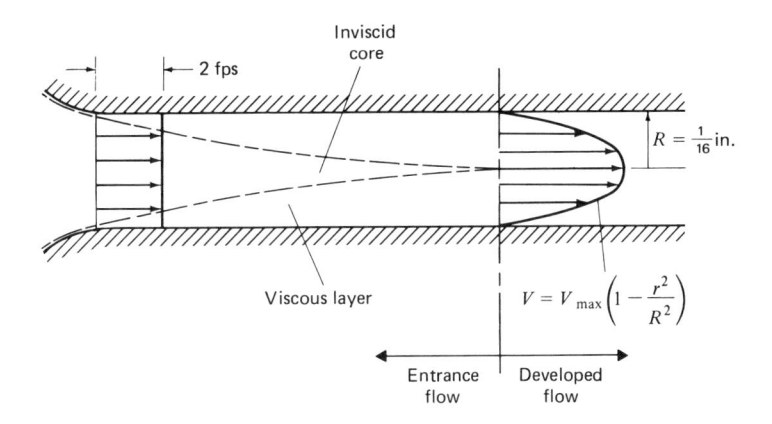

Prob. 2.31

2.32 Air flows from a source through a pipe, impacts a flat disc which weighs 2 lb, and exits radially to the atmosphere. Determine the spacing h at which the disc will just be raised. Assume that no losses occur as the air moves radially outward so that Bernoulli's equation is applicable. The flow is essentially uniform from a radius of 2 in. out to the edge; neglect the force on this small inner area. At low speeds the air is incompressible, with $\rho = 0.0024$ slug/ft³. Explain what physical effect limits the magnitude of the weight which could be raised.

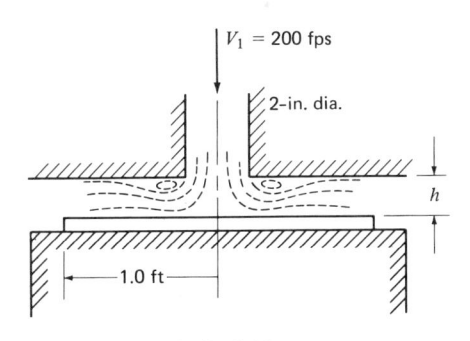

Prob. 2.32

2.33 A lawn sprinkler has four horizontal arms with nozzles oriented at 90° to the 8-in.-long arms and at 45° to the ground. The total flow rate out the

$\frac{1}{4}$-in.-dia. nozzles is 0.3 cfs. Determine the rotational speed if friction is neglected.

2.34 Water flows out of the $\frac{1}{4}$-in. slot with uniform velocity distribution as shown. Determine the limiting angular velocity for the freely rotating arm by considering (a) a stationary reference frame and (b) a rotating reference frame.

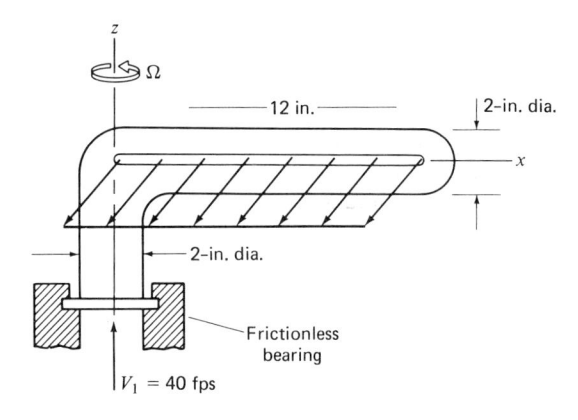

Prob. 2.34

2.35 A simplified dishwasher arm is constructed as shown. The water jets are at 45° with the horizontal. If the water exits at 50 fps, calculate the rotational speed Ω, assuming a frictionless bearing.

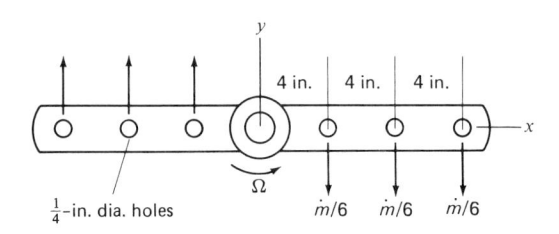

Prob. 2.35

2.36 A motor rotates an impeller in a built-in vacuum system at 5000 rpm. It draws 8 amp on the 110-volt line. Assume the motor to be 90-percent efficient. Estimate the flow rate if the impeller geometry is as shown. The tangential component of velocity V_θ is the same as that of the impeller, that is, $V_\theta = R\Omega$. Assume the air to be incompressible.

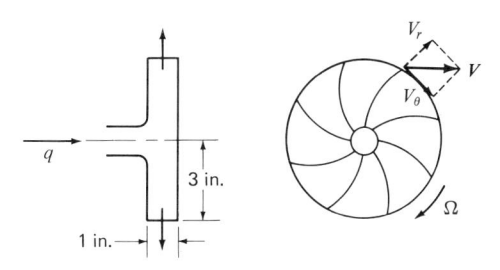

Prob. 2.36

2.37 The hydraulic jump described in Prob. 2.16 is often used as an energy dissipator. How much energy is dissipated (as a percentage of the incoming energy) for the particular hydraulic jump of that problem?

2.38 The bottom of the open channel shown is raised 1 ft. What is the depth of water on the raised section? Assume uniform profiles and no losses.

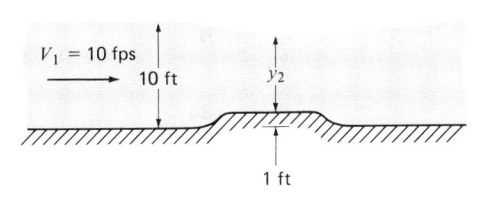

Prob. 2.38

2.39 As H increases, the exit velocity from the reservoir shown increases until a critical elevation is reached and cavitation results. Determine this critical H. Assume no losses. A vapor pressure of 0.5 psia may be assumed.

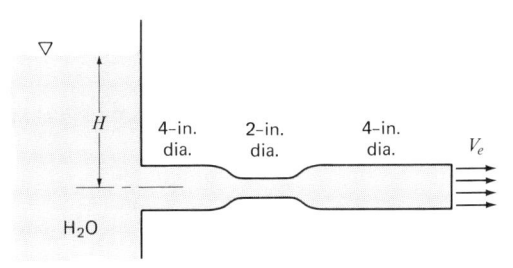

Prob. 2.39

2.40 Water is supplied in a 2-in. dia. pipe at the rate of 2 cfs to a 10-hp, 80-percent efficient pump. Determine the pressure rise if the exit pipe is of (a) 2-in. diameter and (b) 3-in. diameter.

2.41 The turbine shown extracts energy from the flow. For a flow rate q of 40 ft^3/sec, determine the horsepower extracted. The density of mercury is 13.6 ρ_{H_2O}. The losses are given by $1.5\, V_2^2/2g$.

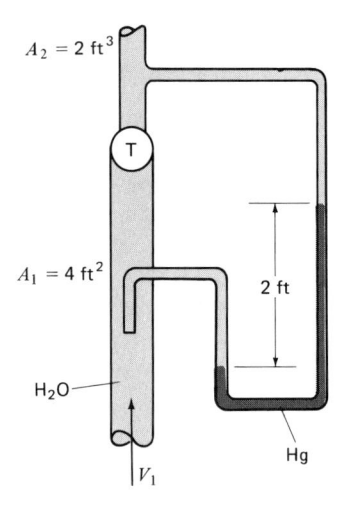

Prob. 2.41

2.42 Water flows from a reservoir through a device and to the atmosphere as shown. For a loss coefficient of $K = 5$, which represents the losses for the entire flow and is based on the average velocity in the pipe, determine the horsepower added to or extracted from the water. There are no losses through the exit nozzle.

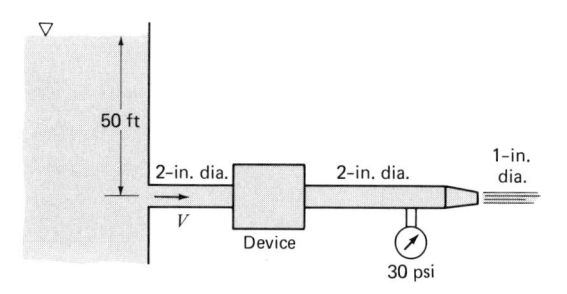

Prob. 2.42

2.43 Water is flowing in a 2-in.-dia. pipe which is on a 30° incline. The pressure at the higher elevation is 20 psi and at the lower elevation it is 35 psi.

Determine the loss coefficient between the two sections which are 100 ft apart if the flow rate is 2 cfs.

2.44 A 70-percent-efficient pump is to be used to provide the necessary energy to deliver 4 cfs of water from a reservoir to a tank at an elevation of 200 ft. The pressure in the tank is 25 psi. Heat is lost at the rate of 6 Btu/sec in the isothermal flow. Determine the necessary pump horsepower.

2.45 For the setup shown, calculate the pump horsepower necessary to cause a pressure $p_A = 30$ psi. The losses up to section A are $0.5 V^2/2$ and for the remainder of the pipe they are $4V^2/2$. Neglect the losses from the pump outlet to the exit. Also determine the pump inlet and exit pressures.

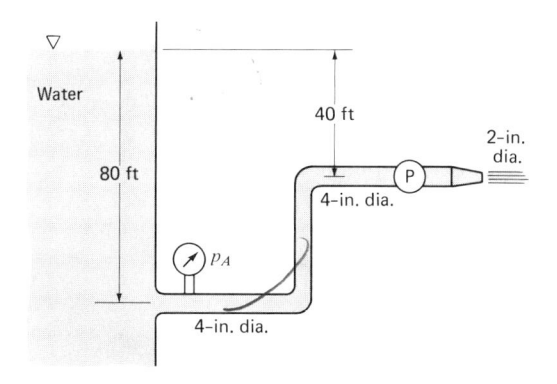

Prob. 2.45

2.46 We wish to express the rate at which kinetic energy is passing a section in a pipe in terms of the average velocity as $\alpha\rho(\overline{V}^3/2)A$ ($= \alpha(\overline{V}^2/2)\dot{m}$), where α is the kinetic-energy-correction factor. For a parabolic profile in a pipe, determine α. Velocity distribution can be written $V = V_{max}[1 - (r^2/r_0^2)]$, where V_{max} is the maximum velocity at the centerline.

2.47 A 500-ft-long, $\frac{1}{2}$-in.-dia. siphon provides water at 60°F from a reservoir to a farm for irrigation. It exits 1 ft below the surface of the reservoir. The velocity distribution exiting from the siphon is $u(r) = 2V[1 - (r^2/r_0^2)]$, where V is the average velocity. Determine the flow rate if the head loss is given by $h_L = f\dfrac{L}{D}\dfrac{V^2}{2g}$ where $f = 64/\text{Re}$ and $\text{Re} = \dfrac{VD\rho}{\mu}$. L is the length of the siphon, D is its diameter, and μ is the viscosity of the water. (The losses are for laminar flow. Actually, a turbulent flow would exist in the siphon and the velocity distribution would not be parabolic.)

*2.48 The plunger in the sketch moves at a constant speed of 2 fps. Estimate the force F necessary if there is leakage past the plunger. Assume no losses so that Bernoulli's equation may be used between the stagnant reservoir and the two exit areas.

Prob. 2.48

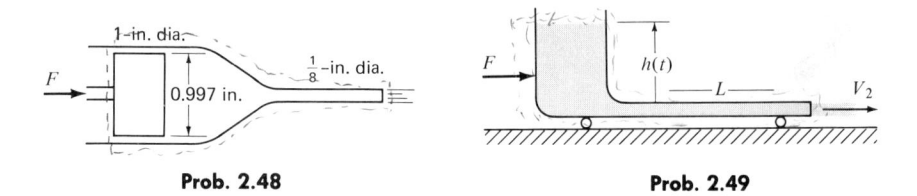

Prob. 2.49

★2.49 The water in the device shown (*above*) is released from rest at $t = 0$. The initial height in the vertical section is h_0, the smaller area is A, the larger area is $4A$, and the mass of the empty device is M. (a) Find an expression for the force $F(t)$ necessary to hold the device stationary. (b) Also, find an expression for the acceleration if the device moves freely with no friction. Assume no losses.

2.50 Hydraulic elevators are often used for buildings of the 10-story category. For these units, it is necessary to extend the shaft into the ground to a depth equal to the maximum height of the elevator rise. A hydraulic piston-cylinder arrangement is used. A 6-in. cylinder is to be used to elevate a maximum load (car and freight) of 6000 lb. Determine the horsepower required to drive the elevator at an acceleration of $0.15g$ with a velocity of 1.0 fps. The losses in the hydraulic line are $f \dfrac{L}{d} \dfrac{V^2}{2g}$. Use $f = 0.015$. Neglect all other losses.

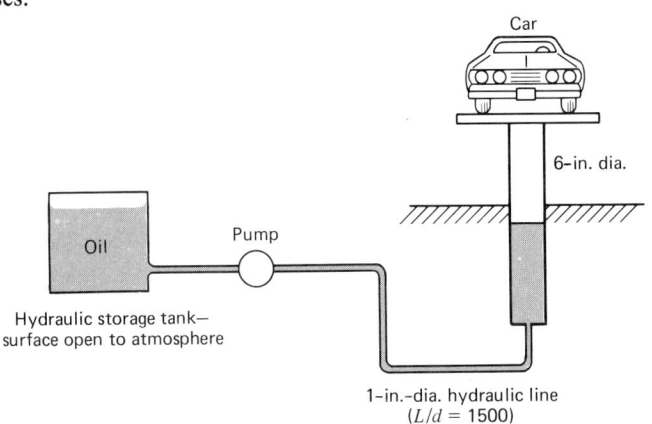

Prob. 2.50

Problems 2.51, 2.52, and 2.53 are difficult, but are included to demonstrate the problem-solving capability made available by the control-volume formulation and a deductive approach. These problems are somewhat open-ended, in that reasoned (but arbitrary) assumptions are required to develop a system of equations from which the desired quantities may be computed. For each problem, the primary answer is a set of differential equations (one or more).

These equations are often quite complex and prohibit a closed-form analytical solution; a numerical solution is therefore suggested.

Note that alternate forms would result if different assumptions were invoked. In one case (Prob. 2.51) the numerical solution has been executed and an experimental confirmation of the numerical results has been obtained. This would be the appropriate procedure to use for the evaluation of the other solutions. For example, the sinking container (Prob. 2.52) would be susceptible to a student's execution of both the computer solution and the experimental evaluation of the height as a function of time.

---

*2.51 The motion of a compressed air-water rocket is to be analyzed. (Devices such as that illustrated are available in toy stores; they gain their propulsion by ejecting a small amount of water with a large velocity.) The rocket has a ballistic trajectory once the water is completely ejected. The mass of the rocket is M_r; the mass of the water at any time t is $M_w(t)$; the other symbols used in the analysis are noted on the figure. An isentropic expansion of the air is assumed. A complete analysis to develop the governing equations is too great a task for assignment as a home problem; therefore, selected terms from the general equations are to be developed.

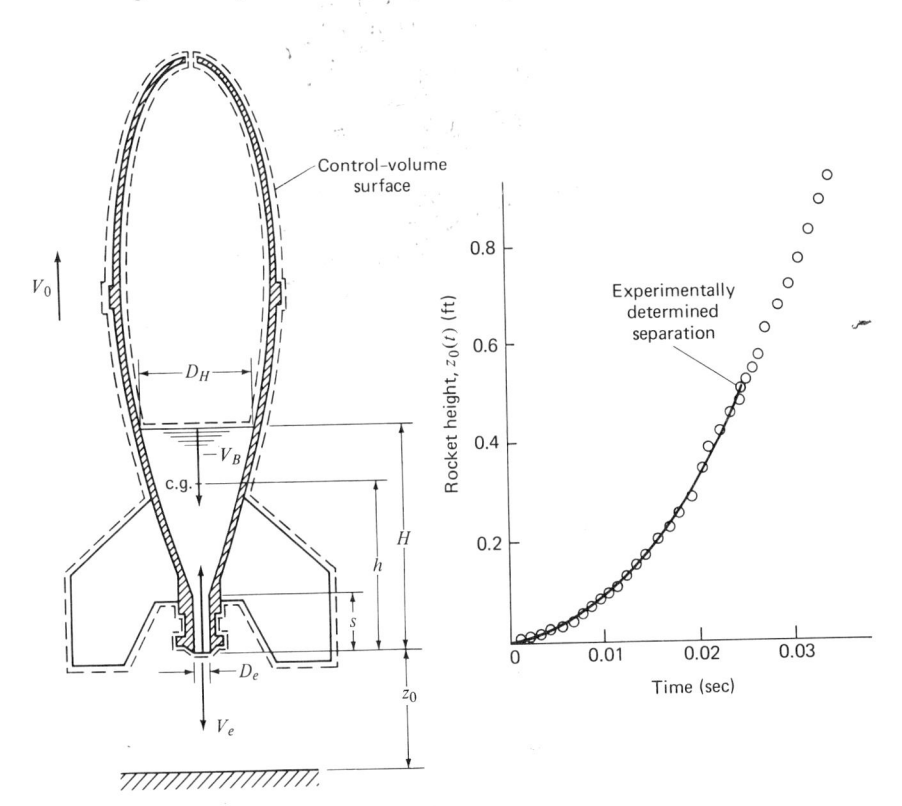

Prob. 2.51.

160   CONTROL-VOLUME FORM OF FUNDAMENTAL LAWS   Ch. 2

The energy equation and the momentum equation may be expressed as two first-order, nonlinear, ordinary differential equations. [The continuity equation has been used to relate $h(t)$ to $V_e(t)$.] The two equations are

$$A\dot{V}_e + B\dot{V}_0 + C = 0$$

$$D\dot{V}_e + E\dot{V}_0 + F = 0$$

where

$$A = \left(s + H\frac{V_0}{V_e} + D_e^2 \int_s^H \frac{dh}{D^2}\right)\rho A_e V_e$$

$$B = \rho H A_e V_e + V_0(M_w + M_R)$$

$$C = \left[\frac{V_e^2}{2}\left(\frac{D_e}{D_H}\right)^4 + V_0 V_e \left(\frac{D_e}{D_H}\right)^2 + gH + \frac{p_a - p_e}{\rho} - \frac{V_e}{2}(2V_0 + V_e)\right]\rho A_e V_e$$

$$+ V_0(M_w + M_R)g + \tfrac{1}{2} C_D \rho V_0^3 A$$

$$D = \rho A_e H$$

$$E = M_w + M_R$$

$$F = V_e\left[\left(\frac{D_e}{D_H}\right)^2 - 1\right]\rho A_e V_e + (M_w + M_R)g + \frac{1}{2} C_D \rho V_0^2 A$$

Note that each term above may be expressed in terms of $V_e(t)$ or $V_0(t)$. For parts a through d below you should start with the indicated term of the original energy or momentum equation and develop the corresponding term for the final equation. You should recognize that such an evaluation for each term in the equations will lead to the given expressions.

a. Explain why the time rate of change of kinetic energy is written as

$$\frac{d}{dt}\int_{c.v.} \frac{\rho}{2}[\mathbf{V}_0 + \mathbf{V}_r]\cdot[\mathbf{V}_0 + \mathbf{V}_r]\,d\mathcal{V}$$

where $\mathbf{V}_r$ is the velocity with respect to the rocket. Further, show that this term can be expanded into

$$\left[V_e \frac{V_H}{2}\left(\frac{D_e}{D_H}\right)^2 + \frac{1}{2}\left(s + D_e^2 \int_s^H \frac{dh}{D^2}\right)V_e\right]\rho A_e V_e$$

$$+ \left[V_0 V_H + H\frac{V_0}{V_e}\dot{V}_e + H\dot{V}_0\right]\rho A_e V_e + \frac{V_0^2}{2}\rho A_e V_e + (M_w + M_R)V_0\dot{V}_0$$

where
$$V_H\left(\frac{\pi D_H^2}{4}\right) = V_e\left(\frac{\pi D_e^2}{4}\right)$$
or
$$V_H = V_e\left(\frac{D_e}{D_H}\right)^2$$

and D_H is the diameter of the liquid surface at time t; the amount of water in the rocket is
$$M_w(t) = M_w(0) + \rho A_e \int_0^t V_e\, dt$$
and, equivalently,
$$M_w(t) = \rho A_e s + \int_s^{H(t)} \rho \frac{\pi D^2}{4}\, dh$$

which defines $H(t)$ in terms of $V_e(t)$.

b. Explain why the time rate of change of potential energy is
$$\frac{d}{dt}\int_{\text{c.v.}} \rho g(z_0 + z_r)\, dV$$

where z_r is the elevation of the water in the rocket with respect to the exit plane and z_0 is the elevation of the rocket. Show that this can be expressed as
$$gz_0 \rho A_e V_e + (M_w + M_R)gV_0 + gH\rho A_e V_e$$

c. Explain why the force term is the surface forces plus
$$\left(-g - \frac{dV_0}{dt}\right)\int_{\text{c.v.}} \rho\, dV$$

and why this term is written as
$$-[g + \dot{V}_0](M_w + M_R)$$

d. Show that the momentum flux term is $(-V_e)(\rho V_e A_e)$.

e. Develop a numerical scheme to evaluate $V_0(t)$, $V_e(t)$, etc., for a generalized compressed air–water rocket; that is, allow $D(h)$ to be a variable quantity and let ρ_w take on various values.

f. Use a polynomial form for $D(h)$ to determine the optimum $D(h)$, using the criterion that V_0 is greatest when the water has been completely ejected from the rocket.

g. Compare the performance for a light oil and a heavy liquid, where $\rho_{\text{oil}} = 0.6\rho_w$ and $\rho_l = 2\rho_w$.

The two ordinary differential equations may be solved numerically, using a predictor-corrector technique. Specifically, the two equations are considered to be algebraic equations for $\dot{V}_e(t)$ and $\dot{V}_0(t)$, where $V_e(t)$, $V_0(t)$, etc., are known. Then $V_e(t + \Delta t)$ and $V_0(t + \Delta t)$ are given by

$$V_e(t + \Delta t) = \int_0^t \dot{V}_e \, dt + \dot{V}_e(t) \Delta t,$$

and similarly for $V_0(t + \Delta t)$. The corrector part of the routine smooths the $t \to (t + \Delta t)$ step based upon the calculation at $t + 2\Delta t$; consult a numerical analysis text for the details. Such a solution has been carried out; in addition, an experimental evaluation of $z_0(t)$ based upon motion pictures at 2000 frames/sec has been used to obtain experimental data for the rocket "firing." The details are discussed in the reference cited.* The results of the numerical evaluation and the comparison with experiment for one case are shown in the

Results of Numerical Solution for Problem 2.51[†]

t (sec)	z_0 (ft)	H (ft)	V_0 (ft/sec)	V_e (ft/sec)	$\dot{V}_0$ (ft/sec^2)	$\dot{V}_e$ (ft/sec^2)
0.001	0.0008	0.1704	1.608	-93.71	1608.	-129.0
0.002	0.0032	0.1675	3.221	-93.71	1613.	9.687
0.003	0.0072	0.1646	4.839	-93.58	1618.	125.6
0.004	0.0128	0.1616	6.462	-93.36	1624.	221.2
0.005	0.0201	0.1587	8.093	-93.06	1631.	297.5
0.006	0.0290	0.1558	9.732	-92.70	1638.	356.4
0.007	0.0396	0.1529	11.38	-92.30	1647.	402.3
0.008	0.0518	0.1496	13.03	-91.86	1656.	443.3
0.009	0.0657	0.1462	14.70	-91.39	1665.	468.4
0.010	0.0812	0.1429	16.374	-90.90	1675.	485.8
0.011	0.0948	0.1391	18.06	-90.40	1685.	503.7
0.012	0.1173	0.1358	19.754	-89.901	1695.	504.0
0.013	0.1379	0.1316	21.46	-89.38	1706.	520.5
0.014	0.1603	0.1279	23.18	-88.86	1717.	514.4
0.015	0.1843	0.1237	24.90	-88.34	1728.	519.
0.016	0.2101	0.1191	26.64	-87.82	1738.	522.4
0.017	0.2376	0.1141	28.39	-87.30	1748.	524.7
0.018	0.2668	0.1091	30.15	-86.78	1756.	518.6
0.019	0.2979	0.1033	31.91	-86.25	1763.	522.0
0.020	0.3307	0.0971	33.68	-85.74	1766.	513.9
0.021	0.3652	0.0900	35.44	-85.24	1764.	498.1
0.022	0.4015	0.0808	37.19	-84.80	1748.	445.8
0.023	0.4396	0.0691	38.89	-84.63	1700.	164.6
0.024	0.4792	0.0458	40.35	-91.43	1460.	-6797.0
0.025	0.5197	-0.000	40.62	-138.6	270.0	-472.0

[†]Initial conditions: $p_a(0) = 4$ atm. gage, $H(0) = 0.1729$ ft, and $V_e(0) = -93.6$ fps.

*J. F. Foss and D. E. Heyer, "The Compressed Air–Water Rocket," *ASEE Journal*, Mechanics Series Monograph M-1. 1969. Also available as a report from the authors. Copies of the 16mm film (2000 frames/sec) are available as well.

accompanying table and figure, respectively. It should be noted that this is an exemplary problem for the analytical method in engineering practice. Once a successful analysis is established (agreement with experimental observation), then engineering decisions are rather easily made. Specifically, answers to such questions as (i) the optimum wall shape, $D(h)$, (ii) the optimum fluid, or (iii) the optimum initial water mass to achieve a specified objective (such as the maximum height or range) can be established from the analysis, as opposed to the often expensive and wasteful "cut and try" approach.

*2.52 A steel container with a large base area (e.g., a "tin can") is punctured and placed in a reservoir of water as shown in the sketch. Determine the equations which govern the time history of the sinking process as a function of d and D. Then, do the experiment with tap water, then with water with detergent added, for several d values. What is the qualitative influence of surface tension? The mass of the container is M_c; $A_c = \pi D^2/4$; $A_H = \pi d^2/4$. Show that the governing differential equations are

$$\frac{d^2 h_w}{dt^2} = g - \rho \frac{g}{M_c} (h_w - h_c) A_c$$

$$2g(h_c - h_w)\left(\frac{A_c}{A_H}\frac{dh_e}{dt} + \frac{dh_w}{dt}\right) = \left(\frac{A_c}{A_H}\frac{dh_c}{dt} + \frac{dh_w}{dt}\right)^3 + \left(\frac{dh_w}{dt}\right)^3\left(\frac{A_c}{A_H}\right)$$

The boundary conditions are

$$h_c = h_w = \frac{dh_w}{dt} = 0 \quad \text{at} \quad t = 0$$

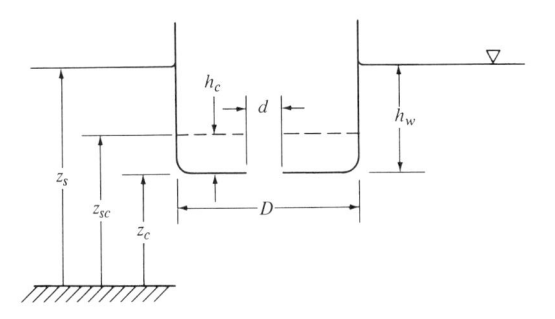

Prob. 2.52

*2.53 The device shown is to modify the effect of a sudden increase in a liquid flow. The desired response is intermediate between that of a continuous pipe and that of a piping system with a pneumatic surge tank. (A surge tank would exist if the piston were removed and the tank closed.) The inlet flow is from a positive-displacement source, that is, the flow rate is independent of the required pressure rise. As the design engineer, you must first establish the general response of this unit to changes in flow rate and then select the free parameters of the problem: size and mass of the piston, cylinder diameter, exit area. Determine the general relationship between the inlet and the outlet flow rates as a function of time. It may be assumed that the cylinder and piston areas are much larger than the inlet and exit areas. As indicated in the sketch, the piston floats at a given depth for steady-state operation.

The problem should be considered for the condition at which the inlet and exit pressures are fixed and a step change in the inlet mass rate of flow is established at time $t = 0$. An important step in the problem is the assumption that the height of the water and the exit velocity V_e are related by the Bernoulli equation. Do not make this assumption until you are forced to do so; and carefully note why this involves the neglect of the term $\int_{h_w}^{0} \frac{\partial V}{\partial t} ds$ and why this might be a reasonable approximation. Show that the governing equations reduce to

$$\dot{m}_e = \dot{m}_i - \frac{dh_w}{dt} \rho(A_w - A_p) - \frac{dh_p}{dt} A_p \rho$$

$$\frac{1}{2} \left(\frac{\dot{m}_e}{\rho A_e} \right)^2 = \left[\frac{1}{2} \left(\frac{dh_w}{dt} \right)^2 + g h_w - \frac{p_e}{\rho} \right]$$

Prob. 2.53

SELECTED REFERENCES

The derivation of the fixed control volume is treated in nearly all recently published texts, for example, in R. W. Fox and A. T. McDonald, *Introduction to Fluid Mechanics*, McGraw-Hill Book Co., New York, 1974. A well-presented derivation of both the nondeformable- and the deformable-control-volume formulations is given by A. G. Hansen, *Fluid Mechanics*, John Wiley & Sons, New York, 1967. The Hansen text is also recommended as a source of insightful review questions which require a (qualitative) control-volume analysis to deduce the behavior of a physical system. Instructive quantitative problems, demonstrating the problem-solving capability of control-volume analysis, are to be found in R. H. Sabersky, A. J. Acosta, and E. G. Hauptman, *Fluid Flow*, 2nd ed., Macmillan Co., New York, 1971. A further source (including solutions for such problems as well) is provided by W. F. Hughes and J. A. Brighton, *Fluid Dynamics*, Schaums Outline Series, McGraw-Hill Book Co., New York, 1967.

3

The Fundamental Laws — Differential Form

3.1 INTRODUCTION

In the previous chapter the fundamental laws were used to establish the integral equations, equations which involve the gross features of a flow, such as lift, pump horsepower, or flow rate. To solve these equations we must have a knowledge of the velocity- and pressure-distribution functions and, for compressible flows, of the density. In the problems posed in Chapter 2 the velocity and pressure distributions were assumed known. An assumption of uniform flow was also commonly made; such an assumption is not only convenient but is often a valid approximation for the computation of global characteristics of the flow. However, for some problems we may not be able to assume velocity and pressure distributions, with sufficient accuracy, or the determination of these distributions may be the subject of the problem. The separation point on an airfoil or a compressor blade (see Art. 5.3) is an example in which the detailed character of the flow is important. The velocity and pressure fields cannot be determined from the integral equations; they are found by solving the differential expressions of the conservation laws.

This chapter will be devoted mainly to derivation of these differential equations; in following chapters, both exact and approximate solutions will be investigated. The derivation of the differential equations is simplified by the use of vector calculus, but since the complete subject may not be part of every reader's mathematical background, we will restrict ourselves to cartesian coordinates.* Thus, we reduce the complexities of a complete exposition of essential concepts and equations. Such an exposition is the basis for advanced analyses of flow fields, remote though it may seem from direct engineering application. It is important that the engineer understand the basis for, and the information presented in, the differential equations; later chapters will reveal many engineering applications for which they are essential; the general strategy to determine the parameters for model studies is also based on them. The derivations in this chapter, along with the integral equations of Chapter 2, constitute the basic equations of fluid mechanics.

The differential form of a conservation law follows directly from the integral form by expressing all quantities as integral quantities (e.g., $\Sigma \mathbf{F} = \int_{c.s.} \tau \, dA + \int_{c.v.} \rho \mathbf{g} \, d\mathcal{V}$) and then transforming the surface integrals into volume integrals with the use of Gauss' theorem. The particular equation can then be written as one integral over the control volume. The integral itself is zero and this requires the integrand to be zero since the control volume may be any arbitrary volume occupied by the fluid. The resulting equation is the differential form of the conservation law. Although the statement of this general technique accurately describes the transformation procedures to develop the differential equation forms, the method cannot be employed for the momentum and energy equations without the use of tensor calculus; the transformation can (and will) however, easily be carried out for the conservation of mass. The differential form of the momentum equation, which is quite important in the analysis of fluid flows, will be derived by an alternate approach, that of applying the basic equations to a fixed elemental volume. A restricted differential form of the energy equation will also be presented. The primary purpose of the energy equation is to provide an equation for the temperature field. For incompressible homogeneous fluids, the velocity field is independent of the temperature field; hence, the temperature is not of interest in most fluid-mechanical engineering applications. Important exceptions are free-convection heat transfer, where density is affected, and flows with temperature differences large enough to significantly change such fluid properties as density, viscosity, and surface tension. The energy equation is, of course, of central importance in heat-transfer studies.

* The basic equations derived in this chapter are presented in cylindrical and spherical coordinates as well in Table 3.1 at the end of this chapter.

The partial differential equations which result naturally require boundary conditions. In fluid mechanics the boundary conditions usually involve the velocity; in viscous fluids the no-slip condition is required so that at a fixed boundary the velocity is zero, and in inviscid flows (fluid flows where the viscous effects may be neglected) the component of velocity normal to a fixed boundary is zero. In problems where free surfaces are involved the pressure becomes a boundary condition; for example, on the oceanic wave–atmosphere interface the pressure is constant.

3.2 THE DIFFERENTIAL EQUATIONS

The fundamental laws are expressed mathematically by the integral equations of Chapter 2. It seems then intuitively reasonable that the differential equations should be derivable from the integral equations since we wish to express the same laws in differential form. This is indeed the case, as will be illustrated for the conservation of mass.

1. Conservation-of-Mass Equation

The integral continuity equation in its most general form* is written as[†]

$$\int_{c.v.} \frac{\partial \rho}{\partial t} \, d\mathcal{V} + \int_{c.s.} \rho \mathbf{V} \cdot \hat{n} \, dA = 0 \tag{3.1}$$

Applying Gauss' theorem, (Eq. 1.42), the integral over the area is transformed into a volume integral, resulting in

$$\int_{c.v.} \frac{\partial \rho}{\partial t} \, d\mathcal{V} + \int_{c.v.} \nabla \cdot (\rho \mathbf{V}) \, d\mathcal{V} = 0 \tag{3.2}$$

which may be written as

$$\int_{c.v.} \left[\frac{\partial \rho}{\partial t} + \nabla \cdot (\rho \mathbf{V}) \right] d\mathcal{V} = 0 \tag{3.3}$$

Because the control volume can be chosen arbitrarily the integrand must be zero, resulting in the *differential continuity equation*

$$\frac{\partial \rho}{\partial t} + \nabla \cdot (\rho \mathbf{V}) = 0 \tag{3.4}$$

---

* For a deformable control volume, we can combine the surface integral involving $\mathbf{V}_b$ with the surface interval involving $\mathbf{V}_r$ by recalling that $\mathbf{V}_b + \mathbf{V}_r = \mathbf{V}$. Equation 3.1 results.

[†] The control volume may be chosen to be any finite volume in a fluid flow; it may have any geometric shape and be fixed in space with no loss in generality.

Art. 3.2 THE DIFFERENTIAL EQUATIONS

which, in rectangular coordinates, is

$$\frac{\partial \rho}{\partial t} + \frac{\partial (\rho u)}{\partial x} + \frac{\partial (\rho v)}{\partial y} + \frac{\partial (\rho w)}{\partial z} = 0 \qquad (3.5)$$

The continuity equation (3.4) may be put in the alternate form

$$\frac{D\rho}{Dt} + \rho \nabla \cdot \mathbf{V} = 0 \qquad (3.6)$$

or written out in rectangular coordinates as

$$\frac{\partial \rho}{\partial t} + u\frac{\partial \rho}{\partial x} + v\frac{\partial \rho}{\partial y} + w\frac{\partial \rho}{\partial z} + \rho\frac{\partial u}{\partial x} + \rho\frac{\partial v}{\partial y} + \rho\frac{\partial w}{\partial z} = 0 \qquad (3.7)$$

For an *incompressible fluid*, one for which $D\rho/Dt = 0$,[†] the continuity equation simplifies to

$$\nabla \cdot \mathbf{V} = 0 \qquad (3.8)$$

or, in cartesian coordinates,

$$\frac{\partial u}{\partial x} + \frac{\partial v}{\partial y} + \frac{\partial w}{\partial z} = 0 \qquad (3.9)$$

We also see that, for an incompressible fluid,

$$\frac{D\rho}{Dt} = \frac{\partial \rho}{\partial t} + u\frac{\partial \rho}{\partial x} + v\frac{\partial \rho}{\partial y} + w\frac{\partial \rho}{\partial z} = 0 \qquad (3.10)$$

showing that it is not necessary that ρ be constant at all points in the flow field. A flow that is incompressible with variable density is a *nonhomogeneous flow* since the property density varies with position. Note that "incompressible" refers to the condition $D\rho/Dt = 0$, which means that the density of a *particle* is not altered as it moves through the flow field. In effect, this is a "Lagrangian concept"; that is, it may be understood using the Lagrangian description. As noted, ρ may be a function of position and time; this is an "Eulerian concept." Examples of incompressible variable-density flows would be the flow of salt water in a channel where the concentration of salt varies with depth, the flow of air over Los Angeles, and the movement of water which is heated from below (convection).

[†] The definition of an incompressible fluid does not demand ρ = constant; however, when we state that the fluid is incompressible we usually mean ρ = constant. The examples of variable-density, incompressible flows are fairly uncommon so that "incompressibility" and "constant density" often are used synonymously.

170 THE FUNDAMENTAL LAWS—DIFFERENTIAL FORM   Ch. 3

Example 3.1

When a fluid flows over a solid surface, the viscous effects in the flow cause the fluid at the surface to be at rest with respect to the surface; this is called the "no-slip" condition. Under certain conditions, the viscous effects are confined to a thin region near the surface; such a region is termed the boundary layer. Figure E3.1A shows a boundary layer on a very thin plate placed in a free stream. Determine the character of the y-component of the velocity in the boundary layer.

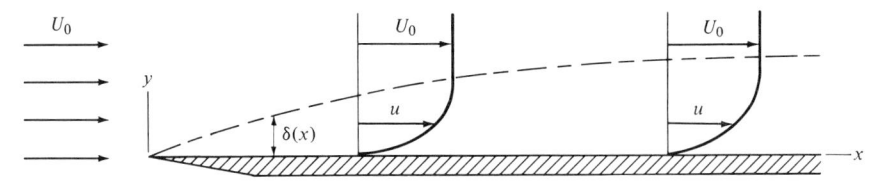

Fig. E3.1A

Solution. From Fig. E3.1A, it is clear that, since $u = U_0$ at $y = \delta$, and since δ increases with respect to x, the x-component velocity at a given y-value decreases with increasing x; that is, $\partial u / \partial x < 0$ for all y in the boundary layer. From the conservation of mass for the incompressible two-dimensional flow, we have

$$\frac{\partial u}{\partial x} + \frac{\partial v}{\partial y} = 0 \quad \text{or} \quad \frac{\partial v}{\partial y} = -\frac{\partial u}{\partial x} \tag{E3.1}$$

The velocity component v may be written as

$$v = \int_0^y \frac{\partial v}{\partial y} \, dy$$

since $v(x, 0) = 0$. In the boundary layer, from Eq. E3.1 $\partial v / \partial y > 0$; hence,

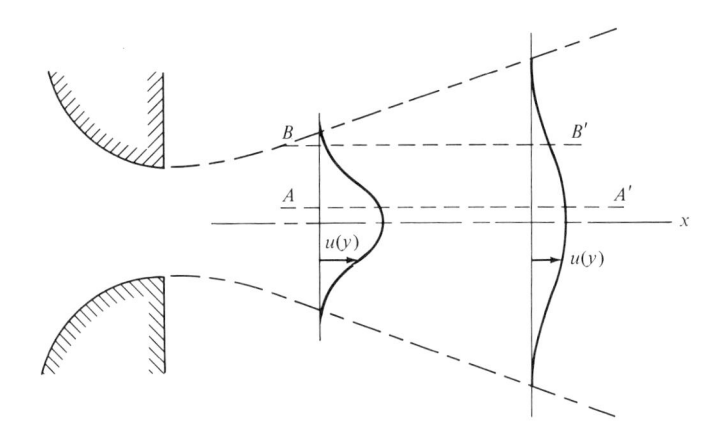

Fig. E.3.1B

Art. 3.2   THE DIFFERENTIAL EQUATIONS   171

$v(x, y) > 0$ and v reaches a maximum at the edge of the boundary layer where $\partial u/\partial x \cong 0$.

Extension 3.1.1. The velocity profiles in a plane jet are as indicated in Fig. E3.1B. Two lines AA' and BB' are indicated on the figure. From the velocity profiles and the behavior of $u(x)$ along these lines, infer the behavior of $v(y)$. You should be able to infer that $v > 0$ near $y = 0$ and $v < 0$ for large enough y. What is the value of $\partial u/\partial x$ where $v(y)$ is a maximum?

2. Momentum Equation

To make use of Gauss' theorem for derivation of the momentum equation in differential form, it is necessary first to transform the surface integral $\int_{c.s.} \tau \, dA \; (= \Sigma \mathbf{F})$ into a volume integral. (Recall this initial step in the derivation of differential continuity equation.) To put it into an appropriate form for this transformation, we must express the stress vector τ as a product of the stress tensor and the unit vector $\hat{n}$; note that all of the area integrals in Gauss' theorem (Eq. 1.42) must include the unit normal $\hat{n}$. This procedure requires tensor calculus, and hence will not be used here. We will employ an alternate method for the derivation of the differential equation.

Consider the infinitesimal cube of Fig. 3.1. It is fixed in the *xyz*-reference frame for a fluid flow, and is thus an infinitesimal control volume through which the fluid passes. Fluid occupies the control volume, and the stresses shown in Fig. 3.1 produce forces which accelerate the infinitesimal mass of fluid occupying the control volume. The relationship between the forces acting on the fluid particle and the acceleration of the particle is*

$$\Sigma \, d\mathbf{F} = \mathbf{a} \, dm \qquad (3.11)$$

In terms of stresses the *x*-component equation becomes

$$\rho g_x dx\, dy\, dz + \left(\tau_{xx} + \frac{\partial \tau_{xx}}{\partial x} dx\right) dy\, dz - \tau_{xx} dy\, dz + \left(\tau_{yx} + \frac{\partial \tau_{yx}}{\partial y} dy\right) dx\, dz$$

$$- \tau_{yx} dx\, dz + \left(\tau_{zx} + \frac{\partial \tau_{zx}}{\partial z} dz\right) dx\, dy - \tau_{zx} dx\, dy$$

$$= a_x \rho \, dx\, dy\, dz \qquad (3.12)$$

---

* The relationship in Eq. 3.11 may appear to be in conflict with that of Eq. 2.42; however, if we write $\Sigma d\mathbf{F} = \dfrac{D}{Dt}(\mathbf{V} \, dm)$ and note that, for a particular particle, dm is constant, this becomes $\Sigma \, d\mathbf{F} = \dfrac{D\mathbf{V}}{Dt} dm$. In Prob. 3.11, the momentum flux in and out of the elemental volume can be shown to give exactly the right-hand side of Eq. 3.12.

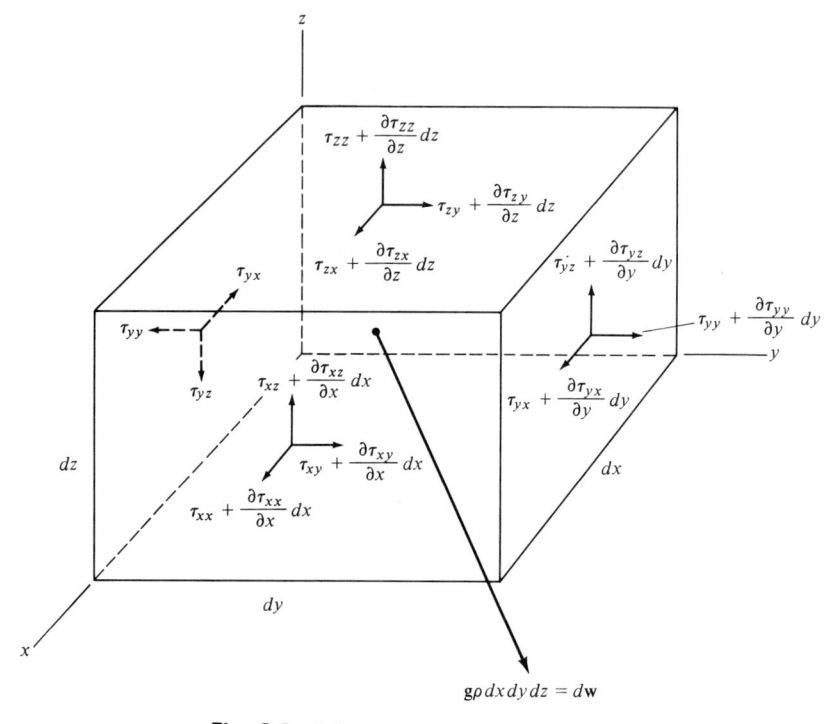

Fig. 3.1. Infinitesimal control volume.

where we recall that τ_{yx} is the component of stress acting on a y-face in the x-direction, or on a negative y-face in the negative x-direction. A negative face is a face oriented by an outward-pointing unit vector in a negative coordinate direction. After subtracting out the appropriate terms in Eq. 3.12 and dividing by $dx\,dy\,dz$, we have

$$\rho g_x + \frac{\partial \tau_{xx}}{\partial x} + \frac{\partial \tau_{yx}}{\partial y} + \frac{\partial \tau_{zx}}{\partial z} = \rho\left(\frac{\partial u}{\partial t} + u\frac{\partial u}{\partial x} + v\frac{\partial u}{\partial y} + w\frac{\partial u}{\partial z}\right) \quad (3.13)$$

where the x-component acceleration a_x is given by Eq. 1.58.

For the y- and z-directions, carrying out the same steps as above, we obtain

$$\rho g_y + \frac{\partial \tau_{xy}}{\partial x} + \frac{\partial \tau_{yy}}{\partial y} + \frac{\partial \tau_{zy}}{\partial z} = \rho\left(\frac{\partial v}{\partial t} + u\frac{\partial v}{\partial x} + v\frac{\partial v}{\partial y} + w\frac{\partial v}{\partial z}\right) \quad (3.14)$$

$$\rho g_z + \frac{\partial \tau_{xz}}{\partial x} + \frac{\partial \tau_{yz}}{\partial y} + \frac{\partial \tau_{zz}}{\partial z} = \rho\left(\frac{\partial w}{\partial t} + u\frac{\partial w}{\partial x} + v\frac{\partial w}{\partial y} + w\frac{\partial w}{\partial z}\right) \quad (3.15)$$

These are the component equations of the differential momentum equation. We cannot, at this point, express them in vector form because of the presence of the stress terms, which cannot be written using vector notation.

The *stress tensor*, with components introduced in the foregoing equations, may be displayed by

$$\tau_{ij} = \begin{bmatrix} \tau_{xx} & \tau_{xy} & \tau_{xz} \\ \tau_{yx} & \tau_{yy} & \tau_{yz} \\ \tau_{zx} & \tau_{zy} & \tau_{zz} \end{bmatrix} \quad (3.16)$$

where, for example, $i = 2, j = 3$ locates the element τ_{yz} in the second row, third column.

In many applications in fluid mechanics the shearing stresses, which are due to the viscosity, are small and may be neglected (Chapter 8 will be devoted to such flows). Also, the normal stresses are nearly equal to $-p$, p being the pressure. Under these conditions the "inviscid" momentum equations result, namely,

$$\rho \frac{Du}{Dt} = -\frac{\partial p}{\partial x} + \rho g_x$$

$$\rho \frac{Dv}{Dt} = -\frac{\partial p}{\partial y} + \rho g_y \quad (3.17)$$

$$\rho \frac{Dw}{Dt} = -\frac{\partial p}{\partial z} + \rho g_z$$

where we have used $D\mathbf{V}/Dt$ to express the acceleration $\mathbf{a}$. These are often referred to as *Euler's equations*. In vector form, Euler's equations may be written as

$$\rho \frac{D\mathbf{V}}{Dt} = -\nabla p + \rho \mathbf{g} \quad (3.18)$$

The boundary conditions on the differential momentum equations are stated in terms of the velocity or the pressure or both. Hence, for flows subject to significant shear-stress effects, Eqs. 3.13, 3.14, and 3.15 are not in acceptable form, since the stresses are included as dependent variables with no accompanying boundary conditions. However, when we make the assumption of zero shear, as in Euler's equations, the dependent variables reduce to u, v, w, and p. Euler's equations, along with continuity Eq. 3.4, then form a tractable set of equations with boundary conditions stated in terms of the velocity and pressure. To put Eqs. 3.13 through 3.15 in solvable form, we must relate the stress components to the velocity and pressure fields. This is done in Art. 3.4.

The observant reader will have noted that the derivation of the differential momentum equations is a subtle blend of concepts which are expressed by both Eulerian (the stress field) and Lagrangian (the acceleration) descriptions. In order to describe the net force on the designated region in space, the Eulerian description is employed to specify the shear stress, as in $\tau_{xx}(x, y, z, t)$. However, the acceleration of a fluid particle (indeed, the concept of the acceleration of anything) is a Lagrangian concept; the acceleration of a region is not under consideration and it is certainly an irrelevant consideration. However, as we have emphasized, the equations of general usefulness in fluid mechanics are those of the Eulerian description. Consequently, it was necessary to make use of the earlier-developed relationship which describes the acceleration using the Eulerian description; see Eq. 1.58.

Example 3.2

Derive Bernoulli's equation from the differential momentum equation and show that it is applicable along a streamline in a steady incompressible flow in which all shearing stresses are zero, and all normal stresses are negatives of the pressure.

Solution. With the foregoing assumptions on the stress tensor, the momentum equations are

$$\rho g_x - \frac{\partial p}{\partial x} = \rho \frac{Du}{Dt}$$

$$\rho g_y - \frac{\partial p}{\partial y} = \rho \frac{Dv}{Dt}$$

$$\rho g_z - \frac{\partial p}{\partial z} = \rho \frac{Dw}{Dt}$$

In vector form these three equations may be written as Eq. 3.18,

$$\rho \mathbf{g} - \nabla p = \rho \frac{D\mathbf{V}}{Dt}$$

The constant-gravity vector $\mathbf{g}$ is conservative and hence can be written as $-\mathbf{g} = g\nabla h$, where h is the vertical distance (if z is vertical then $h = z$, $\nabla h = \hat{k}$, and $\mathbf{g} = -g\hat{k}$).

From Eq. 1.61, the acceleration may be written as

$$\frac{D\mathbf{V}}{Dt} = \frac{\partial \mathbf{V}}{\partial t} + \nabla\left(\frac{V^2}{2}\right) + (\nabla \times \mathbf{V}) \times \mathbf{V}$$

For the steady flow under consideration, $\partial \mathbf{V}/\partial t = 0$, and the momentum equation may be written as

$$\rho \nabla \left(\frac{V^2}{2} \right) + \nabla p + \rho g \nabla h = -\rho (\nabla \times \mathbf{V}) \times \mathbf{V}$$

or, for an incompressible (constant-density) flow,

$$\nabla \left(\frac{V^2}{2} + \frac{p}{\rho} + gh \right) = -(\nabla \times \mathbf{V}) \times \mathbf{V}$$

Since the quantity on the right, $(\nabla \times \mathbf{V}) \times \mathbf{V}$, is a vector normal to the velocity vector, and hence to the streamline, and since the gradient of the scalar function $\left(\frac{V^2}{2} + \frac{p}{\rho} + gh \right)$ is normal to a constant-$\left(\frac{V^2}{2} + \frac{p}{\rho} + gh \right)$ surface, the streamline must be in a constant-$\left(\frac{V^2}{2} + \frac{p}{\rho} + gh \right)$ surface (see Fig. E3.2). Hence $\frac{V^2}{2} + \frac{p}{\rho} + gh =$ constant along a streamline. We have invoked the theorem which states that the gradient of a scalar function ϕ is always normal to a constant-ϕ surface.

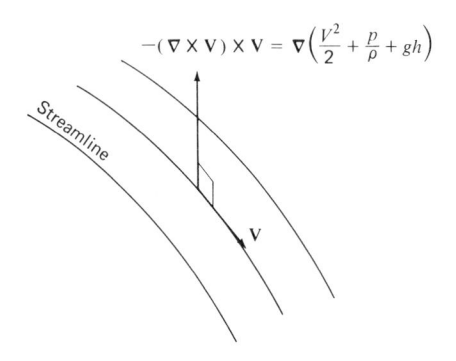

Fig. E3.2

We have, of course, arrived at the same Bernoulli's equation derived in Chapter 1 on the assumption of a steady, incompressible, shearless flow with conservative body forces. (Magnetic body forces for example, are not conservative.)

Extension 3.2.1. If the velocity field were a conservative vector field, how would the development above be different? Would Bernoulli's equation change? Determine what would be different.

Extension 3.2.2. For a nonconservative (rotational) body-force field, would Bernoulli's equation be applicable? If not, why not?

3. Moment-of-Momentum Equation

The stress tensor contains nine components. The equations will be simplified if we can show that it is symmetric. To do this, let us apply the principle of conservation of angular momentum to the infinitesimal cube of Fig. 3.1. Moments about a line passing through the center of the cube and parallel to the z-axis may be expressed in terms of the stresses by assuming that the forces act at the center of a face. The net moment is equated to zero* to give

$$-\frac{dy}{2}\left[\tau_{yx} + \frac{\partial \tau_{yx}}{\partial y} dy\right] dx\, dz - \frac{dy}{2} \tau_{yx}\, dx\, dz$$

$$+ \frac{dx}{2}\left[\tau_{xy} + \frac{\partial \tau_{xy}}{\partial x} dx\right] dy\, dz + \frac{dx}{2} \tau_{xy}\, dy\, dz = 0 \quad (3.19)$$

Dividing by $dx\, dy\, dz$, and keeping only the terms which do not contain a differential quantity, there results

$$-\tau_{yx} + \tau_{xy} = 0 \quad \text{or} \quad \tau_{yx} = \tau_{xy} \quad (3.20)$$

Similarly, we can find that

$$\tau_{yz} = \tau_{zy} \quad \text{and} \quad \tau_{xz} = \tau_{zx} \quad (3.21)$$

thereby showing that the stress tensor is symmetric. This is the contribution of the moment-of-momentum differential equation. The stress tensor (Eq. 3.16) may therefore be displayed in the symmetric form

$$\tau_{ij} = \begin{bmatrix} \tau_{xx} & \tau_{xy} & \tau_{xz} \\ \tau_{xy} & \tau_{yy} & \tau_{yz} \\ \tau_{xz} & \tau_{yz} & \tau_{zz} \end{bmatrix} \quad (3.22)$$

The symmetric second-order stress tensor possesses three principal stresses, associated with three principal directions. The principal stresses include the maximum normal stress that exists at the point where the stress components act. In a solid, failure may occur if this normal stress exceeds a yield criterion for the solid in question;

---

* The summation of moments should be equated to $(dm) \times$ (radius of gyration)^2 $\times$ (angular acceleration); but this is of higher order, since the radius of gyration is an infinitesimal quantity, and thus the term becomes negligibly small as the element is allowed to shrink in size.

4. Energy Equation

Our interest will center around the integral equations and the differential continuity and momentum equations. We will not solve problems for which the differential energy equation is needed; that type of problem is one of the central concerns of a modern heat-transfer course. It is, however, of interest in problems involving temperature gradients (nonisothermal fluid flows) so a somewhat restricted form will be derived here for incompressible flows.

The work-rate term in the integral energy equation results from stresses acting on the control surface. The general stress vector can be written as $\tau = \tau_N \hat{n} + \tau_S \hat{s}$, where $\tau_N \cong -p$. The shearing stress τ_S results from viscosity and is generally quite small and often negligible when considering applications of the energy equation. Let us neglect it in this derivation; this enables us to derive a restricted form of the differential energy equation from the integral energy equation without the use of tensor calculus. Neglecting radiation and viscous shear, the integral energy equation is

$$\int_{c.s.} k\nabla T \cdot \hat{n} \, dA - \int_{c.s.} p\mathbf{V} \cdot \hat{n} \, dA$$

$$= \int_{c.v.} \frac{\partial}{\partial t}\left[\rho\left(\frac{V^2}{2} + \tilde{u} + gz\right)\right] dV + \int_{c.s.} \rho\left[\frac{V^2}{2} + \tilde{u} + gz\right]\mathbf{V} \cdot \hat{n} \, dA \quad (3.23)$$

(the z-axis is vertical) where we have assumed the heat-transfer rate, $\dot{Q} = \int k\nabla T \cdot \hat{n} \, dA$, to result from conduction (no radiation) across the control surface and the work rate to be $\dot{W} = \int p\mathbf{V} \cdot \hat{n} \, dA$ (no shear); k represents conductivity. Converting all the surface integrals to volume integrals by the use of Gauss' theorem results in

$$\int_{c.v.} \left\{\nabla \cdot (k\nabla T) - \nabla \cdot (p\mathbf{V}) - \nabla \cdot \left[\rho \mathbf{V}\left(\frac{V^2}{2} + \tilde{u} + gz\right)\right]\right.$$

$$\left. - \frac{\partial}{\partial t}\left[\rho\left(\frac{V^2}{2} + \tilde{u} + gz\right)\right]\right\} dV = 0 \quad (3.24)$$

For an arbitrary control volume the integrand must be zero, that is,

$$\nabla \cdot (k\nabla T) = \nabla \cdot (\nabla p) + \nabla \cdot \nabla \left[\rho \left(\frac{V^2}{2} + \tilde{u} + gz \right) \right] + \frac{\partial}{\partial t} \left[\rho \left(\frac{V^2}{2} + \tilde{u} \right) \right]$$

(3.25)

where we have used $\nabla \cdot (\phi \mathbf{V}) = \mathbf{V} \cdot (\nabla \phi) + \phi \nabla \cdot \mathbf{V} = \mathbf{V} \cdot \nabla \phi$ since $\nabla \cdot \mathbf{V} = 0$ for an incompressible flow. Also, since z and t are independent $\partial z / \partial t = 0$. Using the definition of the substantial derivative, and assuming k to be constant, Eq. 3.25 may be written as

$$k \nabla^2 T = \mathbf{V} \cdot \nabla p + \rho \frac{D}{Dt} \left(\frac{V^2}{2} + \tilde{u} \right) + \rho \mathbf{V} \cdot \nabla (gz)$$

$$= \mathbf{V} \cdot \nabla p + \rho \mathbf{V} \cdot \frac{D\mathbf{V}}{Dt} - \rho \mathbf{V} \cdot \mathbf{g} + \rho \frac{D\tilde{u}}{Dt} \quad (3.26)$$

where $\mathbf{g} = -\nabla(gz)$. The first three terms on the right-hand side are simply $\mathbf{V}$ dotted with the inviscid momentum equation (3.18) so that the energy equation reduces to

$$\rho \frac{D\tilde{u}}{Dt} = k \nabla^2 T \quad (3.27)$$

for an incompressible flow with zero shear stresses and constant conductivity.

If the effects of viscosity are included, the differential energy equation is

$$\rho \frac{D\tilde{u}}{Dt} = k \nabla^2 T + \tau_{xx} \frac{\partial u}{\partial x} + \tau_{yy} \frac{\partial v}{\partial y} + \tau_{zz} \frac{\partial w}{\partial z}$$

(3.28)

$$+ \tau_{xy} \left[\frac{\partial u}{\partial y} + \frac{\partial v}{\partial x} \right] + \tau_{xz} \left[\frac{\partial u}{\partial z} + \frac{\partial w}{\partial x} \right] + \tau_{yz} \left[\frac{\partial v}{\partial z} + \frac{\partial w}{\partial y} \right]$$

which is applicable for a compressible or an incompressible flow. This equation could be derived by starting with an elemental control volume as was done with the component momentum equations.

The differential energy equation would be necessary if the fluid properties were dependent on temperature; this dependence exists when compressibility affects the flow, but it may also exist when there is a temperature gradient in a liquid flow. For a low-speed air flow (i.e., an incompressible flow) with constant fluid properties, the temperature field is not pertinent to the description of the flow field and the differential energy equation is not of interest. Studies in heat transfer, however, would make use of the differential energy equation.

3.3 VELOCITY GRADIENTS

An examination of differential equations 3.9, 3.13, 3.14, and 3.15 of the previous article indicates that the number of unknowns exceeds the number of basic equations. For the case of constant-density flow the unknowns are u, v, w, τ_{xx}, τ_{yy}, τ_{zz}, τ_{yz}, τ_{xy}, and τ_{xz}, that is, nine unknowns in four equations. To form a tractable problem we must relate the stress components to the velocity field. This may be done rigorously with the use of tensor analysis; since this mathematical background is not assumed of the reader, the relationship between the stress components and the rate-of-strain components—the constitutive equations—will only be stated in this text.

Some general observations should be made with regard to the development of the constitutive equations before proceeding with the detailed considerations. As noted, the differential equations involve nine unknowns in the four equations; this implicitly establishes our motivation for reducing the equations to a tractable form for determining either the velocity or the stress field. Expressing the stress in terms of the velocity indicates our desire to describe the flow field in terms of the velocity since the boundary conditions typically involve the velocity and not the stress. The final result, of course, can be used in either sense: to deduce a velocity distribution from a stress field, or vice versa. Whether or not the stress and velocity fields are related is a matter for experimental determination. The historical motivation for investigating whether or not such a relationship exists was based on the prior success of relating the stress field to the strain field for solid materials. The moment-of-momentum considerations have shown that the stress distribution is described by a symmetric tensor (Eq. 3.22). Therefore, the most general relationship would relate the stress tensor to a symmetric tensor involving the velocity field. We will now consider the velocity field in detail.

If one focuses attention on an individual fluid particle in a flowing fluid, the behavior of the particle may be described as the superposition of three effects: the particle may translate, rotate, or deform; in general, it will do all three. It is only the deformation rate of the particle which is related to the stresses and thus it is necessary to separate the deformation from the rotation and translation. The deformation rate in a fluid is a symmetric tensor, as will now be shown.

At a particular time t the velocity of a fluid particle is $\mathbf{V}$ and that of a neighboring particle is $\mathbf{V} + d\mathbf{V}$, where

$$d\mathbf{V} = \frac{\partial \mathbf{V}}{\partial x} dx + \frac{\partial \mathbf{V}}{\partial y} dy + \frac{\partial \mathbf{V}}{\partial z} dz \qquad (3.29)$$

The quantities $\partial \mathbf{V}/\partial x$, $\partial \mathbf{V}/\partial y$, $\partial \mathbf{V}/\partial z$ are *velocity gradients*. They result from distortion and rigid-body rotation of the particle. It is the distortion of the fluid particle which accounts for shearing stresses in the fluid, and thus its importance is obvious; the rotation of fluid particles is also of considerable interest. Hence, let us study the velocity gradients in some detail.

The gradient $\partial \mathbf{V}/\partial x$ may be written in component form as

$$\frac{\partial u}{\partial x} = \frac{1}{2}\left(\frac{\partial u}{\partial x} + \frac{\partial u}{\partial x}\right) + \frac{1}{2}\left(\frac{\partial u}{\partial x} - \frac{\partial u}{\partial x}\right) \equiv \epsilon_{xx} + \omega_{xx}$$

$$\frac{\partial v}{\partial x} = \frac{1}{2}\left(\frac{\partial v}{\partial x} + \frac{\partial u}{\partial y}\right) + \frac{1}{2}\left(\frac{\partial v}{\partial x} - \frac{\partial u}{\partial y}\right) \equiv \epsilon_{yx} + \omega_{yx} \qquad (3.30)$$

$$\frac{\partial w}{\partial x} = \frac{1}{2}\left(\frac{\partial w}{\partial x} + \frac{\partial u}{\partial z}\right) + \frac{1}{2}\left(\frac{\partial w}{\partial x} - \frac{\partial u}{\partial z}\right) \equiv \epsilon_{zx} + \omega_{zx}$$

where the ϵ-quantities are the first terms in the middle equations and the ω-quantities are the second terms. The other two gradients are, similarly, written as

$$\frac{\partial u}{\partial y} = \frac{1}{2}\left(\frac{\partial u}{\partial y} + \frac{\partial v}{\partial x}\right) + \frac{1}{2}\left(\frac{\partial u}{\partial y} - \frac{\partial v}{\partial x}\right) \equiv \epsilon_{xy} + \omega_{xy}$$

$$\frac{\partial v}{\partial y} = \frac{1}{2}\left(\frac{\partial v}{\partial y} + \frac{\partial v}{\partial y}\right) + \frac{1}{2}\left(\frac{\partial v}{\partial y} - \frac{\partial v}{\partial y}\right) \equiv \epsilon_{yy} + \omega_{yy} \qquad (3.31)$$

$$\frac{\partial w}{\partial y} = \frac{1}{2}\left(\frac{\partial w}{\partial y} + \frac{\partial v}{\partial z}\right) + \frac{1}{2}\left(\frac{\partial w}{\partial y} - \frac{\partial v}{\partial z}\right) \equiv \epsilon_{zy} + \omega_{zy}$$

and

$$\frac{\partial u}{\partial z} = \frac{1}{2}\left(\frac{\partial u}{\partial z} + \frac{\partial w}{\partial x}\right) + \frac{1}{2}\left(\frac{\partial u}{\partial z} - \frac{\partial w}{\partial x}\right) \equiv \epsilon_{xz} + \omega_{xz}$$

$$\frac{\partial v}{\partial z} = \frac{1}{2}\left(\frac{\partial v}{\partial z} + \frac{\partial w}{\partial y}\right) + \frac{1}{2}\left(\frac{\partial v}{\partial z} - \frac{\partial w}{\partial y}\right) \equiv \epsilon_{yz} + \omega_{yz} \qquad (3.32)$$

$$\frac{\partial w}{\partial z} = \frac{1}{2}\left(\frac{\partial w}{\partial z} + \frac{\partial w}{\partial z}\right) + \frac{1}{2}\left(\frac{\partial w}{\partial z} - \frac{\partial w}{\partial z}\right) \equiv \epsilon_{zz} + \omega_{zz}$$

Art. 3.3 VELOCITY GRADIENTS

Equations 3.30, 3.31, and 3.32 define the *strain-rate components* and *vorticity tensor components*. We see that $\epsilon_{xy} = \epsilon_{yx}$, $\epsilon_{xz} = \epsilon_{zx}$, $\epsilon_{yz} = \epsilon_{zy}$, and that $\omega_{xx} = \omega_{yy} = \omega_{zz} = 0$. The *rate-of-strain tensor* is given by the symmetric array

$$\epsilon_{ij} = \begin{bmatrix} \epsilon_{xx} & \epsilon_{xy} & \epsilon_{xz} \\ \epsilon_{xy} & \epsilon_{yy} & \epsilon_{yz} \\ \epsilon_{xz} & \epsilon_{yz} & \epsilon_{zz} \end{bmatrix} \qquad (3.33)$$

and accounts for the distortion rate. The *vorticity tensor* is given by the skew-symmetric array

$$\omega_{ij} = \begin{bmatrix} 0 & \omega_{xy} & \omega_{xz} \\ -\omega_{xy} & 0 & \omega_{yz} \\ -\omega_{xz} & -\omega_{yz} & 0 \end{bmatrix} \qquad (3.34)$$

and accounts for the rigid-body rotation. The discussion which follows (up to Eq. 3.39) verifies that ϵ_{ij} accounts for the distortion rate and that ω_{ij} accounts for the rotation rate.

Consider the displacement of a rectangular fluid element in the xy-plane, shown in Fig. 3.2. Associated with P are the velocity components u and v; with Q, the velocity components $u + (\partial u/\partial x)\,dx$ and $v + (\partial v/\partial x)\,dx$; and with R, the components $u + (\partial u/\partial y)\,dy$ and $v + (\partial v/\partial y)\,dy$. Multiplying these velocities by the time increment dt results in the distances which these points move. Q' moves above P' by the amount $(\partial v/\partial x)\,dx\,dt$ and R' to the right of P' by the amount $(\partial u/\partial y)\,dy\,dt$. The strain of the line element PQ is $(P'Q' - PQ)/PQ$ and is given by

$$\frac{\left[\left(u + \frac{\partial u}{\partial x}\,dx\right)dt + dx - u\,dt\right] - dx}{dx} = \frac{\partial u}{\partial x}\,dt \qquad (3.35)$$

Thus, the rate at which the element is straining in the x-direction is $\partial u/\partial x$, which from Eq. 3.30 is simply ϵ_{xx}. The rate at which a line element in the y-direction is elongating would be ϵ_{yy}, and in the z-direction, ϵ_{zz}. These three are normal strain rates and affect the normal stresses.

To observe the physical significance of ϵ_{xy}, let us evaluate the rate at which the angle α between the line elements is changing. To this end,

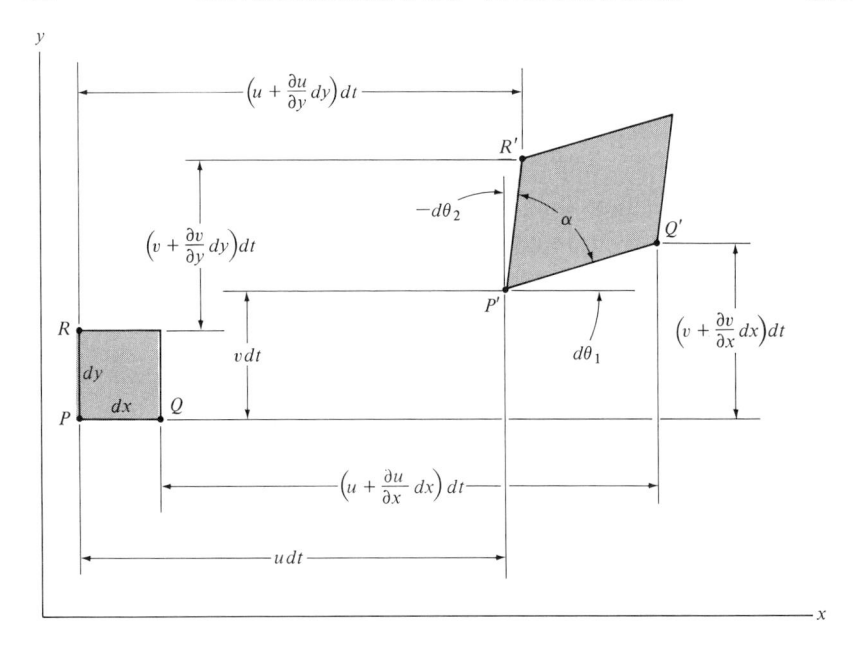

Fig. 3.2. Displacement of an element due to velocity components u and v.

consider the angles $d\theta_1$ and $d\theta_2$. The angle $d\theta_1$ is given by

$$d\theta_1 = \frac{\left(v + \frac{\partial v}{\partial x} dx\right) dt - v\, dt}{dx} = \frac{\partial v}{\partial x} dt \tag{3.36}$$

and $d\theta_2$ by

$$d\theta_2 = -\frac{\left(u + \frac{\partial u}{\partial y} dy\right) dt - u\, dt}{dy} = -\frac{\partial u}{\partial y} dt \tag{3.37}$$

where θ_1 and θ_2 are measured from the x-axis and y-axis, respectively, with the counterclockwise sense positive. There now results

$$\begin{aligned}\epsilon_{xy} &= \frac{1}{2}\frac{d}{dt}(\theta_1 - \theta_2) \\ \omega_{yx} &= \frac{1}{2}\frac{d}{dt}(\theta_1 + \theta_2)\end{aligned} \tag{3.38}$$

The angle α is given by $(\pi/2 - \theta_1 + \theta_2)$; thus ϵ_{xy} represents half the rate at which the angle α decreases with time, accounting for the distortion of the element. Likewise, we can see that the rate at which the element is rotating about the z-axis is ω_{yx}. The inclusion of the

z-component of velocity would not alter this analysis, since its influence is realized only on higher-order terms, which are negligible. Similar physical interpretations can be applied to ϵ_{xz}, ϵ_{yz}, ω_{xz}, and ω_{yz}.

The case of pure rotation occurs when all components of the rate-of-strain tensor are zero, as shown in the xy-plane in Fig. 3.3a. An angle traced in the fluid would not change its magnitude; the lines forming the angle would rotate equally. Pure deformation occurs when the vorticity tensor is zero; two perpendicular lines, as shown in Fig. 3.3b, may rotate an equal amount but in opposite directions to give zero rotation to the fluid element. A small cork placed in the flow would rotate at rate ω_{yx} in Fig. 3.3a and would not rotate at all in the flow of Fig. 3.3b.

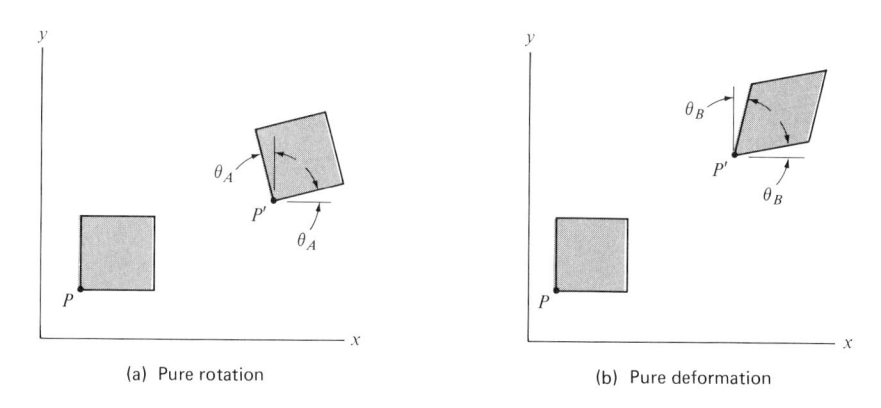

Fig. 3.3. Examples of the motion of perpendicular lines identified in a fluid flow for pure rotation and for pure deformation.

The vorticity tensor is skew-symmetric and thus has only three distinct components. These components may be represented by the *vorticity vector*, commonly referred to as the vorticity, defined by

$$\omega = \nabla \times V \tag{3.39}$$

The cartesian components, ξ, η, and ζ, of the vorticity are related to the components of the vorticity tensor (see Eqs. 3.30, 3.31, and 3.32) by

$$\xi = \frac{\partial w}{\partial y} - \frac{\partial v}{\partial z} = 2\omega_{zy}$$

$$\eta = \frac{\partial u}{\partial z} - \frac{\partial w}{\partial x} = 2\omega_{xz} \tag{3.40}$$

$$\zeta = \frac{\partial v}{\partial x} - \frac{\partial u}{\partial y} = 2\omega_{yx}$$

The vorticity is interpreted physically as twice the rate of rotation of a particle at a point.

A line is a *vortex line* if at each point on the curve the vorticity vector is tangent to the curve. A tube whose wall contains vortex lines is a *vortex tube*. The considerations and several other aspects of vorticity are presented at length in Chapter 5.

Flows for which the vorticity is zero are of special interest; they are known as *irrotational flows*. They usually occur in regions away from the influence of a solid boundary, as in the case of flow away from an airfoil. They will be considered in detail in Chapter 8.

Example 3.3

A velocity field is given by $\mathbf{V} = 10(4y - y^2)\hat{i}$. Is this an irrotational velocity field? If not, determine the vorticity vector. Also determine the rate-of-strain tensor. How rapidly would a particle be rotating at the point (4, 3)? Why is it that we do not have a primary interest in principal strain rates and principal directions, as was the case in your study of solids?

Solution. The vorticity ω is

$$\omega = \left(\frac{\partial w}{\partial y} - \frac{\partial v}{\partial z}\right)\hat{i} + \left(\frac{\partial u}{\partial z} - \frac{\partial w}{\partial x}\right)\hat{j} + \left(\frac{\partial v}{\partial x} - \frac{\partial u}{\partial y}\right)\hat{k}$$

$$= -10(4 - 2y)\hat{k}$$

The flow possesses vorticity; hence, by definition it is not irrotational. At $y = 3$, a particle would rotate about the z-axis with rate

$$\omega_{yx} = \zeta/2$$

$$= 10 \text{ rad/sec}$$

This result is positive; so if the xy-plane were in the plane of the book page, with x-direction to the right, then the right-hand rule would indicate a counterclockwise rotation.

Using Eqs. 3.30, 3.31, and 3.32, the rate-of-strain tensor would be

$$\epsilon_{ij} = \begin{bmatrix} 0 & 5(4-2y) & 0 \\ 5(4-2y) & 0 & 0 \\ 0 & 0 & 0 \end{bmatrix}$$

In solids, failure occurs along the plane of maximum shear or along the plane of maximum tension, depending on the type of material, and for many materials failure is even more complicated; it is important that the maximum stresses, as well as the possible planes of failure, be evaluated. In fluids, however, we are not concerned about failure except for the special phenomenon of cavitation, which takes place when the pressure in a liquid flow drops too low (usually to a point approaching absolute zero). We seldom have interest in maximum normal and shearing stresses and their accompanying principal planes.

3.4 THE CONSTITUTIVE EQUATIONS

We now return to our discussion of the relationship of stresses to the velocity field and relate the components of the symmetric stress tensor to the components of the strain-rate tensor. The general relationship represents six equations for which each stress component may be a function of the six rate-of-strain components. Assuming a linear relationship (a Newtonian fluid), the stress component τ_{xy} would be related to the rate-of-strain components by

$$\tau_{xy} = C_1\epsilon_{xx} + C_2\epsilon_{xy} + C_3\epsilon_{xz} + C_4\epsilon_{yy} + C_5\epsilon_{yz} + C_6\epsilon_{zz} \quad (3.41)$$

The expressions for the other five stress components would involve thirty additional constants. Consequently, even for the linear assumption, thirty-six material-property constants are introduced. The assumption of an *isotropic** medium reduces this number to two since many of the coefficients are identically zero and the others are related to the two material properties needed to describe the material. (Books on continuum mechanics, using tensor calculus, present this derivation.)

For the isotropic Hookean solid the linear stress-strain relationship would similarly include two material constants, expressed in terms of the modulus of elasticity and Poisson's ratio. For fluids, the two material coefficients are the *viscosity* μ and the *second coefficient of viscosity* λ. For the Newtonian, isotropic fluid, the stress–strain-rate relations are

$$\tau_{xx} = -p + 2\mu\epsilon_{xx} + \lambda(\epsilon_{xx} + \epsilon_{yy} + \epsilon_{zz}) \qquad \tau_{xy} = 2\mu\epsilon_{xy}$$

$$\tau_{yy} = -p + 2\mu\epsilon_{yy} + \lambda(\epsilon_{xx} + \epsilon_{yy} + \epsilon_{zz}) \qquad \tau_{xz} = 2\mu\epsilon_{xz} \quad (3.42)$$

$$\tau_{zz} = -p + 2\mu\epsilon_{zz} + \lambda(\epsilon_{xx} + \epsilon_{yy} + \epsilon_{zz}) \qquad \tau_{yz} = 2\mu\epsilon_{yz}$$

---

* The condition of isotropy exists if a material property is found to be independent of direction. As an example, the viscosity of pure water is independent of direction (isotropic). On the other hand, the presence of "chains" in polymers means that polymers are anisotropic, that is, that the properties along the chain are different from the properties normal to the chain.

Note that p is pressure, which is included since we know that when the velocity field vanishes the negative of the normal stress is the pressure. Equations 3.42 are linear relationships between stresses and strain rates, involving two material properties μ and λ, which are constants for homogeneous fluids and which are functions of position for nonhomogeneous fluids. In terms of the velocity gradients, using Eq. 3.30, 3.31, and 3.32, we have

$$\tau_{xx} = -p + 2\mu \frac{\partial u}{\partial x} + \lambda \nabla \cdot \mathbf{V} \qquad \tau_{xy} = \mu\left(\frac{\partial u}{\partial y} + \frac{\partial v}{\partial x}\right)$$

$$\tau_{yy} = -p + 2\mu \frac{\partial v}{\partial y} + \lambda \nabla \cdot \mathbf{V} \qquad \tau_{xz} = \mu\left(\frac{\partial u}{\partial z} + \frac{\partial w}{\partial x}\right) \qquad (3.43)$$

$$\tau_{zz} = -p + 2\mu \frac{\partial w}{\partial z} + \lambda \nabla \cdot \mathbf{V} \qquad \tau_{yz} = \mu\left(\frac{\partial v}{\partial z} + \frac{\partial w}{\partial y}\right)$$

where we have used, as is easily verified,

$$\epsilon_{xx} + \epsilon_{yy} + \epsilon_{zz} = \nabla \cdot \mathbf{V} \qquad (3.44)$$

The sum of the three normal stresses is

$$\tau_{xx} + \tau_{yy} + \tau_{zz} = -3p + (2\mu + 3\lambda)\nabla \cdot \mathbf{V} \qquad (3.45)$$

For an incompressible fluid, where $\nabla \cdot \mathbf{V} = 0$, this relationship shows that the pressure is the negative average of the three normal stresses, that is,

$$p = -\frac{\tau_{xx} + \tau_{yy} + \tau_{zz}}{3} \qquad (3.46)$$

We also observe from the constitutive equations that the second coefficient of viscosity λ does not influence the flow of liquids. This does not, however, mean that $\lambda = 0$ for liquids. The determination of λ for liquids is subject to much controversy, but it appears that λ is positive for liquids whereas in gases it is negative. For monatomic gases the kinetic theory of gases may be used to show that $\lambda = -\frac{2}{3}\mu$. For air, which is a mixture of gases, $\lambda \cong -\frac{2}{3}\mu$; and for most applications in compressible flows, we may safely use $\lambda = -\frac{2}{3}\mu$, often referred to as *Stokes' hypothesis*. From Eq. 3.45 we again see that, in gas flows, for which $\lambda = -\frac{2}{3}\mu$, the pressure is the negative average of the three normal stresses. For an incompressible fluid and for gases in which Stokes' hypothesis is acceptable, μ is the only material property in the stress-strain rate relations.

Art. 3.4 THE CONSTITUTIVE EQUATIONS

We have now introduced six constitutive equations (Eq. 3.43) and one additional unknown, p. Hence, a tractable set of equations results for an in-compressible fluid: ten unknowns and ten equations. For a compressible fluid, ρ would also be unknown, so an equation of state involving temperature and pressure would be necessary. The energy equation (Eq. 3.28), which relates the temperature to other flow properties, would then be introduced and an additional relationship, for example, one between internal energy and temperature, $\tilde{u} = c_v T$, would be required. We would then have thirteen unknowns and thirteen equations, also a tractable set.

Example 3.4

Calculate the normal stress τ_{xx} at the stagnation point (Fig. E3.4) if, along the x-axis, $u = 30 - 120/x^2$. Assume viscous effects are negligible from the free stream, where the pressure is 5 psi, to the stagnation point. Water is flowing with $\mu = 10^{-5}$ lb-sec/ft^2.

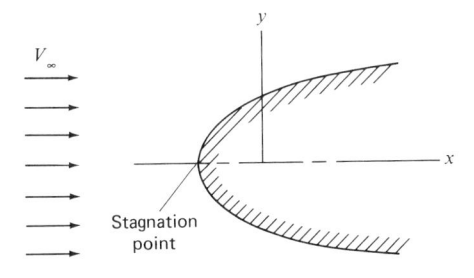

Fig. E3.4

Solution. From the constitutive equations (3.43) the normal stress τ_{xx} is

$$\tau_{xx} = -p + 2\mu \frac{\partial u}{\partial x}$$

since for water $\nabla \cdot \mathbf{V} = 0$. From the given velocity distribution along the x-axis, as $x \to -\infty$, $u \to 30$, so that $V_\infty = 30$ fps. The free-stream pressure is given as 5 psi, so that, if we neglect viscous effects from the free stream to the stagnation point, Bernoulli's equation is (note that the stagnation point is at $x = -2$ ft)

$$\frac{V_\infty^2}{2} + \frac{p_\infty}{\rho} = \frac{p_0}{\rho}$$

or

$$p_0 = 5 \times 144 + \frac{30^2}{2} \times 1.94 = 1593 \text{ psf}$$

The normal stress τ_{xx} is then

$$\tau_{xx} = -1593 + 2 \times 10^{-5}\left(\frac{240}{x^3}\right)$$

$$= -1593 - 6 \times 10^{-4}$$

$$\cong -1593 \text{ psf}$$

Thus, we see that the normal stress is very well approximated by the negative pressure, an approximation we have used previously. The viscosity is so small for most fluids that the viscous term in the normal-stress expressions (Eq. 3.43) is simply too small to be significant.

3.5 THE GOVERNING DIFFERENTIAL EQUATIONS

If the constitutive equations of the preceding article are substituted in the differential equations of motion, Eqs. 3.13, 3.14, and 3.15, the governing differential equations for an *incompressible*, homogeneous fluid become, after algebraic manipulation,

$$\frac{\partial u}{\partial x} + \frac{\partial v}{\partial y} + \frac{\partial w}{\partial z} = 0 \tag{3.47}$$

$$\rho\left[\frac{\partial u}{\partial t} + u\frac{\partial u}{\partial x} + v\frac{\partial u}{\partial y} + w\frac{\partial u}{\partial z}\right]$$
$$= -\frac{\partial p}{\partial x} + \rho g_x + \mu\left[\frac{\partial^2 u}{\partial x^2} + \frac{\partial^2 u}{\partial y^2} + \frac{\partial^2 u}{\partial z^2}\right] \tag{3.48}$$

$$\rho\left[\frac{\partial v}{\partial t} + u\frac{\partial v}{\partial x} + v\frac{\partial v}{\partial y} + w\frac{\partial v}{\partial z}\right]$$
$$= -\frac{\partial p}{\partial y} + \rho g_y + \mu\left[\frac{\partial^2 v}{\partial x^2} + \frac{\partial^2 v}{\partial y^2} + \frac{\partial^2 v}{\partial z^2}\right] \tag{3.49}$$

$$\rho\left[\frac{\partial w}{\partial t} + u\frac{\partial w}{\partial x} + v\frac{\partial w}{\partial y} + w\frac{\partial w}{\partial z}\right]$$
$$= -\frac{\partial p}{\partial z} + \rho g_z + \mu\left[\frac{\partial^2 w}{\partial x^2} + \frac{\partial^2 w}{\partial y^2} + \frac{\partial^2 w}{\partial z^2}\right] \tag{3.50}$$

where u, v, w, and p are the four unknowns in the four equations. The three momentum equations are known as the *Navier-Stokes equations*.

The differential equations may be written in vector form as (refer to Eqs. 1.37, 1.58, and 1.59)

$$\nabla \cdot \mathbf{V} = 0 \qquad (3.51)$$

$$\frac{\partial \mathbf{V}}{\partial t} + (\mathbf{V} \cdot \nabla)\mathbf{V} = -\frac{1}{\rho}\nabla p - g\nabla h + \nu \nabla^2 \mathbf{V} \qquad (3.52)$$

where ν is the kinematic viscosity, $\nu = \mu/\rho$, and we have used $\mathbf{g} = -g\nabla h$.

It is reasonable to assume that the student will have noticed the numerous detailed considerations required to arrive at these equations. The equations are for the simplest, nontrivial relationship between the stress and the rate of strain: the Newtonian, isotropic, homogeneous, incompressible fluid. The derivation is important to establish the restrictions on the equations, and the details of the velocity gradients are important to recognize the character of the constitutive equations.

The primary utility of the equations is that they provide, in principle, a means to determine the velocity and the pressure fields *without recourse to experimental evaluation*. However, this utilization is realized only if we can execute their solution for the problem of interest. If we could determine the details of the flow field, $\mathbf{V}(x, y, z, t)$ and $p(x, y, z, t)$, then the question of the separation point for the airfoil or the amount of flow in each branch of the ducting system would be answered without recourse to experiment. The contribution to the engineering task in terms of the ability to obtain optimum configurations is immediately obvious. The design of streamlined bodies and lubrication systems are two examples in which this situation is realized to a considerable degree; however, it is not as simple as the foregoing presentation would make it seem.

The equations are nonlinear (e.g., $u\, \partial u/\partial x$), and this greatly complicates their solution. The nonlinearity also means that the solution may not be unique. An apparently steady-flow situation may involve significant unsteady effects, known as turbulence, for which the character of the flow field is radically altered. The flow may not follow the solid contours of interest (i.e., separation may occur). Exact solutions can be obtained in certain restrictive situations involving simple geometries or steady flow, or both; some of these will be considered in a later chapter. Other flows are of such a nature as to allow ordinary differential equations to be formed.

In fluid mechanics, boundary conditions are given on velocity or pressure, yielding a boundary-value problem which is well posed. If boundary conditions were partly on stress, as is the case in solid mechanics, a mixed set of boundary conditions would occur, resulting in an extremely complicated boundary-value problem. This would also lead to the necessity of compatibility relationships.

Boundary conditions and initial conditions on the time-dependent variables are, of course, extremely important in the solution of differential equations. It is in this context that we can best understand the difference between *laminar* and *turbulent* flow. *Laminar flow* has the following primary characteristic: the effects of viscosity, realized through the viscous terms in the Navier-Stokes equations, are significant with respect to the other quantities (pressure gradient, gravity) influencing the region of flow. Many laminar flows are well behaved, that is, the time plot of a velocity component (or pressure) would be a smooth curve, as shown in Fig. 3.4a. A steady flow ($\partial/\partial t = 0$) is an example of a well-behaved function of time; hence, a steady flow for which the viscous terms are nonzero is laminar. An unsteady laminar flow may be an erratic function of time; the flow associated with the stirring of a mixture of heavy oils is an example of this.

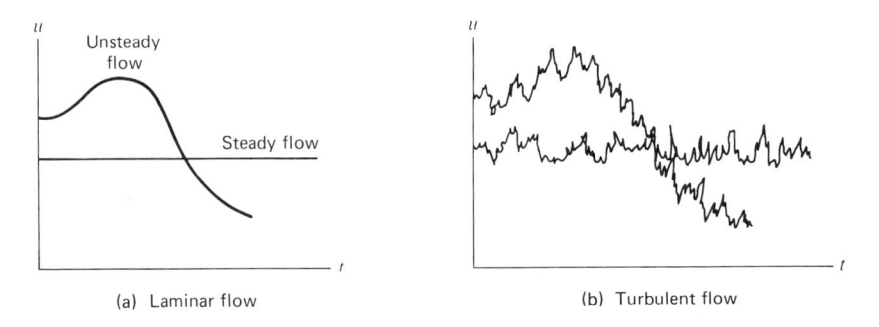

Fig. 3.4 Examples of laminar and turbulent flows for the x-component of velocity.

Turbulent flow displays this important characteristic: its velocity field, with all three components nonzero, is a fluctuating, erratic function of time. There are other characteristics of turbulent flow, which will be presented in a later chapter; now we wish only to note its erratic unsteadiness. We may never neglect the $\partial/\partial t$ terms in the governing equations for turbulent flow. Consequently, we need an initial condition, that is, we must prescribe the velocity at all points in the region of flow at some time, say $t = 0$, as well as the boundary conditions, the values of the velocity on the boundary enclosing the region. To state the initial condition is, of course, impossible even for very simple turbulent flows, for example, that of fully developed pipe flow. Consequently, we will not solve the differential equations (3.47–3.50) for turbulent flow. It is important to note, however, that the equations are applicable for both laminar and turbulent flow. With some general assumptions, it is possible to extract some information from the equations for turbulent flow; this will be done later.

It is important to recognize that the erratic fluctuations of the velocity and pressure values at all points in a region are a necessary but not a sufficient condition for the region of flow to be turbulent. Laminar flows, as noted above, may have this character. Similarly, a *free-stream flow* in which the velocity field of interest is negligibly influenced by shear effects (i.e., dominated by inertial and pressure effects) may also fluctuate erratically. A flow around a body is a free-stream flow away from the body (and outside the wake region) and a laminar or a turbulent flow near the body. Erratic fluctuations in the free stream are called *free-stream fluctuations* or, alternately, free-stream turbulence.

To summarize, we have assumed a linear relationship between the stress tensor and the rate-of-strain tensor. By further assuming that the fluid is isotropic, two material constants μ and λ are introduced. Only the viscosity μ is of importance in incompressible flows and most compressible flows (for which Stokes' hypothesis is valid). Obviously, many arbitrary steps have occurred, starting with the assumption that the stress and rate-of-strain field can be related. The final arbiter of the value of these assumptions is the degree to which the final results are useful in describing the behavior of real fluids. Fortunately, many fluids (e.g., air, oil, and water) can be found for which the assumptions are valid. For our purposes, the validity of these equations is assumed for all fluids unless otherwise specified. They are statements of the conservation of mass; relationships between force and the time rate of change of momentum; and heat, work, and energy. These basic principles are fundamentally established for systems, but we have used mathematical methods to express them as Eulerian equations for the distributions of the velocity, pressure, and temperature fields.

In concluding this article we will note the equations which result for a compressible, homogeneous fluid, equations which would be necessary if velocities exceeded 30 percent of the speed of sound. If we substitute the constitutive equations in the momentum equations and the energy equation, the following set of equations governs the flow of a perfect gas:

Continuity:
$$\frac{D\rho}{Dt} + \rho \nabla \cdot \mathbf{V} = 0 \tag{3.53}$$

Momentum:
$$\rho \frac{D\mathbf{V}}{Dt} = -\nabla p - \rho g \nabla h + \mu \nabla^2 \mathbf{V} + \frac{\mu}{3} \nabla(\nabla \cdot \mathbf{V}) \tag{3.54}$$

Energy:
$$\rho c_v \frac{DT}{Dt} = k \nabla^2 T - p \nabla \cdot \mathbf{V} + \Phi \tag{3.55}$$

State:
$$p = \rho R T \tag{3.56}$$

We have assumed $D\tilde{u}/Dt = c_v(DT/Dt)$ and $\lambda = -\frac{2}{3}\mu$. In the energy equation the *dissipation function* Φ is introduced, given in cartesian coordinates by

$$\Phi = 2\mu \left[\left(\frac{\partial u}{\partial x}\right)^2 + \left(\frac{\partial v}{\partial y}\right)^2 + \left(\frac{\partial w}{\partial z}\right)^2 \right.$$
$$\left. + \frac{1}{2}\left(\frac{\partial u}{\partial y} + \frac{\partial v}{\partial x}\right)^2 + \frac{1}{2}\left(\frac{\partial v}{\partial z} + \frac{\partial w}{\partial y}\right)^2 + \frac{1}{2}\left(\frac{\partial u}{\partial z} + \frac{\partial w}{\partial x}\right)^2 \right] \tag{3.57}$$

For a compressible flow the six unknowns u, v, w; p, ρ, T are contained in the six component equations included in the above.

We have now completed the development of both the integral and differential equations of fluid mechanics. We have applied the integral equations to typical problems encountered in various fields of engineering to express macroscopic or gross quantities such as pump horsepower, total thrust of a rocket engine, or total flow rate through a conduit. Although the differential equations express the same fundamental principles, their application is often directed toward a different objective, namely, to provide a continuous solution for the velocity and pressure distribution throughout a region of interest, for example, $\mathbf{V}(x, y, z, t)$. Such information is often necessary in the solution of engineering problems. The integral equations require $\mathbf{V}(x, y, z, t)$, $p(x, y, z, t)$, and so on, and it is not always satisfactory to approximate such information. The analytical prediction of separation phenomena (where the flow leaves a surface), determining regions of low pressure in which vapor bubbles (i.e., cavitation) may form, and many other problems require solutions of the differential equations. If we could simply solve the differential equations our job would be completed; however, because the equations are so complex and because there are so many different boundary conditions which can be imposed (various geometries, injection of fluids through boundaries, heating of boundaries, etc.) we must continually solve new problems. The later chapters of this text are devoted to providing the student with the problem-solving tools that are characterized by a thorough grasp of these differential equations and their various forms.

Example 3.5

A steady laminar flow exists in a conduit, with all streamlines parallel to the walls. Determine the simplified differential equations which govern the flow.

Solution. If all streamlines are parallel to the wall (and in the x-direction) then $u \neq 0$, and $v = w = 0$. The equation of continuity

$$\frac{\partial u}{\partial x} + \cancelto{0}{\frac{\partial v}{\partial y}} + \cancelto{0}{\frac{\partial w}{\partial z}} = 0$$

shows that $u \neq u(x)$; that is, $u = u(y, z)$ where y and z are normal to the walls. The velocity profile is independent of x. The momentum equations (3.48, 3.49, and 3.50) take the forms

$$0 = -\frac{\partial p}{\partial x} + \rho g_x + \mu \left(\frac{\partial^2 u}{\partial y^2} + \frac{\partial^2 u}{\partial z^2} \right)$$

$$0 = -\frac{\partial p}{\partial y} + \rho g_y$$

$$0 = -\frac{\partial p}{\partial z} + \rho g_z$$

Consider the conduit to be horizontal so that $g_x = 0$, and neglect the pressure variation normal to the conduit (this is permissible if the conduit is small); then we would have

$$\frac{dp}{dx} = \mu \left(\frac{\partial^2 u}{\partial y^2} + \frac{\partial^2 u}{\partial z^2} \right)$$

With the appropriate boundary conditions this can be solved.

3.6 BOUNDARY CONDITIONS

In the preceding articles we have referred to the boundary conditions which accompany the various differential equations. In this article the boundary conditions will be discussed more specifically. We will illustrate the various conditions that occur by considering flow past a cylinder, shown in Fig. 3.5.

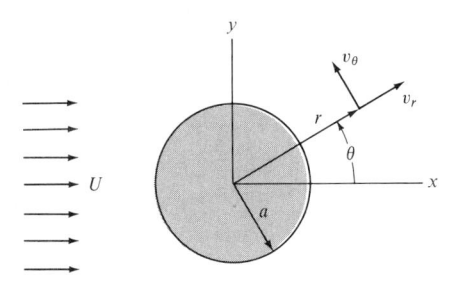

Fig. 3.5. Flow past a cylinder.

The most common boundary condition that occurs for viscous fluid flows is the *no-slip condition*.* The no-slip condition requires that the fluid velocity at each boundary point be the same as the velocity of the boundary at the point. For a stationary boundary the fluid velocity at the boundary would be zero. For the cylinder this would be

$$v_r = v_\theta = 0 \quad \text{at} \quad r = a \quad (3.58)$$

Far from the cylinder we would require that the velocity approach some prescribed value; for a uniform flow we would have

$$v_r = U \cos \theta \quad v_\theta = -U \sin \theta \quad \text{at} \quad r = \infty \quad (3.59)$$

Often, it is possible to completely neglect the viscous term $\nu \nabla^2 \mathbf{V}$ in the differential momentum equation and obtain an approximate solution to the problem for the region of the flow field not influenced by viscous shear. This is often valid for flows around bodies, especially streamlined bodies such as airfoils. This neglect of the viscous terms reduces the order of the differential equations (see Eqs. 3.48–3.50), and thus the boundary conditions must be relaxed. (The number of

---

* This is a most difficult effect to observe with a typical velocity-measuring instrument such as a pitot probe or a hot-wire anemometer. For an enlightening discussion of the no-slip condition see Art. 1.3.2 of *Laminar Boundary Layers*, edited by L. Rosenhead, Oxford University Press, London (1963).

boundary conditions must be reduced.) For this *inviscid flow* the fluid velocity *normal* to a boundary point is set equal to the normal component of the velocity of the boundary. This is a no-penetration condition. For the cylinder we would have

$$v_r = 0 \quad \text{at} \quad r = a \tag{3.60}$$

and the boundary condition at $r = \infty$ would remain the same as for the viscous-flow problem.

Another type of boundary condition occurs at a porous surface where we control the velocity at the surface, usually the normal component only. For a uniform suction on the cylinder in a viscous flow, the boundary conditions at the cylinder would be

$$v_r = -K(\theta) \quad v_\theta = 0 \quad \text{at} \quad r = a \tag{3.61}$$

where K is determined by the magnitude of the suction provided.

A pressure boundary condition is encountered whenever an interface between two fluids occurs; this includes the *free surface*, a liquid-gas interface. On a free surface, the pressure has a prescribed value, for example, for surface waves the gage pressure is zero at the free surface. An interface separates two different liquids flowing in contact with each other. Across the interface the pressure in each liquid is the same, except when surface tension is important; then the pressure difference is related to the surface tension. The difficulty in prescribing the pressure on an interface is that the location of the interface is not known; the interface location is included as part of the solution and this makes problems involving interfaces quite difficult to solve. In addition to the pressure condition, the component of velocity normal to the interface is the same in both fluids.

In flows involving thermal effects the energy equation (3.55) requires that boundary conditions on the temperature be stated. One condition requires that the temperature of the fluid at the boundary must equal the temperature of the boundary. The heat-transfer rate through the boundary prescribes the temperature gradient at the boundary since the heat-transfer rate for a Fourier-Biot material is equal to $-k(\partial T/\partial n)$, where n is normal to the surface. If the cylinder in the uniform flow were insulated (no heat transfer) and held at constant temperature T_b, the thermal boundary conditions would be

$$T = T_b \quad \frac{\partial T}{\partial r} = 0 \quad \text{at} \quad r = a \tag{3.62}$$

The condition far away from the body would also need to be prescribed, for example,

$$T = T_\infty \quad \text{at} \quad r = \infty \tag{3.63}$$

With the differential equations derived and with the boundary conditions stated, various phenomena for which the differential analysis is required may now be investigated.

TABLE 3.1
Fundamental Laws for Incompressible Flows

Continuity

CARTESIAN

$$\frac{\partial u}{\partial x} + \frac{\partial v}{\partial y} + \frac{\partial w}{\partial z} = 0$$

CYLINDRICAL

$$\frac{1}{r}\frac{\partial}{\partial r}(rv_r) + \frac{1}{r}\frac{\partial v_\theta}{\partial \theta} + \frac{\partial v_z}{\partial z} = 0$$

SPHERICAL

$$\frac{1}{r^2}\frac{\partial}{\partial r}(r^2 v_r) + \frac{1}{r \sin \theta}\frac{\partial}{\partial \theta}(v_\theta \sin \theta) + \frac{1}{r \sin \theta}\frac{\partial v_\phi}{\partial \phi} = 0$$

Momentum

CARTESIAN

$$\frac{Du}{Dt} = -\frac{1}{\rho}\frac{\partial p}{\partial x} + g_x + \nu \nabla^2 u$$

$$\frac{Dv}{Dt} = -\frac{1}{\rho}\frac{\partial p}{\partial y} + g_y + \nu \nabla^2 v$$

$$\frac{Dw}{Dt} = -\frac{1}{\rho}\frac{\partial p}{\partial z} + g_z + \nu \nabla^2 w$$

$$\frac{D}{Dt} = \frac{\partial}{\partial t} + u\frac{\partial}{\partial x} + v\frac{\partial}{\partial y} + w\frac{\partial}{\partial z}$$

$$\nabla^2 = \frac{\partial^2}{\partial x^2} + \frac{\partial^2}{\partial y^2} + \frac{\partial^2}{\partial z^2}$$

Table 3.1 (Continued)

CYLINDRICAL

$$\frac{Dv_r}{Dt} - \frac{v_\theta^2}{r} = -\frac{1}{\rho}\frac{\partial p}{\partial r} + g_r + \nu\left[\nabla^2 v_r - \frac{v_r}{r^2} - \frac{2}{r^2}\frac{\partial v_\theta}{\partial \theta}\right]$$

$$\frac{Dv_\theta}{Dt} + \frac{v_r v_\theta}{r} = -\frac{1}{\rho r}\frac{\partial p}{\partial \theta} + g_\theta + \nu\left[\nabla^2 v_\theta + \frac{2}{r^2}\frac{\partial v_r}{\partial \theta} - \frac{v_\theta}{r^2}\right]$$

$$\frac{Dv_z}{Dt} = -\frac{1}{\rho}\frac{\partial p}{\partial z} + g_z + \nu\nabla^2 v_z$$

$$\frac{D}{Dt} = \frac{\partial}{\partial t} + v_r\frac{\partial}{\partial r} + \frac{v_\theta}{r}\frac{\partial}{\partial \theta} + v_z\frac{\partial}{\partial z}$$

$$\nabla^2 = \frac{\partial^2}{\partial r^2} + \frac{1}{r}\frac{\partial}{\partial r} + \frac{1}{r^2}\frac{\partial^2}{\partial \theta^2} + \frac{\partial^2}{\partial z^2}$$

SPHERICAL

$$\frac{Dv_r}{Dt} - \frac{v_\theta^2 + v_\phi^2}{r}$$
$$= -\frac{1}{\rho}\frac{\partial p}{\partial r} + g_r + \nu\left[\nabla^2 v_r - \frac{2v_r}{r^2} - \frac{2}{r^2}\frac{\partial v_\theta}{\partial \theta} - \frac{2v_\theta \cot\theta}{r^2} - \frac{2}{r^2 \sin\theta}\frac{\partial v_\phi}{\partial \phi}\right]$$

$$\frac{Dv_\theta}{Dt} + \frac{v_r v_\theta - v_\phi^2 \cot\theta}{r}$$
$$= -\frac{1}{\rho r}\frac{\partial p}{\partial \theta} + g_\theta + \nu\left[\nabla^2 v_\theta + \frac{2}{r^2}\frac{\partial v_r}{\partial \theta} - \frac{v_\theta}{r^2 \sin^2\theta} - \frac{2\cos\theta}{r^2 \sin^2\theta}\frac{\partial v_\phi}{\partial \phi}\right]$$

$$\frac{Dv_\phi}{Dt} + \frac{v_\phi v_r}{r} + \frac{v_\theta v_\phi \cot\theta}{r}$$
$$= -\frac{1}{\rho r \sin\theta}\frac{\partial p}{\partial \phi} + g_\phi + \nu\left[\nabla^2 v_\phi - \frac{v_\phi}{r^2 \sin^2\theta} + \frac{2}{r^2 \sin^2\theta}\frac{\partial v_r}{\partial \phi} + \frac{2\cos\theta}{r^2 \sin^2\theta}\frac{\partial v_\theta}{\partial \phi}\right]$$

TABLE 3.1 (Continued)

Momentum (Continued)

SPHERICAL (Continued)

$$\frac{D}{Dt} = \frac{\partial}{\partial t} + v_r \frac{\partial}{\partial r} + \frac{v_\theta}{r} \frac{\partial}{\partial \theta} + \frac{v_\phi}{r \sin \theta} \frac{\partial}{\partial \phi}$$

$$\nabla^2 = \frac{1}{r^2} \frac{\partial}{\partial r}\left(r^2 \frac{\partial}{\partial r}\right) + \frac{1}{r^2 \sin \theta} \frac{\partial}{\partial \theta}\left(\sin \theta \frac{\partial}{\partial \theta}\right) + \frac{1}{r^2 \sin^2 \theta} \frac{\partial^2}{\partial \phi^2}$$

Energy

CARTESIAN

$$\rho \frac{D\tilde{u}}{Dt} = k \nabla^2 T + 2\mu \left[\left(\frac{\partial u}{\partial x}\right)^2 + \left(\frac{\partial v}{\partial y}\right)^2 + \left(\frac{\partial w}{\partial z}\right)^2 + \frac{1}{2}\left(\frac{\partial u}{\partial y} + \frac{\partial v}{\partial x}\right)^2 \right.$$

$$\left. + \frac{1}{2}\left(\frac{\partial v}{\partial z} + \frac{\partial w}{\partial y}\right)^2 + \frac{1}{2}\left(\frac{\partial u}{\partial z} + \frac{\partial w}{\partial x}\right)^2 \right]$$

CYLINDRICAL

$$\rho \frac{D\tilde{u}}{Dt} = k \nabla^2 T + 2\mu \left[\left(\frac{\partial v_r}{\partial r}\right)^2 + \left(\frac{1}{r} \frac{\partial v_\theta}{\partial \theta} + \frac{v_r}{r}\right)^2 + \left(\frac{\partial v_z}{\partial z}\right)^2 \right.$$

$$\left. + \frac{1}{2}\left(\frac{1}{r} \frac{\partial v_z}{\partial \theta} + \frac{\partial v_\theta}{\partial z}\right)^2 + \frac{1}{2}\left(\frac{\partial v_r}{\partial z} + \frac{\partial v_z}{\partial r}\right)^2 + \frac{1}{2}\left(\frac{1}{r} \frac{\partial v_r}{\partial \theta} + \frac{\partial v_\theta}{\partial r} - \frac{v_\theta}{r}\right)^2 \right]$$

SPHERICAL

$$\rho \frac{D\tilde{u}}{Dt} = k \nabla^2 T + 2\mu \left[\left(\frac{\partial v_r}{\partial r}\right)^2 + \left(\frac{1}{r} \frac{\partial v_\theta}{\partial \theta} + \frac{v_r}{r}\right)^2 + \left(\frac{1}{r \sin \theta} \frac{\partial v_\phi}{\partial \phi} + \frac{v_r}{r} + \frac{v_\theta \cot \theta}{r}\right)^2 \right]$$

$$+ \mu \left[\left(\frac{1}{r \sin \theta} \frac{\partial v_\theta}{\partial \phi} + \frac{\sin \theta}{r} \frac{\partial}{\partial \theta}\left(\frac{v_\phi}{\sin \theta}\right)\right)^2 \right.$$

$$\left. + \left(\frac{1}{r \sin \theta} \frac{\partial v_r}{\partial \phi} + r \frac{\partial}{\partial r}\left(\frac{v_\phi}{r}\right)\right)^2 + \left(r \frac{\partial}{\partial r}\left(\frac{v_\theta}{r}\right) + \frac{1}{r} \frac{\partial v_r}{\partial \theta}\right)^2 \right]$$

TABLE 3.1 (Continued)

Stresses

CARTESIAN

$$\tau_{xx} = -p + 2\mu \frac{\partial u}{\partial x} \qquad \tau_{xy} = \mu\left(\frac{\partial u}{\partial y} + \frac{\partial v}{\partial x}\right)$$

$$\tau_{yy} = -p + 2\mu \frac{\partial v}{\partial y} \qquad \tau_{yz} = \mu\left(\frac{\partial v}{\partial z} + \frac{\partial w}{\partial y}\right)$$

$$\tau_{zz} = -p + 2\mu \frac{\partial w}{\partial z} \qquad \tau_{xz} = \mu\left(\frac{\partial u}{\partial z} + \frac{\partial w}{\partial x}\right)$$

CYLINDRICAL

$$\tau_{rr} = -p + 2\mu \frac{\partial v_r}{\partial r} \qquad \tau_{r\theta} = \mu\left[r\frac{\partial}{\partial r}\left(\frac{v_\theta}{r}\right) + \frac{1}{r}\frac{\partial v_r}{\partial \theta}\right]$$

$$\tau_{\theta\theta} = -p + 2\mu\left(\frac{1}{r}\frac{\partial v_\theta}{\partial \theta} + \frac{v_r}{r}\right) \qquad \tau_{\theta z} = \mu\left[\frac{\partial v_\theta}{\partial z} + \frac{1}{r}\frac{\partial v_z}{\partial \theta}\right]$$

$$\tau_{zz} = -p + 2\mu \frac{\partial v_z}{\partial z} \qquad \tau_{rz} = \mu\left[\frac{\partial v_r}{\partial z} + \frac{\partial v_z}{\partial r}\right]$$

PROBLEMS

3.1. State the boundary conditions for a viscous flow (a) between parallel plates, (b) on top of a harmonically oscillating plate, and (c) around a cylinder. For an inviscid flow, state the boundary conditions (d) around a cylinder, and (e) for honey flowing from a hole.

3.2. Express the general integral forms of $\dot{W}_s$, $\dot{Q}$, and $\Sigma \mathbf{M}$ which appear in the integral equations.

3.3 Starting with a small fluid element of volume $dx\,dy\,dz$, derive the continuity equation (Eq. 3.4) in rectangular cartesian coordinates.

3.4. Show that Eq. 3.6 follows from Eq. 3.5 and verify that, for an isothermal flow of air, $-(1/p)(Dp/Dt) = \nabla \cdot \mathbf{V}$.

3.5. Velocity measurements in the channel shown are made along the axis at points A, B, C, and D, spaced 1 in. apart. The velocities measured are 50, 56, 66, and 80 ft/sec, respectively. Estimate the y-component of velocity v at a point located 1/4 in. above C.

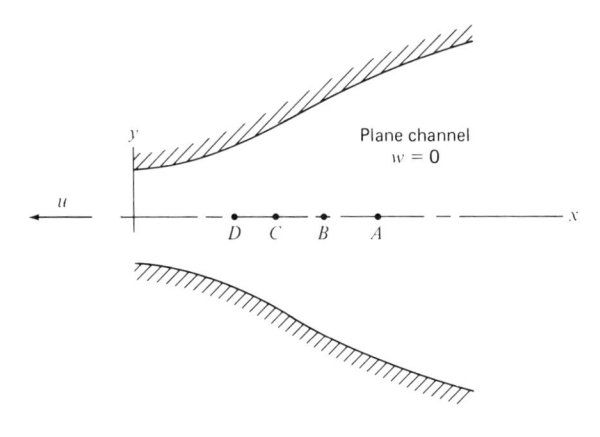

Prob. 3.5

3.6. Determine the y-component of velocity of an incompressible plane flow if the x-component is given by $u = 10 - 2xy$. Along the x-axis, $v = 0$.

3.7. An incompressible fluid flows so that

$$v_\theta = 100\left(1 + \frac{4}{r^2}\right)\sin\theta - \frac{200}{r}$$

Determine v_r if v_r is zero at $r = 2$ for all θ. The flow does not depend on z.

3.8. The velocity in a one-dimensional compressible pipe flow is given by $u = c_1(1 - e^{-c_2 x})$, where c_1 and c_2 are positive constants. Determine the expression for the density.

3.9. An incompressible fluid is flowing in a region in which $u = 10(x^2 - y^2)$ and $v(1, 0) = 0$. The density varies in the region and $\partial\rho/\partial y = -0.2$ at a particular point $(1, -2)$. Find $\partial\rho/\partial x$ at the point. Assume $w = 0$ everywhere.

3.10. Write the simplified differential momentum equations for a two-dimensional, plane, steady, horizontal flow ($w = 0$). Also, verify that $\mathbf{g} = -g\nabla h$, where h is measured vertically upward by assuming the xy-plane to be horizontal.

3.11. In the development of the momentum equations, the summation of the infinitesimal forces could have been equated to

$$\int_{c.v.} \frac{\partial(\rho\mathbf{V})}{\partial t} dV + \int_{c.v.} \rho\mathbf{V}(\mathbf{V}\cdot\hat{n}) dA$$

for the infinitesimal element. Assume only a two-dimensional flow with $w = 0$ for simplicity, and show that the right-hand sides of Eqs. 3.13 and 3.14, with $w = 0$, would result by using the above integrals. The momentum flux quantities which contribute to $\int_{c.s.} \rho\mathbf{V}(\mathbf{V}\cdot\hat{n}) dA$ are shown on the sketch.

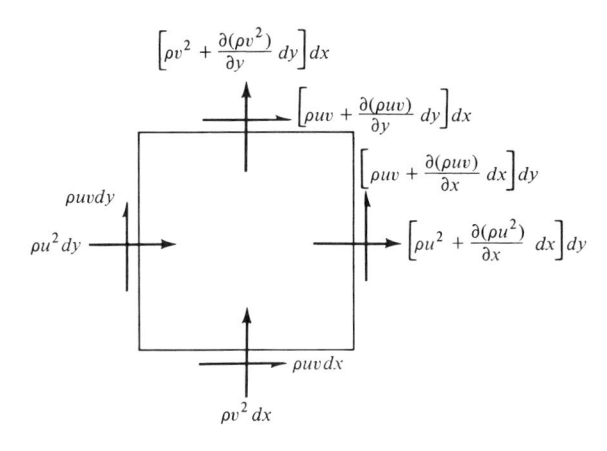

Prob. 3.11

$$\left[\rho v^2 + \frac{\partial(\rho v^2)}{\partial y} dy\right] dx$$

$$\left[\rho uv + \frac{\partial(\rho uv)}{\partial y} dy\right] dx$$

$$\left[\rho uv + \frac{\partial(\rho uv)}{\partial x} dx\right] dy$$

$$\rho uv\, dy$$

$$\rho u^2\, dy$$

$$\left[\rho u^2 + \frac{\partial(\rho u^2)}{\partial x} dx\right] dy$$

$$\rho uv\, dx$$

$$\rho v^2\, dx$$

3.12. Show that if the velocity field is everywhere zero, the normal stresses at a point are all equal. Use an infinitesimal triangular element and sum forces. The shearing stresses vanish as the velocity field vanishes.

3.13. For a compressible flow, the differential energy equation, with zero shear, may be written as

$$\rho \frac{D\tilde{u}}{Dt} = k \nabla^2 T - p\left(\frac{\partial u}{\partial x} + \frac{\partial v}{\partial y} + \frac{\partial w}{\partial z}\right)$$

Show that this can also be put in the form ($h = p/\rho + \tilde{u}$)

$$\rho \frac{Dh}{Dt} = k \nabla^2 T + \frac{Dp}{Dt}$$

3.14. Derive the differential energy equation (Eq. 3.27) for a non-viscous, incompressible fluid, starting with an infinitesimal element instead of using Gauss' theorem. Assume that the normal stresses are negative the pressure. Explain why this approach would be necessary in deriving the energy equation if the shearing stresses were included.

3.15. Determine the angular velocity of a cork in the flow of water between two parallel walls if $u(y) = 4(y - y^2)$ fps with y measured normal to a wall. The cork is placed $\frac{1}{4}$ ft from the wall.

3.16. Substitute the constitutive equations in the equations of motion (the three momentum equations) and derive the governing equations for compressible flow, using Stokes' hypothesis. Then show that Eqs. 3.48, 3.49, and 3.50 result for an incompressible flow. Assume homogeneity throughout.

3.17. For a gas flow, where compressibility effects are important, density is not constant. Let the negative average of the three normal stresses be designated by $\bar{p}$ and let the pressure in Eq. 3.45 be the thermodynamic pressure given by an equation of state. Determine an expression for $(p - \bar{p})$ in terms of the two coefficients of viscosity and the density, and observe that in general $p - \bar{p} \neq 0$ except when $\lambda = -\frac{2}{3}\mu$, by Stokes' hypothesis.

3.18. (a) An unsteady laminar flow exists in a conduit so that the streamlines are all parallel to the walls (blood flow through an artery, for example). Sketch $u(t)$ and indicate a typical path of a particle. Now consider the flow to be turbulent; sketch $u(t)$ and show a typical path of a particle. (b) A laminar flow exists between rotating cylinders and a certain velocity field is measured. The cylinders are stopped and then brought back up to the same speed; now the laminar flow is different from that previously measured even though the equations and boundary conditions are identical. How can this be? Explain by referring to the equations.

3.19. A temperature gradient occurs in a liquid flow (the motion of oil in a bearing) so that $\mu = \mu(T)$. This is a nonhomogeneous flow. How would the differential momentum equations change to account for $\mu(T)$?

3.20. Water flows between two parallel plates for which the width w is much greater than the spacing h. Determine the simplified set of differential equations governing the laminar flow and write the necessary boundary conditions if at the entrance the velocity is uniform as shown and at 15 ft downstream, where the streamlines are again parallel to the wall, the velocity profile is $u = 4U_{max}[(y/h) - (y^2/h^2)]$. Also determine U_{max}.

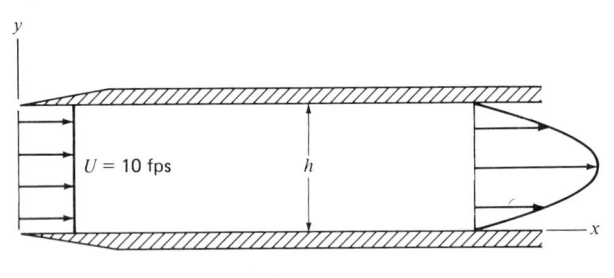

Prob. 3.20

3.21. Write the governing differential equation for laminar flow and state the boundary conditions if the figure represents (a) long concentric cylinders and (b) concentric spheres.

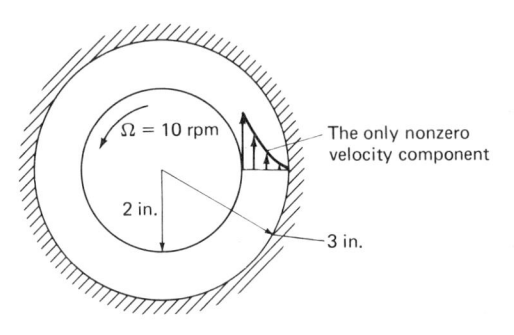

Prob. 3.21

PROBLEMS

3.22. A large flat plate oscillates beneath a fluid as shown. Write the governing differential equations and boundary conditions if the laminar flow moves in laminae parallel to the plate. The plate is heated and insulated.

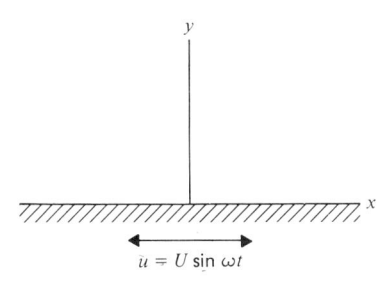

$u = U \sin \omega t$

Prob. 3.22

3.23. A laminar flow over a step is as shown. It is observed that the flow near the lower surface separates ahead of the step, runs into the step, and then proceeds downstream. At point B the fluid leaves the surface, so $\partial u/\partial y|_{y=0} = 0$. Respond to the following: (a) Consider a point a distance $\delta y (\delta y \ll h)$ above point B. Estimate the value of u at $(x_B, \delta y)$. (*Hint*: Use Taylor's series.) Is $\partial u/\partial x$ at $(x_B, \delta y) > 0$? (b) Estimate the magnitude of v at $(x_B, \delta y)$ if $\partial u/\partial x$ is assumed known. (c) If $u(y)$ along AB is known, estimate $\partial p/\partial x$ at $(x_B, \delta y)$. Is $\partial p/\partial x > 0$? (d) In the vicinity of point A, is $\partial p/\partial x > 0? < 0? = 0$? (e) Estimate the value of τ_{xy} at $(x_B, \delta y)$. At $(x_B, \delta y)$ is $\partial \tau_{xy}/\partial y > 0? < 0? = 0$? Near point A, is $\partial \tau_{xy}/\partial y > 0? < 0? = 0$?

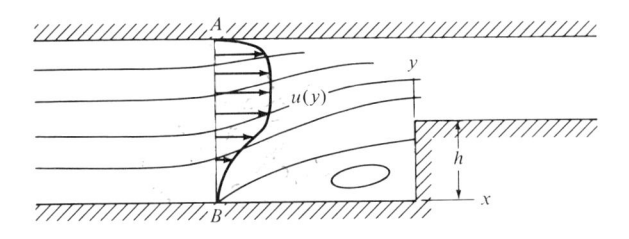

Prob. 3.23

3.24. For the wake of a circular cylinder, as shown in the sketch: (a) Plot the velocity along the center line. (b) For the x-location shown, plot the magnitude of v as a function of y. (c) From experimental data, it is known that along the centerline the longitudinal velocity gradient is numerically equal to $|\Delta u/\Delta x| = 35$ sec^{-1}. Estimate the magnitude of v at $y = -0.12$ inches below the wake centerline. (d) A particle is shown at x_0. Show the location and shape of the particle at x_2. (e) Consider a particle at $y = y_1$. Is the normal strain rate ϵ_{xx} greater than, less than, or equal to zero? What of ϵ_{yy} at y_1? (f) Determine the sign of $\partial^2 u/\partial y^2$, at $y = 0$ and at y_1.

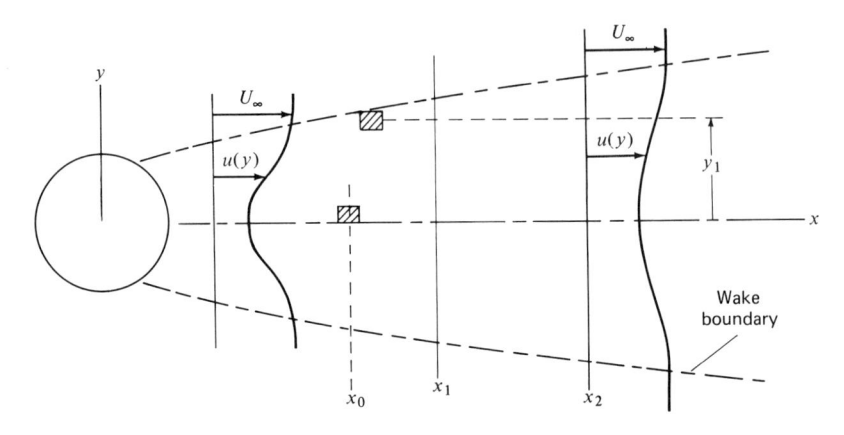

Prob. 3.24

SELECTED REFERENCES

The motion picture *Deformation of a Continuous Media* ((No. 21608); J. L. Lumley, film principal) may serve as a helpful supplement to this chapter.

The approach taken in developing the field equations is a key to the sophistication required of readers in our subject. Four texts which approximate our own level of sophistication are these: J. W. Daily and D. R. F. Harleman, *Fluid Dynamics*, Addison-Wesley Publishing Company, Reading, Mass., 1968; Y. S. Yuan, *Foundations of Fluid Mechanics*, Prentice-Hall, Englewood Cliffs, N.J., 1967; R. H. Sabersky, A. J. Acosta, and E. G. Hauptman, *Fluid Flow*, 2nd ed., Macmillan Co., New York, 1971. A somewhat more complete presentation of the second coefficient of viscosity and the details of the overall derivation is that of H. Schlichting, *Boundary Layer Theory*, 6th ed., McGraw-Hill Book Co., New York, 1968.

Familiarity with tensor notation permits a more generalized derivation. Such derivations are provided in the following: S. Whitaker, *Introduction to Fluid Mechanics*, Prentice-Hall, Englewood Cliffs, N. J., 1968; G. K. Batchelor, *An Introduction to Fluid Dynamics*, Cambridge University Press, London, 1967; J. A. Owczarek, *Introduction to Fluid Mechanics*, International Textbook Co., Scranton, Pa., 1968.

4

Similitude

4.1 INTRODUCTION

The principles of mass conservation, Newton's second law, and the first law of thermodynamics have been developed to describe the behavior of the physical world about us. Because of their apparently universal validity, we sometimes say that these principles "govern" the behavior of the material universe. In this sense, the control-volume equations of Chapter 2 and the field equations of Chapter 3 are simply different ways to express the laws that govern fluid motions. Since the basic engineering tasks are design and prediction, these equations represent the basic tools for the engineer involved with fluid-mechanical problems. If it were possible to solve the full set of equations for every configuration and condition of interest, the engineering task would be quite straightforward (if somewhat unchallenging). In fact, the opposite situation more often prevails. The governing equations are so complex that very few problems of engineering relevance can be solved directly. However—and this is the central and quite important point of this chapter—the governing equations are of great importance to the engineering task even if a solution cannot be found for the problem under consideration. The nature of this important contribution is best presented in the context of a particular example.

4.2 SIMILITUDE

Flowing fluids are often used to transfer thermal energy; an automobile radiator, a gas stove, or a cooling breeze are examples from common experience. High-temperature devices such as the combustion chamber of an internal combustion engine or the first stages of a gas turbine are cases where cooling is necessary or the device will fail mechanically, owing to excessive stresses in the high-temperature condition. Some engine designs use the liquid fuel as an energy-absorbing agent by passing it through internal passages; this not only cools the structural material but preheats the fuel for better thermal efficiency. It is not difficult to realize that the design of these passages is of great importance: the total rate of cooling fuel flow must not only be sufficient but the internal flow patterns must be such that no local hot-spots can develop which would cause an explosion hazard.

In the design of such a device, a model-study phase is clearly appropriate to evaluate the design of the critical regions. Because fuel passages are often small, a larger-scale model will allow a more detailed study. Since the explosion hazard is always present when a volatile substance is the working fluid it would be desirable to use a different fluid for the laboratory study. However, if a different fluid and a different size are to be investigated, then how can the results of this different study be related to the problem of concern? In this chapter we we will consider how to treat this—and any other model or simulation study—in order to extract the most general and most reliable information.

The internal passages for the flow of the cooling fuel represent the geometric boundaries for the solution of the "boundary-value problem" for the velocity $\mathbf{V}(x, y, z, t)$ and pressure $p(x, y, z, t)$ distributions inside the passages. Basically, the strategy to make the model accurately reflect the conditions of the actual device is quite simple: *we will force the boundary-value problem for the model to be exactly* (or as close as we can make it) *the same as that for the prototype* (actual device). *Then, identical solutions must result from identical equations, with identical boundary and initial conditions.* But how is this possible if the model is larger and uses a different fluid? It will be shown that the dimensionless (or normalized) problem can be made identical even though the dimensional variables of the problem differ between prototype and model. Since this problem deals with a liquid and since the flow of a confined[†] incompressible, isotropic, Newtonian fluid is the simplest case we will consider, the derivations below will cover all such examples. Extensions to more complex cases will be treated following this example.

We will begin the normalization by geometrically scaling (dividing) all the lengths in the problem by a characteristic length. "Characteristic length," as the term implies, is the length which best represents the problem. In the present example we could use the hydraulic diameter d_H as the characteristic length.[‡] The velocity is normalized by use of a characteristic velocity U, usually defined as the average velocity in the passage.

Although it need not be so in reality, the inlet pressure can be specified as that value which would result if the inlet flow came from a reservoir in which the total pressure is $p_0 = p_s + \rho U^2/2$ and this total pressure can be used as the reference pressure. (Note that this form of the total pressure is only valid for incompressible flow, and will not be used if compressibility is important.) Since only the pressure gradient appears in the equation, the pressure need only be specified to within an additive constant, hence let $p_s = 0$; therefore, all pressures may be normalized by ρU^2 where the 1/2 has been omitted for simplicity. Finally, we note for generality in this problem, and for reference in other problems, that there is no independent characteristic time (e.g., that of an oscillation) associated with the phenomenon under consideration. Therefore, the time t can be normalized by using the characteristic time L/U, which is the time necessary for a fluid particle traveling with velocity U to move the distance L. Summarizing, we have the dimensionless variables (the asterisk [*] will be used to identify the dimensionless variable for the remainder of this chapter)

$$x^* = \frac{x}{L} \quad y^* = \frac{y}{L} \quad z^* = \frac{z}{L} \quad \mathbf{V}^* = \frac{\mathbf{V}}{U} \quad p^* = \frac{p}{\rho U^2} \quad t^* = \frac{tU}{L}$$

(4.1)

Actually, the foregoing may be thought of as a simple transformation

[†]A confined flow is one which has no free surfaces or interfaces. The flow is confined to move within a specified region and the boundary conditions are only expressed in terms of the velocity. Confined flows include flow around airfoils, turbine blades, and submarines, along with the obvious conduit flows.
[‡]The hydraulic diameter is defined as $d_H = 4 \times \text{area}/P$ where P is the perimeter of the solid in contact with the fluid. For a circular pipe $d_H = d$; for very narrow rectangles where $w \gg h$, $d_H \simeq 2h$.

of variables. In vector form, the dimensional Navier-Stokes equation is

$$\rho\left[\frac{\partial \mathbf{V}}{\partial t} + (\mathbf{V}\cdot\nabla)\mathbf{V}\right] = -\nabla p - \rho g \nabla h + \mu \nabla^2 \mathbf{V} \qquad (4.2)$$

or, combining the pressure and body-force terms,

$$\rho\left[\frac{\partial \mathbf{V}}{\partial t} + (\mathbf{V}\cdot\nabla)\mathbf{V}\right] = -\nabla p_k + \mu \nabla^2 \mathbf{V} \qquad (4.3)$$

where the *kinetic pressure* p_k has been introduced. It results from the dynamic effects as expressed in Eq. 4.3. Comparing Eq. 4.2 with Eq. 4.3, we have

$$p_k = p + \rho g h \qquad (4.4)$$

It is introduced because the body-force term does not influence the velocity field of confined-flow problems and thus it is "hidden." To nondimensionalize Eq. 4.3 we use the transformations of Eq. 4.1 and, noting $\nabla^* = L\nabla$, arrive at

$$\rho \frac{U^2}{L}\left[\frac{\partial \mathbf{V}^*}{\partial t^*} + (\mathbf{V}^*\cdot\nabla^*)\mathbf{V}^*\right] = -\rho \frac{U^2}{L}\nabla^* p_k^* + \frac{\mu U}{L^2}\nabla^{*2}\mathbf{V}^* \qquad (4.5)$$

or

$$\frac{\partial \mathbf{V}^*}{\partial t^*} + (\mathbf{V}^*\cdot\nabla^*)\mathbf{V}^* = -\nabla^* p_k^* + \frac{\mu}{UL\rho}\nabla^{*2}\mathbf{V}^* \qquad (4.6)$$

which is the *dimensionless* Navier-Stokes equation. The equation for the conservation of mass in dimensionless form is

$$\nabla^* \cdot \mathbf{V}^* = 0 \qquad (4.7)$$

Only one parameter appears in these two equations. The dimensionless group

$$\mathrm{Re} = \frac{\rho U L}{\mu} \qquad (4.8)$$

is the *Reynolds number*[†] and includes fluid properties and the length and velocity scales for the problem. Consequently, if this parameter is the same for the model and the prototype flows, then the governing equations are the same. It is then only necessary to specify that the dimensionless boundary conditions are the same in order to guarantee

---

[†]Named for Osborne Reynolds, an English engineer, who recognized the importance of this group of parameters in his studies of the breakdown of laminar flow in tubes.

that the solution to the boundary-value problem must be the same for the two nondimensional flow fields. The geometry of the model and the prototype differ by a constant scale factor (geometric similarity); then, since the no-slip condition at a solid wall insures that **V** (boundary) = 0, the velocity boundary conditions are identical at the boundaries. The inlet boundary condition requires that

$$\left(\frac{\mathbf{V}_{inlet}}{U}\right)_{model} = \left(\frac{\mathbf{V}_{inlet}}{U}\right)_{prototype} \qquad (4.9)$$

which means, for example, that if a parabolic profile occurs at the inlet in the prototype, it must also be parabolic in the model at the inlet. It also means that if a turbulent flow exists in the prototype at the inlet it must also be turbulent in the model at the inlet. Thus, it may be necessary to model the turbulent structure that exists in a turbulent motion.

Consequently, the measured data from the model, such as the internal velocity distributions and pressure measurements along the walls, which represent the solution for the above-described boundary-value problem, are identical[†] with those which will exist in the prototype when both are expressed in dimensionless form. *Such a condition is defined as similitude between model and prototype.*

We can now easily set the requirements for the model study, namely Reynolds-number and boundary-condition equivalence, expressed as

$$\left(\frac{\rho U L}{\mu}\right)_{model} = \left(\frac{\rho U L}{\mu}\right)_{prototype} \qquad (4.10)$$

$$\left(\frac{\mathbf{V}}{U}\right)_{\substack{model \\ inlet}} = \left(\frac{\mathbf{V}}{U}\right)_{\substack{prototype \\ inlet}} \qquad (4.11)$$

along with the no-slip condition on *identical* dimensionless boundaries of the model and prototype.

With the exception of uniqueness questions, the principle that identical equations and boundary and initial conditions result in identical solutions, and therefore a condition of similitude, has led us to a very straightforward way to establish the conditions for model studies. Before proceeding with considerations of similitude, we will identify several important questions often encountered in model studies.

[†]Some rather delicate questions of uniqueness enter these considerations at a level more advanced than considered here. In effect, we will assume that for given geometry and boundary conditions there is only one possible flow field.

A subtle and difficult-to-answer question can be posed for the fuel-cooling example: Is it possible to make a model for which the boundary and initial conditions are "identical"? Forcing the equations to be identical is not operationally difficult; we simply make the Reynolds numbers the same. But, is it possible to exactly reproduce the inlet velocity distribution which exists in the prototype? More importantly, is it necessary? The prototype device might have been formed by metal casting; in that event the inside walls will be rough whereas the walls of a clear plastic model are smooth. Does this difference cause any important effects in the structure of the flow field? The answers to these questions depend upon the detailed nature of any particular problem. They are also much easier to ask than to answer. The insight to ask and to seek answers for such questions comes from enlightened experience. The remaining chapters of the book are to help provide this enlightenment. The formal method of extracting the similarity parameters from the governing equations relies on the strategy of developing a general framework for our analysis; one that is itself complete and sufficient. Then, real problems with their attendant limitations and uncertainties can be dealt with in an inclusive and general framework.

Example 4.1

The flow around a model of a turbine blade is investigated. The model is five times larger than the prototype. A maximum pressure of 4 psi is measured at the leading edge, a maximum velocity of 30 fps is measured near the top of the blade, and a small device attached to the surface measures a shearing stress of 0.02 psi at a particular location. Determine the associated quantities to be expected on the prototype. Water is the fluid for both model and prototype.

Solution. An important consequence of similarity is that, in dimensionless form, all quantities in the model at a particular point are the same as those in the prototype at the corresponding point. That is, at the leading edge

$$p_M^* = p_P^*$$

or, from expressions 4.1,

$$\left(\frac{p}{\rho U^2}\right)_M = \left(\frac{p}{\rho U^2}\right)_P$$

Since the density is the same for the fluid in both the model and prototype, the pressure at the leading edge of the prototype blade is

$$p_P = p_M \frac{U_P^2}{U_M^2}$$

Art. 4.2 **SIMILITUDE**

We find the velocity ratio from the Reynolds-number equivalence, namely,

$$\left(\frac{UL}{\nu}\right)_M = \left(\frac{UL}{\nu}\right)_P$$

or

$$\frac{U_P}{U_M} = \frac{L_M}{L_P} = 5$$

The pressure is then

$$p_P = p_M \frac{U_P^2}{U_M^2} = 4 \times 5^2 = 100 \text{ psi}$$

At the top of the blade we have

$$v_M^* = v_P^*$$

or

$$\left(\frac{v}{U}\right)_M = \left(\frac{v}{U}\right)_P$$

The velocity on the prototype is then

$$v_P = v_M \frac{U_P}{U_M} = 30 \times 5 = 150 \text{ fps}$$

The dimensionless shearing-stress relationship is

$$\tau_M^* = \tau_P^*$$

or, since shearing stress has the same units as pressure,

$$\left(\frac{\tau}{\rho U^2}\right)_M = \left(\frac{\tau}{\rho U^2}\right)_P$$

Finally,

$$\tau_P = \tau_M \frac{U_P^2}{U_M^2} = 0.02 \times 5^2 = 0.5 \text{ psi}$$

We have predicted the quantities to be expected on the actual turbine blade by investigating the flow around a model blade.

Extension 4.1.1. Determine the velocity at the top of the prototype blade corresponding to 30 fps on the model if air at 80°F were used in the model study. Use $\rho_{\text{air}} = 0.0024$ slug/ft^3 and $\nu_P = 10^{-5}$ ft^2/sec. *Ans.* 8.88 fps

Example 4.2

Consider that the prototype discussed in Art. 4.2 uses kerosene as the working fluid, with $\nu = 3 \times 10^{-5}$ ft^2/sec and a characteristic length (the diameter at the circular inlet) of 0.4 in. A minimum flow rate of 0.5 cfm is to be examined. Develop the design considerations for a model to be made of clear plastic to allow flow-visualization studies and quantitative data to be obtained. Although water or air could be used for the visualization phase, air is preferred because the quantitative measurements are often more easily obtained in an air flow. For the air flow, use $\nu_{air} = 1.6 \times 10^{-4}$ ft^2/sec.

Solution. The characteristic velocity for the prototype is the average velocity at the inlet, namely,

$$U_P = \text{volume flow rate/Area}$$

$$= \left(\frac{0.5}{60}\right) / \frac{\pi(0.4)^2}{4 \cdot 144} = 9.5 \text{ fps}$$

The Reynolds number for the problem is

$$(Re)_P = \left(\frac{UL}{\nu}\right)_P = \frac{(9.5)(0.4)/12}{3 \times 10^{-5}}$$

$$= 10.6 \times 10^3$$

and therefore, for air at 68°F as the working fluid,

$$(Re)_M = \left(\frac{UL}{\nu}\right)_M = \frac{U_M L_M}{1.6 \times 10^{-4}} = 10.6 \times 10^3$$

or

$$U_M L_M = 1.7 \text{ ft}^2/\text{sec}$$

At this point, the designer is forced to make the best selection based on his estimate of the optimum trade-offs. A large model will make the visualization study most revealing; however, the higher the velocity, the more easily are velocity measurements and wall static measurements made. Experience is, of course, of great help in making these decisions. Velocities of the order of 10 to 20 fps can be reliably measured with hot-wire anemometers; these velocities represent minimum values for reasonably reliable pitot-tube measurements. Selecting $U_M = 15$ fps, L_M is calculated to be 1.36 inches. Although it will depend upon the size of the regions of critical importance, this length would seem to be large enough for visualization work; we might even make L_M, the diameter of the model inlet, smaller and U_M larger to facilitate the quantitative measurements.

Extension 4.2.1. If the inlet velocity profile is uniform for the prototype, state the dimensionless boundary condition u^* at the inlet plane.

Ans. $u^* = 1.0$

Extension 4.2.2. If water were used as the fluid for the model study and if flow visualization studies could best be carried out for velocities of the order of 5 fps, what physical size should be used for the model inlet diameter?

Ans. 0.305 in.

4.3 INTEGRAL QUANTITIES

Although the control-volume, or integral, equations will not introduce any new parameters, they explicitly involve quantities that are often of great interest in engineering applications, quantities such as a drag force, a twisting moment, and pump horsepower. We know from our considerations that *if similarity exists then the dimensionless velocity field is identical for the dimensionless prototype and the dimensionless model; the pressure fields, the shear stress distributions, and the temperature fields are also the same for the model and prototype*. Since the integral quantities (force, moment, work) involve integrals of identical quantities on the dimensionless model and prototype, it follows that *all dimensionless, or normalized, integral quantities are the same for model and prototype*:

$$F_M^* = F_P^* \quad M_M^* = M_P^* \quad \dot{W}_M^* = \dot{W}_P^* \quad \dot{Q}_M^* = \dot{Q}_P^* \quad (4.12)$$

The integral momentum equation involves the integral $\int_{c.s.} \rho \mathbf{V}(\mathbf{V} \cdot \hat{n}) dA$, which in dimensionless form is $\rho U^2 L^2 \int_{c.s.*} \mathbf{V}^*(\mathbf{V}^* \cdot \hat{n}) dA^*$. Since this is equal to $\Sigma \mathbf{F}$, we see that forces are normalized by dividing by $\rho U^2 L^2$. From the moment-of-momentum equation we would conclude that moments are normalized by dividing by $\rho U^2 L^3$, and from the energy equation the work rate and heat-transfer rate would be nondimensionalized by dividing by $\rho U^3 L^2$. Hence, in dimensional terms, Eq. 4.12 is

$$\frac{F_M}{\rho_M U_M^2 L_M^2} = \frac{F_P}{\rho_P U_P^2 L_P^2} \quad \frac{M_M}{\rho_M U_M^2 L_M^3} = \frac{M_P}{\rho_P U_P^2 L_P^3}$$
$$\frac{\dot{W}_M}{\rho_M U_M^3 L_M^2} = \frac{\dot{W}_P}{\rho_P U_P^3 L_P^2} \quad \frac{\dot{Q}_M}{\rho_M U_M^3 L_M^2} = \frac{\dot{Q}_P}{\rho_P U_P^3 L_P^2} \quad (4.13)$$

This type of relationship allows us to predict integral quantities associated with the prototype from measurements made on the model. The net towing force on the model of a ship can be used to predict the propulsive force needed for the actual ship. The measured lift and pitching moment on an airfoil in a wind tunnel allow the actual lift and pitching moment to be estimated.

Example 4.3

A model of a large air fan is used to test a new design proposal. It is one-tenth full size. The model is tested in a wind tunnel at 200 fps and a net force of 400 lb is noted. What prototype speed does this correspond to and what force is required to support the large air fan?

Solution. Reynolds number is used for the model study, so

$$\left(\frac{UL}{\nu}\right)_M = \left(\frac{UL}{\nu}\right)_P$$

or

$$U_P = U_M \frac{L_M}{L_P} = 200 \times \frac{1}{10} = 20 \text{ fps}$$

From Eq. 4.13 the force can be modeled as

$$\left(\frac{F}{\rho U^2 L^2}\right)_M = \left(\frac{F}{\rho U^2 L^2}\right)_P$$

so the force to be expected on the fan is

$$F_P = F_M \frac{U_P^2}{U_M^2} \frac{L_P^2}{L_M^2} = 400 \times \frac{20^2}{200^2} \times 10^2 = 400 \text{ lb}$$

The forces are identical for both model and prototype. This is always the case if Reynolds number is the governing parameter and if the fluid is the same for both model and prototype.

Extension 4.3.1. If Reynolds number is used to model the flow over the rotating fan, determine the ratio of the torque on the fan to the torque measured on the model. *Ans.* 10

Extension 4.3.2. The velocity necessary in the model study was 200 fps to simulate 20 fps of the prototype. This is quite large. Possibly the situation could be improved if the model study were done in a water channel. Would you recommend a water channel study? Approximately what force on the model in the water channel could be expected if 400 lb exists on the prototype?
Ans. 1820 lb

4.4 PARAMETERS FOR MORE COMPLEX FLOWS

Although the Reynolds number was the only parameter to appear in the fuel-cooling problem, it must be remembered that this problem involved a nonfluctuating, incompressible flow of a Newtonian fluid with no surface effects. We will now systematically develop additional parameters which appear in the mathematical description of more complicated problems. Although we will consider these parameters somewhat independently, the student should appreciate that several parameters may be important for similitude considerations of a given problem.

1. Froude Number

The pressure term of Eq. 4.6 was written in the form which absorbs the body force of the conservative field in the definition of the pressure; namely,

$$p_k = p + \rho g h \tag{4.14}$$

where h is the height above a horizontal datum plane.

When the boundary conditions involve only the velocity (as in the examples considered), the boundary-value problem involving Eqs. 4.6 and 4.7 can be solved with no reference to Eq. 4.14. However, if a boundary is given in terms of the pressure p then Eq. 4.14 would be necessary. Examples for which a pressure-boundary condition is necessary are shown in Fig. 4.1. In dimensionless form Eq. 4.14 is

$$p_k^* = p^* + \left(\frac{gL}{U^2}\right)h^* \tag{4.15}$$

where the asterisk again denotes a dimensionless quantity. In order to introduce the dimensionless pressure p^* in a boundary condition, and insure identical boundary conditions on both model and prototype, we must make the dimensionless group gL/U^2 the same for model and prototype. This group is termed the *Froude number*[†] and is most often expressed as

$$\text{Fr} = \frac{U^2}{gL} \tag{4.16}$$

(a) Pressure is zero on surface of waves

(b) Pressure is zero on free surface

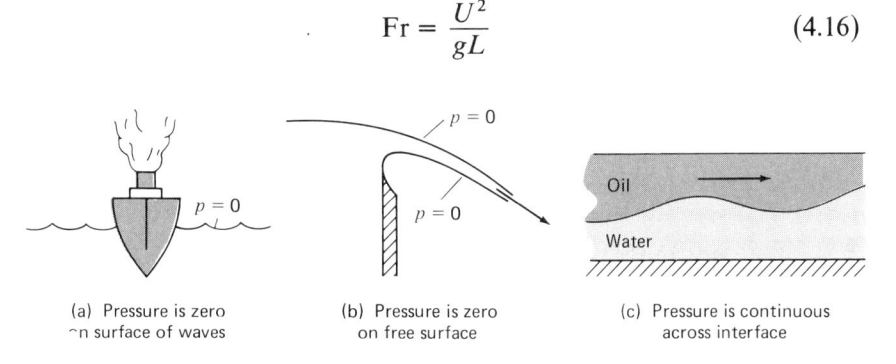

(c) Pressure is continuous across interface

Fig. 4.1. Examples showing pressure as a boundary condition.

[†]A somewhat uncommon extension of the Froude number is the *densiometric Froude number*. If the density is not constant throughout the flow and we use $\Delta\rho$ to normalize ρ and $\rho_0 U^2$ to normalize the pressure, then $\text{Fr}_d = \rho_0 U^2/(\Delta\rho g L)$ is the appropriate dimensionless parameter. This parameter describes the buoyancy effects associated with the flow up a chimney, the fresh water–salt water boundary of an estuary, and the dispersal of gasoline vapors in air.

The design of a hydrofoil, shown in Fig. 4.2, provides an interesting case where the Froude number plays an important role. For low-speed operation, the hydrofoil is sufficiently submerged that the flow field around the hydrofoil is effectively independent of the free surface and, hence, the Froude number is unimportant; the Reynolds number is the important similarity parameter. For a hydroplaning condition, the distortion of the free surface by the hydrofoil will strongly affect the pressure distribution over the hydrofoil surface and must be accounted for by the Froude number in any model studies.

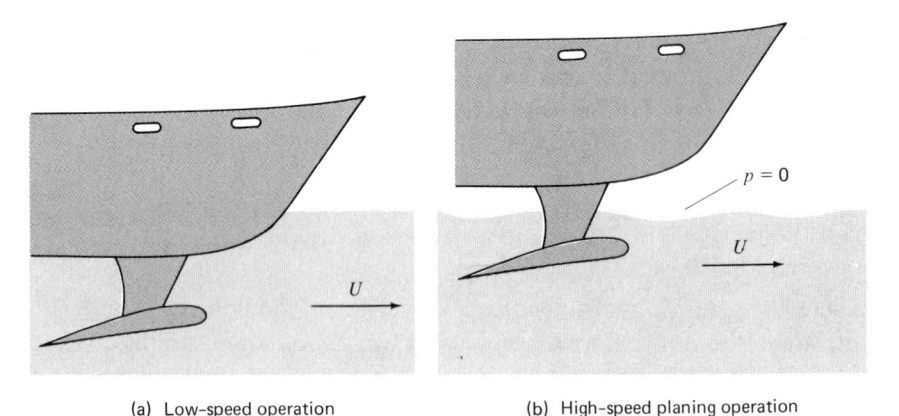

(a) Low-speed operation   (b) High-speed planing operation

Fig. 4.2. A hydrofoil example for Froude-number effects.

2. Weber Number

An easily observed example of surface-tension effects is the formation of droplets in a water jet issuing slowly from a faucet. It is intuitively obvious that a liquid jet will accelerate under the influence of gravity with a consequent decrease in the jet diameter. When the jet diameter becomes small enough, surface tension effects act in such a way as to continue to reduce the diameter at those positions where a chance fluctuation in the diameter might occur. As a result, the jet disintegrates into individual droplets. In terms of the momentum equation such an effect is expressed in terms of the pressure-boundary condition. The surface tension σ is such that the force parallel to the surface is

$$F = \sigma l \tag{4.17}$$

When important, the surface tension will enter the problem as a force-boundary condition which will balance a pressure times an area.

This can be expressed as
$$pA = \sigma l \quad (4.18)$$

Since ρU^2 and L^2 are used to normalize pressure and area respectively, Eq. 4.18 is normalized as

$$p^*A^* = \left(\frac{\sigma}{\rho U^2 L}\right)l^* \quad (4.19)$$

This introduces the dimensionless group known as the *Weber number*:

$$\text{We} = \frac{\rho U^2 L}{\sigma} \quad (4.20)$$

It is an important similarity parameter whenever surface tension is significant.

3. Strouhal Number

Consider a flow field which is influenced by a time-varying boundary condition, for example, the flow in the inlet or exhaust manifolds of an internal combustion engine or the flow over a blade in a turbine as it passes behind a row of stator (guide) vanes. In order to satisfactorily model such a flow, some provision must be made for these unsteady fluctuations. If there is a periodic variation in a boundary condition with frequency n, then the reference time used may be chosen as the time per cycle, $1/n$. The *Strouhal number* is then introduced to account for such effects and is defined as

$$\text{St} = \frac{nL}{U} \quad \overset{RAD}{s} \quad (4.21)$$

It would appear in the Navier-Stokes equation† as

$$\text{St}\,\frac{\partial \mathbf{V}^*}{\partial t^*} + (\mathbf{V}^* \cdot \boldsymbol{\nabla}^*)\mathbf{V}^* = -\boldsymbol{\nabla}^* p_k^* + \frac{1}{\text{Re}}\,\nabla^{*2}\mathbf{V}^* \quad (4.22)$$

If viscous effects were important in such an oscillatory flow then the Reynolds number would also play an important role.

†The Strouhal number would be introduced in the boundary conditions if the time were normalized by $t^* = tU/L$. The boundary condition for the velocity on the periodically varying part of the boundary would then contain the Strouhal number St, e.g., the dimensional boundary condition $u = r\Omega$ would become $u^* = \frac{nL}{U}\,r^*\Omega^*$ where $\Omega^* = \Omega/n$.

Time-dependent effects may be forced or they may be an intrinsic part of the flow field. The Strouhal number is not limited to the former. The important natural phenomenon of vortex shedding (see Fig. 10.2) occurs in a flow past a cylindrical body such as a TV tower. This subject is discussed in detail in Chapter 10.

4. High-Reynolds-Number Flows

There are flows which exhibit a rather interesting characteristic, namely, that no parameter is introduced by the describing equations. This is the situation for many flows for which the Reynolds number is sufficiently large that the viscous effects do not play a discriminating role in the behavior of the flow field. For example, in flow around a body the viscous effects will always result in the condition that $\mathbf{V} = 0$ at the surface of the body. However, if the flow follows the contour of the body or if it separates from the contour of the body at a location which is insensitive to the Reynolds number, then the character of the flow field is governed by nonviscous effects. When these conditions are met, a necessary condition for similitude is that the Reynolds number simply be above a certain value.[†] It is not necessary that the model Reynolds number be the same as the prototype Reynolds number; this may allow other, more significant parameters to be matched, such as the Mach number or Froude number.

Example 4.4

A 16-to-1 model of a submarine is built. We wish to determine certain characteristics of the submarine when it is traveling at 20 knots on the surface (where we neglect viscous effects) and when it is traveling at 20 knots far below the surface. At what speed would we tow the model, both on the surface and below the surface? (Water is used in the model study.)

Solution. On the surface the viscous effects in the water flowing around the submarine are expected to be less important than the influence of the surface waves. The Reynolds number and the Froude number cannot both be used for the same model study (see Extension 4.4.1); hence, the Froude number is used as the similarity parameter. The waves created by the submarine involve a pressure-boundary condition so we must have equal Froude numbers, that is,

$$\left(\frac{U^2}{gL}\right)_P = \left(\frac{U^2}{gL}\right)_M$$

[†]Characteristic examples are given for fully developed pipe flow in Art. 7.4 (Fig. 7.13) and flow around a sphere in Art. 10.3 (Fig. 10.10).

or
$$U_M = U_P\sqrt{L_M/L_P}$$
$$= 20 \times \tfrac{1}{4} = 5.0 \text{ knots}$$

The model must be towed at 5.0 knots.

Below the surface, the viscous drag is the only parameter since no waves are present; hence the Reynolds numbers must be equal, requiring that

$$\left(\frac{UL}{\nu}\right)_P = \left(\frac{UL}{\nu}\right)_M$$

or

$$U_M = U_P \frac{L_P}{L_M} = 20 \times 16 = 320 \text{ knots}$$

This speed is much too high to be practical, so we should either change the model size or the model fluid; or possibly the Reynolds number is sufficiently large so that viscous effects may be ignored.[†] Consider a submarine to be 10 ft in diameter. Let us use a Reynolds number of 10^5 for the model.[‡] Then,

$$\frac{U_M L_M}{\nu_M} = 10^5$$

or, using $L_M = 10/16 = 0.624$ ft,

$$U_M = \frac{10^5 \nu_M}{L_M} = \frac{10^5 \times 1.2 \times 10^5}{0.624} = 1.92 \text{ fps}$$

A model velocity greater than 2 fps appears to be acceptable. This would be the case providing the flow is separated (this would be the situation for a short submersible); and the engineer could choose any velocity greater than 2 fps to meet his instrumentation and/or facility requirements.

We have solved the problem, but you may react to the neglect of the viscous effects for the surface study. Why did we not satisfy both the Reynolds number and the Foude number? If we use water in the model study the Froude number demands a towing velocity of 5.0 knots, and the Reynolds number demands 320 knots; hence it is impossible to satisfy both criteria if we use the same fluid for both the model and the prototype. But it is also impractical to choose a fluid other than water in which to tow a typical model. Thus we are

---

[†]It may be improper to ignore viscous effects since the drag may be primarily due to viscous drag and not the form drag of a separated flow. For a streamlined submarine with little separation Reynolds number must be the same on model and prototype.

[‡]This magnitude of the Reynolds number is selected on the basis of experimental data for flow around bodies of various shapes, some of which are presented in Chapter 10. The value of 10^5 appears to be sufficiently large for bodies that have rough surfaces. Sandpaper on the front of the model could provide the roughness.

forced to use only the Froude number in model studies involving surface phenomena since the wave effects are greater than the viscous effects. The results must then be empirically corrected to account for the viscous effects.

Extension 4.4.1. For the model in this example, suppose that both the Froude number and the Reynolds number must be satisfied for the surface study. What kinematic viscosity would the model fluid be required to have for similarity to exist? *Ans.* 2.81×10^{-7} ft^2/sec

Extension 4.4.2. If an air tunnel were used for the underwater study, what should the wind tunnel velocity be? Would this be acceptable? What do you propose?

4.5 COMPRESSIBLE FLOW

Compressible flow is the term for the situation in which the density variations become significant enough to affect the velocity and pressure distributions for a given flow problem. At the introductory level of this text, it is not too restrictive to consider only perfect gases as the fluid for such flows, thus greatly simplifying the thermodynamic considerations which along with the mass, momentum, and energy equations, govern the fluid motion. The governing equations in dimensional form are:

Momentum:
$$\frac{\partial \mathbf{V}}{\partial t} + (\mathbf{V} \cdot \nabla)\mathbf{V} = -\frac{\nabla p}{\rho} + \nu \nabla^2 \mathbf{V} + \frac{\nu}{3} \nabla (\nabla \cdot \mathbf{V}) \qquad (4.23)$$

Energy:
$$\rho c_v \frac{DT}{Dt} = k \nabla^2 T - p \nabla \cdot \mathbf{V} + \Phi \qquad (4.24)$$

Mass:
$$\frac{\partial \rho}{\partial t} + \nabla \cdot (\rho \mathbf{V}) = 0 \qquad (4.25)$$

State:
$$p = \rho RT \qquad (4.26)$$

The characteristic length and velocity may again be identified as L and U. If, for example, the problem under consideration were that of an aircraft wing, the chord length could be used for L and the aircraft speed would be the characteristic velocity U. The characteristic pressure, temperature, and density values are taken as the approach conditions[†] in the undisturbed atmosphere, namely p_0, T_0, ρ_0. Alternately, these values are the pressure, temperature, and density which would exist in a reservoir supplying the compressible flow. For a perfect gas the speed of sound a_0 is recalled to be $a_0 = \sqrt{\gamma R T_0}$. Consequently, although we shall not show the detailed steps, the dimensional equations may be reformulated in dimensionless form (neglecting the gravity term, since body forces are negligible in gas flows) as follows:

Momentum:

$$\rho^*\left[\frac{\partial \mathbf{V}^*}{\partial t^*} + (\mathbf{V}\cdot\nabla)\mathbf{V}^*\right] = -\frac{1}{M_0^2\gamma}\nabla^* p^* + \frac{1}{\text{Re}}\left[\nabla^{*2}\mathbf{V}^* + \frac{1}{3}\nabla^*(\nabla^*\cdot\mathbf{V}^*)\right]$$
(4.27)

Energy:

$$\rho^*\frac{D^*T^*}{Dt^*} = \frac{\gamma}{\text{PrRe}}\nabla^{*2}T^* - (\gamma-1)p^*\nabla^*\cdot\mathbf{V}^* + \gamma(\gamma-1)\frac{M_0^2}{\text{Re}}\Phi^*$$
(4.28)

Mass: $$\frac{\partial \rho^*}{\partial t^*} + \nabla^*\cdot(\rho^*\mathbf{V}^*) = 0 \qquad (4.29)$$

State: $$p^* = \rho^* T^* \qquad (4.30)$$

The *Mach number*

$$M_0 = U/a_0 \qquad (4.31)$$

appears explicitly in these equations, as well as the *specific heat ratio* γ and the *Prandtl number*

$$\text{Pr} = \mu C_P/k \qquad (4.32)$$

If heat-transfer and energy-dissipation effects are not important, then the Prandtl and Reynolds numbers are not important in specifying the flows unless there are significant shear effects in the momentum equations. For example, a laminar boundary layer exists over a significant length on a supersonic airfoil because the characteristic thickness of the layer is exceedingly small even though the free-stream velocity is greater than that of sound. The Reynolds number would be important

---

[†]The characteristic pressure is sometimes chosen as ρU^2 even for a compressible flow. This is considered inconsistent with the "demand" that a characteristic quantity be a physically realizable quantity.

222   SIMILITUDE   Ch. 4

in describing such a flow. Conversely, if the flow is effectively inviscid as in a nozzle, the shear effects are negligible and hence the Reynolds number is not important. The Mach number and the specific heat ratio then effectively determine the behavior of the flow.

The assumption of incompressible flow has the direct result that the density which appears in the governing equations is treated as a constant. We can quantitatively identify the effects of compressible flows in terms of the effect on the magnitude of the density. In Art. 1.8 it was demonstrated that a Mach number of 0.25 resulted in a 3-percent change in the density from reservoir conditions. The commonly accepted "limit" between compressible and incompressible flow is at a Mach number of 0.3. Consequently, for air flow near standard conditions the velocity has to exceed approximately 350 fps before compressibility is important. Up to this speed only the Reynolds number governs the flow.

It is important to realize that each dimensionless group which has been identified expresses the effect of a term in either the governing equations or the boundary conditions. If the term represents an important physical effect, then the parameter will be important for establishing a condition of similitude.

Example 4.5

A rather common request for a wind tunnel study involves the aerodynamic testing of a scale-model automobile. Determine the requirements for such a test in terms of a 4-in.-wide scale model of a 70-in.-wide prototype. Prototype speeds of 50 to 120 mph are of interest.

Solution. The prototype will operate in the incompressible regime; consequently, the Reynolds number is the relevant dimensionless parameter. We can assume that both the model and prototype will operate in standard air; therefore, the Reynolds-number similarity requirement can be used to determine the velocity for the prototype as

$$\left(\frac{UL}{\nu}\right)_M = \left(\frac{UL}{\nu}\right)_P$$

or

$$U_M = U_P(L_P/L_M) = U_P(70/4)$$

$$= 875 \text{ to } 2100 \text{ mph}$$

The reader will note that this velocity range is supersonic: the speed of sound is about 750 mph for air at standard conditions. Since a supersonic flow is needed to satisfy the Reynolds-number similarity requirement, similarity cannot be achieved using air for the model study. However, if one restricts the velocity to approximately 210 mph, the compressibility effects would be eliminated, although the Reynolds-number criterion cannot be met exactly. The error in the Reynolds number would be 4- to 10-fold but some of the phenomena of interest may be sufficiently insensitive to Reynolds number to allow this to be a good representation. Note that this example shows that even though the Mach number is not important for the prototype it is important that the model Mach number is sufficiently low to make it unimportant there also.

Extension 4.5.1. A large water tunnel is available. Could the necessary similarity be obtained using this unit and the 4-inch model?

Extension 4.5.2. Identify the boundary conditions from the prototype that should be duplicated in the model study. Note that the model will be mounted in a wind tunnel of dimension w.

4.6 DIMENSIONLESS VARIABLES

The preceding articles in this chapter have established the conditions for similitude. When these conditions are met, information obtained from a model study can be used to predict the behavior of a prototype. The formulation of nondimensional equations and initial and boundary conditions is itself an important contribution. The dimensionless parameters included in the equations and conditions are of particular interest. When an investigator performs an experiment or seeks an analytical solution, the study will be most useful if dimensionless variables are properly used. This is best illustrated by an example.

Let us determine the pressure drop Δp due to a fluid flowing in a horizontal pipe. We must anticipate the quantities that can influence Δp and make certain that all such quantities are included in the study. The quantities will include the length L and the diameter D of the pipe, the viscosity μ and the density ρ of the fluid, the roughness size e of the pipe walls, and the average velocity V of the fluid. This dependence of the pressure drop on the other variables is expressed as

$$\Delta p = f(L, D, \mu, \rho, e, V) \qquad (4.33)$$

Suppose that an analytical solution is not feasible so that an experimental investigation is necessary. Experiments would be done in which all variables except one would be held constant and the remaining one varied. The resulting pressure drop could then be plotted against the varying quantity. To do this for all the variables would be very time-consuming and expensive. In fact, it would be quite difficult to hold the viscosity constant and vary the density since one would vary the density by choosing different fluids which themselves possess varying viscosities. Difficulty would also arise for other investigators who wanted to use the results of the experiments, or who wanted to compare results.

Fortunately, the relationship (Eq. 4.33) can be simplified by considering dimensionless variables. In order for a quantity to influence the pressure drop it must be included in the describing equations or the boundary conditions. The differential equation which describes the pressure variation in the pipe is the x-component Navier-Stokes equation,

$$\frac{D^*u^*}{Dt^*} = -\frac{\partial p^*}{\partial x^*} + \frac{\mu}{\rho D V} \nabla^{*2} u^* \tag{4.34}$$

where we have used the average velocity V and the diameter D as characteristic quantities. A boundary condition requires that the velocity be zero on the wall of the pipe. For a smooth pipe, this is stated as

$$u^* = 0 \quad \text{at } r^* = 0.5 \tag{4.35}$$

where r^* is the dimensionless radial dimension. For a rough wall, we can invoke the same condition but we must recognize that the boundary is not at $r^* = 0.5$ because of the roughness. Consequently, the presence of the roughness must be recognized by a suitable description, such as e, the root-mean-square height of the protrusions from the nominal radius. The dimensionless parameter introduced in the modified boundary condition would be e/D. A statement of the velocity profile at the entrance to the pipe is also required as a boundary condition. Assuming that the entrance profile is uniform this condition is expressed as

$$u^* = 1 \quad \text{at } x^* = 0 \tag{4.36}$$

where x^* is measured from the entrance to the pipe. No parameters are introduced in this condition, but if the profile had not been uniform a parameter would have appeared. It should also be noted that if the pipe were long the entrance effects could be neglected,[†] so that the condition (4.36) would not be necessary; then the entrance profile would introduce no new parameters.

[†]The flow in the pipe evolves into a state independent of the entrance condition, that is, the flow becomes fully developed (see Art. 6.1).

The dimensionless pressure drop would be obtained from

$$\Delta p^* = \int_0^{L/D} \frac{\partial p^*}{\partial x^*} dx^* \qquad (4.37)$$

If an analytical solution could be obtained, the pressure drop would be easily established. Many problems, though, cannot be solved analytically; however, the governing equations and boundary conditions will provide valuable information. Specifically, it is known that the dimensionless dependent variables depend only on the parameters introduced in the dimensionless boundary conditions and the describing equations. In the example being considered the differential equation and boundary conditions introduce the two parameters $VD\rho/\mu$ and e/D. The expression for the pressure drop introduced L/D. Hence, we can express the dimensionless pressure drop as

$$\Delta p^* = f\left(\frac{VD\rho}{\mu}, \frac{e}{D}, \frac{L}{D}\right) \qquad (4.38)$$

The final expression being sought is found by writing the dimensionless pressure in dimensional terms. It is

$$\frac{\Delta p}{\rho V^2} = f\left(\frac{VD\rho}{\mu}, \frac{e}{D}, \frac{L}{D}\right) \qquad (4.39)$$

There is a significant difference between this relationship of dimensionless terms and the dimensional relationship (Eq. 4.33). Specifically, the quantity $\Delta p/\rho V^2$ depends on only three dimensionless variables, whereas the pressure drop Δp depended on six variables. This greatly reduces the number of experimental cases to be considered. In addition, we do not have the complexity of using different fluids in the study; we can vary the Reynolds number $VD\rho/\mu$ simply by varying the flow rate (the average velocity V). The final experimental results would be expressed in terms of the dimensionless variables. This would allow quick usage of the results and easy comparison with other data.

The reduction of the number of variables influencing a quantity of interest, as illustrated in the foregoing example, is a direct result of writing the describing equations and boundary conditions (in unsteady flows initial conditions are also necessary) in dimensionless forms. The variables in the example of this article were reduced by three, from six to three. This reduction by three always occurs when all three basic dimensions, mass, length, and time, are contained in the dimensions of the variables. If only two dimensions, such as length and time, are involved in the problem then the variables would be reduced by two. This could be demonstrated by considering the example of a freely falling body, neglecting viscous drag, with the appropriate initial conditions.

The method of *dimensional analysis* is an alternate approach to the results presented in this article. In dimensional analysis the Buckingham π-theorem is used in the formation of the dimensionless terms of Eq. 4.39; however, certain variables thought to influence the quantity of interest may not enter the describing equations and boundary conditions. Hence, they should not be considered in the study. The describing equations and boundary conditions introduce the relevant quantities.

Example 4.6

The drag force F_D on an immersed axisymmetric body depends on the maximum diameter D, length L, and speed U of the body and the viscosity μ, and density ρ of the fluid. Extract a functional relationship between these parameters by considering the differential equation and boundary conditions.

Solution. The appropriate differential equation, written in nondimensional form, is

$$\frac{D^*\mathbf{V}^*}{Dt^*} = -\nabla p_d^* + \frac{\mu}{\rho D U} \nabla^{*2}\mathbf{V}^*$$

The only parameter introduced is the Reynolds number $\rho DU/\mu$, where D and U have been chosen as characteristic quantities. The boundary conditions on the dimensionless axisymmetric body shown in Fig. E4.6 are

$$u^* = 1 \quad v^* = 0 \quad \text{at } y^* = \pm\infty$$

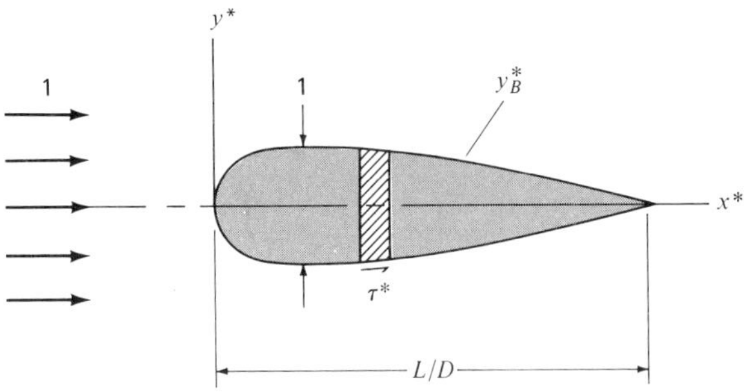

Fig. E4.6

and
$$u^* = 0 \quad v^* = 0 \quad \text{at } y^* = y_B^*(x^*)$$

The dimensionless drag force is determined from
$$F_D^* = \int_{A^*} \tau^* \cos\theta \, dA^*$$
$$= \int_0^{L/D} \tau^* \cos\theta \, 2\pi y_B^* \, dx^*/\cos\theta$$

This introduces the parameter L/D. The dimensionless drag is thus written as
$$F_D^* = f\left(\frac{UD\rho}{\mu}, \frac{L}{D}\right)$$
or
$$\frac{F_D}{\rho U^2 D^2} = f\left(\frac{UD\rho}{\mu}, \frac{L}{D}\right)$$

Instead of the drag force depending on five quantities, the dimensionless drag depends on two dimensionless parameters.

Extension 4.6.1. If L had been chosen as the characteristic length, state the functional relationship that would have resulted.

$$\text{Ans.} \quad \frac{F_D}{\rho U^2 L^2} = f\left(\frac{UL\rho}{\mu}, \frac{L}{D}\right)$$

Extension 4.6.2. It has been observed that the roughness ϵ of the body can substantially affect the drag force. How would the roughness enter the problem?

Example 4.7

Air flows radially outward between two parallel discs, as shown in Fig. E4.7. It is assumed that $R \gg d$ so that the three-dimensional flow at the entrance to the parallel discs causes a negligible influence on the outward flow. The physical device is constructed so that the radius R is fixed but the gap height $h \, (\ll R)$ and the flow rate q can both vary. Write the describing equations and boundary conditions and conclude from the dimensionless boundary conditions that two characteristic lengths should be used in order for the flow for two different gap heights to be similar. (Ignore the flow near the center hole.)

Solution. The dimensional describing equation, written in cylindrical coordinates, is
$$v_r \frac{\partial v_r}{\partial r} = -\frac{1}{\rho}\frac{\partial p}{\partial r} + \nu\left(\frac{\partial^2 v_r}{\partial z^2} - \frac{v_r}{r^2}\right)$$

where we have assumed $v_\theta = v_z = 0$ and $\partial^2 v_r/\partial z^2 \gg \partial^2 v_r/\partial r^2$. The boundary

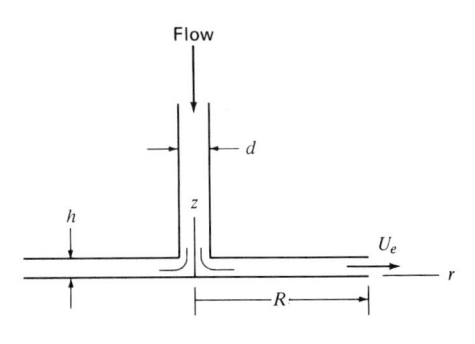

Fig. E4.7

conditions necessary are

$$v_r = 0 \quad \text{at } z = 0$$

$$v_r = 0 \quad \text{at } z = h$$

$$\bar{v}_r = \bar{U}_e \quad \text{at } r = R$$

$$p = 0 \quad \text{at } r = R$$

where $\bar{U}_e$ is the exiting average velocity. The boundary conditions suggest that we normalize z with h and r with R; that is,

$$z^* = z/h \qquad r^* = r/R$$

Choosing $\bar{U}_e$ as the characteristic velocity, the dimensionless boundary conditions become

$$v_r^* = 0 \quad \text{at } z^* = 0$$

$$v_r^* = 0 \quad \text{at } z^* = 1$$

$$\bar{v}_r^* = 1 \quad \text{at } r^* = 1$$

$$p^* = 0 \quad \text{at } r^* = 1$$

The dimensionless describing equations become

$$v_r^* \frac{\partial v_r^*}{\partial r} = -\frac{\partial p^*}{\partial r^*} + \frac{R\nu}{h^2 \bar{U}_e} \left(\frac{\partial^2 v_r^*}{\partial z^{*2}} - \overset{\text{neglect}}{\cancel{\frac{h^2}{R^2} \frac{v_r^*}{r^{*2}}}} \right)$$

where $(h^2/R^2) \ll 1$ so that the last term is negligible. For the dimensionless boundary conditions and equations to be identical it is only necessary that the Reynolds number $h^2 \bar{U}_e / R\nu$ be the same for the two different gap heights. In

Art. 4.6   DIMENSIONLESS VARIABLES   229

terms of the flow rate q

$$\text{Re} = h^2 \overline{U}_e / R\nu = qh / 2\pi R^2 \nu$$

where $\overline{U}_e 2\pi R h = q$. Hence for $R = $ constant, in order for Re to be the same it is necessary to require that

$$(qh)_1 = (qh)_2$$

for the two runs. Suppose that $h_2 = h_1/2$; then, if we let $q_2 = 2q_1$ the two dimensionless solutions would be identical and the two flows would be similar. (Near the inlet hole the flow would not be similar, but this flow region is not significant.)

Extension 4.7.1. Suppose that only one characteristic length (h or R) is used in the transformation to dimensionless variables. Show that for $h_2 = h_1/2$, in order for identical dimensionless solutions to exist, $R_2 = R_1/2$ and $q_2 = q_1/2$.

Example 4.8

The average velocity V of water flowing from a short pipe of length L attached to an open tank depends on the vertical distance H of the pipe outlet below the water surface, the viscosity μ, the density ρ, the acceleration of gravity g, and the diameter D of the pipe. Relate V to the other variables in the problem.

Solution. The average velocity V and the diameter D are chosen as the characteristic velocity and length. The Reynolds number $VD\rho/\mu$ is the parameter introduced from the dimensionless Navier-Stokes equation

$$\frac{D^* \mathbf{V}^*}{Dt^*} = -\nabla^* p_k^* + \frac{\mu}{VD\rho} \nabla^{*2} \mathbf{V}^*$$

where $p_k^* = p^* + (gD/V^2)h^*$. The boundary condition $p^* = 0$ on the free surface requires that on the free surface $p_k^* = (gD/V^2) \times (H/D)$ at $h^* = H/D$. The pressure at the exit is $p^* = 0$ at $x^* = L/D$. The dimensionless relation is then

$$V^* = f\left(\frac{VD\rho}{\mu}, \frac{V^2}{gD}, \frac{H}{D}, \frac{L}{D} \right)$$

The average velocity in the pipe has been chosen as the characteristic velocity; thus $V^* = 1$. The preceding relationship is then

$$1 = f\left(\frac{VD\rho}{\mu}, \frac{V^2}{gD}, \frac{H}{D}, \frac{L}{D} \right).$$

This relationship may be written as

$$\frac{V^2}{gD} = f_2\left(\frac{VD\rho}{\mu}, \frac{H}{D}, \frac{L}{D} \right)$$

where f_2 represents a function different from the function f. In an investigation the results would be presented in terms of the dimensionless parameters.

Example 4.9

A study is to be made of a proposed STOL (short take off and landing) airfoil by investigating the flow characteristics of a model. The objective is to obtain the optimum design to give the maximum lift at a particular speed. The quantities which can be varied in the model study are the position x_E of the engines, angle θ orienting the engines, boundary-layer control-slot width w, location x_s of the slot, wind tunnel speed U, length L of the airfoil from leading to trailing edge, temperature T of the air, angle of attack α of the airfoil, and engine mass flux $\dot{m}$. Propose how the study should be organized.

Solution. The important parameter from the governing equations is the Reynolds number. The study is conducted in a wind tunnel so it is assumed that the viscosity for model fluid and prototype fluid are equal. The characteristic length is the airfoil length L and the characteristic speed is the aircraft speed U. In dimensionless terms the lift depends on the various parameters in the equations and boundary conditions written in dimensionless form, that is,

$$\frac{\text{Lift}}{\rho_0 U^2 L^2} = f\left(\text{Re}, \frac{x_E}{L}, \frac{x_s}{L}, \frac{w}{L}, \theta, \alpha, \frac{\dot{m}}{\rho_0 U L^2}\right).$$

We are primarily interested in the variation of dimensionless lift with Reynolds number; hence we would collect data for the study by varying one parameter at a time. First, the results for $\text{Lift}/\rho_0 U^2 L^2$ vs. Re would be plotted for various engine locations, holding all other parameters constant. Then, for a particular engine location, the slot location would be varied, next, the slot width, the angle of attack, and so forth. The best possible design would then follow from an analysis of the data and the proposed STOL airfoil configuration could then be evaluated.

4.7 CLOSURE

This chapter was intended to deliver two messages, one explicit and one implicit. Explicitly, we have shown that the Reynolds number is the governing parameter provided by the equations for an incompressible, nonoscillatory flow and that the Mach number, the Prandtl number, and the specific heat ratio are the additional parameters which are introduced by the equations for a compressible flow of a perfect gas. Additional parameters may be introduced (possibly through the boundary conditions): the Froude number when a free surface or buoyancy effects are important in the description of the problem, the Weber number when surface-tension effects are important, the Strouhal number for oscillatory flows.

Implicitly, we have attempted to stress the idea that the governing equations provide the rational framework or structure from which these important effects may be deduced. The above-noted parameters do, in fact, provide a basis for similitude studies. In this sense, they could have been proposed as elements of "divine revelation" or abstractions from the intuition of the "technological priesthood" to be simply used by the "laity." For straightforward problems, this blind-faith utilization would be sufficient to obtain useful answers. However, the engineering challenges are not to be found in straightforward problems. In many cases exact similitude is not possible; one must assess whether the available conditions can be made "close enough." Very often, the assessment of the proper boundary conditions requires considerable skill and judgment. The hoped-for implicit learning experience will mean that the student can approach these more sophisticated questions from the rational framework of the appropriate equations and boundary conditions. In the pages that follow, we will consider many special cases of the field equations. The particular characteristics of inviscid flows, laminar flows, turbulent flows, compressible flows, and others will be investigated in detail. If we are jointly successful, the reader will see that these individual topics are not only appropriate for the solution of specific problems but they will also provide the insight to allow him to deal successfully with the challenging types of problems we have noted above.

This chapter could have been expanded to include many ramifications of the matters of similitude and the use of the governing equations. Instead, these remaining concepts will be considered in relationship to the special kinds of flows to be treated in detail. For example, the description of turbulent flow requires a recognition of the time-varying character of the flow field even if it is "steady" in the sense of a long-term average. For a given geometry, the magnitude of the Reynolds number may be used to characterize the nature of an incompressible flow; this is examined in detail in Chapter 5.

A specific example will provide an appropriate terminus for this chapter. In the early days of wind tunnel testing, the measurement of the drag on a sphere was evolved as a standard measurement that each laboratory could perform to establish the accuracy of its instrumentation and techniques. The drag measurement on the sphere was selected because dimensional reasoning indicated that it should depend only upon the diameter D, approach velocity U, density ρ, and viscosity μ; that is, from what we have shown in this chapter,

$$\frac{F_D}{\rho U^2 D^2} = f\left(\frac{\rho U D}{\mu}\right) \qquad (4.40)$$

When the detailed measurements by highly respected laboratories were compared, certain details of the function f did not agree. It was not until some time later that it was discovered that the level of turbulent velocity fluctuations in the oncoming stream was also an important factor in the details of the f function. That is, these fluctuations represent an aspect of the upstream boundary conditions for the boundary-value problem of the flow around a sphere which does have an observable influence on the solution for the flow around the body. A boundary-value problem approach would not guarantee that one would recognize the importance of these fluctuations *but*, unlike a purely dimensional-reasoning approach, it does provide a framework in which the observation may be rationally understood and perhaps predicted.

PROBLEMS

4.1 Express the following quantities in non-dimensional form: the shearing stress τ, the flow rate q, the torque T on an impeller, and the power input P necessary to operate a device. Use U and D as characteristic quantities.

4.2 Non-dimensionalize the Navier-Stokes equation (Eq. 4.2), keeping the body-force term in the equation directly. Do not introduce the dynamic pressure. Show that the parameter gL/U^2 appears as a coefficient in the equation.

4.3 A model study is to be made on a large water pump. The pressure at the outlet of the pump is 500 psi. What pressure should exist at the pump model outlet if water is used in the 10:1 scale model?

4.4 A model is constructed of a STOL (short-take-off-and-landing) aircraft airfoil at a scale ratio of 4:1. The flow at low speeds, take-off and landing, is being investigated by experimenting with the model in a wind tunnel at atmospheric temperature and pressure. (a) Determine the wind tunnel speed that should be used to simulate landing at 100 fps. (b) The velocity at the top of the airfoil is predicted by a computer analysis to be 125 fps. What velocity is expected on the model? (c) A drag force of 20 lb is measured on the model. Determine the corresponding drag force on the STOL airfoil.

4.5 A model of a hydroelectric turbine blade is constructed ten times larger than the prototype blade. The flow around the model is investigated in a wind tunnel with air at 70°F and 14.7 psia. The water is jetted from nozzles at the blades at a velocity of 200 fps and a temperature of 50°F. (a) What velocity should be used in the wind tunnel? (b) What force on the prototype corresponds to 10 lb measured on the model?

4.6 A 16:1 scale model of a dam spillway is to be used to study flow conditions to be expected when the actual dam is constructed. If 50,000 gallons per minute of water flow over the prototype, what should the flow rate of water be over the model?

4.7 A 25:1 model of a ship is used to predict the effect of a proposed design change. What towing speed should be used for the model to simulate a ship speed of 30 knots? A drag force of 4 lb is measured on the model. What drag is expected on the ship?

4.8 If a droplet falls from an eyedropper, it undergoes a natural oscillation as it falls. To study the flow characteristics around a droplet, a 5:1 scale model (the model is larger than the droplet) is constructed so that velocity measurements can be made. The model, capable of undergoing the same type of oscillation as observed in the droplet, is studied in a vertical flow wind tunnel. Determine the frequency ratio of the model and droplet. We are interested in the flow as terminal velocity is reached. Air is the fluid flowing around both model and prototype.

4.9 The equations considered in this chapter have been written for an inertial reference frame. Consider a generalized problem in which the similarity parameters are desired for an incompressible flow field observed in a reference frame that is accelerating linearly with a magnitude $\ddot{X}$ and undergoing a rotational motion described by Ω and $\dot{\Omega}$. What parameters are introduced by these non-inertial effects? At time $t = 0$, the rotational speed is Ω_0 and the acceleration is $\ddot{X}_0$.

4.10 A golf ball manufacturer wishes to study the effects of the dimple shapes for the control of lateral pressure forces which influence the seriousness of a hook or slice. (The purpose of the dimples is to make the flow turbulent near the surface of the ball, not only reducing the size of the wake but making it more nearly symmetric in the presence of the spin.) In order to carry out detailed studies, a model ball which is three times standard size will be investigated in the wind tunnel. What parameters must be controlled to successfully model the hook or slice phenomenon? What should be the velocity in the wind tunnel if we wish to simulate a golf ball speed of 60 fps, and what rotational speed should it be rotated at if the prototype golf ball rotates at 40 rad/sec?

4.11 A model of a supersonic aircraft is to be studied with the objective of obtaining a lift-vs.-speed relationship. The aircraft is designed to fly at 60,000 ft, where $T = -70°F$ and $p = 2$ in. of mercury. A 20:1 scale model is used in a wind tunnel and the lifting force, measured at 2000 mph, is 900 lb. What force would this correspond to on the aircraft? Neglect viscous effects. The pressurized wind tunnel operates at 10 psia with a temperature of 40°F.

4.12 A ship is cruising at 30 knots and the passengers experience smoke pollution on the deck from stack exhaust. A 20:1 scale model is to be studied to provide a hopeful solution. Should a water channel or a wind tunnel be used? What velocity should be used in the model study?

4.13 A model of a swimming microorganism found in water is to be studied. If the tiny organism is 10^{-3} in. long, choose a possible model size and then determine the velocity which would be necessary in a wind tunnel. The organism swims at a speed of 0.1 body lengths per second.

4.14 A model of a large 5-ft-dia. fan is to be studied in order that the operating characteristics of the prototype fan may be predicted. The model is one-fifth size. In operation it moves air at the rate of 200 cfs while rotating at 2500 rpm. (a) Predict the flow rate and rotational speed of the prototype fan. (b) Also determine the ratio of power required to run the prototype compared to that required to run the model.

4.15 A fan jet cruises at 30,000 ft at a speed of 600 mph. A model study is proposed to determine an engine's performance after certain changes are made in the jet engine. The model engine, built to one-sixth scale, is tested in a wind tunnel at 70°F and 14.7 psia. What wind tunnel speed should be used? What rotational speed of the compressor and turbine should be used to simulate rotational speeds of 15,000 rpm? Determine the ratios of the captured air mass flux $\dot{m}$, of thrusts, and of fuel rates. Neglect viscous effects.

4.16 Starting with the momentum and energy equations (Eqs. 3.54 and 3.55, show that the dimensionless Eqs. 4.27 and 4.28 result.

4.17 The torque T necessary to rotate a disc of radius R a distance h from a flat plate depends on the rotational speed ω and the fluid viscosity μ. Relate T to the appropriate variables.

4.18 The power P necessary to drive a fan depends on the rotational speed ω of the fan, the fan diameter D, the viscosity μ, and density ρ of the fluid. Relate P to the other parameters.

4.19 The average velocity V of water flowing from a hole in the side of an open tank depends on the height h of the hole below the water surface and the acceleration of gravity g. Show that $V = \text{const}\sqrt{gh}$ by considering the dimensionless differential equations and boundary conditions. It is postulated that the density influences V. Is this reasonable?

4.20 The performance of a centrifugal pump measured in terms of the pressure rise Δp depends on a number of parameters, including the inlet velocity V and the inlet diameter d, the impeller diameter D, the rotational speed ω of the impeller, the viscosity μ and density ρ of the fluid, and the outlet area A. Express the pressure rise in terms of the other parameters.

4.21 Water flows from a reservoir out a pipe of diameter D and length L. The describing equation is

$$0 = -\frac{1}{\rho}\frac{\partial p}{\partial z} + \nu\left(\frac{\partial^2 v_z}{\partial r^2} + \frac{1}{r}\frac{\partial v_z}{\partial r}\right)$$

The boundary conditions are $v_z = 0$ at $r = D/2$, v_z is finite at $r = 0$, and $p = 0$ at $z = L$. The average velocity is V. Normalize the equations, using V as the characteristic velocity, and select the appropriate characteristic lengths from the boundary conditions. A flow rate of 2 cfs is used during a particular study. The pipe is then doubled in length with the same diameter. Determine the new flow rate in order that identical dimensionless solutions result.

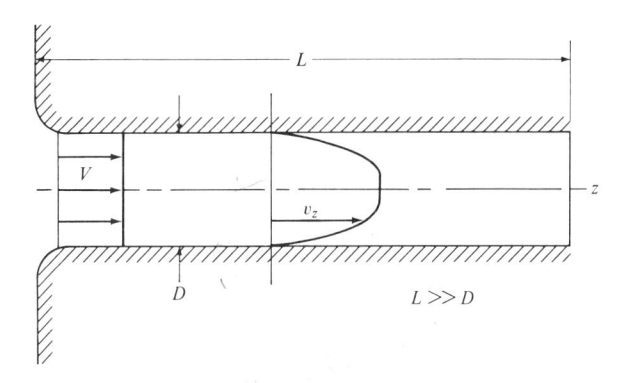

Prob. 4.21

4.22 It is possible for a flow situation to occur in which no characteristic velocity or length occurs naturally. The flow caused by an infinite flat disc rotating about an axis normal to the disc is an example. In this problem the kinematic viscosity ν and the rotational speed ω are the only parameters which influence the solution. "Invent" a characteristic velocity, length, and time, non-dimensionalize the momentum equations and boundary conditions, and determine any dimensionless numbers which appear.

4.23 The design of the agitator in a washing machine is the object of an experimental study. One phase of the study deals with the relationship between the maximum torque T necessary to oscillate the agitator shaft and the associated parameters, which include the oscillation frequency f, the angular velocity ω with which the agitator rotates during the oscillation, the diameter D of the agitator, the height h and the width w of the paddles, and the number N of paddles. Relate the maximum torque to the other variables in the problem. Viscous effects do not influence the flow. The effect of the free surface is also unimportant.

4.24 Two fluid flows with total flow rate q are dumped into a large mixing chamber of diameter D, are mixed thoroughly by a paddle wheel of diameter d rotating at speed ω, and then exit through a common pipe of radius R. Rectangular baffle plates with dimensions $h \times w$ are positioned in the chamber to enhance the mixing process. A model study is performed to determine the optimum dimensions on the several components of the mixer. To perform the study one must choose the various parameters to be varied. The viscous mixing process dominates the fluid flow with the through-flow having only a small effect. Indicate what parameter governs the flow and then propose how the study should be carried out. It is assumed that the quality of mixing can be measured in the exit pipe.

SELECTED REFERENCES

A discussion on extracting the similitude parameters from the governing equations that parallels that of the present text is provided by J. W. Daily and D. R. F. Harleman, *Fluid Dynamics*, Addison-Wesley Publishing Co., Reading, Mass., 1968. A more comprehensive presentation, as well as a critique of dimensional analysis methods is that by S. J. Kline, *Similitude and Approximation Theory*, McGraw-Hill Book Co., New York, 1965. Nearly all texts provide a discussion of the Buckingham pi theorem and dimensional analysis. For a lucid discussion of these points, see D. G. Shepard, *Elements of Fluid Mechanics*, Harcourt Brace and World, New York, 1965.

5

Fluid-Flow Phenomena

5.1 INTRODUCTION

In this chapter, many of the phenomena found in fluid flows will be introduced. Our objective is to give a short description of each phenomenon, with appropriate definitions, so that these concepts may be utilized in considering subsequent applications. With a conceptual grasp of these phenomena, with the basic equations in integral and differential form, and with the capacity to extract the governing parameters from the appropriate equations, the basis for consideration of many fluid mechanics problems will have been established.

The phenomena to be considered, which will be described in terms of their occurence in natural or technological flows, will be related to the governing equations. These equations serve as a useful reference for the description of the phenomena since the physical occurrences are characterized by the neglect or dominance of certain terms or boundary conditions. Examining these phenomena in terms of the equations will also allow an easier transition from qualitative descriptions to quantitative problem solving using these concepts.

In general, the governing equations are nonlinear; typically, they possess no unique solutions. Their nonlinearity prohibits the very powerful technique of superimposing various solutions to achieve the solution to a more complicated problem. (Recall that superposition was possible for fluid statics since the governing equation was linear.) Since the physical phenomena may be described in terms of special forms of the governing equations, it has been natural for various special analytical techniques to be developed for these cases. Some of these analytical considerations will be presented with the detailed considerations in subsequent chapters.

Our attention will be confined to Newtonian fluids involving isotropic, homogeneous fluid properties. Various other, often very interesting phenomena occur because of non-Newtonian fluids, non-isotropic, or nonhomogeneous properties. These include non-Newtonian "elastic water" and the nonhomogeneous regions in fluid flows which account for the smog concentration in the Los Angeles basin.

5.2 PHENOMENA ASSOCIATED WITH REYNOLDS NUMBER

For an incompressible flow, the nondimensional momentum equation is

$$\frac{\partial \mathbf{V}}{\partial t} + (\mathbf{V} \cdot \nabla)\mathbf{V} = -\nabla p_k + \frac{1}{\text{Re}} \nabla^2 \mathbf{V} \tag{5.1}$$

where the symbols represent the dimensionless quantities of Eq. 4.5. (Asterisks are omitted for convenience.)

The left side of the equation is the Eulerian statement of the acceleration; the right side results from the net force acting on a fluid particle, that is, the fluid particle occupying the Eulerian location at the instant of interest. Recall that this form of the equation is most appropriate for those physical situations without a free surface or interface, where gravity effects are not important.

Equation 5.1 represents a balance between the force caused by the pressure distribution, the force caused by the shear stresses, and acceleration (or the d'Alembert "inertial force"). The phenomena to be described below may be considered to be special cases for which one or more of these physical effects is negligible. It will be convenient to consider these special cases in terms of an increasing Reynolds number.

1. Stokes' Flow

When the viscous shear effects are so large as to balance the driving effect of the pressure, a *"creeping"* or *Stokes' flow* condition exists; for this type of flow the acceleration is negligible. Such a condition will exist if the Reynolds number is small, that is,

$$\text{Re} = \frac{UL}{\nu} \ll 1 \tag{5.2}$$

where U and L are representative quantities in the particular flow situation. The governing equation, found by neglecting the appropriate terms in Eq. 5.1, is

$$\nabla p_k = \frac{1}{\text{Re}} \nabla^2 \mathbf{V} \qquad (5.3)$$

Examples of a low-Reynolds-number flow may be found in a glacier (small U), the drag on liquid droplets in a sneeze (small L), and the motion of "molasses in January" (large ν). The engineering applications of this type of motion are quite limited; hence Stokes' flows will not be considered in detail.

2. Laminar Flow

Consider fluid to be flowing in a round pipe such that the flow rate is independent of time (Fig. 5.1a). If the output signal from a hot-wire anemometer (a device that produces a signal proportional to the velocity component normal to the wire) were displayed on an oscilloscope, for Reynolds numbers below 2000 it would appear as shown in Fig. 5.1b. The velocity at a particular location would be steady except for very small disturbances in the fluid due to vibrations of the pipe

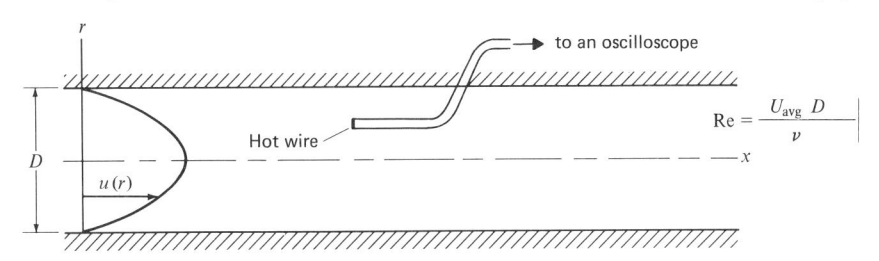

(a) Schematic for hot-wire measurement

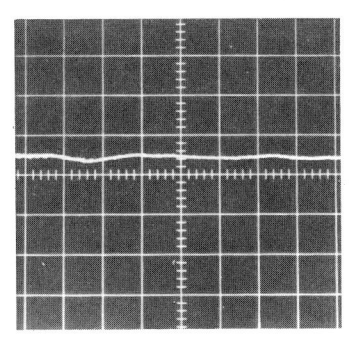

(b) Oscilloscope output

Fig. 5.1. Laminar flow in a pipe.

caused by the impeller blades of a pump or fan, or other such effects. A flow possessing this well-behaved, nonfluctuating character is a *laminar* flow; however, all laminar flows are not steady flows. A flow could fluctuate and yet be laminar.

Flow in a geometry other than a circular pipe can be laminar if the Reynolds number is sufficiently small. The flow of an incompressible, isotropic, homogeneous Newtonian fluid is governed by the full Navier-Stokes equations, along with the continuity equation; in cartesian coordinates, these are

$$\frac{\partial u}{\partial x} + \frac{\partial v}{\partial y} + \frac{\partial w}{\partial z} = 0$$

$$\frac{\partial u}{\partial t} + u\frac{\partial u}{\partial x} + v\frac{\partial u}{\partial y} + w\frac{\partial u}{\partial z} = -\frac{\partial p_d}{\partial x} + \frac{1}{\text{Re}}\left(\frac{\partial^2 u}{\partial x^2} + \frac{\partial^2 u}{\partial y^2} + \frac{\partial^2 u}{\partial z^2}\right)$$

$$\frac{\partial v}{\partial t} + u\frac{\partial v}{\partial x} + v\frac{\partial v}{\partial y} + w\frac{\partial v}{\partial z} = -\frac{\partial p_d}{\partial y} + \frac{1}{\text{Re}}\left(\frac{\partial^2 v}{\partial x^2} + \frac{\partial^2 v}{\partial y^2} + \frac{\partial^2 v}{\partial z^2}\right)$$

$$\frac{\partial w}{\partial t} + u\frac{\partial w}{\partial x} + v\frac{\partial w}{\partial y} + w\frac{\partial w}{\partial z} = -\frac{\partial p_d}{\partial z} + \frac{1}{\text{Re}}\left(\frac{\partial^2 w}{\partial x^2} + \frac{\partial^2 w}{\partial y^2} + \frac{\partial^2 w}{\partial z^2}\right)$$

(5.4)

All terms—including the nonlinear acceleration terms—in these equations may be important for laminar flow. Only in certain simple geometries, typically those in which the nonlinear terms vanish, can analytic solutions be found for laminar flow. Some of these solutions will be presented in Chapter 6.

Examples of laminar flow include "smooth-surfaced" water flow from a faucet, flow near the leading edge of an airfoil, flow in a small-diameter conduit, and flow of oil in a bearing. All Stokes' flows are also examples of laminar flows. Unsteady laminar flows include start-up of fluid flow in a conduit, flow around a body falling from rest for small elapsed time, and flow near the vanes for a slowly rotating turbine. Non-Newtonian fluids like ketchup, blood, and mayonnaise may exhibit a liminar-flow character although they may not be strictly defined in terms of a Reynolds number since a single material property to describe the relationship between the stress and the rate of strain cannot be defined.

3. Instability, Transition, and Turbulent Flows

As the Reynolds number for a given flow is increased, disturbances of the regular streamline pattern are often observed. For example, in the pipe flow of Fig. 5.1 the streamlines in the pipe can be observed to show an undulatory pattern at Reynolds numbers of approximately 2000. An experiment on this point was carried out by Reynolds.[*] Although his study may be criticized on certain technical grounds, it was most insightful and is generally recognized as the first examination of disturbances in laminar flow and an assessment of the conditions for growth and decay. The dimensionless group $\rho U D/\mu$ was recognized by Reynolds as the parameter determining the response of the flow to a disturbance. Figure 5.2 shows a schematic drawing of Reynolds' experiment and his results.

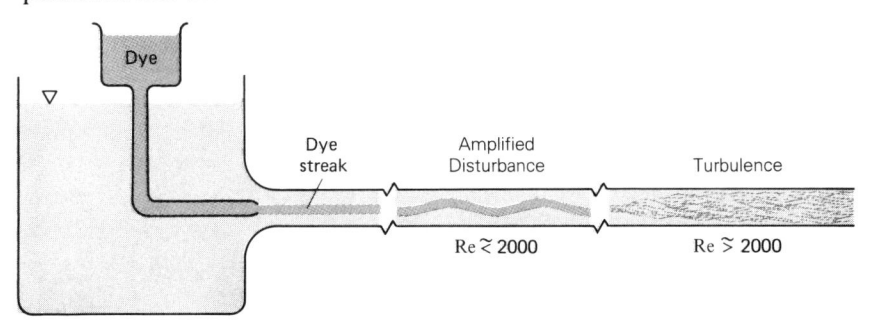

Fig. 5.2. The Reynolds experiment for flow in a circular tube.

What Reynolds observed may be termed "laminar-flow breakdown." It occurs when the Reynolds number exceeds a critical value. Its occurrence is not easily calculated analytically. The linear stability of many flows has been examined by the process of (1) starting with a solution to the Navier-Stokes equations for a given flow, (2) postulating a regular or ordered disturbance function, (3) assuming small disturbances so that terms involving the product of disturbances can be neglected, and (4) determining the minimum Reynolds number for which the postulated disturbance will grow. In general, this technique will give a critical Reynolds number which is an upper bound for the experimentally determined critical Reynolds number of the physical

---

[*]"An Experimental Investigation of the Circumstances which Determine Whether the Motion of Water in Channels Shall Be Direct or Sinuous," *Philosophical Transactions of the Royal Society of London,* 1883, Vol. 174, pp. 935–982.

problem. A second method, which examines the stability of a disturbance on the basis of energy considerations, provides a lower bound for the observed critical Reynolds number. For a circular pipe, linear stability calculations show that the flow is stable at all Reynolds numbers, no matter how large; energy considerations result in a critical Reynolds number of 1620 and typical experimental results show a critical value near 2000 with a maximum of 40,000 recorded in an experiment with an exceedingly small disturbance level.

A laminar flow in an unstable condition may undergo a catastrophic breakdown to a nonperiodic motion where the streamlines are irregular and strongly time-dependent (a turbulent motion) or it may develop into a second laminar-flow state comprising a primary and a secondary motion. The Taylor cells of the flow between concentric cylinders are a classic example of a laminar condition with a primary and secondary flow; see Fig. 5.3. Many additional types of flow have been observed in this geometry; they are obtained by various combinations of inner and outer cylinder-rotation speeds.

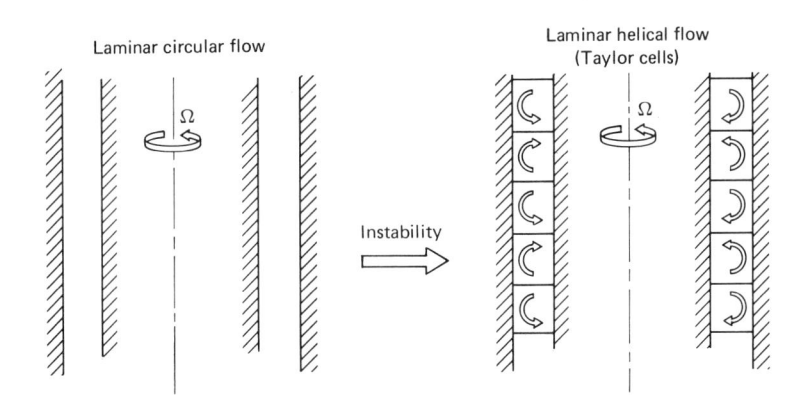

Fig. 5.3. Flow between concentric cylinders.

This rather brief description of a rather vast arena for fluid-mechanical research (stability theory) is primarily to serve as an introduction to the phenomenon of transition. "Transition" is a general term which can be used to signify either the change of a laminar flow to another, more complex one or its breakdown and development into a turbulent flow. As will be noted subsequently, the transition phenomenon is of great importance for the study of high-Reynolds-number flows over bodies.

4. Description of a Turbulent Flow

For the pipe flow of Fig. 5.1a, let the flow rate increase so that the Reynolds number increases. The disturbances, which are ever present, will, as the critical Reynolds number is exceeded, be amplified in the pipe. As they are swept by the fixed probe an intermittent signal as shown in Fig. 5.4a would be recorded. The flow will be intermittently laminar and turbulent. As the Reynolds number is increased still further a completely turbulent flow results.* Parts b and c of the figure show a fully turbulent signal with a high frequency and a lower frequency from such a probe. The high-frequency signal results from a relatively large Reynolds number for which there are "small"-scale motions in the flow being convected "rapidly" past the probe.

If one would contemplate a solution to the governing equations, the time-dependent Navier-Stokes equations, for this pipe flow it would be

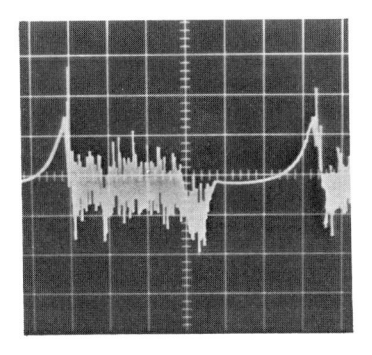

(a) Intermittent signal

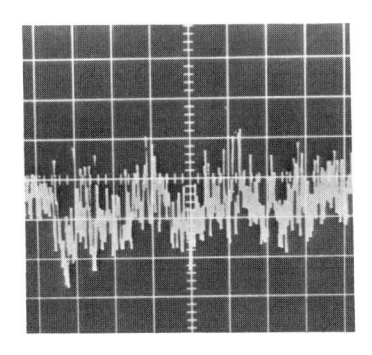

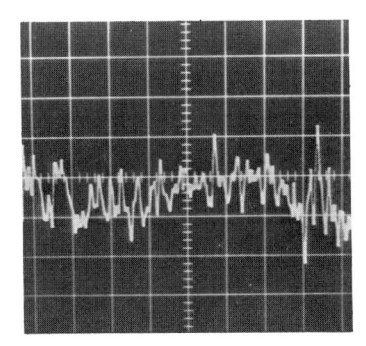

(b) High-frequency signal   (c) Low-frequency signal

Fig. 5.4. Turbulent-flow velocity-signal outputs.

*Turbulent flow is discussed in Chapter 7.

necessary to solve an initial-value problem. This requires the velocity at all points of the flow field at some initial instant to be known. Such information cannot be expected. More importantly, the complete solution, which is characterized by the trace of Fig. 5.4b, for only one point in the flow field, contains far more information than is useful for an engineering task. That is, if one knew the time mean (average) value and possibly some of the statistical features of the indicated $u(t)$, one could determine the various quantities of interest.

In order to obtain some insight into the problem we invent a different way of looking at a turbulent flow. Consider the velocity components at a point in a turbulent flow to be given by

$$u = \bar{u} + u' \quad v = \bar{v} + v' \quad w = \bar{w} + w' \qquad (5.5)$$

where $\bar{u}$ is the x-component of the *time-averaged velocity* and the fluctuating velocity component u' is, by definition, the difference between the instantaneous and the time-averaged velocity. The x-component of the velocity would appear as shown in Fig. 5.5a. Such a time record could be generated from a probe in the pipe of our mental experiment. The average velocity $\bar{u}$ is

$$\bar{u} = \frac{1}{T} \int_0^T u \, dt \qquad (5.6)$$

In many problems such as the pipe flow it is not difficult to decide upon a suitable time T over which the integration process should occur; one simply lets T grow large enough so that repeated samples (i.e., for different $t = 0$ conditions) yield sufficiently similar results so that the computed value is tending toward a stationary condition (mathematically, $d\bar{u}/dT = 0$). There are, however, cases where $\bar{u}$ is a function of time for which the determination of a sensible T value is a matter for the analyst or experimentalist to arbitrarily decide based upon the character of the problem. The velocity in the exhaust of a finite-volume pressurized container, and the wind velocity on a gusty day, are examples of unsteady flows for which a suitably defined average would be useful, as the reader may verify from his own experience. For turbulent flow like that in a washing machine with an oscillating agitator, it is more useful to construct averages from repeated measurements that are synchronized with the periodic forcing function than to determine long-term time averages. These usually include flows for which the magnitude of the fluctuations are larger than the magnitude of the time-averaged quantities.

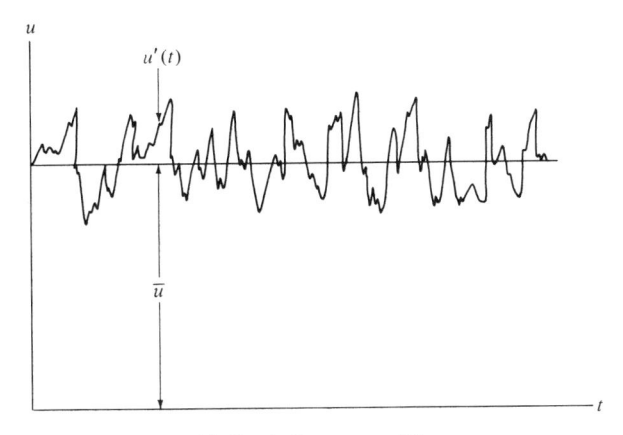

(a) Steady time-averaged flow

(b) Unsteady time-averaged flow

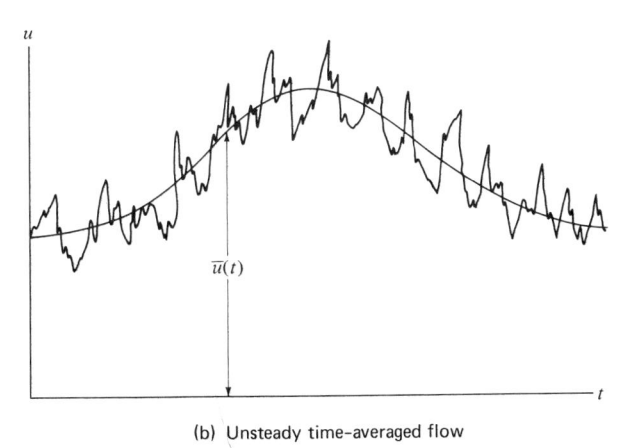

Fig. 5.5. A turbulent velocity component.

5.3 BOUNDARY-LAYER FLOW, INVISCID FLOW, AND SEPARATION

Whenever the coefficient of the highest-order term in a differential equation is small, compared to the other coefficients in the equation, a *boundary-layer solution* may result; that is, there may exist a narrow region, in which the dependent variable possesses large gradients making this highest-order term important. The highest-order derivative term is negligible beyond this narrow region, which is called a *boundary layer*. Boundary-layer behavior occurs in all areas of engineering; in fluid mechanics the highest-order term in the momentum equation is the term which represents the viscous effect and whose coefficient is Re^{-1} (see Eq. 5.1). We can thus expect boundary-layer behavior for large-Reynolds-number flows. The boundary layer, outside of which the viscous term is negligible, exists next to a solid surface. It was this characteristic which led L. Prandtl to coin the name "boundary layer" (originally "grenschict"). The boundary-layer thickness δ usually increases as the fluid flows along the boundary. Boundary-layer theory will be presented in detail in Chapter 9.

From a physical viewpoint, the fluid particles stick to the surface; consequently, their velocity is equal to the surface velocity. This is known as the no-slip condition and is valid for all flows for which the continuum approximation is valid. For large Reynolds numbers, the viscous effects are manifest near the surface in the region where the viscous term is important with respect to the other terms. Beyond this narrow boundary layer, the viscous effects become negligible and the fluid behaves as if it possesses no viscosity. This outer region is termed a region of inviscid flow.

Boundary layers exist in both external and internal flows (Fig. 5.6). Since a boundary layer is invariably thin compared to the body dimensions and since the boundary-layer region always exists underneath the inviscid flow, a good approximation to a flow around a body is to assume that the boundary layer does not exist. This approximation allows an inviscid solution to be determined from which the pressure distribution on the body can be calculated. This pressure distribution must be known before the boundary-layer equations can be solved. Formal mathematical techniques are available to "solve" the boundary-layer equations for the flow near the body and to join this solution to that of the inviscid flow. This is shown in Fig. 5.7 for point B on the airfoil surface of Fig. 5.6a. We see that the boundary-layer flow matches the inviscid flow at the edge of the boundary layer. The inviscid flow solution is thus extremely important for the analyses of flows around bodies. Inviscid flows are discussed in Chapter 8.

(a) Flow around an airfoil

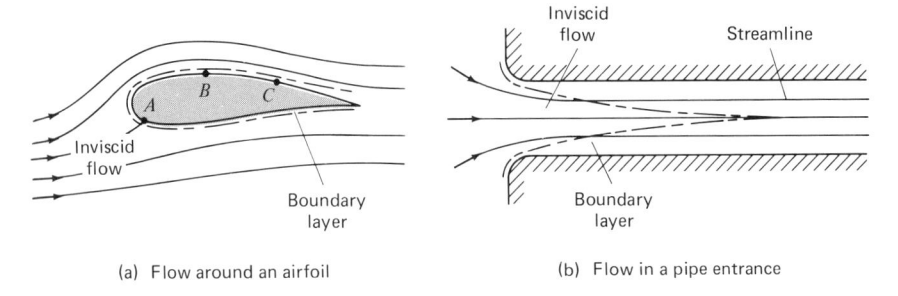

(b) Flow in a pipe entrance

Fig. 5.6. Examples of boundary layers in an external flow and in an internal flow.

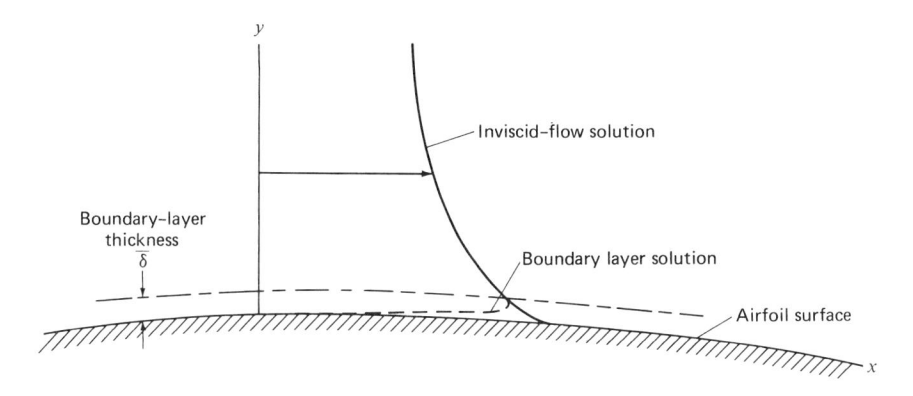

Fig. 5.7. Inviscid-flow and boundary-layer solution matching.

At point A on the airfoil in Fig. 5.6a a hot-wire sensor would indicate a laminar-flow boundary layer for all Reynolds numbers. At point C a turbulent flow would be indicated in the boundary layer for sufficiently large Reynolds numbers. This, of course, implies a transition from laminar to turbulent flow in the boundary layer between points A and C. The transition phenomenon is quite complex. Because it is a three-dimensional random occurrence involving an instability phenomenon (which in turn is influenced by surface roughness, external disturbances, etc.), it is not easily predicted. However, since such quantities as the shear stress and surface heat or mass transfer act quite differently in laminar as opposed to turbulent flow, its prediction is of considerable technical interest. This laminar-transition-turbulent behavior of a boundary layer occurs in all boundary layers of sufficient length (Fig. 5.8). For a flat plate aligned parallel to the free stream, as in Fig. 5.8, the *transition length* x_T may be influenced by several

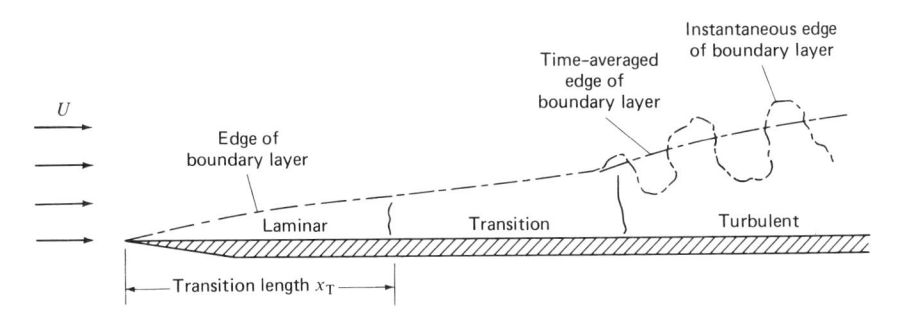

Fig. 5.8. Transition to turbulent flow in a boundary layer on a flat plate.

factors. It may be computed from $Ux_T/\nu \cong 3 \times 10^5$ as a lower limit. If the body is shorter than the transition length then transition will not occur. As well as on the Reynolds number, the transition length is dependent on the pressure gradient, the intensity of free-stream fluctuations, wall roughness, and wall temperature.

Occasionally, transition occurs from turbulent to laminar flow; a turbulent boundary layer can be forced into a laminar state by being exposed to a large favorable pressure gradient ($\partial p/\partial x < 0$). This process is referred to as *relaminarization*. Transition from turbulent to laminar flow also occurs as the Reynolds number associated with a turbulent flow is reduced. This transition usually occurs at a lower Reynolds number than that associated with the laminar-to-turbulent transition process.

One of the most striking features of a fluid flow is that the main flow does not have to (and often does not) follow the contour of the solid geometry. "*Separation*" is the term used to describe this phenomenon; the term "boundary-layer separation" is often used to describe the pattern denoted in Fig. 5.9 since it is the presence of the lower velocities in the boundary layer, in this case, which causes the separa-

(a) Separation on an airfoil

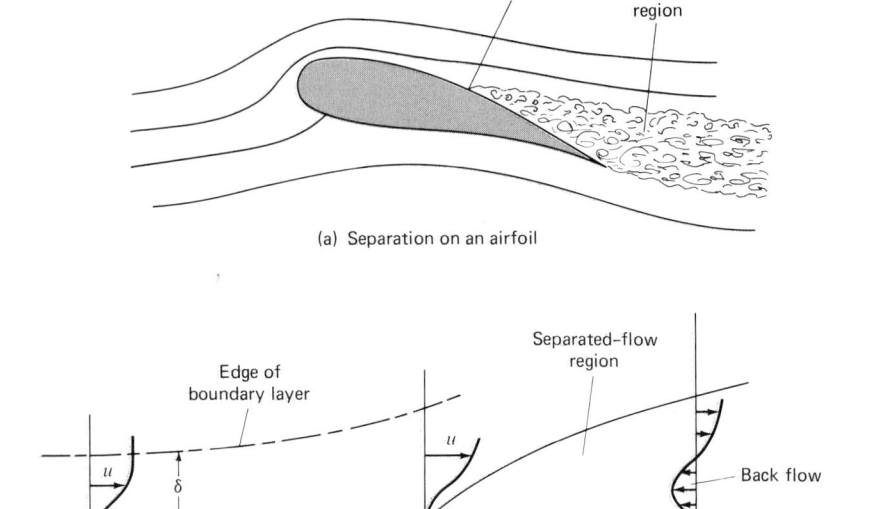

(b) Detail of region near a point of separation

Fig. 5.9. Boundary-layer separation.

Art. 5.3 BOUNDARY-LAYER FLOW, INVISCID FLOW, SEPARATION

tion phenomenon. Separation may result from a streamwise pressure increase or a radius-of-curvature effect. The *separation point* is that location where $\partial u/\partial y = 0$ at the boundary. Downstream of separation a back flow exists. A separation streamline divides the main flow from the recirculating (separated) flow and, on the time average, fluid does not pass from the main flow into the separated region. However, turbulent mixing of fluid does occur between the two regions. Separated flow is responsible for stall of an aircraft, the high drag of blunt objects moving at high Reynolds numbers, and the oscillatory wind forces acting on a tower. The separated region typically contains large-scale turbulent eddies.

Flow over a body at various Reynolds numbers is shown in Fig. 5.10. At low Reynolds number no separation is observed. As the Reynolds number increases, separation is observed in part b of the figure and several large-scale eddies in part d. Note the small angle between the separation streamline and the surface of the body, contrary to the way it is often sketched (e.g., Fig. 5.9b) for purposes of instruction.

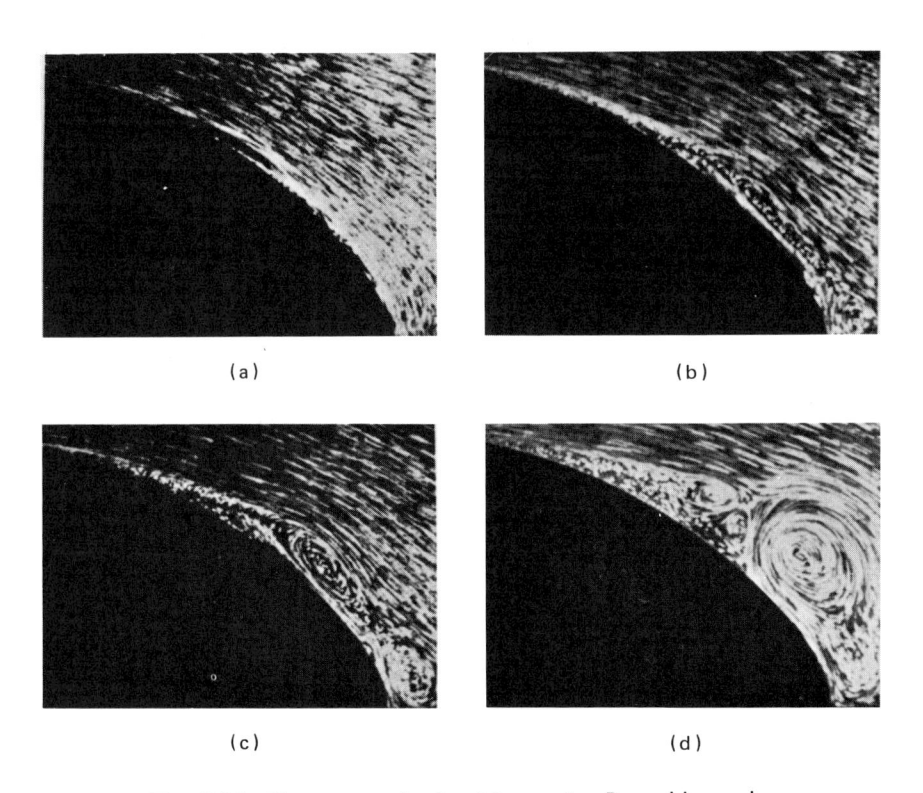

Fig. 5.10. Flow over a body at increasing Reynolds number.

5.4 VORTICITY AND CIRCULATION

1. Fundamentals

Vorticity is a vector quantity which is proportional to the angular momentum of a fluid element; like velocity, it is a pointwise vector function. Mathematically, vorticity is defined as

$$\omega = \nabla \times \mathbf{V} \tag{5.7}$$

and was introduced in Art. 3.3. The notion of vorticity is reintroduced here because it is quite useful to describe certain fluid-flow phenomena. The basic relationships involving vorticity will first be developed; examples of the pertinent phenomena will then be noted.

By means of the Stokes theorem and the definition of vorticity, we may equate the flux of vorticity through a surface area to the *circulation* Γ about a curve surrounding the area. A schematic representation of this is shown in Fig. 5.11 and is presented in equation form as

$$\Gamma = \oint_C \mathbf{V} \cdot d\mathbf{s} = \int_A \nabla \times \mathbf{V} \cdot \hat{n} \, dA$$

$$= \int_A \omega \cdot \hat{n} \, dA \tag{5.8}$$

where the curve represented by C is a reducible* curve. The enclosed area is always to the left as curve C is traversed.

An important property of the vorticity vector can be developed from its definition. Since the divergence of the curl of a vector quantity is identically zero we have

$$\nabla \cdot \omega = 0 \tag{5.9}$$

which is of the same form as the incompressible conservation-of-mass differential equation. The analogy with the conservation of mass is instructive because one can consider a *vortex tube*, a tube composed of a bundle of vortex lines, to be the equivalent of a streamtube. In a streamtube the mass flux through any cross section of the tube is independent of the orientation of the area or the position along the tube. A similar result is obtained for the vortex tube; that is, the flux of

---

*A reducible curve is one which can be collapsed continuously until it forms a point without passing outside the area. A curve encircling a hole would not be reducible.

Art. 5.4 VORTICITY AND CIRCULATION 251

(a) General definition of terms

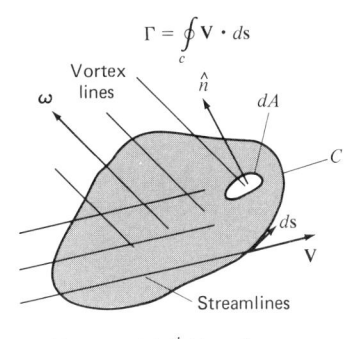

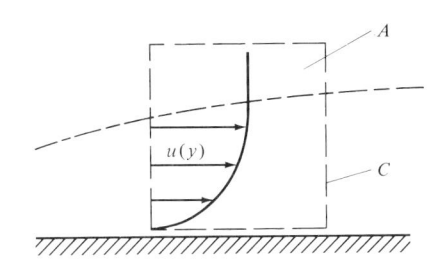

(b) Vortex lines are oriented normal to streamlines in this boundary-layer flow

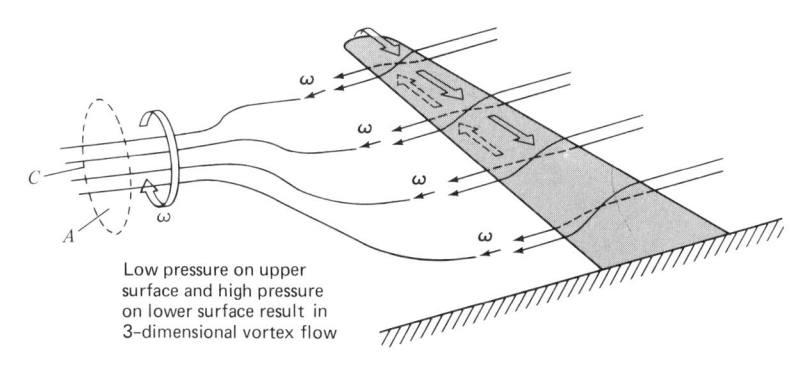

(c) Concentrated wing-tip vortex

Fig. 5.11. Vorticity and circulation.

vorticity is independent of the area orientation and the position along the tube. Consider the circulation around the curve shown in Fig. 5.12. The area surrounded by the curve is the lateral area of the vortex tube; $\omega \cdot \hat{n} = 0$ for this area. The circulation around this curve must then be zero. It is written as

$$\Gamma = \oint_C \mathbf{V} \cdot d\mathbf{s} = \int_①^② \mathbf{V} \cdot d\mathbf{s} + \int_②^③ \mathbf{V} \cdot d\mathbf{s} + \int_③^④ \mathbf{V} \cdot d\mathbf{s} + \int_④^① \mathbf{V} \cdot d\mathbf{s} = 0 \quad (5.10)$$

In the limit, as ① → ④ and ② → ③, the integral from ① to ② cancels the integral from 3 to 4 ; consequently,

$$\int_②^③ \mathbf{V} \cdot d\mathbf{s} = \int_①^④ \mathbf{V} \cdot d\mathbf{s} \quad (5.11)$$

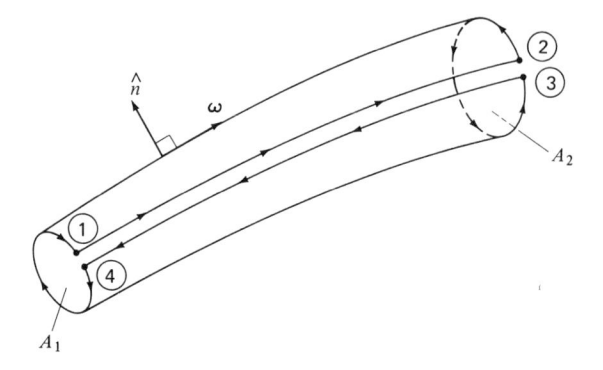

Fig. 5.12. Segment of a vortex tube.

where we have interchanged the limits of integration on the last integral. Equation 5.11 may be written (see Eq. 5.8)

$$\int_{A_1} \omega \cdot \hat{n} \, dA = \int_{A_2} \omega \cdot \hat{n} \, dA \tag{5.12}$$

which is of the same form as the integral continuity equation for an incompressible flow in a streamtube. If the vorticity is constant over the area then $\omega_1 A_1 = \omega_2 A_2$.

We can infer from Eq. 5.12 that a vortex tube (and in the limit, as the cross-sectional area of the vortex tube shrinks to zero, a vortex line) cannot originate or terminate in the fluid since this would require an infinite vorticity, a physical impossibility.

2. Vorticity Transport Equation

The Navier-Stokes equations can be used to derive two important relationships for vorticity and circulation. By forming the curl of the full Navier-Stokes equation, one obtains the vorticity transport equation,

$$\nabla \times \left[\frac{D\mathbf{V}}{Dt} = -\frac{\nabla p}{\rho} - g\nabla h + \nu \nabla^2 \mathbf{V} \right] \tag{5.13}$$

which can be written as

$$\frac{D\omega}{Dt} = (\omega \cdot \nabla)\mathbf{V} + \nu \nabla^2 \omega \tag{5.14}$$

where the curl of the pressure and body-force terms is zero and the first term on the right-hand side results from the curl operation on the convective acceleration.* This equation has several important consequences. Since $D\omega/Dt$ is the time rate of change of the vorticity of a fluid particle, an element possessing no vorticity will become vortical only through the action of viscosity for the conditions of the derivation; namely, for an incompressible flow and for an inertial reference frame.[†] Because of this, many flows which originate from rest, such as the flow of air in a wind tunnel and the flow around bodies in the atmosphere or a body of water, may be considered to be without vorticity (irrotational) for the bulk of the flow. Note that the viscous term, and thus the vorticity, is important near the boundary.

The boundary-layer phenomenon introduced in the preceding article is an excellent example of this. The fluid approaching the boundary layer is nonvortical (assuming it starts from a uniform or zero velocity condition). In a fundamental sense, the boundary-layer fluid is that which has become vortical. This results in the presence of a velocity gradient $\partial u/\partial y$. The outward growth of the boundary layer is governed by the viscous diffusion[‡] of vorticity, represented by the term $\nu \nabla^2 \zeta$ for a two-dimensional flow; this outward diffusion attempts to "contaminate" the outer fluid with vorticity. Recognizing that a boundary layer is fundamentally a region of concentrated vorticity

---

*The curl of the acceleration is expressed as

$$\nabla \times \frac{D\mathbf{V}}{Dt} = \nabla \times [\partial \mathbf{V}/\partial t + \tfrac{1}{2}\nabla V^2 - \mathbf{V} \times (\nabla \times \mathbf{V})]$$

$$= \partial \omega/\partial t + \tfrac{1}{2}\nabla \!\!\!\!/\, \nabla V^2 - \nabla \times (\mathbf{V} \times \omega)$$

$$= \partial \omega/\partial t + (\mathbf{V} \cdot \nabla)\omega - (\omega \cdot \nabla)\mathbf{V}.$$

See the expression for acceleration and the vector identies in Appendix C.

[†]The Navier-Stokes equations (Eq. 5.13 and, hence, Eq. 5.14) are restricted by the assumptions of an incompressible flow in an inertial reference frame. Additional vorticity-producing effects exist if the flow is not barotropic, that is, where $\rho \neq \rho(p)$, or if there are body forces which do not act through the center of mass of an element. An example of the latter is the Coriolis body force associated with a noninertial reference frame to describe the motion. These additional aspects are more readily accounted for in the context of the rate of change of circulation; hence, their introduction will be deferred until the rate of change of circulation is considered, in the next section.

[‡]Any physical process which is described by the equation $dC/dt = \alpha \nabla^2 C$ is called a *diffusive* process; α is a material property and C is a concentration of some physical quantity. Physical examples of diffusive processes are common in your experience: for example, the hot handle of a frying pan and water absorption by a paper tissue.

makes it much easier to understand the behavior of a boundary layer starting from rest. Figure 5.13 shows sequential time values for an impulsively started flow on a flat plate. The growth of the vortical region is easily recognized to result from the diffusion of vorticity into an ever larger region of the flow until the convective and diffusive effects at a point are in balance. For plane flow on a flat plate ($w = 0$ and $\omega = \zeta \hat{k}$)

$$\frac{\partial \zeta}{\partial t} + (\mathbf{V} \cdot \nabla)\zeta = \nu \nabla^2 \zeta \tag{5.15}$$

When the diffusive term $\nu \nabla^2 \zeta$ balances the convective term $(\mathbf{V} \cdot \nabla)\zeta$ the vorticity at a point ceases to change with time and the boundary layer reaches a steady-state condition.

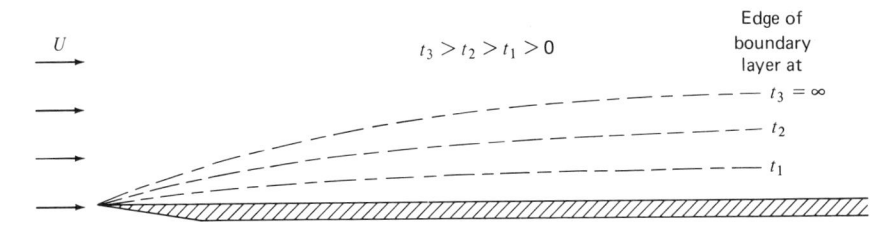

Fig. 5.13. Impulsively started flow on a flat plate. Boundary-layer thickness increases with time.

The "edge" of the boundary layer is located at the y value where the vorticity diffusion to larger y values is negligible. The asymptotic decrease of this diffusion effect means that an arbitrary definition of the boundary-layer thickness must be invoked, e.g. $u = 0.99 U$. An alternate definition could be based upon an arbitrarily small value for the vorticity.

The first term on the right-hand side of Eq. 5.14 is termed the "stretching" or "production" term since it can (1) amplify (or attenuate) the vorticity by the elongation (or shortening) of a vortex tube element and (2) produce, for example, x-component vorticity by reorientating the vortex tube segment which originally contained only y-component vorticity. As an example of the former, consider a vortex tube element approaching a contraction as shown in Fig. 5.14. The x-component of the vorticity vector is amplified as the fluid moves through the contraction since $\xi \frac{\partial u}{\partial x} > 0$ and $\frac{D\xi}{Dt} = \xi \frac{\partial u}{\partial x}$.

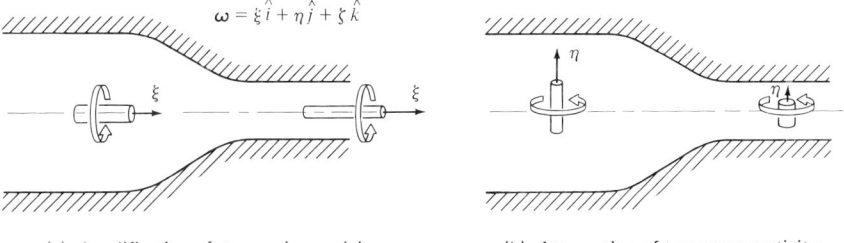

(a) Amplification of streamwise vorticity

(b) Attenuation of transverse vorticity

Fig. 5.14. Effect of a contraction on vorticity.

An example of the importance of production by reorientation of a vortex structure is provided by a turbulent flow near a wall. (See Fig. 5.15.) Although its time-averaged value would be zero, some y-component vorticity η will exist near the wall as a result of random fluctuations in the flow. These vertical vorticity elements are reoriented by the strong velocity gradients $\partial u/\partial y$, and their consequent scrubbing motions on the plate result in the streaking which is characteristic of the turbulent wall layer.

(a) Velocity profile in a turbulent flow near a wall

(b) Reorientation of vortex tube to produce ξ ($D\xi/Dt = \eta\,\partial u/\partial y$)

(c) Dye streaking the wall beneath a turbulent wall layer

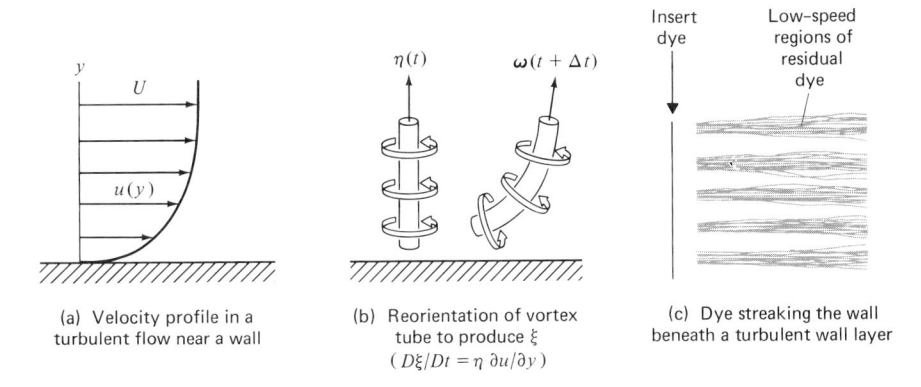

Fig. 5.15. Production of vorticity by vortex tube reorientation.

3. Time Rate of Change of Circulation

The time rate of change of the circulation about a material curve* is given by

$$\frac{D\Gamma}{Dt} = \frac{D}{Dt}\oint_C \mathbf{V}\cdot d\mathbf{s} = \oint_C \frac{D\mathbf{V}}{Dt}\cdot d\mathbf{s} + \oint_C \mathbf{V}\cdot \frac{D}{Dt}d\mathbf{s} \quad (5.16)$$

*A contiguous set of material particles which are free to compress, elongate, or otherwise distort, subject only to the restriction that their mutual boundaries remain in contact.

where the term $\dfrac{D}{Dt}d\mathbf{s}$ represents the rate of change of a fluid line element's length. From the definition of a derivative, we may write

$$\frac{D}{Dt}d\mathbf{s} = \lim_{\Delta t \to 0} \frac{d\mathbf{s}(t + \Delta t) - d\mathbf{s}(t)}{\Delta t} \tag{5.17}$$

The incremental distance $d\mathbf{s}$ is shown graphically in Fig. 5.16. From the vector polygon we see that

$$d\mathbf{s}(t) + \mathbf{V}(s + ds)\Delta t = \mathbf{V}(s)\Delta t + d\mathbf{s}(t + \Delta t)$$

or

$$d\mathbf{s}(t + \Delta t) - d\mathbf{s}(t) = [\mathbf{V}(s + ds) - \mathbf{V}(s)]\Delta t \tag{5.18}$$

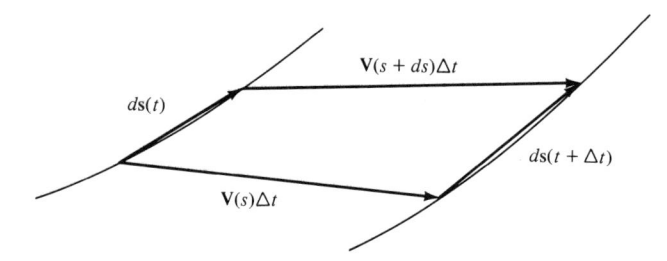

Fig. 5.16. Integration element $d\mathbf{s}$ at time t and at time $t + \Delta t$.

This allows us to write (see Eq. 5.17)

$$\frac{D}{Dt}d\mathbf{s} = \mathbf{V}(s + ds) - \mathbf{V}(s)$$

$$= d\mathbf{V} \tag{5.19}$$

Hence, the last integral in Eq. 5.16 becomes

$$\oint_C \mathbf{V} \cdot \frac{D}{Dt} d\mathbf{s} = \oint_C \mathbf{V} \cdot d\mathbf{V} = \oint_C d\left(\frac{V^2}{2}\right) = 0 \tag{5.20}$$

Using the Navier-Stokes equation to substitute for $D\mathbf{V}/Dt$, Eq. 5.16 may be written as*

$$\frac{D\Gamma}{Dt} = -\oint_C \mathbf{a}_I \cdot d\mathbf{s} - \oint_C \frac{\nabla p}{\rho} \cdot d\mathbf{s} + \oint_C \nu \nabla^2 \mathbf{V} \cdot d\mathbf{s} \tag{5.21}$$

*Equation 5.21 forms the basis for a significant portion of the excellent film *Vorticity* (film principal A. H. Shapiro), of the NCFMF Series of motion pictures depicting fluid-flow phenomena. The films are distributed by the Encyclopaedia Britannica Educational Corp. The student is encouraged to view this film if it is available.

where $\mathbf{a}_I$ represents the additional acceleration terms due to the choice of a noninertial reference frame. The gravitational body force is not included in Eq. 5.21 since $g\,\nabla h \cdot d\mathbf{s} = g\,dh$ and the cyclic integral of such a term is zero. Other body-force effects, such as those resulting from a magnetic field, may be circulation-producing, but are excluded here. The cyclic integral of the pressure term is retained so that examples in which $\rho \neq$ constant may be considered.

It is often more useful to examine the rate of change of circulation about a curve fixed in space. The development for this is similar to that for a material curve, as shown by the following. (The notation $d\Gamma/dt$ is appropriate to form the rate of change of circulation since the curve is fixed; we are not following a material curve.) The infinitesimal length element $d\mathbf{s}$ is fixed in space; hence, the rate of change of circulation is given by

$$\frac{d\Gamma}{dt} = \frac{d}{dt}\oint_C \mathbf{V}\cdot d\mathbf{s}$$

$$= \oint_C \frac{\partial \mathbf{V}}{\partial t}\cdot d\mathbf{s} \qquad (5.22)$$

Using the Navier-Stokes equations to substitute for $\partial \mathbf{V}/\partial t$, this may be put in the form

$$\frac{d\Gamma}{dt} = -\oint_C (\mathbf{V}\cdot\nabla)\mathbf{V}\cdot d\mathbf{s} - \oint_C \frac{dp}{\rho} - \oint_C \mathbf{a}_I\cdot d\mathbf{s} + \oint_C \nu\nabla^2\mathbf{V}\cdot d\mathbf{s} \qquad (5.23)$$

where we have used $\nabla p \cdot d\mathbf{s} = dp$. The last three terms are of the same form as for the material curve. The cyclic integral of the convective acceleration is more easily accounted for if it is expressed as an area integral. Using $(\mathbf{V}\cdot\nabla)\mathbf{V} = \boldsymbol{\omega}\times\mathbf{V} + \nabla(V^2/2)$, the transformation gives

$$\oint_C (\mathbf{V}\cdot\nabla)\mathbf{V}\cdot d\mathbf{s} = \int_A [\nabla\times(\boldsymbol{\omega}\times\mathbf{V})]\cdot\hat{n}\,dA \qquad (5.24)$$

since $\nabla\times\nabla(V^2/2) = 0$. The resulting expression for $d\Gamma/dt$ is

$$\frac{d\Gamma}{dt} = -\int_A \nabla\times(\boldsymbol{\omega}\times\mathbf{V})\cdot\hat{n}\,dA - \oint_C \mathbf{a}_I\cdot d\mathbf{s} - \oint_C \frac{dp}{\rho} + \oint_C \nu\nabla^2\mathbf{V}\cdot d\mathbf{s}$$

$$(5.25)$$

where curve C surrounds area A.

As an example of the viscous-term contribution and as an example of a problem for which the fixed curve is quite useful, consider the flow over a thin flat plate placed parallel to a free stream flow, that is, a boundary layer as shown in Fig. 5.17. The time rate of change of circulation will be examined for a material curve and for a fixed curve. The flow is incompressible and is viewed from an inertial reference frame; hence, $\mathbf{a}_I = 0$ and $\oint_C dp/\rho = 0$. For the contour shape shown, the viscous term is significant only along the surface where $\nabla^2 \mathbf{V} \cdot d\mathbf{s} = (\partial^2 u/\partial y^2)dx$. The full Navier-Stokes equations, evaluated at the wall, where $u = v = w = 0$, allow us to write

$$\nu \left.\frac{\partial^2 u}{\partial y^2}\right|_{y=0} = \frac{1}{\rho}\left.\frac{\partial p}{\partial x}\right|_{y=0} \tag{5.26}$$

The rate of change of circulation for a material curve is then

$$\frac{D\Gamma}{Dt} = \oint_C \nu \nabla^2 \mathbf{V} \cdot d\mathbf{s} = \int_{x_1}^{x_2} \nu \left(\frac{\partial^2 u}{\partial y^2}\right)_{y=0} dx = \int_{x_1}^{x_2} \frac{1}{\rho}\left(\frac{\partial p}{\partial x}\right)_{y=0} dx$$
$$= \frac{p_2 - p_1}{\rho} \tag{5.27}$$

This expression has quite an interesting consequence. A material curve, as shown by the dashed line, will be convected downstream by the flow, except at the surface, where $\mathbf{V} = 0$. During the time period of the convection, the circulation about it will remain constant ($D\Gamma/Dt = 0$) if $p_2 = p_1$. However, if the x_1-location is chosen near (or at) the leading edge of the plate, then $p_1 > p_2$ and the circulation about the material

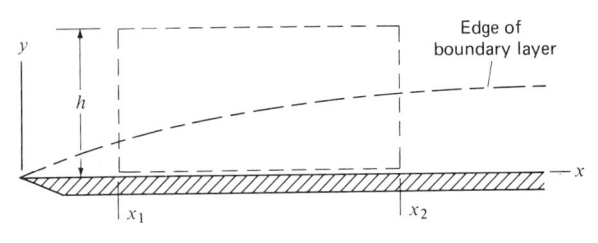

(a) Material and fixed curve at time t

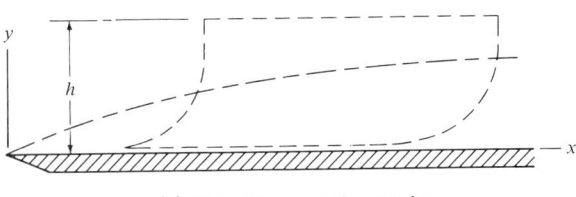

(b) Material curve at time $t + \Delta t$

Fig. 5.17. Curves for circulation evaluation in a boundary layer.

Art. 5.4   VORTICITY AND CIRCULATION   259

curve decreases. This is equivalent to introducing negative vorticity into the area bounded by the curve.

The physical behavior of the circulation rate of change for this and many other problems is more easily visualized in terms of a fixed rather than a material curve. The relationship between the $\nu \nabla^2 \mathbf{V}$ term and the pressure term remains the same; hence,

$$\frac{d\Gamma}{dt} = -\int_A \nabla \times (\boldsymbol{\omega} \times \mathbf{V}) \cdot \hat{n} \, dA + \frac{p_2 - p_1}{\rho} \tag{5.28}$$

For a steady flow, $d\Gamma/dt = 0$. However, if one considers x_1 to be located at the leading edge—a stagnation point—the magnitude of $p_2 - p_1$ is equal to

$$\frac{p_2 - p_1}{\rho} = -\frac{U^2}{2} \tag{5.29}$$

assuming p_2 is equal to the free-stream static pressure. The area integral must balance this term since $d\Gamma/dt = 0$. For this two-dimensional flow, the area is in the xy-plane and consequently the z-component of $\nabla \times (\boldsymbol{\omega} \times \mathbf{V})$ is the pertinent term for this integral. There results, with $\hat{n} = \hat{k}$ and $\xi = \eta = 0$,

$$\nabla \times (\boldsymbol{\omega} \times \mathbf{V}) \cdot \hat{k} = \frac{\partial}{\partial x}(\zeta u) + \frac{\partial}{\partial y}(\zeta v) \tag{5.30}$$

The area integral term can then be written as

$$\int_A \nabla \times (\boldsymbol{\omega} \times \mathbf{V}) \cdot \hat{n} \, dA = \int_0^h \int_{x_1}^{x_2} \frac{\partial}{\partial x}(\zeta u) \, dx \, dy + \int_{x_1}^{x_2} \int_0^h \frac{\partial}{\partial y}(\zeta v) \, dy \, dx$$

$$= \int_0^h (u\zeta)_{x=x_2} dy - \int_0^h (u\zeta)_{x=x_1} dy + \int_{x_1}^{x_2} (v\zeta)_{y=h} dx - \int_{x_1}^{x_2} (v\zeta)_{y=0} dx$$

$$\tag{5.31}$$

Consequently, with h located outside the boundary layer,

$$\frac{p_2 - p_1}{\rho} = \int_0^h (u\zeta)_{x=x_2} dy - \int_0^h (u\zeta)_{x=x_1} dy \tag{5.32}$$

The contribution of the area integral term is now quite apparent; the four integrals in Eq. 5.31 represent the net flux of vorticity transported by the streamwise and transverse motions, respectively. Mechanistically, the vorticity introduced by the viscous shear at the wall would result in a net increase of the circulation Γ were it not for the convective transport, which carries it downstream and out of the region. If x_1 is located at a point where $p_1 = p_2 = p_{atm}$, then no new vorticity is introduced into the boundary layer and the net convection of vorticity is zero. It is a rather interesting, and certainly not an obvious, result that all of the vorticity in a flat-plate boundary layer is introduced in the region of the nose of the plate and is independent of the shape of the nose. Note that the total vorticity flux past any x = constant section in the zero-pressure-gradient region is $-U^2/2$, by using Eq. 5.29.

The presence of a high (or low) atmospheric-pressure center is always associated with a characteristic swirl pattern of atmospheric motion. The cloud formations as seen by weather satellites clearly demonstrate this effect. The swirling motion is seen by an observer on the earth; his reference frame is noninertial since it is rotating at an angular velocity of one revolution each day. Consider a low-pressure center of a particular location. The surrounding air responds to this radial pressure gradient with a centrally directed motion. The coriolis acceleration $2\Omega \times \mathbf{V}$, and hence, $\oint_C \mathbf{a}_I \cdot d\mathbf{s}$ leads to a production of circulation, $D\Gamma/Dt \neq 0$. Additional circulation-producing effects will be presented in Examples 5.1, 5.2, and 5.3.

Example 5.1

Use the vorticity transport equation (5.14) and explain the presence of the secondary flow downstream of a bend in a river or creek. Can the typical eroded outer bank and the deposition of stones on the inner bank be explained by such a secondary flow? The velocity profile prior to the bend is as shown in Fig. E5.1. It is known that the flow speeds up on the inside of the bend, as shown at section ②.

Solution. The vorticity at the start of the bend is in the y-direction near the bottom ($\omega = \eta \hat{j}$) and in the z-direction on the inner side wall ($\omega = \zeta k$), assuming the river to form a rectangular cross-section. The flow on the inside of the bend accelerates (a fact which will be discussed in Chapter 8) resulting, at section ②, in a non-zero $\partial u/\partial y$ for the bottom flow and a non-zero $\partial u/\partial z$ for the side wall flow. Note that $\partial u/\partial y = 0$ near the bottom and $\partial u/\partial z = 0$ on the sidewall at section ①. The x-component vorticity equation (see Eq. 5.14) is

$$\frac{D\xi}{Dt} = \xi \frac{\partial u}{\partial x} + \eta \frac{\partial u}{\partial y} + \zeta \frac{\partial u}{\partial z} + \nu \nabla^2 \xi$$

Thus we see that $\eta \partial u/\partial y$ from the bottom flow results in a positive $D\xi/Dt$ and hence a positive ξ at section ②. Likewise, $\zeta \partial u/\partial z$ from the sidewall results in a positive ξ. The sense of ξ is to cause a net flow from the outer bank to the

Art. 5.4 VORTICITY AND CIRCULATION

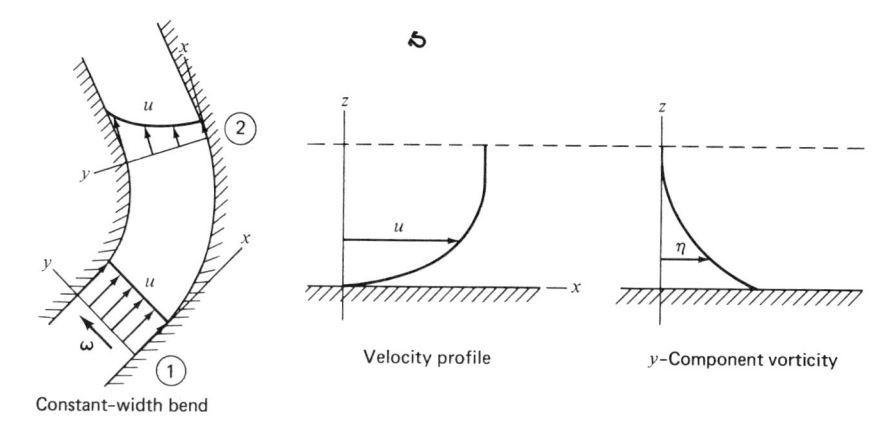

Fig. E5.1

Constant-width bend — Velocity profile — y-Component vorticity

inner bank near the bottom of the stream. (If your right thumb points in the direction of the vorticity your curled fingers will indicate the vortical motion.) This is the agent which transports stones to the inner bank. The ξ-vorticity component also causes the surface flow to move toward the outer bank after causing errosion near the stream surface. (This outward flow may be observed by dropping a leaf on the water surface upstream of the bend.)

Example 5.2

The second circulation-producing term in Eq. 5.21, $-\oint_C (dp/\rho)$, is nonzero if the density is a variable and is not related uniquely to the pressure, that is, if there is a nonbarotropic relationship between ρ and p. [It will be left as an

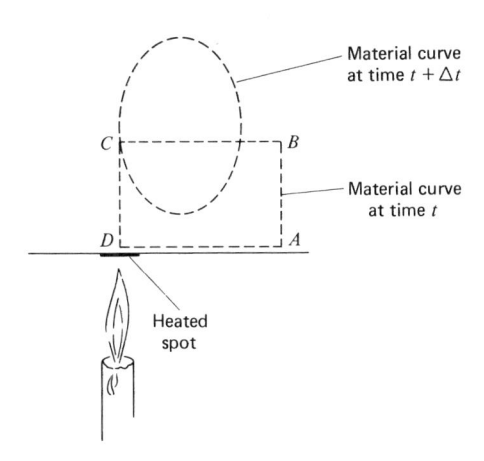

Fig. E5.2

exercise for the student to show that $\oint_C (dp/\rho) = 0$ for a barotropic flow, i.e., $\rho = \rho(p)$. Note that ρ = constant satisfies this condition, as do isentropic and isothermal flows.] A nonbarotropic circulation-producing condition is the localized heating of a fluid such as would occur above a stove burner, a light bulb, or a fire. Such a situation is demonstrated in Fig. E5.2. Explain how the circulation is produced.

Solution. A closed contour of fluid particles (a material curve as shown on the figure), will experience a change in its circulation as a result of the localized heating effect. The density of the fluid above the heated spot will decrease in the region where the fluid has been heated; hence the hydrostatic-pressure variation, from some distance above the heated spot down to it, would be less than the hydrostatic-pressure change from the same initial height to the level of the spot. As a result of this, there is a pressure gradient from D to A ($p_D < p_A$) and hence a flow from A to D and from D to C. Hence, the fluid of the material curve will be carried upward. Since the pressure variation from C to D is nonbarotropic (the upward flow from D to C implies that $\partial p/\partial z > \rho g$ along CD) there will be a net contribution to $D\Gamma/Dt$ from this term. Consequently, a circulation around the material curve will be developed as the fluid of the curve rises.

Example 5.3

A relatively large boundary layer exists at the inlet to a contraction. It is observed that the boundary layer at the exit of the contraction is smaller than at the inlet (see Fig. E5.3); explain this observation.

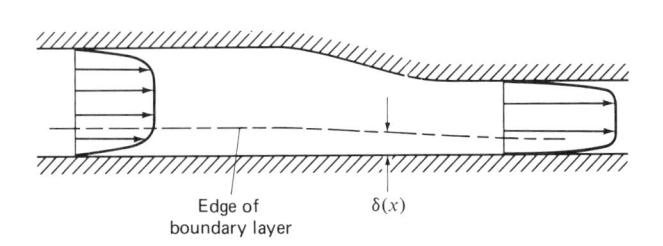

Fig. E5.3

Solution. Consider a region in space as shown on the sketch. The edge of the boundary layer is taken as the location where the outward diffusion of vorticity is negligible. For the accelerating flow, the free-stream velocity has a component directed toward the lower surface; consequently, the edge of the boundary layer will be located at the y-location where the convective transport bringing non-vortical fluid into a spatial region from above is balanced by the outward diffusion from below. For this example the boundary layer decreases in thickness with the x-location since the convective transport exceeds the viscous diffusion in the contracting region.

5.5 COMPRESSIBLE-FLOW PHENOMENA

For a given fluid and a given geometry, we have seen that as the characteristic velocity is increased from zero the flow progresses through the conditions described as Stokes', laminar, and turbulent flow. The flow may also include boundary layers and an inviscid motion in the region at some distance from a solid surface. As will be established in Chapter 7, the flow grows more and more independent of the Reynolds number as the Reynolds number increases, that is, the normalized results asymptotically approach given values for a given problem.

With this information we must not be led to the conclusion that nothing new will be observed from further increases in the velocity. In fact, two additional phenomena occur as the velocity is increased: the compressibility effects discussed in this article, and cavitation, discussed in the next.

All fluids are compressible; however, our interest in compressibility effects will be here confined to gases. Basically, what happens is that large changes in the velocity of the gas are associated with pressure changes sufficiently large to lead to significant density changes. If the Mach number remains below a value of 0.3 at all points in the flow the density does not undergo significant change, and the gas behaves as an incompressible fluid; see Art. 1.8. The equations which describe the compressible flow of a gas are those for continuity, momentum, and energy, and the equation of state. Invoking Stokes' hypothesis, and restricting our attention to a steady flow, the equations are

$$\nabla \cdot (\rho \mathbf{V}) = 0$$

$$(\mathbf{V} \cdot \nabla)\mathbf{V} = -\frac{\nabla p}{\rho} + \nu \nabla^2 \mathbf{V} + \frac{\nu}{3} \nabla(\nabla \cdot \mathbf{V})$$

$$\rho c_v (\mathbf{V} \cdot \nabla) T = k \nabla^2 T - (\mathbf{V} \cdot \nabla) p + \Phi \qquad (5.33)$$

$$p = \rho R T$$

This set of equations is extremely difficult to solve even for the simplest of geometries. Thus, we will consider only flows in which viscous effects may be neglected (except for gas flow in a pipe); further aspects will be presented in Chapter 11.

Perhaps the most spectacular phenomenon in compressible gas flow is the shock wave, which changes the fluid properties discontinuously; that is, there is a jump in the pressure, temperature, velocity, etc. in a distance of several mean free paths of the molecules. The Mach number must be greater than unity for a shock wave to occur. Shock waves generated by supersonic aircraft (Fig. 5.18a have made the "boom" familiar to many. The "boom" is actually a sudden pressure rise, which always accompanies a shock wave.

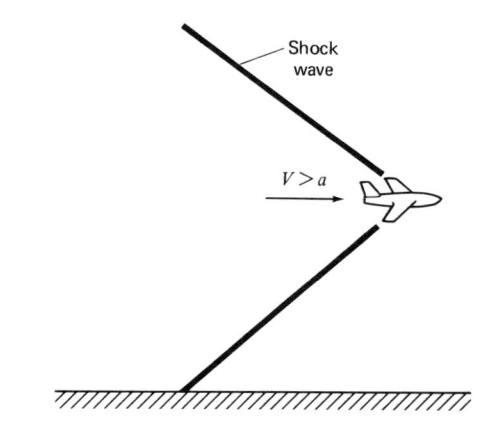

(a) A shock wave propagating from an aircraft

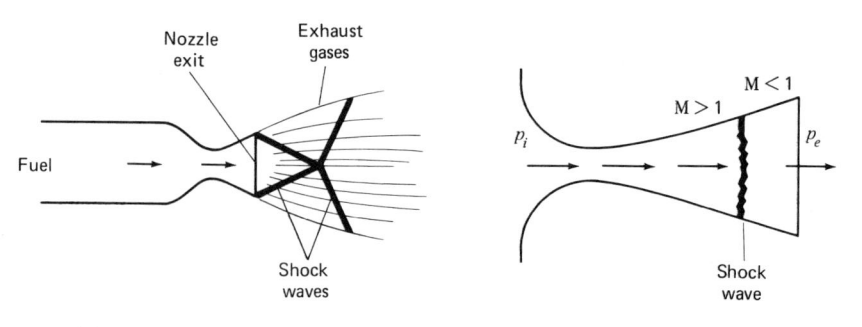

(b) A converging–diverging exhaust nozzle on a rocket engine

(c) A shock wave occurring in a converging–diverging nozzle.

Fig. 5.18. Examples of shock waves and supersonic flows.

To achieve supersonic flow in a conduit a converging-diverging nozzle must be used. At appropriate values of the pressure ratio p_i/p_e, a shock wave will occur in the diverging section (Fig. 5.18b). The nozzles on rocket engines (such as those on the Saturn rocket used in the United States space program), are converging-diverging nozzles which allow supersonic exhaust flows with corresponding large thrusts. The interaction of the supersonic exhaust flow with the ambient atmosphere produces a very complex flow situation, as shown in Fig. 5.18c. The shock wave and other compressible flow phenomena will be included in Chapter 11, on compressible flow.

5.6 CAVITATION

Cavitation is the dynamic phenomenon which occurs in a liquid when it is subjected to pressure at or below the vapor pressure: the formation of bubbles, or cavities, as the liquid vaporizes, and their subsequent collapse if pressure decreases below that point. The force exerted as liquid rushes into a cavity gives rise to a very high localized pressure that may damage surfaces in the region of cavitation; such erosion—in turbomachinery, for example—has stimulated most of the interest in the subject. For flows involving liquids the pressure should be higher than the vapor pressure at all points; therefore, Bernoulli's equation indicates that the velocity is limited to that value associated with the vapor pressure if cavitation is to be avoided. This limits the performance of machinery where high velocities are otherwise desireable.

Four types of cavitation have been identified: (1) traveling cavitation, (2) fixed cavitation, (3) vortex cavitation, and (4) vibratory cavitation; the first three are sketched in Fig. 5.19. *Traveling cavitation* occurs when individual bubbles or cavities are formed, transported with the liquid (usually a very short distance), and suddenly collapse. *Fixed cavitation* results when a cavitation cavity detaches from a body to form a fixed cavitation pocket. It may reattach to the body or be closed by the main liquid stream downstream of the body. This latter case has become known as *supercavitation*. *Vortex cavitation* is found in the high-velocity core of a vortex, a region of low pressure. The most remarkable example is that of "tip" cavitation from a propeller, shown in Fig. 5.19c. Vortex cavitation sustains itself much longer than the other types since high-speed vortices usually penetrate a considerable distance into the flow before gradual dissipation. *Vibratory cavitation* may occur in a liquid at rest. It is an unsteady phenomenon resulting from pressure waves induced in the flow. A pressure wave would consist of a high pressure followed by a low pressure; obviously if the low-pressure part of the wave (or vibration) is below the vapor pressure cavitation results.

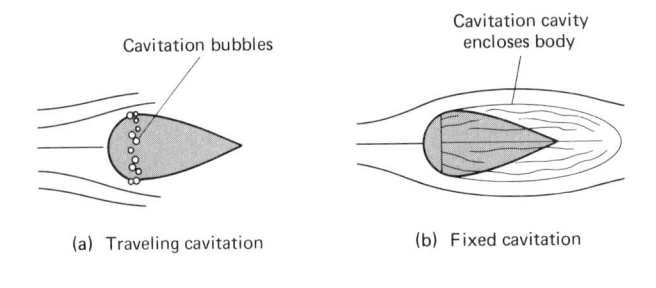

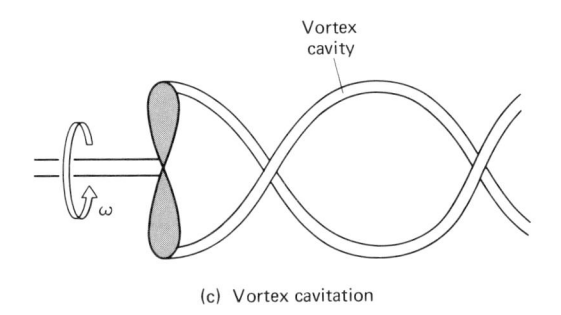

Fig. 5.19. Examples of cavitation.

To study cavitation experimentally we would choose as the characteristic pressure the pressure drop from free-stream pressure p_∞ to vapor pressure p_v, namely, $2(p_\infty - p_v)$ where the factor 2 is used to conform with accepted practice. Neglecting viscous effects and body forces, the dimensionless differential equations of motion become

$$\frac{D\mathbf{V}}{Dt} = -\frac{p_\infty - p_v}{\frac{1}{2}\rho U^2} \nabla p \tag{5.34}$$

The cavitation number K is defined by the equation

$$K = \frac{p_\infty - p_v}{\frac{1}{2}\rho U^2} \tag{5.35}$$

For a particular flow the cavitation number at which cavitation inception occurs is designated by K_i. As K decreases below K_i cavitation becomes more "advanced" and supercavitation may result. If $K > K_i$, no cavitation results.

Cavitation has two major effects; it modifies the flow field and it causes damage. The modified flow is usually accompanied by increased drag and poorer performance of mechanical devices. Propellors experience reduced thrust, turbines reduced power output, and pumps reduced efficiency. Damage to solid surfaces occurs in turbines, pumps, valves, nozzles, on hydraulic structures, and propellors. Hydraulic systems in which liquids are subjected to low pressures are usually designed so that cavitation does not result.

PROBLEMS

5.1 For an incompressible Stokes' flow, show that $\nabla^2 p_k = 0$. Also, for a plane Stokes' flow, show that $\nabla^2 \nabla^2 \psi = 0$ where ψ is the stream function defined by $u = \partial \psi / \partial y$ and $v = -\partial \psi / \partial x$.

5.2 Water is used in a simulated heart-artery system. It is important that the flow everywhere be laminar. A probe which senses the velocity is placed in the artery, and the signal is displayed on an oscilloscope screen. Sketch the expected signal display for a laminar flow. Should a turbulent flow develop, sketch the expected signal display. Would a time-averaged velocity component be useful for the turbulent flow? If so, what T should be chosen?

5.3 An engineer states that water flowing with an average speed of 0.1 fps in a 6-in.-dia. sewage pipe is a laminar flow. Comment on the plausibility of his statement. Determine the speed below which the flow would always be laminar.

5.4 A river is flowing rather placidly; and it is of interest to determine if it is a laminar flow. The critical Reynolds number for a relatively wide river is approximately 2000, based on average velocity and depth. The river of interest is 10 ft deep. Determine the maximum speed for laminar flow and comment on the plausibility of laminar flow in any river.

5.5 The weather man states that the average wind velocity is 5 mph from the north. He does not mention, however, that the fluctuations from this wind velocity reach 4 mph. Sketch the wind velocity from the north as a function of time. Also plot the wind velocity from the east vs time, assuming the vertical velocity component is small.

5.6 Starting with the definition of an average quantity, verify the following:

(a) $\overline{\dfrac{\partial \bar{u}}{\partial x}} = \dfrac{\partial \bar{u}}{\partial x}$ (c) $\overline{\bar{v} \dfrac{\partial u'}{\partial y}} = 0$

(b) $\overline{\bar{u} u'} = 0$ (d) $\overline{\bar{u} \dfrac{\partial \bar{u}}{\partial x}} = \bar{u} \dfrac{\partial \bar{u}}{\partial x}$

5.7 Two pistons are moving in a pipe at constant velocity. To create a steady-flow problem, we fix the pistons and move the pipe. Sketch the boundary layer regions if the pistons are close together and if the pistons are far apart. Sketch a typical velocity profile between the pistons. Assume no leakage by the pistons.

5.8 Consider airflow over and through an automobile. State where you would expect boundary layers to exist and where you would expect separated regions. Would the flow be laminar or turbulent?

5.9 The process of transpiration from a leaf is being simulated in the laboratory. Air at 10 fps blows over the leaf aligned parallel to the air velocity. The experimenter wonders whether the boundary layer over most of the leaf is laminar or turbulent. What should the experimenter conclude?

5.10 The differential equation $10^{-4}\dfrac{d^2u}{dy^2} + \dfrac{du}{dy} - u = -2$ is of the boundary-layer type. For boundary conditions $u = du/dy = 0$ at $y = 0$, the solution is approximately $u = 2[1 - e^y - 10^{-4}e^{-10^4 y}]$. Plot this between $y = \pm 1$ and identify the boundary-layer region. Outside the boundary layer in the region $y > 0$, what is the approximate solution? What would be the approximate equation in this region $y > 0$? Discuss the order of this approximate equation and the necessary boundary condition.

5.11 Consider flow around a power plant smoke stack. Identify the laminar and turbulent boundary-layer regions, the separated region, and the inviscid-flow region.

5.12 A thin-walled container filled with a hot liquid is immersed in a cool environment. A circulation pattern is noted from the movement of bubbles on the top of the liquid. Make a sketch of the velocity pattern and identify the mechanism for its origin. For the reverse problem, that is, a container filled with cold liquid placed in a hot environment, deduce the circulation pattern and identify the causal mechanism.

5.13 Consider a jet approaching a plane wall as shown. (This is the geometry of a VTOL [vertical-take-off-and-landing] aircraft.) Identify the vortex structures in the approach jet as $\omega = \omega_\theta \hat{\theta}$. Consider a vortex loop near the stagnation streamline and trace its path as the flow spreads along the plate. If this material loop appears in the indicated boundary-layer fluid, what has happened to its vorticity? How has this change occurred? What about the magnitude of ω_θ as the material loop flows away from the stagnation point?

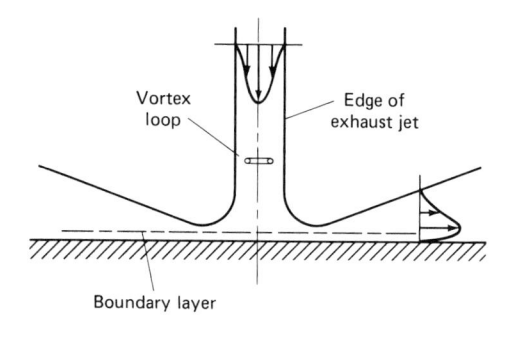

Prob. 5.13

5.14 For Prob. 5.13, consider a contour with boundaries along the jet axis, in the surface of the plate and a closing loop which is everywhere perpendicular to the velocity. What is the time rate of change of Γ about this loop? What is the value of $\oint_C \nu \nabla^2 \mathbf{V} \cdot d\mathbf{s}$ for this loop? The value of $\int_A \nabla \times (\boldsymbol{\omega} \times \mathbf{V}) \cdot \hat{n}\, dA$ for the loop?

5.15 A baratropic flow is one for which $\rho = \rho(p)$; the density is related uniquely to the pressure. Show that

$$\oint_C \frac{\nabla p}{\rho} \cdot d\mathbf{s} = 0$$

for a baratropic flow. Note that $\nabla p \cdot d\mathbf{s} = dp$, so that this integral can be written as $\oint_C dp/\rho$.

5.16 The cavitation number at which cavitation inception occurs on a submarine is 0.5. Determine the maximum speed at which a submarine at a depth of 500 ft can travel without inducing cavitation.

5.17 From your observational experience (or from your creative imagination), give two examples of each of the following types of flow: (a) Stokes flow, (b) steady laminar flow, (c) unsteady laminar flow, (d) separated laminar flow, (e) non-separated turbulent flow, (f) separated turbulent flow, (g) cavitated flow, (h) incompressible airflow, and (i) compressible air flow.

SELECTED REFERENCES

The motion picture *Vorticity* (in two parts, No. 21605 and No. 21606; A. H. Shapiro, film principal) is a helpful supplement to this chapter.

The flow phenomena discussed in this chapter are also considered in the introductory chapter of J. A. Owczarek, *Introduction to Fluid Mechanics*, International Textbook Co., Scranton, Pa., 1968.

The material on vorticity is covered at somewhat greater depth and breadth by G. K. Batchelor, *An Introduction to Fluid Dynamics*, Cambridge University Press, London, 1967. An exceptionally lucid and informative treatment of vorticity is that by M. J. Lighthill in Section II of *Laminar Boundary Layers*, edited by L. Rosenhead for Oxford University Press, Oxford, 1963.

6

Laminar Flows

6.1 INTRODUCTION

In Chapter 3, the equations governing the flow of an incompressible fluid were developed. They include the Navier-Stokes equations and the equation of continuity. We assumed a Newtonian, isotropic, homogeneous fluid and arrived at the four equations with the four unknowns u, v, w, and p. For laminar-flow problems the boundary and initial conditions are often easily identified and for many simple geometries solutions can be found. The purpose of this chapter is to obtain analytical solutions for several of these simple laminar flows; this will adequately demonstrate the nature and contribution of such solutions.

The Navier-Stokes equations and the continuity equation are, in dimensional form,

$$\frac{\partial u}{\partial t} + u\frac{\partial u}{\partial x} + v\frac{\partial u}{\partial y} + w\frac{\partial u}{\partial z} = -\frac{1}{\rho}\frac{\partial p_k}{\partial x} + \nu\left(\frac{\partial^2 u}{\partial x^2} + \frac{\partial^2 u}{\partial y^2} + \frac{\partial^2 u}{\partial z^2}\right)$$

[STEADY FLOW → GOES TO 0]

$$\frac{\partial v}{\partial t} + u\frac{\partial v}{\partial x} + v\frac{\partial v}{\partial y} + w\frac{\partial v}{\partial z} = -\frac{1}{\rho}\frac{\partial p_k}{\partial y} + \nu\left(\frac{\partial^2 v}{\partial x^2} + \frac{\partial^2 v}{\partial y^2} + \frac{\partial^2 v}{\partial z^2}\right)$$

$$\frac{\partial w}{\partial t} + u\frac{\partial w}{\partial x} + v\frac{\partial w}{\partial y} + w\frac{\partial w}{\partial z} = -\frac{1}{\rho}\frac{\partial p_k}{\partial z} + \nu\left(\frac{\partial^2 w}{\partial x^2} + \frac{\partial^2 w}{\partial y^2} + \frac{\partial^2 w}{\partial z^2}\right)$$

$$\frac{\partial u}{\partial x} + \frac{\partial v}{\partial y} + \frac{\partial w}{\partial z} = 0 \qquad (6.1)$$

As noted, this is a set of nonlinear equations. Because of the nonlinearity there exists the possibility of nonunique solutions; that is, when we obtain a solution for a given set of boundary conditions we have no guarantee that the solution we obtain analytically will actually be realized experimentally.

An example of this is a liquid flow between two concentrically rotating cylinders (see Fig. 6.1). One solution of the Navier-Stokes equation and continuity equation for the velocity between the cylinders is, in cylindrical coordinates,

$$v_\theta = \frac{A}{r} + Br \qquad (6.2)$$

which describes the velocity field shown in Fig. 6.1a. A and B are constants involving r_1, r_2, Ω_1, and Ω_2. This solution (see Art. 6.4 for its derivation) is valid for Ω_1 and Ω_2 only within a certain range of values. For various other values of Ω_1 and Ω_2, flows such as those shown in the other parts of Fig. 6.1 have been observed in the laboratory. These flows correspond to other solutions, more complicated than the one given in Eq. 6.1: in part b, the flow appears as regularly spaced cells (doughnuts); in parts c and d with a superimposed wavy motion; and in part e as a turbulent flow. All these flows have the same governing equations and identical boundary conditions; the boundary conditions are

$$v_\theta = r_2 \Omega_2 \quad \text{at} \quad r = r_2 \quad \text{and} \quad v_\theta = r_1 \Omega_1 \quad \text{at} \quad r = r_1 \qquad (6.3)$$

but the flows are all different. (Actually, researchers have observed about thirty different kinds of flow in the rotating cylinders.) The foregoing example gives a good physical meaning to the mathematical term "nonunique."

Actually to determine when a solution such as Eq. 6.2 is valid is usually quite difficult analytically; we must often resort to experiment for the answer. In this chapter we will be concerned only with obtaining such a solution, not in determining when (for what Reynolds-number range) it is valid.

The flow field shown in Fig. 6.1a is one example of a general state of fluid motion called "fully developed." A fully developed flow is one in which the velocity profile is not changing with respect to the longitudinal (streamwise) position. The shear stresses are distributed

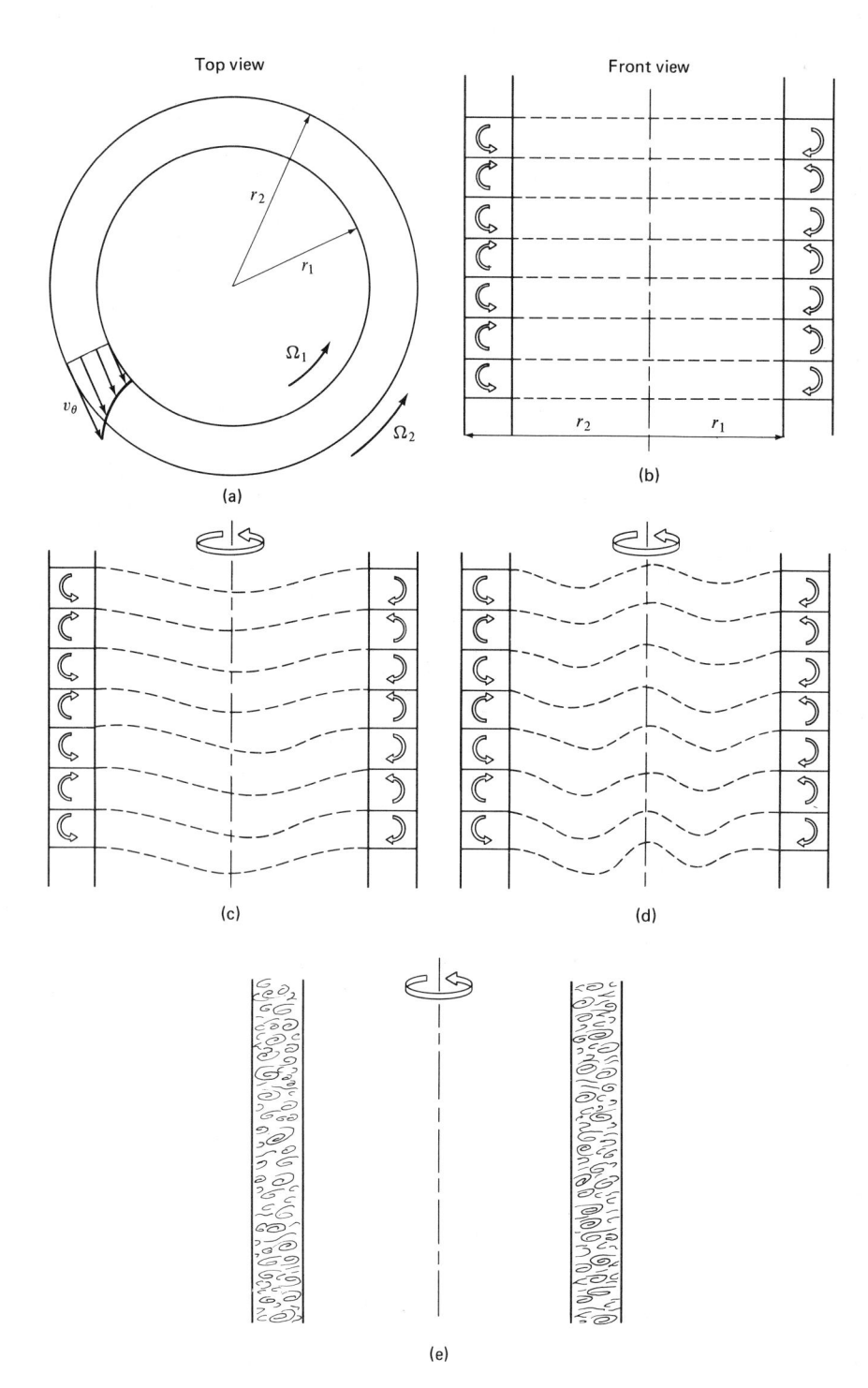

Fig. 6.1. Flow between concentric cylinders.

throughout such a flow. The flow in Fig. 6.1a is also a steady flow since it is time-independent.

The fully developed flow condition, which requires the velocity profile to be independent of the streamwise coordinate, can be expressed as

$$\frac{\partial \mathbf{V}}{\partial x} = 0 \qquad (6.4)$$

where the x-direction is aligned parallel to the flow. The incompressible continuity equation then reduces to

$$\frac{\partial v}{\partial y} + \frac{\partial w}{\partial z} = 0 \qquad (6.5)$$

We will be interested in a class of problems in which the streamlines are parallel to the wall; for such flows $v = w = 0$ and the continuity equation is satisfied. The acceleration of a fluid element is then

$$\frac{Du}{Dt} = \underbrace{\cancel{\frac{\partial u}{\partial t}}}_{\text{steady flow}} + \underbrace{u\cancel{\frac{\partial u}{\partial x}}}_{\text{fully developed}} + \underbrace{\cancel{v\frac{\partial u}{\partial y}} + \cancel{w\frac{\partial u}{\partial z}}}_{\text{parallel to wall}} = 0 \qquad (6.6)$$

so that the nontrivial Navier-Stokes equation becomes, for a steady, laminar, fully developed flow, in a conduit with parallel walls,

$$0 = -\frac{1}{\rho}\frac{\partial p_k}{\partial x} + \nu\left(\frac{\partial^2 u}{\partial y^2} + \frac{\partial^2 u}{\partial z^2}\right) \qquad (6.7)$$

This is a linear second-order, partial differential equation and expresses the physical condition that the net force acting on a particle is zero. From this equation, it is clear that in such a fully developed flow, there is a balance between the pressure force on the particle and the effect due to stress. The flow can be driven by the stress as in Fig. 6.1, or it can be driven by the pressure as in a pipe flow.

The boundary conditions for Eq. 6.7 are the no-slip velocity conditions at the walls of the channel and, if the pressure is desired, the pressure at some station in the flow.

A turbulent flow in a rectangular channel is an example of a fully developed flow for which the acceleration of a fluid particle is nonzero. The conservation of mass

$$\frac{\partial \bar{v}}{\partial y} + \frac{\partial \bar{w}}{\partial z} = 0 \qquad (6.8)$$

is satisfied where $\bar{v}$ and $\bar{w}$ are time-averaged velocities but they are nonzero since a spiraling motion exists in the channel; Fig. 6.2 shows this condition. However, beyond the entrance length all velocities are independent of x; hence by our definition this flow is fully developed.

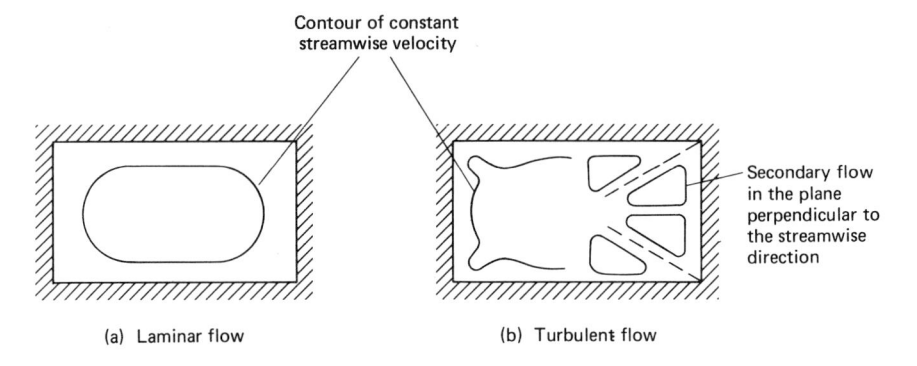

Fig. 6.2. Flow in a rectangular duct.

A parallel-wall configuration that is clearly not fully developed is that of radial flow between parallel discs (see Example 4.7). In this situation the flow cannot become fully developed since the velocity continually decreases as r increases. This is not associated with nonzero values for v_θ and v_z since these both may be zero; it is associated with the expanding area for the flow.

6.2 LAMINAR FLOW IN A PIPE

The fully developed flow in a circular tube is a comparatively simple yet important flow. For actual flows, it is usually found that a Reynolds number (Re = UD/ν) of the order of 2000 is sufficient for transition to occur. If the flow system is carefully isolated from external disturbances this critical value can be made markedly larger; under carefully controlled circumstances, experimental Reynolds-number values of the order of 40,000 have been obtained without transition to turbulence.

In this article, we shall derive the velocity profile for the fully developed region of a circular pipe flow. The solution for the velocity field in the entrance section of the pipe (see Fig. 6.3) will not be considered here; it involves a much more difficult analysis because the nonlinear convective accelerations are nonzero. The considerations

Art. 6.2 LAMINAR FLOW IN A PIPE

which would need to be included in such an analysis are introduced in Chapter 9, on boundary-layer flows.

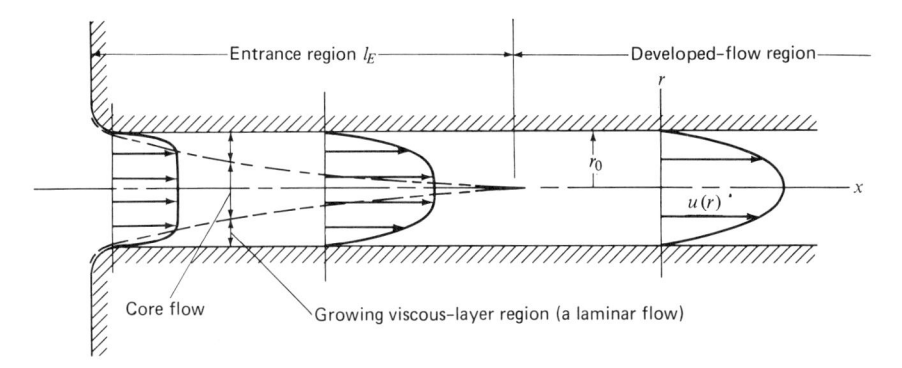

Fig. 6.3. Flow in a pipe.

The length l_E of the laminar entrance region is related to the Reynolds number by $l_E/D = 0.065\text{Re}$. The entrance region terminates where viscous effects completely penetrate the flow in the pipe.

We will designate the cylindrical velocity components as u, v, w in the x, r, θ directions, respectively. For the developed-flow region and for streamlines parallel to the wall, $v = w = 0$. The x-momentum equation, expressed in cylindrical coordinates (see Table 3.1), is

$$\frac{\partial u}{\partial t} + u\frac{\partial u}{\partial x} + v\frac{\partial u}{\partial r} + \frac{w}{r}\frac{\partial u}{\partial \theta}$$
$$= -\frac{1}{\rho}\frac{\partial p_k}{\partial x} + \nu\left[\frac{1}{r}\frac{\partial}{\partial r}\left(r\frac{\partial u}{\partial r}\right) + \frac{1}{r^2}\frac{\partial^2 u}{\partial \theta^2} + \frac{\partial^2 u}{\partial x^2}\right] \quad (6.9)$$

For steady, fully developed flow this equation reduces to

$$0 = -\frac{\partial p_k}{\partial x} + \frac{\mu}{r}\frac{\partial}{\partial r}\left(r\frac{\partial u}{\partial r}\right) \quad (6.10)$$

The r-component equation results in $\partial p_k/\partial r = 0$. We have assumed all quantities to be independent of θ. Hence, p_k is only a function of x: $p_k = p_k(x)$; also, $u = u(r)$.

We may then write Eq. 6.10, using ordinary derivatives,* as

$$\frac{1}{r}\frac{d}{dr}\left(r\frac{du}{dr}\right) = \frac{1}{\mu}\frac{dp}{dx} \quad (6.11)$$

*Note that the subscript k on the pressure is usually dropped for convenience; however, it is emphasized that the pressure p used throughout this and the remaining sections is $p_k = p + \rho gh$. Elevation changes may have a very significant effect on the pressure p but these are dynamically insignificant unless there is a free surface in a liquid flow or unless there is a variable density field.

The left-hand side is at most a function of the radius r since $u = u(r)$; the right-hand side is a function of the axial dimension x. Since either can be varied independently of the other, each side is necessarily a constant; that is

$$\frac{1}{r}\frac{d}{dr}\left(r\frac{du}{dr}\right) = \frac{1}{\mu}\frac{dp}{dx} = \lambda \qquad (6.12)$$

where λ = constant. We can solve the velocity part of Eq. 6.12 to yield

$$\frac{du}{dr} = \lambda\frac{r}{2} + \frac{C_1}{r} \qquad (6.13)$$

and, integrating again,

$$u = \frac{\lambda r^2}{4} + C_1 \ln r + C_2 \qquad (6.14)$$

The boundary conditions are

$$u = 0 \quad \text{at} \quad r = r_0$$

$$u = \text{finite} \quad \text{at} \quad r = 0 \qquad (6.15)$$

hence, $C_1 = 0$ and $C_2 = -\lambda r_0^2/4$. Finally, the velocity distribution is

$$u = \frac{\lambda}{4}(r^2 - r_0^2) \quad \text{PARABALOID OF REVOLUTION}$$

$$= \frac{1}{4\mu}\frac{dp}{dx}(r^2 - r_0^2) \qquad (6.16)$$

where we have introduced the pressure gradient from Eq. 6.12. In the developed region the velocity distribution is parabolic.

The shear stress ($\tau = -\tau_{rx}$; see Table 3.1) in this developed laminar flow is

$$\tau = -\mu\left(\frac{\partial u}{\partial r} + \cancelto{0}{\frac{\partial v}{\partial x}}\right)$$

$$= -\mu\frac{\lambda r}{2}$$

$$= -\frac{r}{2}\frac{dp}{dx} \qquad (6.17)$$

That is, the stress varies linearly from a zero value at the center line to a maximum value at the surface of the pipe. The wall shear stress τ_0 is the value of τ at $r = r_0$. From Eq. 6.17,

$$\tau_0 = -\frac{dp}{dx}\frac{r_0}{2} \qquad (6.18)$$

Art. 6.2 LAMINAR FLOW IN A PIPE

Since the pressure gradient, $dp/dx = \Delta p/L$, is constant, the pressure drop Δp in a given length of pipe L may be written as

$$\Delta p = -\frac{2\tau_0 L}{r_0} \tag{6.19}$$

using Eq. 6.18. A more convenient form can be arrived at by algebraic manipulation: specifically, using $D = 2r_0$,

$$-\frac{\Delta p}{\rho g} = \left(\frac{8\tau_0}{\rho V^2}\right)\frac{L}{D}\frac{V^2}{2g} \tag{6.20}$$

where $-\Delta p/\rho g$ is the head loss (the head loss was introduced in Section 7 of Art. 2.4; note that it has the dimensions of length). The coefficient in parentheses on the right-hand side is termed the *friction factor f*, defined by

$$f = \frac{8\tau_0}{\rho V^2} \tag{6.21}$$

where V is the average velocity in the pipe.

If total-pressure (pitot) tubes were placed at two locations in a pipe, the difference in the readings between the two locations would represent the loss effects in the flow. (The absence of loss effects would result in the same Bernoulli constant and hence the same total pressure.) If the total pressure were measured by a water column in a manometer tube then one could refer to the pressure difference as a head loss. Since the earliest identification of such an effect was made by hydraulic engineers concerned with water distribution systems, the name "head loss" was adopted as a standard description of such a loss effect and has been retained for the more general cases of gas and other liquid flows. From Eq. 6.20, the head loss for steady, incompressible, horizontal, constant-area flow is

$$h_L = -\frac{\Delta p}{\rho g} \tag{6.22}$$

Using the earlier-defined friction factor f,

$$h_L = f\frac{L}{D}\frac{V^2}{2g} \tag{6.23}$$

for the constant-area pipe. The average velocity V in the pipe may be expressed as

$$V = \frac{1}{A}\int_0^{r_0} u(r)2\pi r\, dr \tag{6.24}$$

or, using Eq. 6.16 for $u(r)$, there results

$$V = \frac{1}{\pi r_0^2} \frac{2\pi}{4\mu} \frac{dp}{dx} \int_0^{r_0} (r^2 - r_0^2) r \, dr$$

$$= -\frac{r_0^2}{8\mu} \frac{dp}{dx} \tag{6.25}$$

This relationship shows that the average velocity is directly proportional to the pressure gradient; conversely, the pressure gradient is related to the average velocity by

$$\frac{dp}{dx} = -\frac{8\mu V}{r_0^2} \tag{6.26}$$

The pressure change over a length L of the pipe is

$$\Delta p = -\frac{8\mu V L}{r_0^2} \tag{6.27}$$

The maximum velocity occurs at the center of the pipe, where $r = 0$. From Eq. 6.16, it is

$$u_{\max} = -\frac{r_0^2}{4\mu} \frac{dp}{dx} \tag{6.28}$$

Combining this with Eq. 6.26 relates the average velocity to the maximum velocity:

$$V = \frac{u_{\max}}{2} \tag{6.29}$$

Finally, it is of interest to relate the friction factor to the average velocity for laminar flow. Using Eq. 6.20 and the definition of f, we have

$$f = -\frac{\Delta p}{\rho g} \frac{D}{L} \frac{2g}{V^2} \tag{6.30}$$

or, by using Eq. 6.27,

$$f = \frac{64\mu}{VD\rho}$$

$$= 64/\text{Re} \tag{6.31}$$

for laminar flow in a pipe. Substituting this in Eq. 6.23, we observe that the head loss in a pipe is directly proportional to the average velocity. This result is, in fact, quite general. For laminar flow, losses are proprotional to the first power of the average velocity.

Example 6.1

Use a control volume of length L and radius r and derive Eq. 6.17. Comment as to the assumption of laminar flow.

Solution. The control volume is shown in Fig. E6.1. Applying the integral momentum equation, we find that

$$\Sigma F_x = 0$$

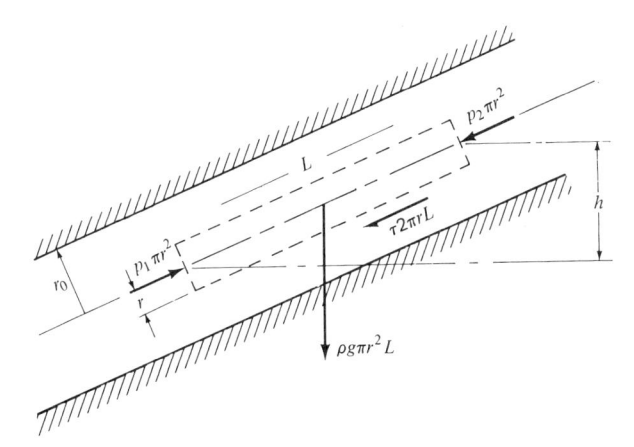

Fig. E6.1

Thus, the force balance requires that

$$(p_1 - p_2)\pi r^2 - \tau 2\pi r L - \rho g \pi r^2 L \times \frac{h}{L} = 0$$

or

$$\tau = -(p_2 - p_1)\frac{r}{2L} - \rho g \frac{r}{2L}$$

$$= -(\Delta p + \rho g h)\frac{r}{2L}$$

$$= -\Delta p_k \frac{r}{2L}$$

Using $\Delta p_k/L = dp_k/dx$, and dropping the subscript k, there results

$$\tau = -\frac{r}{2}\frac{dp}{dx}$$

We have nowhere made the assumption of laminar flow; consequently, this result is also applicable for turbulent flow in a pipe. Obviously, Eqs. 6.18 through 6.24 are also applicable to both laminar and turbulent flow. Eq. 6.25 uses the parabolic velocity distribution, which is a result of the laminar flow assumption; hence the remaining equations (6.25–6.31) are valid only for laminar flow.

Extension 6.1.1. Sketch the shear-stress distribution in a pipe for either laminar or turbulent flow.

Extension 6.1.2. State where the assumption of laminar flow was made in the solution to the pipe flow problem leading to Eq. 6.16.

Example 6.2

Apply the integral energy equation between two points in a pipe and show that $\Delta p/\rho g$ is indeed a loss. Assume no shaft-work and no heat transfer between the points. Also, assume that the velocity profile does not change along the pipe.

Solution. For steady flow in a pipe the energy equation (2.75) becomes

$$\cancel{\dot{Q}}^{0} - \cancel{\dot{W}_s}^{0} = \int_{c.s.} \left[\frac{V^2}{2} + \frac{p}{\rho} + gh + \tilde{u} \right] \mathbf{V} \cdot \hat{n} \rho \, dA$$

The velocity distribution is the same at both sections. Hence,

$$0 = \int_{A_1} \left(\frac{p}{\rho} + \tilde{u} \right) \rho \mathbf{V} \cdot \hat{n} \, dA + \int_{A_2} \left(\frac{p}{\rho} + \tilde{u} \right) \rho \mathbf{V} \cdot \hat{n} \, dA$$

where the kinetic pressure (see Eq. 4.4) has been introduced. (For convenience again drop the subscript k.) Since the pressure p and internal energy $\tilde{u}$ are constant over each section, we have

$$0 = \left(\frac{p_1}{\rho} + \tilde{u}_1 \right)(-\rho VA) + \left(\frac{p_2}{\rho} + \tilde{u}_2 \right)(\rho VA)$$

or

$$\frac{\Delta p}{\rho g} = \frac{(\tilde{u}_2 - \tilde{u}_1)}{g}$$

This shows that the pressure drop between two points in a pipe is related directly to the internal energy change by increasing the temperature of the fluid, and is considered a loss. It represents a degradation of mechanical energy to unusable thermal energy.

Extension 6.2.1. Assume isothermal flow, so that there is no internal-energy change. Show that the pressure drop is again a loss. Do not assume that the heat transfer is zero.

Example 6.3

Derive the equation for the velocity profile in the annulus shown in Fig. E6.3. The steady flow is fully developed and $v = w = 0$.

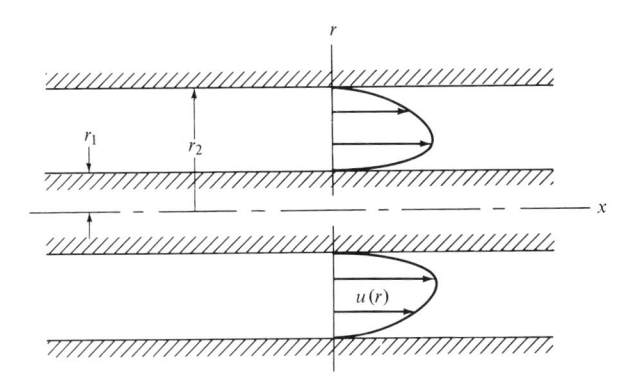

Fig. E6.3

Solution. The velocity components in the r and θ-directions are v and w, respectively. For the fully developed flow $\partial u/\partial x = 0$. The describing equations are identical to those which describe pipe flow. The $\dot x$-momentum equation (6.9), with $v = w = 0$, is

$$0 = -\frac{dp}{dx} + \frac{\mu}{r}\frac{d}{dr}\left(r\frac{du}{dr}\right)$$

The general solution is

$$u = \frac{\lambda r^2}{4} + C_1 \ln r + C_2$$

where $\lambda = (1/\mu)(dp/dx)$. The boundary conditions are $u = 0$ at $r = r_1$ and r_2. This results in

$$0 = \frac{\lambda r_1^2}{4} + C_1 \ln r_1 + C_2$$

$$0 = \frac{\lambda r_2^2}{4} + C_1 \ln r_2 + C_2$$

Solving for C_1 and C_2 yields

$$C_1 = \frac{\lambda}{4}\frac{r_1^2 - r_2^2}{\ln(r_2/r_1)}$$

$$C_2 = -\frac{\lambda}{4}\frac{(r_1^2 - r_2^2)\ln r_2}{\ln(r_2/r_1)} - \frac{\lambda r_2^2}{4}$$

The velocity distribution is then

$$u = \frac{1}{4\mu}\frac{dp}{dx}\left[(r^2 - r_2^2) + \frac{r_2^2 - r_1^2}{\ln(r_1/r_2)}\ln(r/r_2)\right]$$

As we let r_1 approach 0, $\ln r_1/r_2$ approaches $-\infty$ and the last term in the brackets drops out, resulting in the pipe flow parabolic velocity profile.

Extension 6.3.1. Find an expression for the average velocity in an annulus flow. *Ans.* $-\dfrac{1}{8\mu}\dfrac{dp}{dx}[r_2^2 + r_1^2 + (r_2^2 - r_1^2)/\ln(r_1/r_2)]$

Extension 6.3.2. Determine the position of maximum velocity in an annulus flow. *Ans.* $[(r_2^2 - r_1^2)/2 \ln(r_2/r_1)]^{1/2}$

Extension 6.3.3. As the inner radius reduces to zero, the inner cylinder has less and less influence on the flow. This means that a wire down the center of a pipe would not significantly influence the flow even though the velocity must be zero at the wire surface. To support this, show that the dragging force on a length L of the inner cylinder approaches zero as $r_1 \to 0$.

6.3 LAMINAR FLOW BETWEEN PARALLEL PLATES

Laminar flow also exists between parallel plates at low Reynolds numbers ($\text{Re} = Vh/\nu$), although not at such high Reynolds numbers as sometimes occurs in a pipe. Laminar flow always exists between parallel plates if the Reynolds number is less than 1500, but it will never be observed at Reynolds numbers greater than 7700 since the flow is unstable in the presence of infinitesimal disturbances at this Reynolds number.* We again are only interested in the developed flow, as shown in Fig. 6.4, where there are no velocity components in the y and z-directions. We will assume the side walls have no influence on the flow in the central portion of the channel, an acceptable assumption if the width $w \gg h$.

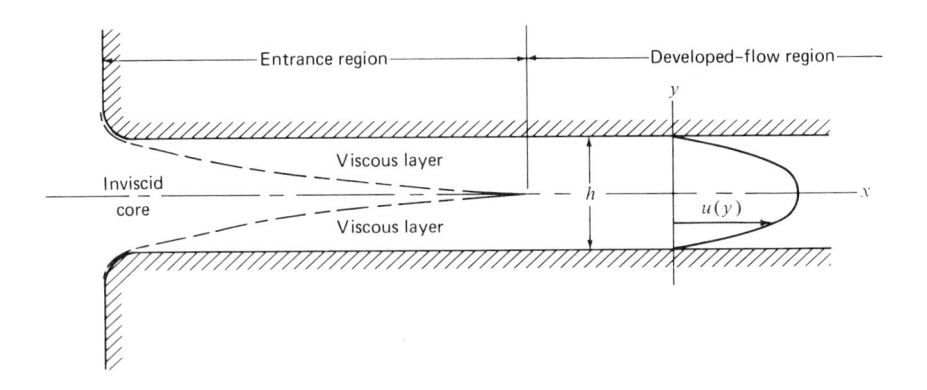

Fig. 6.4. Flow between parallel plates.

*A Reynolds number of 7500 has been observed in a wide channel where great care was taken to isolate the flow from disturbances. (See the footnote on page 296.)

Art. 6.3 LAMINAR FLOW BETWEEN PARALLEL PLATES

The continuity equation, for our incompressible flow, with $\partial u/\partial x = 0$ and $v = w = 0$, is identically satisfied. For $w \gg h$, $\partial u/\partial z = 0$, so that $u = u(y)$. The second Navier-Stokes equation (see Eq. 6.1) reduces to

$$0 = -\frac{\partial p}{\partial y} \tag{6.32}$$

so that $p = p(x)$. The first Navier-Stokes equation is then

$$0 = -\frac{dp}{dx} + \mu \frac{d^2 u}{dy^2} \tag{6.33}$$

since the substantial derivative

$$\frac{Du}{Dt} = 0 \tag{6.34}$$

We again see that

$$\frac{d^2 u}{dy^2} = \frac{1}{\mu}\frac{dp}{dx} = \lambda \tag{6.35}$$

where λ = constant. The solution is

$$\frac{du}{dy} = \lambda y + C_1 \tag{6.36}$$

Integrating again, there results

$$u = \frac{\lambda y^2}{2} + C_1 y + C_2 \tag{6.37}$$

The boundary conditions, $u = 0$ at $y = \pm h/2$, require that $C_1 = 0$ and $C_2 = -\lambda h^2/8$, so that

$$u = \frac{\lambda}{2}(y^2 - h^2/4) \tag{6.38}$$

Introducing the pressure gradient from Eq. 6.35, we have

$$\boxed{u = \frac{1}{2\mu}\frac{dp}{dx}(y^2 - h^2/4)} \tag{6.39}$$

The head loss is again ⟶ BETWEEN PARALLEL PLATES

$$h_L = -\Delta p/\rho g \tag{6.40}$$

between any two sections.

Following the same procedure as for the circular pipe flow, we find that the shearing stress is

$$\tau = -y\frac{dp}{dx} \tag{6.41}$$

and
$$\Delta p = -\frac{2\tau_0 L}{h} \tag{6.42}$$

The average velocity is related to the pressure gradient and the maximum velocity by
$$V = \frac{-h^2}{12\mu}\frac{dp}{dx} \tag{6.43}$$

and
$$V = \tfrac{2}{3} u_{max} \tag{6.44}$$

The derivation of this relationship is the subject of Prob. 6.13.

If we again define the friction factor for flow between parallel plates as
$$f = \frac{8\tau_0}{\rho V^2} \tag{6.45}$$

then
$$h_L = f\frac{L}{2h}\frac{V^2}{2g} \tag{6.46}$$

and
$$f = \frac{48}{\text{Re}} \tag{6.47}$$

Example 6.4

Determine the pressure drop in the entrance region of the channel flow (Fig. E6.4), assuming an entrance length equal to $l_E/h = 0.04$ Re. The entrance-velocity profile is uniform. The Reynolds number of the flow is 1500.

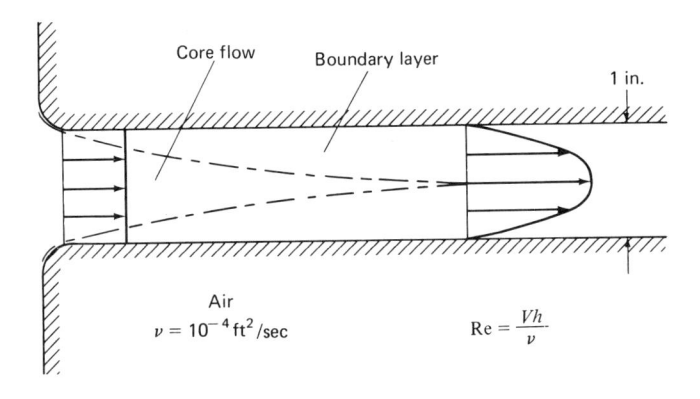

Air
$\nu = 10^{-4}$ ft^2/sec

$\text{Re} = \dfrac{Vh}{\nu}$

Fig. E6.4

Solution. Then entrance length is

$$l_E = 0.04 \, \text{Re} \times h$$
$$= 0.04 \times 1500 \times \tfrac{1}{12}$$
$$= 5.0 \text{ ft}$$

This entrance-length calculation is not necessary for the determination of the pressure drop; it is presented as a matter of interest. The average velocity is found from the definition of the Reynolds number.

$$V = \frac{\text{Re} \, \nu}{h}$$
$$= \frac{1500 \times 10^{-4}}{1/12}$$
$$= 1.8 \text{ fps}$$

In a channel the maximum velocity at the centerline at the end of the entrance length where the velocity profile is parabolic is, by Eq. 6.44,

$$u_{\max} = \tfrac{3}{2} V$$
$$= 2.7 \text{ fps}$$

Now, the flow is inviscid in the core region, so that Bernoulli's equation is applicable along the various streamlines. The air that flows down the streamline at the center of the channel has a velocity of 2.7 fps just at the end of the inviscid core. Hence, Bernoulli's equation allows us to write

$$\frac{V_1^2}{2} + \frac{p_1}{\rho} = \frac{V_2^2}{2} + \frac{p_2}{\rho}$$

or

$$\Delta p = \frac{V_1^2 - V_2^2}{2} \rho$$
$$= \frac{1.8^2 - 2.7^2}{2} \times 0.0024$$
$$= -0.0049 \text{ psf}$$

a very small pressure difference to measure with any type of instrument.

Extension 6.4.1. Compute the pressure drop for the same entrance region, but use water with $\nu = 10^{-5} \text{ ft}^2/\text{sec}$. *Ans.* -0.0393 psf

Extension 6.4.2. Explain why Bernoulli's equation can be used only along the center streamline, as the fluid flows from the entrance to the beginning of the developed flow.

6.4 LAMINAR FLOW BETWEEN ROTATING CYLINDERS

Another flow which, because of its simple geometry, can be solved exactly is that between concentrically rotating cylinders, as shown in Fig. 6.5, and discussed in Art. 6.1. At low speeds, a laminar flow will exist, and again only one component of the velocity remains, namely, v_θ. Fully developed flow requires that

$$\frac{\partial v_\theta}{\partial \theta} = 0 \qquad (6.48)$$

giving $v_\theta = v_\theta(r)$. The Navier-Stokes equations in cylindrical coordinates (neglecting the body force) give, in the r-direction,

$$\frac{\rho v_\theta^2}{r} = \frac{\partial p_k}{\partial r} \qquad (6.49)$$

and (assuming $p_k \neq p_k(\theta)$), in the θ-direction,

$$\frac{d^2 v_\theta}{dr^2} + \frac{d}{dr}\left(\frac{v_\theta}{r}\right) = 0 \qquad (6.50)$$

(For nonconcentric cylinders there would be a nonzero $\partial p_k/\partial \theta$.) Integrating once gives

$$\frac{dv_\theta}{dr} + \frac{v_\theta}{r} = C_1$$

or

$$\frac{1}{r}\frac{d}{dr}(rv_\theta) = C_1 \qquad (6.51)$$

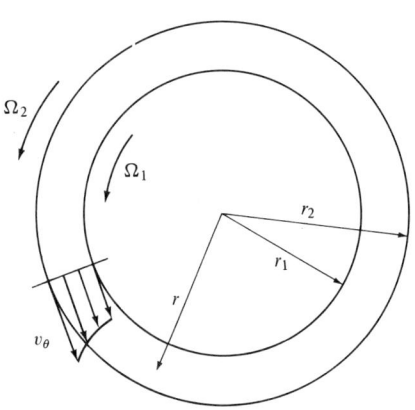

Fig. 6.5. Flow between rotating cylinders.

A second integration yields

$$v_\theta = \frac{C_1}{2} r + \frac{C_2}{r} \tag{6.52}$$

The boundary conditions are $v_\theta = r_1 \Omega_1$ at $r = r_1$ and $v_\theta = r_2 \Omega_2$ at $r = r_2$. The constants of integration are thus

$$C_1 = 2 \frac{\Omega_2 r_2^2 - \Omega_1 r_1^2}{r_2^2 - r_1^2} \qquad C_2 = \frac{r_1^2 r_2^2 (\Omega_1 - \Omega_2)}{r_2^2 - r_1^2} \tag{6.53}$$

This solution corresponds to that of the flow shown in Fig. 6.1a. The critical Reynolds number varies, depending on Ω_1 and Ω_2; however, if $\Omega_2 = 0$, as is often the case in actual situations, the critical Reynolds number ($\mathrm{Re} = r_1^2 \Omega_1 / \nu$) is 1708. Above 1708 the flow in Fig. 6.1b develops. For the flows in Fig. 6.1c and d, Ω_2 must be smaller than Ω_1 and rotating with the same sense as Ω_1. For sufficiently large Ω_1, Ω_2, or Ω_1 and Ω_2, a turbulent flow will result.

Once the velocity profile is known, shear stresses, drag on the surfaces, and power input to the fluid can be found. The normal stresses are all equal to the pressure. The $\tau_{\theta z}$ and τ_{rz} shear stresses (axially directed) are zero. The only remaining shear stress is azimuthal (around the cylinders). Letting $\tau = -\tau_{r\theta}$ (see Table 3.1), resulting in

$$\tau = -\mu \left[r \frac{\partial}{\partial r} \left(\frac{v_\theta}{r} \right) + \frac{1}{r} \cancel{\frac{\partial v_r}{\partial \theta}}^{0} \right]$$

$$= -\mu r \frac{d}{dr} \left(\frac{C_1}{2} + \frac{C_2}{r^2} \right) \tag{6.54}$$

or

$$\tau = \frac{2\mu C_2}{r^2} \tag{6.55}$$

which can be evaluated at any r. In particular, consider a bearing problem where $\Omega_2 = 0$; then $C_1 = -2\Omega_1 r_1^2/(r_2^2 - r_1^2)$ and $C_2 = r_1^2 r_2^2 \Omega_1/(r_2^2 - r_1^2)$. The shear stresses on the cylinders at r_1 and r_2 are

$$\tau_1 = \frac{2\mu r_2^2 \Omega_1}{r_2^2 - r_1^2} \qquad \text{and} \qquad \tau_2 = \frac{2\mu r_1^2 \Omega_1}{r_2^2 - r_1^2} \tag{6.56}$$

The shear stresses act in the direction opposite to that of the velocity.

The total drag on the journal ($r = r_1$) and the bearing ($r = r_2$) is in each case the product of the stress by the area over which it acts.

Torques necessary to drive the journal and restrain the bearing can be found by multiplying by the appropriate moment arm. The journal surface exerts a force on the fluid (oppositely directed from the force of the fluid on the journal), causing the fluid to move in the direction of the force. Thus there is a power input to the fluid,

$$\dot{W}_{in} = \mathbf{F} \cdot \mathbf{V} = [A_{\text{surface}} \tau v_\theta]_{r=r_1}$$

$$= 2\pi r_1 L \frac{2\mu r_2^2 \Omega_1}{r_2^2 - r_1^2} r_1 \Omega_1 \tag{6.57}$$

or

$$\dot{W}_{in} = \frac{4\pi L \mu r_1^2 r_2^2 \Omega_1^2}{r_2^2 - r_1^2} \tag{6.58}$$

where L is the axial length of the bearing. This power is all used to overcome the retarding effect of the shear stresses and is thus dissipated as an increase in internal energy of the fluid. In the steady state, application of the integral energy equation indicates that all of this power input results in a net heat-transfer rate from the bearing.

It is interesting to note that the force acting on the bearing is stationary, since $\Omega_2 = 0$; hence, there is no work rate at this surface. Consequently, the work-rate input calculated above is also the net work associated with a control volume defined to encompass the fluid only.

Example 6.5

Show that as the gap between concentric cylinders becomes small, with only the inner one rotating, (Fig. E6.5), the velocity distribution approaches a linear function of r. This is called a *Couette flow*. It approximates the flow between parallel plates between which there is a zero pressure gradient and one plate is moving.

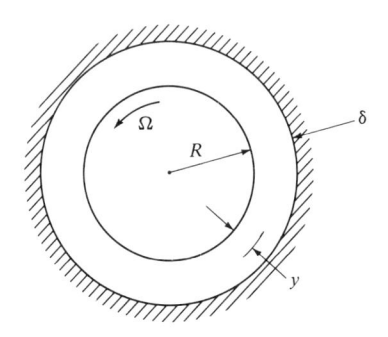

Fig. E6.5

Art. 6.5   SINUSOIDAL OSCILLATIONS OF A FLAT PLATE   289

Solution. The gap width, $r_2 - r_1 = \delta$, is small, so the sum $r_2 + r_1$ may be approximated by $2R$. The velocity distribution (Eq. 6.52), with $\Omega_2 = 0$, is

$$v_\theta = -\frac{\Omega r_1^2}{r_2^2 - r_1^2}\left(r - \frac{r_2^2}{r}\right)$$

with C_1 and C_2 replaced. This may be written as

$$v_\theta = -\frac{\Omega r_1^2}{(r_2 - r_1)(r_2 + r_1)}\left(\frac{r^2 - r_2^2}{r}\right)$$

Using the approximation $r_2 + r_1 \cong 2R$, we have

$$v_\theta = -\frac{\Omega R}{2\delta}\frac{(r - r_2)(r_2 + r)}{r}$$

where $\Omega = \Omega_1$. Now, let $r = R + y$ where y is measured from the inner cylinder, as shown. Then, for $y \ll R$,

$$\frac{r_2 + r}{r} \cong \frac{2R + y}{R + y} \cong 2$$

The velocity distribution, finally, is

$$v_\theta = \frac{-\Omega R(y + R - r_2)}{\delta}$$

$$= \frac{\Omega R}{\delta}(\delta - y)$$

This is a linear distribution in the small gap. It is a good approximation to the flow whenever $\delta \ll R$.

Extension 6.5.1. Approximate the torque necessary to rotate the inner cylinder and the power input to the fluid for the journal-bearing problem assuming the gap to be small.   Ans. $2\pi\mu\Omega R^3 L/\delta$, $2\pi\mu\Omega^2 R^3 L/\delta$

6.5 SINUSOIDAL OSCILLATIONS OF A FLAT PLATE

The examples thus far presented in the chapter have all involved steady flow. A simple problem in laminar, unsteady flow that can be solved exactly is the case of a fluid above an infinite flat plate which is oscillating sinusoidally. This condition is as shown in Fig. 6.6. We will express the velocity at the wall as

$$u_w = Ue^{i\omega t}$$

$$= U[\cos \omega t + i \sin \omega t] \tag{6.59}$$

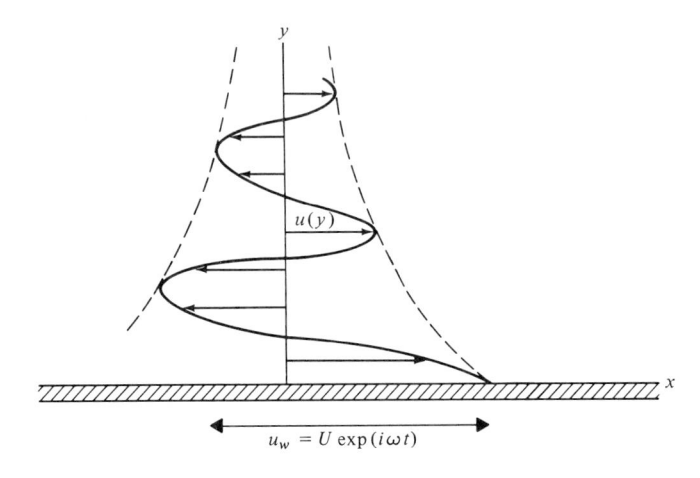

Fig. 6.6. An oscillating flat plate.

where only the real part is used to represent the physical velocity of the wall. The frequency is ω and U is the amplitude of the oscillation velocity. In this problem the velocity components in the y and z-directions are zero (and continuity shows that $\partial u/\partial x = 0$); hence the governing momentum equation becomes

$$\frac{\partial u}{\partial t} = \nu \frac{\partial^2 u}{\partial y^2} \tag{6.60}$$

where we must keep the partial derivatives since $u = u(y, t)$.

The pressure-gradient term can be shown, with the use of the y and z-component equations, to be at most a function of x. We will maintain the pressure the same at $x = \pm \infty$; hence, the constant-pressure gradient dp/dx must therefore be zero. This is a reasonable condition for a sufficiently large fluid space.

We can solve Eq. 6.60 by using separation of variables, with boundary conditions

$$u = Ue^{i\omega t} \text{ at } y = 0 \quad \text{and} \quad u = 0 \text{ at } y = \infty \tag{6.61}$$

Assume that $u(y, t)$ separates into a function of t multiplied by a function of y:

$$u = T(t)Y(y) \tag{6.62}$$

Substitution in Eq. 6.60 gives

$$YT' = \nu TY'' \tag{6.63}$$

Art. 6.5   SINUSOIDAL OSCILLATIONS OF A FLAT PLATE

where the primes denote differentiation ($T' = dT/dt$, $Y'' = d^2Y/dy^2$). This equation may be written as

$$\frac{T'}{T} = \nu \frac{Y''}{Y} \tag{6.64}$$

and again we have a function of t on the left and a function of y on the right; hence, both sides must be equal to a constant. Choose this constant to be the imaginary quantity $i\lambda$.* This results in the two ordinary differential equations

$$T' - i\lambda T = 0 \tag{6.65}$$

and

$$Y'' - \frac{i\lambda}{\nu} Y = 0 \tag{6.66}$$

Their solutions are

$$T = C_1 \exp[i\lambda t] \tag{6.67}$$

and

$$Y = C_2 \exp\left[\sqrt{\frac{\lambda}{2\nu}} (1+i)y\right] + C_3 \exp\left[-\sqrt{\frac{\lambda}{2\nu}} (1+i)y\right] \tag{6.68}$$

where we have used $\sqrt{i} = (1+i)/\sqrt{2}$. The general solution is then

$$u(y, t) = TY$$

$$= C_1 \exp(i\lambda t)\left\{C_2 \exp\left[\sqrt{\frac{\lambda}{2\nu}} (1+i)y\right]\right.$$

$$\left. + C_3 \exp\left[-\sqrt{\frac{\lambda}{2\nu}} (1+i)y\right]\right\} \tag{6.69}$$

Using the second boundary condition in (6.61), $C_2 = 0$. The solution may then be written

$$u(y, t) = C_4 \exp[i\lambda t] \exp\left[-\sqrt{\frac{\lambda}{2\nu}} (1+i)y\right] \tag{6.70}$$

where $C_4 = C_1 C_3$. Using the other boundary condition we find $C_4 = U$

---

*An imaginary constant allows $T(y)$ to be an oscillating time function whereas a real constant would result in an exponential function. Certainly the time dependence should be oscillatory.

and $\lambda = \omega$, giving the final solution as

$$u(y, t) = U \exp\left[i\left(\omega t - \sqrt{\frac{\omega}{2\nu}}\, y\right)\right] \exp\left[-\sqrt{\frac{\omega}{2\nu}}\, y\right] \quad (6.71)$$

with the real part being the actual, physical velocity. Recalling that $e^{i\theta} = \cos\theta + i\sin\theta$, we write

$$u(y, t) = U \cos\left(\omega t - \sqrt{\frac{\omega}{2\nu}}\, y\right) \exp\left(-\sqrt{\frac{\omega}{2\nu}}\, y\right) \quad (6.72)$$

This solution represents shear waves which propagate outward at the rate $\sqrt{2\omega\nu}$ and which are damped as they move outward. (See Example 6.6.)

The wall velocity could have been specified as

$$u_w = \tfrac{1}{2} U[e^{i\omega t} + e^{-i\omega t}] \quad (6.73)$$

which is a real quantity. The solution would then have been

$$u(y, t) = \frac{U}{2} \exp\left[-\sqrt{\frac{\omega}{2\nu}}\, y\right]\left[e^{i\left(\omega t - \sqrt{\frac{\omega}{2\nu}}\, y\right)} + e^{-i\left(\omega t - \sqrt{\frac{\omega}{2\nu}}\, y\right)}\right]$$
$$(6.74)$$

which is also a real quantity, and is, in fact, the same as Eq. 6.72. The development leading to Eq. 6.71 follows because we can use superposition, since Eq. 6.60 is linear. Because of this linearity we carry out the solution as was done with the separation-of-variables technique, using only the real part. If the governing equation were nonlinear we would have to use a wall condition like Eq. 6.73 since superposition would no longer be possible.

Example 6.6

Show that the solution (Eq. 6.72) represents shear waves which propagate outward at the rate $\sqrt{2\omega\nu}$.

Solution. Let us calculate the quantity h_n, shown in Fig. E6.6, which identifies the y-coordinates at which $u = 0$. It is found from Eq. 6.72 by letting

$$\cos\left(\omega t - \sqrt{\frac{\omega}{2\nu}}\, h_n\right) = 0$$

so that
$$\omega t - \sqrt{\frac{\omega}{2\nu}}\, h_n = \frac{\pi}{2}(2n-1) \qquad n = 1, 2, 3, \ldots$$
or
$$h_n = \sqrt{2\nu\omega}\, t - \frac{\pi}{2}\sqrt{\frac{2\nu}{\omega}}\,(2n-1)$$

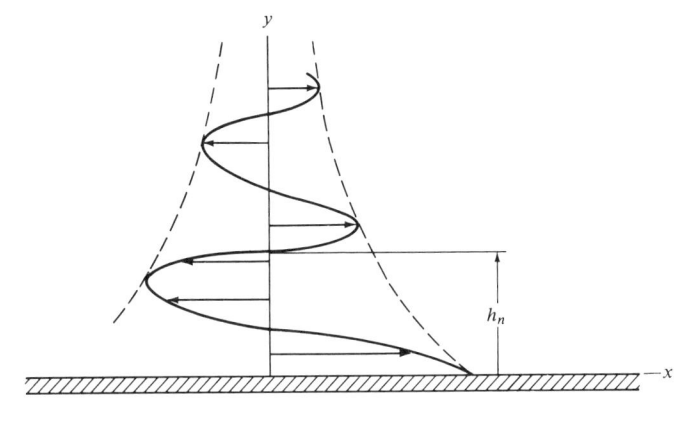

Fig. E6.6

At any particular instant, say t_0, the various points where $u = 0$ would be found by letting $n = 1, 2, 3, \ldots$ to yield $h_1, h_2, h_3, \ldots$ respectively. The zero points located by h_n move outward at the rate dh_n/dt, given by

$$\frac{dh_n}{dt} = \sqrt{2\nu\omega}$$

These waves which move outward at this rate are referred to as shear waves since they are caused by the action of viscosity through the shearing stresses.

6.6 GENERAL OBSERVATIONS ON LAMINAR FLOWS

A number of problems involving laminar flows have been solved in this chapter. Although several more examples might be included, the number of laminar-flow problems for which an exact solution can be obtained is quite small (about 15). However, based on the solutions we have obtained, we can make certain general observations about laminar flows.

1. Superposition

The solutions which have been obtained are for those physical situations that can be described by a linear form of the Navier-Stokes equations and that have allowed the use of the method of separation of variables. An important feature of linear equations is that the individual solutions of these equations can be added to obtain the solution to the problem represented by the combination of the separate problems. For example, if an axial pressure gradient exists for the problem of the rotating cylinders (Fig. 6.5) it would be possible to generate the solution for the axial flow, similar to that in Example 6.3, and add it to the azimuthal velocity v_θ given by Eq. 6.52. A helical velocity field, $\mathbf{V} = v_\theta \hat{e}_\theta + v_z \hat{e}_z$, would result, where v_z is the velocity given in Example 6.3.

2. Steady Conduit Flows

For the fully developed flow region in a pipe or between parallel plates (see Eqs. 6.16 and 6.39), we observed that the pressure gradient is directly proportional to both the velocity and the viscosity and inversely proportional to the square of the characteristic length. Thus, if the velocity or the viscosity is doubled or the length is decreased by $\sqrt{2}$, the pressure gradient doubles. It is postulated, and has been verified experimentally for several geometries, that for conduit flows in the fully developed region it is always true that the pressure gradient is directly proportional to the velocity and the viscosity and inversely proportional to the square of the characteristic length of the flow field.

These considerations can be more carefully examined in relation to the dimensionless form of the Navier-Stokes equations. For the fully developed condition in a conduit, the convective acceleration is zero and a balance exists between the pressure gradient and the shear-stress gradient. As noted for the several cases that have been examined, the pressure gradient is a constant for such conditions. A characteristic velocity such as the average velocity V in the conduit and a characteristic length such as the hydraulic diameter d_H ($4 \times$ area/circumference) of the conduit can be introduced as the appropriate scales for the problem. The normalized stress-gradient term is related to the normalized pressure-gradient term, for our conduit flow, as

$$\nabla^* p^* = \frac{\mu}{\rho d_H V} \nabla^{*2} \mathbf{V}^* \tag{6.75}$$

where the normalized pressure $p^* = p/\rho V^2$. As in Chapter 4, the Reynolds number is the parameter that the various dimensionless quantities of interest depend on for the conduit flow. For example, the pressure gradient $\Delta p/L$ is of particular importance in conduit flow. Eq. 6.75 would indicate that the dimensionless pressure gradient depends on the Reynolds number, that is,

$$\frac{\Delta p^*}{L^*} = f(\text{Re}) \tag{6.76}$$

For a pipe flow (see Eq. 6.27), we know that

$$\frac{\Delta p}{L} = \text{const}\ \frac{\mu V}{D^2} \tag{6.77}$$

which, in dimensionless form, is

$$\frac{\Delta p^*}{L^*} = \frac{\text{const}}{\text{Re}} \tag{6.78}$$

or the dimensionless pressure gradient is inversely proportional to the Reynolds number. Thus, we see that the relationship of Eq. 6.76, with the results from our study in pipe flow, leads to the conclusion in Eq. 6.78. This is generally valid for laminar flow in conduits, D being the hydraulic diameter.

This linear behavior has been put to effective use for fluidic elements (i.e., no moving part, logic, control, and sensing devices) in order to make a *linear fluidic resistor*, shown in Fig. 6.7. "Linear" means that the pressure drop is directly proportional to the flow rate. The pressure drop Δp is taken to be the analog of the electrical voltage drop.

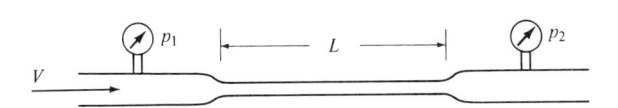

Fig. 6.7. Fluidic linear resistor. (The small-bore passage may be of any shape.)

6.7 LAMINAR-FLOW INSTABILITY

The laminar flows considered in this chapter are several examples of flows for which the velocity profiles can be solved exactly. They all involve simple geometries and simple boundary conditions. Usually, in order to find a laminar-flow solution for a more complex flow situation, flow in the entrance region of a conduit or flow around a sphere for example, we must use approximate techniques or numerical methods. Having found a laminar flow, either exactly or with some other technique, the immediate question raised is, "For what Reynolds number range is this solution valid?" For the problems considered in this chapter, the describing equation is Eq. 4.6, so that Re is the only parameter. We observe experimentally that at a large enough Reynolds number, the laminar flow becomes unstable and the flow undergoes a transition into a "secondary," more complex laminar flow, as discussed in Art. 6.1, or into a turbulent flow. The Reynolds number at which this transition occurs, the critical Reynolds number, is of much interest and the theory used to determine it is called "hydrodynamic stability theory." The critical Reynolds number for a particular type of flow (i.e., flow in a pipe and flow between rotating cylinders) depends on the size of disturbances inherent in the flow, the frequency of these disturbances, the roughness of the boundaries, the temperature field, boundary flexibility, and so on.

If we were to observe a flow between parallel plates at Re = 10,000, as shown in Fig. 6.8, we would observe at various x-locations an oscilloscope output as shown for $u'(t)$. At A the flow is laminar; at B the flow is unstable and fluctuations appear as fairly regular sinusoidal disturbances; at C the disturbances "burst" into turbulence (no stable secondary laminar flows occur between parallel plates or in a pipe). This is a typical transition-to-turbulence process which occurs in the entrance regions of parallel plates and pipes. Between parallel plates this process may occur at Reynolds numbers as low as 1500 if large disturbances occur in the flow or hypothetically at Re = 7700 if all disturbances can be eliminated. The figure of Re = 7500 (below which no turbulence occurs) has been attained* for disturbances of 0.12-percent fluctuation intensity, where

$$\text{Fluctuation intensity} = \frac{\sqrt{\overline{u'^2}}}{V_{\text{avg}}} \times 100 \qquad (6.79)$$

$u'(t)$ is the velocity fluctuation at the entrance.

For flow in pipes the critical Reynolds number has been observed as high as 40,000 when extreme care has been used to eliminate disturbances. There is no reason to believe it could not be higher, if even greater care were used in eliminating disturbances. As an experimental

---

*This has been attained in the fluid mechanics laboratories of the Mechanical Engineering Department at Michigan State University in a 70-to-1 aspect-ratio channel developed by Drs. M. A. Karnitz and N. A. Feliss.

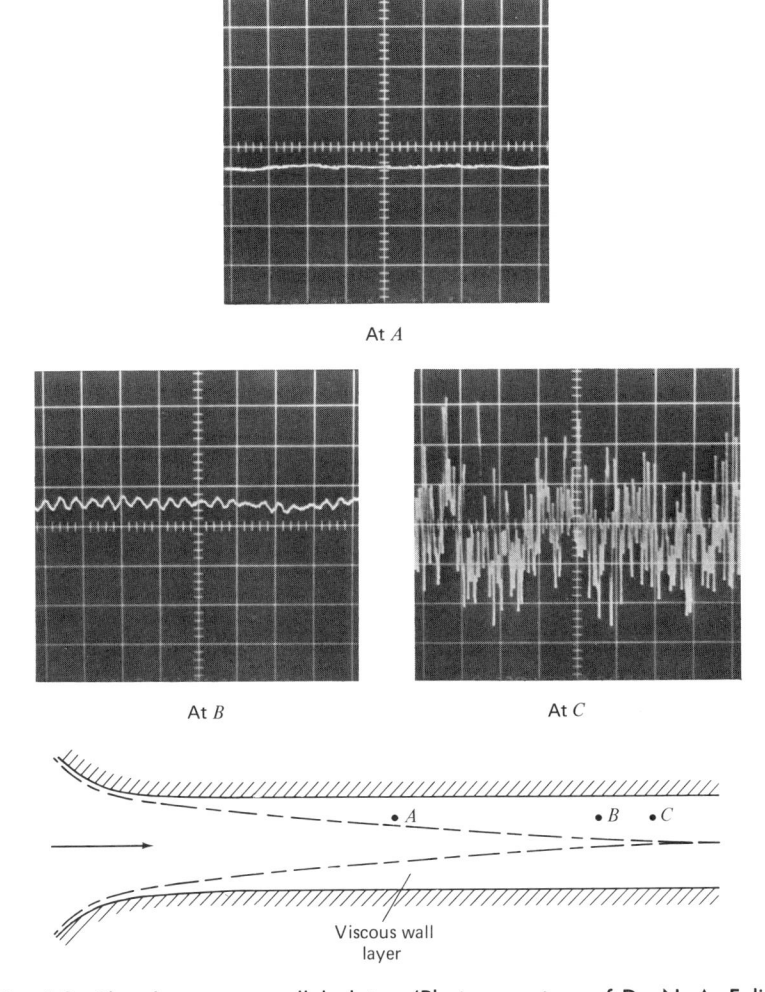

Fig. 6.8. Flow between parallel plates. (Photos courtesy of Dr. N. A. Feliss)

point of interest, the flow facility must be isolated from the building containing the facility at these high Reynolds numbers so that disturbances which exist in the structure do not "trigger" the transition process.

The reason that such high critical Reynolds numbers are possible in pipe flow—this can be shown analytically—is that the disturbances are damped at higher and higher Reynolds numbers as the disturbance amplitude is decreased, with zero-amplitude disturbances having an infinite critical Reynolds number. However, in flow between parallel plates the critical Reynolds number for zero-amplitude disturbances is Re = 7700; hence, we can only hope to approach these limits as the disturbances are decreased in strength.

The frequency f or wave number α of the disturbance is also of importance. A large disturbance of one frequency may not cause transition whereas a small disturbance of another frequency may. If we plot wave number α vs Reynolds number for controlled frequency disturbances of a particular amplitude we would obtain a *neutral stability curve*, as shown in Fig. 6.9. The solid curve would represent data for a set of disturbances with larger amplitude than those associated with the dotted line.

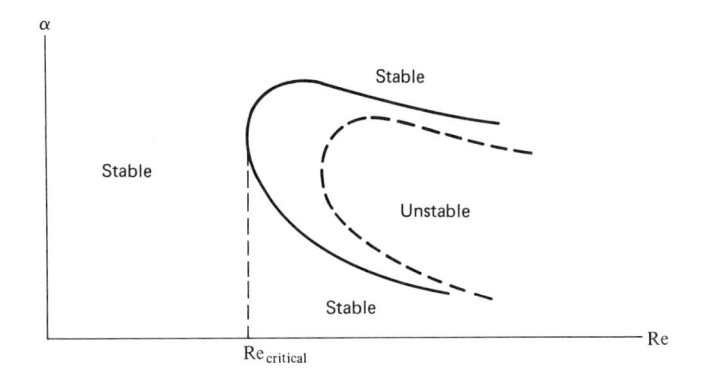

Fig. 6.9. Neutral stability curve.

The critical Reynolds number shown in the figure is the minimum Reynolds number of the neutral stability curve. For a flow in an experimental apparatus we assume that all wave numbers are represented by the natural disturbances which exist in the flow. The natural disturbances are usually quite random with some natural frequencies prevailing, but the whole spectrum of wave numbers is present.

Other interesting instabilities are shown in Fig. 6.10. Many of these problems have never been solved, only observed. Stability theory is usually presented in a graduate course and may be found in advanced texts on fluid mechanics.

(a) Honey piling up

(b) Cigarette smoke

(d) Boundary-layer instability lending to turbulent flow

(e) Surface wave instability

(c) Flow up a concave wall

(f) Convective thermal instability

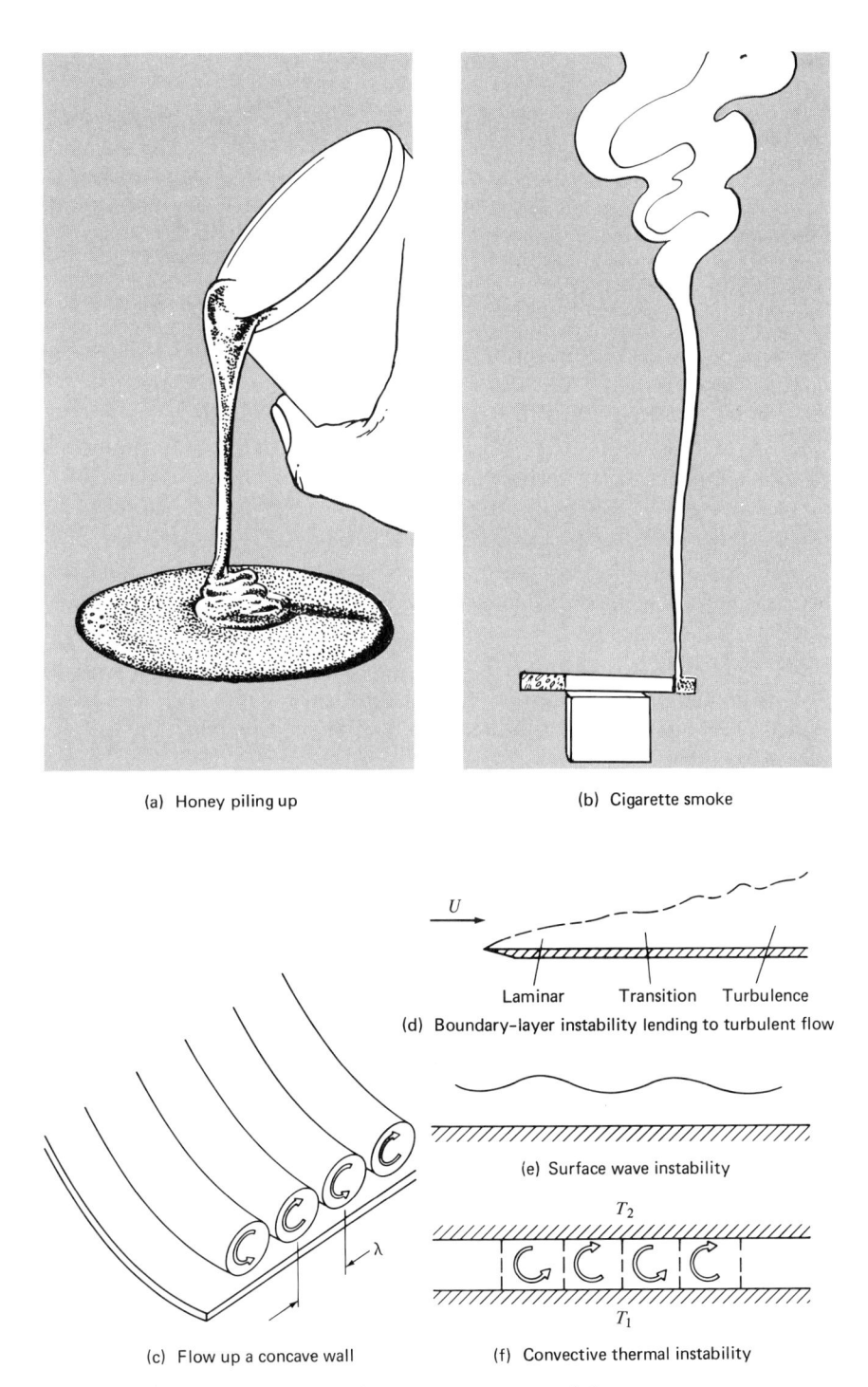

Fig. 6.10. Some interesting instabilities.

PROBLEMS

6.1 For steady laminar flow in a channel, Eq. 6.7 is the governing equation. Why is this equation not acceptable for turbulent flow? (Do not consider time-averaging.) What terms would be necessary to include for a turbulent flow? How would the boundary or initial conditions be affected?

6.2 An element of a fluidic circuit, which performs a function similar to that of an electrical resistance, requires that the pressure drop be proportional to the first power of the average velocity. The device has a $\frac{1}{64}$-in.-dia. hole through which air at 70°F flows. Determine the maximum pressure drop allowed if the air passage is 10 in. long. Use $\rho = 0.0024$ slug/ft^3.

6.3 A pressure drop of 10 psi occurs in a section of a 1-in.-dia. pipe at a Reynolds number of 1000. How long is the section of horizontal pipe? Water flows with $\nu = 10^{-5}$ ft^2/sec.

6.4 Water at 80°F flows down a slight decline in a $\frac{1}{2}$-in.-dia. pipe. For a Reynolds number of 2000, determine the slope such that no pressure drop occurs. (Note that the p used in the equations in Art. 6.2 is actually the kinetic pressure $p_k = p + \rho gh$.)

6.5 Determine the shearing stress at the wall and the dragging force on 100 ft of $\frac{1}{4}$-in.-dia. pipe if a pressure drop of 2 psi occurs over the 100-ft length. Water at 60°F is flowing. Also determine the Reynolds number and the head loss.

6.6 The friction factor for water at 60°F flowing in a $\frac{1}{2}$-in.-dia. pipe is 0.06. Find the pressure drop for the laminar flow in 1000 ft of pipe.

6.7 A laminar flow of water exists in a 10-ft-long, $\frac{1}{4}$-in.-dia. pipe at a Reynolds number of 2000. Calculate the pressure drop in the pipe. The velocity distribution at the entrance to the 10 ft length is uniform and $\nu = 10^{-5}$ ft^2/sec.

6.8 Laminar flow exists in a pipe between two pistons moving at a constant speed U. To create a steady flow, superimpose a velocity $(-U)$ on the whole flow field. Sketch the "entrance" regions and the developed-flow region and determine the velocity profile for the developed flow if the pistons are far apart.

6.9 Estimate the flow rate through the pipe shown. Assume that a laminar flow exists.

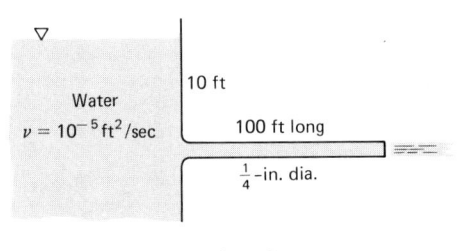

Water
$\nu = 10^{-5}$ ft^2/sec
10 ft
100 ft long
$\frac{1}{4}$-in. dia.

Prob. 6.9

6.10 Water flows in an annulus as shown. For a pressure gradient of -0.01 psf/ft, find the shearing stress on the inner cylinder assuming a laminar flow.

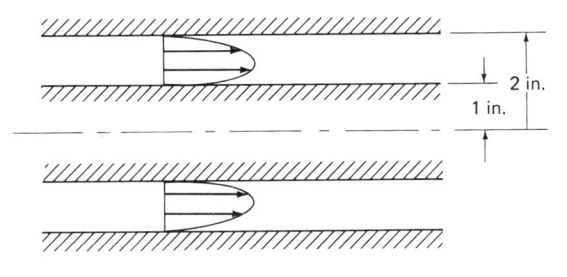

Prob. 6.10

6.11 The inner cylinder of Prob. 6.10 has been heated so that a temperature gradient exists in a liquid flowing in the annulus. The viscosity is highly sensitive to the temperature. It is assumed that $\mu(T)$ is known. What differential equation would describe this non-homogeneous flow? State the boundary conditions.

6.12 A small device is constructed to measure the shearing stress at the wall of a channel. Determine an expression which relates the average velocity to the wall shearing stress for a laminar flow.

6.13 Verify the relationship for the average velocity given in Eq. 6.44. Also verify that for flow between parallel plates, $f = 48/\text{Re}$.

6.14 Calculate the flow rate in a channel $\frac{1}{2} \times 20$ in. if water is flowing at a Reynolds number of 2000 and $\nu = 10^{-5}$ ft²/sec. Assume laminar flow.

6.15 A channel is designed so that air at room temperature can flow in a laminar state at an average velocity of 15 fps. This magnitude allows easy measurement of the velocity distribution. For laminar flow at a Reynolds number of 5000, what should be the distance between the parallel plates?

6.16 A slot $\frac{1}{32} \times 2$ in. is machined in the side of a 4-in.-thick pressure vessel. What maximum pressure can exist in the vessel if laminar flow exists in the slot? Assume the entrance length is negligible; $\nu = 1.6 \times 10^{-4}$ ft²/sec and $\rho = 0.0024$ slug/ft³. Now consider the entrance flow to be significant; what influence would this have on the flow assuming the same pressure difference as that calculated?

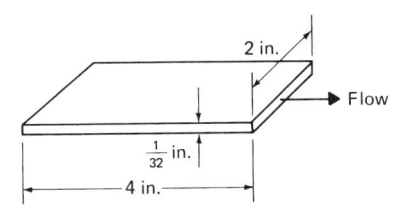

Prob. 6.16

6.17 A large flat plate is constrained to move with a constant velocity U at a fixed distance h above a plane surface. The gap is filled with a fluid of density ρ and kinematic viscosity ν. Formulate the Navier-Stokes equations for this problem, assuming a developed flow with zero pressure gradient. Show that the velocity distribution in the developed region is $u = Uy/h$, where y is normal to the plates and measured from the stationary plate.

6.18 Let the physical situation described in Prob. 6.17 be modified by the addition of a flow system such that the fluid is driven by a pressure differential as well as the shearing action of the upper plate. The pressure drop is fixed at the value dp/dx. This part of the problem has been solved in Art. 6.3. Since the governing equations are linear, the solution to Prob. 6.17 may be added to the earlier solution to obtain the velocity distribution between the plates. Let y be the coordinate normal to the plates and measured from the lower plate. Determine $u(y)$ for the laminar flow.

6.19 A flat belt is moved parallel to a fixed wall at 10 fps as shown. (a) Calculate the pressure gradient required to cause a zero flow rate. (b) Calculate the pressure gradient required for a zero shear on the wall. (c) Calculate the flow rate for a zero shear on the belt.

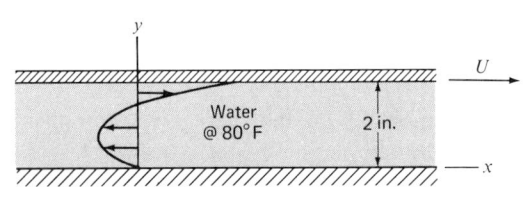

Prob. 6.19

6.20 Water flows down a long inclined plane at the rate of 10 cfs per ft of width. Determine the depth of flow and the shearing stress at the wall for a 10° incline, assuming a laminar flow.

6.21 Water at 50°F is used as the lubricant for the propeller shaft of a large ship. The water is pumped through the gap between the shaft and bearing. Assuming a laminar flow, approximate the flow rate if the pressure drop is 10

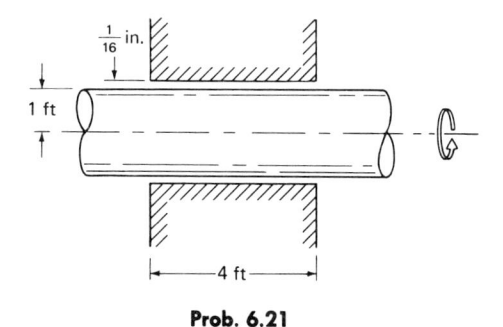

Prob. 6.21

lb/ft². Is the laminar-flow assumption reasonable? (It is acceptable to assume that the flow in the gap approximates flow between parallel plates. Because the governing equations are linear, the flow in the gap may be solved independently of the rotational flow.)

6.22 A shaft, at a certain time in an operation cycle, is moving at a velocity of 20 fps as shown. The pressure difference across the bearing is $p_2 - p_1 = 10$ psi. Calculate the flow rate of the lubricating oil if $\mu = 10^{-3}$ lb-sec/ft².

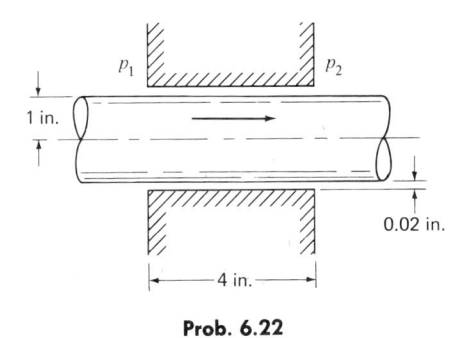

Prob. 6.22

6.23 Approximate the torque necessary to rotate the disc with an angular velocity Ω. A liquid with viscosity μ fills the small gap of depth δ between the disc and the fixed container.

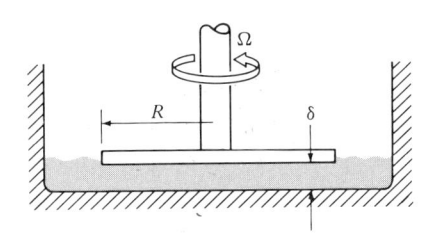

Prob. 6.23

6.24 A flat plate is moved parallel to an upper fixed plate with a velocity U. The temperatures of the upper and lower plates are T_2 and T_1, respectively. The energy differential equation yields a temperature distribution of approximately $T(y) = T_1 + (T_2 - T_1)(y/h)$, where y is measured from the lower plate. The viscosity-temperature relation is $\mu(T) = k_1 e^{k_2 T}$, where k_1 and k_2 are constants. Assuming no pressure gradient, determine $u(y)$ for the developed flow.

6.25 Determine the velocity distribution in the fluid for a cylinder of radius R, rotating with angular velocity Ω in a large container of liquid. (*Hint*: In Art. 6.4, let $\Omega_2 = 0$ and $r_2 \to \infty$.) Calculate the torque per unit length on such a cylinder if $R = 2$ in. and $\Omega = 20{,}000$ rpm. The liquid is water at 70°F.

6.26 Calculate the torque necessary to rotate an inner cylinder at $\Omega_1 = 10$ rad/sec if the outer one is fixed. The two radii are 2 in. and 2.01 in. Water fills the gap between the 4-in-long cylinders, and $\nu = 10^{-5}$ ft²/sec.

6.27 The first instability occurs in the flow of Prob. 6.26 at Re = 1700, where Re = $r_1\Omega_1\delta/\nu$ and $\delta = r_2 - r_1$. Determine the value Ω_1 at which the flow illustrated in Fig. 6.1b would be expected to occur.

6.28 An instrument designed to measure the viscosity of a liquid is composed of two concentric cylinders with the inner one suspended from a wire as shown. The outer one rotates at a prescribed angular speed. The torque characteristics of the wire are so calibrated that a certain angular deflection of the wire corresponds to a particular torque. Find an expression for the viscosity μ in terms of the measured torque T and the other appropriate quantities.

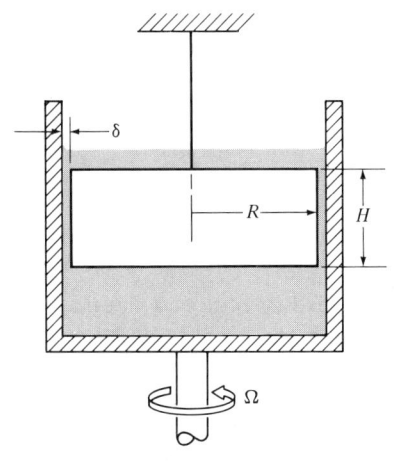

Prob. 6.28

6.29 A viscosity pump is constructed as shown. Its main feature is a stiff membrane which is in slip contact with the inner cylinder and across which a

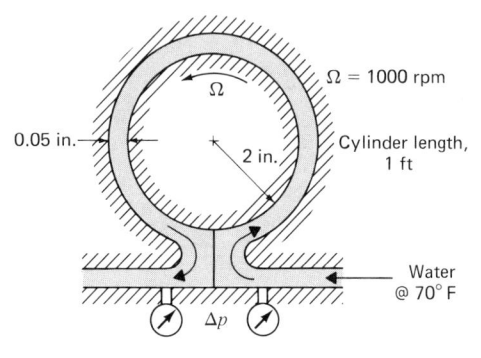

Prob. 6.29

pressure difference may occur. This pressure difference Δp is controlled by external means. Determine the flow rate q for (a) $\Delta p = -10$ in. water and (b) $\Delta p = 0$; (c) also determine the value of Δp which is necessary for $q = 0$. Assume the narrow-gap approximation (see Example 6.5).

6.30 In deriving $\partial u/\partial t = \nu\, \partial^2 u/\partial y^2$, (Eq. 6.60), the pressure gradient $\partial p/\partial x$ must be shown to be zero if the same pressure is maintained at $x = \pm\infty$. By starting with the y- and z-component momentum equations, show that $\partial p/\partial x = 0$ for the flow above an oscillating flat plate.

6.31 To solve Eq. 6.60, the separation-of-variables technique was used. The constant in the separation process was assumed to be $i\lambda$. Instead of the imaginary quantity $i\lambda$, assume the constant to be the real quantity λ and show that the boundary conditions cannot be satisfied.

6.32 Calculate the shear stress as a function of y at $t = \pi/2$ sec in a fluid above a large flat plate oscillating in its own plane with oscillation velocity $20 \cos 4t$. The fluid is water at $60°F$.

6.33 Determine the velocity distribution in the fluid contained between two plates a distance h apart if the lower plate oscillates in its own plane with frequency ω.

6.34 Consider the problem of a plate being accelerated parallel to itself at the exponential rate $e^{0.01t}$. Determine the solution $u(y, t)$. The footnote referenced between Eqs. 6.64 and 6.65 will be helpful for this. Use $\nu = 10^{-4}$ ft^2/sec.

6.35 How far from the leading edge of a flat plate can we expect a laminar flow of air if $V_\infty = 600$ mph and $\nu = 2 \times 10^{-4}$ ft^2/sec? The critical Reynolds number for a boundary-layer flow is 4×10^5, where $\text{Re} = V_\infty x_T/\nu$. This answer would be an approximation of the length of the laminar region on the airfoil of a commercial passenger jet.

6.36 A wave is traveling at a speed of 10 ft/sec with a frequency of 200 cps. Determine the wave number if $\alpha = 2\pi/\lambda$, where λ is the wavelength.

SELECTED REFERENCES

The motion pictures *Low Reynolds Number Flows* (No. 21617; G. I. Taylor, film principal) and *Flow Instabilities* (No. 21619; E. L. Mollo Christenson, film principal) are recommended as supplements to this chapter.

For an alternative discussion of the Navier-Stokes solutions at a level compatible with that of the present text, see Y. S. Yuan, *Foundation of Fluid Mechanics*, Prentice-Hall, Englewood Cliffs, N. J., 1967; and, for a somewhat more advanced, comprehensive treatment, see R. B. Bird, W. E. Stewart, and E. N. Lightfoot, *Transport Phenomena*, John Wiley & Sons, New York, 1960, and H. Schlichting, *Boundary Layer Theory*, 6th ed., McGraw-Hill Book Co., New York, 1968. Numerous laminar-flow solutions (utilizing tensor notation) and stability analyses are provided by G. K. Batchelor, *An Introduction to Fluid Dynamics*, Cambridge University Press, London, 1967, and by C. S. Yih, *Fluid Mechanics*, McGraw-Hill Book Co., New York, 1969.

7

Turbulence

7.1 INTRODUCTION

When we realize that the vast majority of flows of natural and technological importance—oceans, rivers, the atmosphere, flows through heat exchangers and combustion chambers and around automobiles and marine propellers—operate in a turbulent-flow regime, the importance of turbulent flow becomes self-evident. That turbulence is randomly time-dependent and strongly nonlinear in its mathematical model makes it a very difficult subject. Both points were emphasized by George Carrier of Harvard University, speaking at a sesquicentennial discussion at the University of Michigan on fluid mechanics research of the future (from the present to the year 2000), who began his address with the observation that "turbulence is the major unsolved problem to be studied." He closed his presentation with the remark: "If, in the year 2000, someone delivers a similar lecture on fluid mechanics research in the next 50 years, he will probably start by noting that turbulence is our major unsolved problem."

In the light of these observations, the material selected for this chapter has been chosen to meet a difficult but rather direct objective. The material is more descriptive than quantitative and is intended to provide the student with insight as to *why* the general turbulence problem is so difficult, *how* the technical community has organized the search for general and specific "solutions" to problems involving turbulent flow, and *what* general knowledge exists to deal with turbulent flows. Although the complexities of turbulent flow are substantial, they do not result from mysterious "ghost forces." The objective of the chapter is to allow the student to comprehend not only the real difficulties but also the productive avenues which exist and, possibly, to stimulate his interest in pursuing this subject to learn how to better apply present knowledge or to generate, through research, new knowledge.

We have so far used the word "turbulence" without attempting its definition. We will follow the suggestion of R. W. Stewart, who, in his introductory film* on the subject, refers to turbulence in terms of three basic characteristics: *disorder (randomness)*; *enhanced mixing*; and *three-dimensional motions involving angular momentum*. We might also add that *turbulence occurs at high Reynolds numbers and is generated in the presence of mean velocity gradients*. Somewhat in common with the word "sky," attempts at formal definitions do more harm than good; hence these characterizations must suffice.

Vorticity, defined and discussed in Chapter 5, plays an important although complicated role in turbulent motions. The phenomenon of vortex stretching, which was introduced in this previous chapter, is strongly related to the randomness, the enhanced mixing, and the three-dimensional motions of turbulent flows. Vortex stretching is also associated with the easily observed tendency of turbulence to develop small-scale and intense motions from large-scale motions. (Observe this when cream is stirred in a coffee cup; the mixing occurs on a scale quite small with respect to the size of the spoon.) The student is encouraged to maintain an awareness of the relationship between vorticity and the physical phenomena discussed in the present chapter.

7.2 THE DESCRIBING EQUATIONS

The field equations which describe a fluid motion, whether laminar or turbulent, are the differential continuity, momentum, and energy equations. We will be concerned with those incompressible flows which are not dependent on the temperature field, so that only the continuity and momentum equations are necessary. The four equations, in cartesian component form, follow immediately on the next page.

*See reference list at the end of the chapter.

$$\frac{\partial u}{\partial x} + \frac{\partial v}{\partial y} + \frac{\partial w}{\partial z} = 0$$

$$\frac{\partial u}{\partial t} + u\frac{\partial u}{\partial x} + v\frac{\partial u}{\partial y} + w\frac{\partial u}{\partial z} = -\frac{1}{\rho}\frac{\partial p_k}{\partial x} + \nu \nabla^2 u$$

$$\frac{\partial v}{\partial t} + u\frac{\partial v}{\partial x} + v\frac{\partial v}{\partial y} + w\frac{\partial v}{\partial z} = -\frac{1}{\rho}\frac{\partial p_k}{\partial y} + \nu \nabla^2 v \qquad (7.1)$$

$$\frac{\partial w}{\partial t} + u\frac{\partial w}{\partial x} + v\frac{\partial w}{\partial y} + w\frac{\partial w}{\partial z} = -\frac{1}{\rho}\frac{\partial p_k}{\partial z} + \nu \nabla^2 w$$

Turbulent motion is always an erratic, unsteady motion; hence, if these equations were to be solved it would be necessary to specify an initial condition, namely, the velocity and pressure at each point in the region of interest at some instant. It is very unlikely that this information would be available except for the simplest of cases. A complete solution to this unsteady-flow problem would involve rather a phenomenal amount of information: specifically, the three velocity components and the pressure at each point in space and time. It is quite unlikely that such detailed information could be effectively used for engineering purposes even if it were available. Rather, we are interested in average quantities, such as average pressure, average shearing stress, average location of separation point, and average lift force.* With this motivation, we separate quantities into time-averaged and fluctuating parts, as shown in Fig. 7.1. (The x-velocity component u is shown as characteristic.) The flow may be "steady," as shown in part a, or "unsteady" as shown in part b. It is useful to think of this as decomposing the instantaneous, actual quantity into two parts. It is important to recognize that neither of these parts exists in the flow; they are merely a mathematical convenience. By "steady" we mean that the time-averaged part is independent of time even though the fluctuating part is unsteady.

To write Eqs. 7.1 in terms of time-averaged quantities, we separate each dependent variable into a time-averaged part and a fluctuating part as

$$u = \bar{u} + u' \qquad v = \bar{v} + v' \qquad w = \bar{w} + w' \qquad p_k = \bar{p}_k + p'_k \quad (7.2)$$

* This is not to imply that the basic mechanics of turbulent motion are revealed solely by such average quantities. Indeed, turbulence research has been rejuvenated in the late 1960s by the recognition that coherent and structured events apparently occur somewhat randomly in space and time.

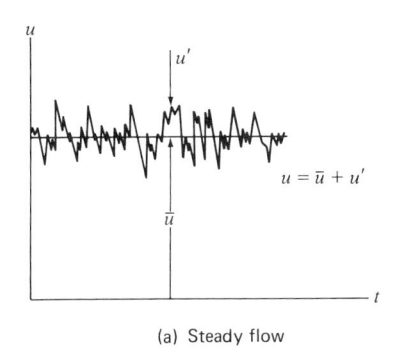

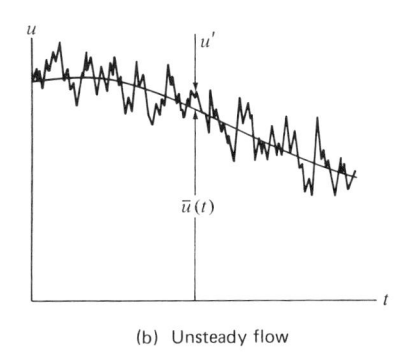

Fig. 7.1. Examples of quantities in a turbulent flow, showing average and fluctuating components.

where the time-averaged quantity $\bar{u}$ is defined by

$$\bar{u} = \frac{1}{T} \int_0^T u \, dt \tag{7.3}$$

In many problems, such as a pipe flow used to extract fluid from a large reservoir under the action of gravity, it is not difficult to decide upon a suitable time T over which the integration process should occur. We simply let T be large enough for repeated samples (i.e., for different $t = 0$ conditions) to yield sufficiently similar results so that the computed value is tending toward a stationary condition (mathematically, $d\bar{u}/dT = 0$). There are, however, cases where $\bar{u}$ is a function of time* for which the determination of a sensible T value is a matter for the analyst or experimentalist to arbitrarily decide, based upon the character of the problem. The velocity at a point in a washing machine, the velocity in the exhaust of a finite-volume pressurized container and the wind velocity on a gusty day are all examples of unsteady turbulent flows, as the reader may verify from his own experience.

The quantities in Eq. 7.2 may be substituted in the continuity equation to obtain

$$\frac{\partial \bar{u}}{\partial x} + \frac{\partial \bar{v}}{\partial y} + \frac{\partial \bar{w}}{\partial z} + \frac{\partial u'}{\partial x} + \frac{\partial v'}{\partial y} + \frac{\partial w'}{\partial z} = 0 \tag{7.4}$$

A time average of the entire equation and use of the "rule" $\overline{A + B} =$

---

*This rather subtle point is perhaps better made in terms of an *ensemble average* in which an experiment is run many times, and the average, at a specified time from the initiation of the experiment, is determined. This would be $\bar{u}(t)$ and the fluctuation can be defined as above. For a single experiment $\bar{u}(t)$ cannot be precisely determined.

$\bar{A} + \bar{B}$ (this can be verified by using the definition of a time-averaged quantity, Eq. 7.3) results in

$$\frac{\partial \bar{u}}{\partial x} + \frac{\partial \bar{v}}{\partial y} + \frac{\partial \bar{w}}{\partial z} = 0 \tag{7.5}$$

In this process we have used a second "rule" of averaging,*

$$\overline{\frac{\partial u'}{\partial x}} = \frac{\partial \overline{u'}}{\partial x} = 0 \tag{7.6}$$

This follows from

$$\overline{\frac{\partial u'}{\partial x}} = \frac{1}{T}\int_0^T \frac{\partial u'}{\partial x}\,dt = \frac{\partial}{\partial x}\left(\frac{1}{T}\int_0^T u'\,dt\right) = \frac{\partial \overline{u'}}{\partial x} = 0 \tag{7.7}$$

which is permissible since the limits on the integration process are independent of x. If we now substitute Eq. 7.5 back into Eq. 7.4 we arrive at the fluctuating continuity equation

$$\frac{\partial u'}{\partial x} + \frac{\partial v'}{\partial y} + \frac{\partial w'}{\partial z} = 0 \tag{7.8}$$

To determine the time-averaged momentum equations it is necessary to time-average the nonlinear acceleration terms, for example, $v\,\partial u/\partial y$. This is done as follows:

$$\overline{v\frac{\partial u}{\partial y}} = \overline{(\bar{v} + v')\frac{\partial(\bar{u}+u')}{\partial y}} = \overline{\bar{v}\frac{\partial \bar{u}}{\partial y}} + \overline{\bar{v}\frac{\partial u'}{\partial y}} + \overline{v'\frac{\partial \bar{u}}{\partial y}} + \overline{v'\frac{\partial u'}{\partial y}}$$

$$= \bar{v}\frac{\partial \bar{u}}{\partial y} + \bar{v}\overline{\frac{\partial u'}{\partial y}} + \overline{v'}\frac{\partial \bar{u}}{\partial y} + \overline{v'\frac{\partial u'}{\partial y}}$$

$$= \bar{v}\frac{\partial \bar{u}}{\partial y} + \overline{v'\frac{\partial u'}{\partial y}} \tag{7.9}$$

Note that $\overline{v'\,\partial u'/\partial y} \neq \overline{v'}\,\overline{\partial u'/\partial y}$. This is obvious if one returns to the definition of the time average, Eq. 7.3. However, it is true that $\overline{\bar{v}\partial u'/\partial y} = \bar{v}\,\overline{\partial u'/\partial y}$ which can be verified by Eq. 7.3. The three time-averaged momentum equations may be put in the form

---

*We will often make use of the fact that $\overline{u'} = 0$. This follows from time-averaging $u = \bar{u} + u'$, which gives $\bar{u} = \bar{u} + \overline{u'}$, resulting in $\overline{u'} = 0$.

$$\frac{\overline{D}\overline{u}}{Dt} = -\frac{1}{\rho}\frac{\partial \overline{p}_k}{\partial x} + \nu \nabla^2 \overline{u} - \frac{\partial \overline{u'^2}}{\partial x} - \frac{\partial \overline{u'v'}}{\partial y} - \frac{\partial \overline{u'w'}}{\partial z}$$

$$\frac{\overline{D}\overline{v}}{Dt} = -\frac{1}{\rho}\frac{\partial \overline{p}_k}{\partial y} + \nu \nabla^2 \overline{v} - \frac{\partial \overline{u'v'}}{\partial x} - \frac{\partial \overline{v'^2}}{\partial y} - \frac{\partial \overline{v'w'}}{\partial z} \quad (7.10)$$

$$\frac{\overline{D}\overline{w}}{Dt} = -\frac{1}{\rho}\frac{\partial \overline{p}_k}{\partial z} + \nu \nabla^2 \overline{w} - \frac{\partial \overline{u'w'}}{\partial x} - \frac{\partial \overline{v'w'}}{\partial y} - \frac{\partial \overline{w'^2}}{\partial z}$$

where

$$\frac{\overline{D}}{Dt} = \overline{u}\frac{\partial}{\partial x} + \overline{v}\frac{\partial}{\partial y} + \overline{w}\frac{\partial}{\partial z} + \frac{\partial}{\partial t}$$

Several important features are implicit in Eq. 7.10; these will be separately examined in the following sections of the article.

1. The Mean Flow Field

It would be possible for an observer looking at a turbulent flow to imagine that two separate flows exist, namely, the mean and fluctuating flows. (Such, of course, is not the case; only the instantaneous flow exists in nature!) Pursuing this arbitrary distinction, our observer now asks: "What equation governs the behavior of the mean velocity flow?" The answer to this question is provided by Eqs. 7.10. Note that the left members are the accelerations as defined by the mean flow. The right members contain the mean pressure gradient and the mean viscous shear stresses plus other terms which exert a stresslike effect on the mean flow. This, of course, is a matter of viewpoint; the terms are actually accelerations. Equations 7.10 and these stresslike terms are named for Osborne Reynolds in recognition of his identifying the terms as components of a stress tensor:

$$-\rho \begin{pmatrix} \overline{u'^2} & \overline{u'v'} & \overline{u'w'} \\ \overline{u'v'} & \overline{v'^2} & \overline{v'w'} \\ \overline{u'w'} & \overline{v'w'} & \overline{w'^2} \end{pmatrix} = \text{Reynolds stress tensor} \quad (7.11)$$

The question of viewpoint is important. Although these terms do have a stresslike effect, which in a jet or the outer portion of a boundary layer completely dominates the viscous shear stress, they are not capable of directly dissipating kinetic energy to the thermal form nor can they transmit vorticity to irrotational fluid; they are simply components of the acceleration. The identification of these terms as stresses means that the shear stress in a plane flow is often given as

$$\bar{\tau}_{xy} = (\bar{\tau}_{xy})_{\text{laminar}} + (\bar{\tau}_{xy})_{\text{turbulent}}$$

$$= \mu \frac{\partial \bar{u}}{\partial y} - \rho \overline{u'v'} \quad (7.12)$$

2. The Correlation Coefficient

The term $\overline{u'v'}$ represents the time average of the product of the fluctuating x- and y-component velocities and could be rewritten as

$$\overline{u'v'} = K_{uv}\sqrt{\overline{u'^2}}\sqrt{\overline{v'^2}} \quad -1 \leq K_{uv} \leq 1 \quad (7.13)$$

(Note that whereas $\overline{u'} = 0$, $\overline{u'^2} \neq 0$ if $u' \neq 0$.) The term K_{uv} is termed a *correlation coefficient* and has the indicated limits. An examination of any flow visualization of a turbulent flow makes it easy to accept the existence of u' or v'; that is, the random motion of the fluid will clearly lead to a fluctuating velocity and hence $\overline{u'^2} > 0$, $\overline{v'^2} > 0$. It is not clear that $K_{uv} \neq 0$. Physically $K_{uv} = 0$ if the effects causing u' are unrelated to those causing v'. However, if the physical mechanism responsible for a v' leads (on the average) to a negative or a positive u' then the fluctuations will be correlated. Consider the flow in the inlet of a rectangular channel, as shown in Fig. 7.2a. The following argument is used to show that, at point A in the channel entrance, u' and v' are correlated.

The turbulent eddying motion is considered to cause, at some instant, an upward motion $v' > 0$ at point A (case *i*, Fig. 7.2b). This sweeps fluid with a lower average momentum up to point A. This lower-momentum fluid results in a lesser magnitude for the instantaneous x-component velocity than for the average value, that is, $u' < 0$. Conversely, a negative v' induces a positive u' by the same argument (case *ii*). Therefore, for point A, $\overline{u'v'} < 0$ (shown in Fig. 7.2c). A similar argument will demonstrate that $\overline{u'v'} = 0$ at B and that $\overline{u'v'} > 0$ at C.

Art. 7.2 THE DESCRIBING EQUATIONS 313

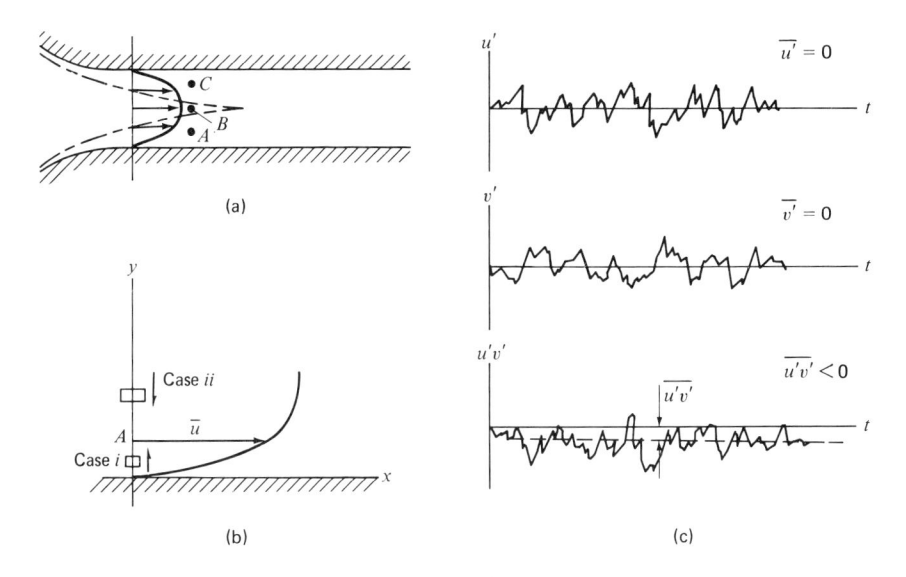

Fig. 7.2. Velocities u' and $\overline{u'v'}$ in a channel entrance.

Turbulent fluctuations are always three-dimensional even when the mean flow structure is two-dimensional. That is, $w' \neq 0$ for $\partial(\ )/\partial z = \overline{w} = 0$. Consequently, $\overline{w'^2} \neq 0$ for any turbulent flow. However, for $\overline{u'w'} = K_{uw}\sqrt{\overline{u'^2}}\sqrt{\overline{w'^2}}$ and $\overline{v'w'} = K_{vw}\sqrt{\overline{v'^2}}\sqrt{\overline{w'^2}}$, the physical effects causing w' are physically unrelated to those causing u' and v' on the average; hence $K_{uw} = K_{vw} = 0$ for a two-dimensional flow.

3. The "Indeterminate" Equations for Turbulent Flow

Since the Reynolds stress terms appear in the momentum (Reynolds) equations, we must count them as unknowns if we wish to solve for $\overline{u}$, $\overline{v}$, $\overline{w}$, and $\overline{p}$. Consequently, in addition to these four unknowns, we have the additional six independent unknowns $\overline{u'^2}$, $\overline{v'^2}$, $\overline{w'^2}$, $\overline{u'v'}$, $\overline{v'w'}$, $\overline{u'w'}$: that is, four equations for ten unknowns! Even for the very simple case of fully developed channel flow (Fig. 7.2a), where $\overline{v} = \overline{w} = \overline{u'w'} = 0$ and $\partial \overline{u'^2}/\partial x = 0$, the equation for $\overline{u}$ retains one unknown Reynolds stress in addition to the mean velocity, specifically,

$$\frac{1}{\rho}\frac{\partial \overline{p}}{\partial x} = \nu \frac{\partial^2 \overline{u}}{\partial y^2} - \frac{\partial \overline{u'v'}}{\partial y} \qquad (7.14)$$

The presence of $\overline{u'v'}$ prevents an analytical solution for $\overline{u}(y)$.

Since the equation for the instantaneous velocity is not indeterminate, we might be tempted to seek a solution for this quantity. However, $\partial u/\partial t \neq 0$ for the instantaneous unsteady motion; hence, for $t = 0$ the velocity component u would have to be specified at every point in the flow field, as would the other dependent variables. This is, of course, extremely difficult if not impossible, therefore, this is not an escape route.

4. Similitude Considerations for Turbulent Flow

Since the Navier-Stokes equations describe the instantaneous turbulent motions and since the time-averaging process does not introduce any terms which are fundamentally different from those considered earlier, similitude considerations will yield the same set of parameters as presented in Chapter 4. Consequently, for a submerged, incompressible flow that is not driven by periodic effects, the Reynolds number is the governing parameter for the flow field providing that the viscous terms are significant. However, the establishment of characteristic lengths and velocities may be rather more complex. For example, consider a stirring action in an industrial blender. The container might have a characteristic dimension of L whereas the flow patterns themselves might be quite independent of L. The flow pattern and hence the mixing characteristics of the blender are dependent upon a characteristic "eddy" size l which reflects the scale of the agitator. Similarly, there may be no characteristic velocity of translation $\bar{u}$, therefore the velocity scale would have to be obtained from the intensity of the velocity fluctuations $(\overline{u'^2})^{1/2}$. Other problems may have natural scales. For example, the Reynolds number based on the average velocity and the diameter for a turbulent pipe flow not only characterizes the mean motion but is also uniquely related to the turbulence characteristics.

For highly turbulent motions the Reynolds stress terms may completely dominate the viscous terms, so that the Reynolds number is no longer a significant parameter in the mean flow. This situation is encountered in high-Reynolds-number flow in rough conduits, in meteorological flows, and in flow around an automobile. For such flows the initial and boundary conditions introduce the important parameters. The conditions may include the size and frequency of the turbulent eddies and the size, spacing, and shape of the wall roughness elements. This type of flow is usually termed a *fully turbulent flow*.

Example 7.1

Use an argument similar to that used for Fig. 7.2b to show that the stress at point C is such that $\overline{u'v'} > 0$.

Solution. A positive v' in the neighborhood of C carries relatively high-velocity fluid toward the upper plate. The upward motion of the fluid in this region is therefore related to a positive u'. Consequently, a positive v' is correlated with a positive u'; hence

$$\overline{u'v'} > 0$$

Extension 7.1.1. Determine the sign of the correlation coefficient K_{uv} for a two-dimensional jet at location A, and at location B (Fig. E7.1), respectively.

Ans. $>0, <0$

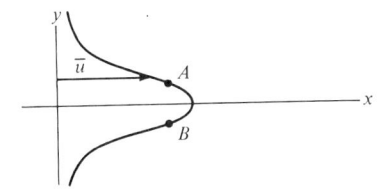

Fig. E7.1.

7.3 APPROACHES TO THE GENERAL TURBULENCE PROBLEM

In determining the character of the mean flow field, the problems introduced by the additional unknown Reynolds stresses must be considered. Furthermore, if one is concerned with thermal-energy transport or mass transport (e.g., heating, cooling, or drying processes) due to the turbulence or if one is concerned with the turbulence structure itself (e.g., its contribution to the accoustic sound from a jet engine exhaust or a ventillating system) one would find that the convective acceleration terms again introduce correlations in these equations which are both important and unknown in the analytical formulation. The response of the technological community concerned with these phenomena has taken three forms, which, in some ways, are mutually supportive and beneficial.

1. To force an analytically complete formulation by various phenomenological theories

2. To compile empirical data on flows of technological significance
3. To conduct research to establish the fundamental phenomena of turbulent motion

These three approaches will be considered in the following sections.

1. Phenomenological Theories

To reiterate the need for these theories: The fluctuating velocities of the convective acceleration terms lead to additional terms for the time-averaged turbulent-flow equations which are not present in the laminar-flow equations. For example, the term $\partial \overline{u'v'}/\partial y$ appears in the x-momentum equation for fully developed channel flow (see Eq. 7.14). An early and quite logical postulate was advanced,* in which an *eddy viscosity* ϵ_M was proposed to serve the same purpose as the kinematic molecular viscosity for laminar flows; that is, we let

$$\epsilon_M \frac{\partial \bar{u}}{\partial y} = -\overline{u'v'} \tag{7.15}$$

so that

$$\bar{\tau}_{xy} = (\bar{\tau}_{xy})_{\text{viscous}} + (\bar{\tau}_{xy})_{\text{turbulent}} = \mu \frac{\partial \bar{u}}{\partial y} + \rho \epsilon_M \frac{\partial \bar{u}}{\partial y} \tag{7.16}$$

Using the eddy viscosity, the momentum equation (7.14) becomes

$$\frac{1}{\rho}\frac{\partial \bar{p}}{\partial x} = \nu \frac{\partial^2 \bar{u}}{\partial y^2} + \epsilon_M \frac{\partial^2 \bar{u}}{\partial y^2} + \frac{\partial \bar{u}}{\partial y}\frac{\partial \epsilon_M}{\partial y} \tag{7.17}$$

for channel flow in which $\bar{u} = \bar{u}(y)$.

An examination of the defining equation (Eq. 7.15) shows that the eddy viscosity may be thought of as merely a proportionally constant, and—if nothing is assumed about its universal character (e.g., it is certainly not a thermodynamic property like μ or ν)—an ϵ_M value could be so defined for any turbulent flow. However, with this interpretation we are simply trading an unknown $\overline{u'v'}$ for an unknown ϵ_M. It is only when ϵ_M can be assumed known that it is of utility.† For example, if one assumes $\epsilon_M = $ constant, an exponential solution is obtained for the

*By Joseph Boussinesq, in 1877.
†There are many proposed eddy viscosity models. One of these which allows a complete profile in a channel is

$$\epsilon_M = \frac{1}{2}\left\{1 + 0.0178\, \text{Re}^2 \frac{\partial p}{\partial x}(1-y^2)^2(1+2y^2)^2\left[1 - \exp\left(\frac{y-1}{26}\text{Re}\sqrt{\frac{\partial p}{\partial x}}\right)\right]^2\right\}^{1/2} - \frac{1}{2}.$$

mean velocity distribution in the wake of a cylinder. Such a solution provides a good description (i.e., good agreement with experimental data) for the mean velocity distributions in the sufficiently far downstream regions of the cylinder wake. However, the constant, which is determined for the wake flow, is not valid for different flows for which ϵ_M = constant provides a good $\bar{u}$ distribution. Furthermore, a constant-eddy-viscosity solution does not provide a good description for a boundary layer or conduit flow. This is not unexpected since it would be asking a great dispensation from nature to allow universal validity for such a simple model.

The term "phenomenological theory" is rather loosely applied to the introduction of an eddy viscosity; the "phenomenon" it appeals to is a postulated physical analogy with the kinematic viscosity, whereby the mean velocity in a turbulent flow is affected by ϵ_M just as ν affects the velocity for a laminar flow.

A more elaborate theory, due to Prandtl,* will be briefly examined. This will both demonstrate the development and establish the nature of a more characteristic phenomenological theory. The derivation of (or the rational for) the *mixing length theory* will be presented in highly abbreviated form. Some of the physical reasoning of Prandtl will thus be lost. However, in considering the "worth" of these phenomenological theories we must remember that their purpose is to provide a way to complete the mean velocity (or temperature) equation, not to provide insight into the structure of turbulent flows.

In terms of Fig. 7.3, Prandtl argued that a fluid mass arriving at location y from a lower position $(-l_1)$ in the shear flow could be assumed to be absorbed by the rest of the fluid at y. In so doing, the initial momentum of the fluid would also be absorbed by the fluid at y, and a velocity fluctuation of a magnitude approximated by

$$u' = \bar{u}(y - l_1) - \bar{u}(y) \tag{7.18}$$

would result. Using the linear Taylor-series expansion, this becomes

$$u' \cong \bar{u}(y - l_1) - \left[\bar{u}(y - l_1) + l_1 \frac{\partial \bar{u}}{\partial y} \right]$$

$$= -l_1 \frac{\partial \bar{u}}{\partial y} \tag{7.19}$$

*Ludwig Prandtl, "Über die ausgebildete Turbulenz," *Zeitschrift für angewandte Mathematik und Mechanik* (*ZAMM*) 5, pp. 136–139, 1925. The basic framework employed by Prandtl was advanced earlier by G. I. Taylor in an analysis of thermal-energy transport.

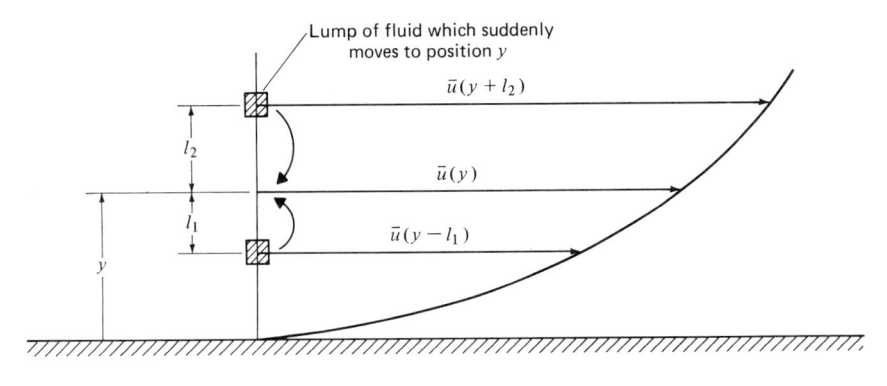

Fig. 7.3. Turbulent motion of a fluid lump.

Note that a similar development applies for fluid arriving from $y + l_2$. The layer at y will, of course, exchange fluid with a wide range of l_1 and l_2 values and it is therefore necessary to introduce an average effective length l_e.

The objective, to characterize the Reynolds stress, can be accomplished by assuming that the magnitude of the y-component velocity fluctuation $\overline{v'^2}$ is roughly proportional to $\overline{u'^2}$. Then, introducing the correlation coefficient,

$$\overline{u'v'} = K_{uv}\sqrt{\overline{u'^2}}\sqrt{\overline{v'^2}}$$

we have

$$\overline{u'v'} = K_{uv}\left(l_e \frac{\partial \bar{u}}{\partial y}\right)\left(c l_e \frac{\partial \bar{u}}{\partial y}\right)$$

$$= -l_M^2 \left|\frac{\partial \bar{u}}{\partial y}\right| \frac{\partial \bar{u}}{\partial y} \quad (7.20)$$

where $l_M^2 = -c l_e^2 K_{uv}$ and the sign of $\overline{u'v'}$ is accounted for by the use of the absolute value for the velocity derivative. The relationship between the eddy viscosity and the mixing length (see Eq. 7.15) is then

$$\epsilon_M = l_M^2 \left|\frac{\partial \bar{u}}{\partial y}\right| \quad (7.21)$$

This last relation reminds us that such theories do not contribute new information, but simply exchange the unknown $\overline{u'v'}$ for an unknown

ϵ_M or l_M. The advantage of the mixing length is that it can be logically expected to be a function of y in some turbulent shear flows, e.g., the distance from the wall in a boundary layer. Other such theories are available for the mean velocity distribution as well as for mean temperature and chemical species. An account of these theories is given by Schlichting.*

The mixing length has been used for both pipe flows and boundary-layer flows even though l_M turns out to be much larger than Eq. 7.19 allows. A restriction to small l_M values is implicit in Eq. 7.19 since we have truncated the Taylor series of this equation by assuming l_M to be small.

By briefly considering these two theories we have attempted to show one response of the technological community to the indeterminate governing equations: the development of engineering-oriented computing equations. An example follows.

Example 7.2

One of the uses of the phenomenological theories is to predict the velocity distribution for a particular geometry. Assuming a linear variation of l_M with y, predict the time-averaged velocity profile for fully developed flow in a circular pipe. State the shortcomings of the predicted profile, if any.

Solution. The mixing length must approach zero at the wall, where the turbulent motions cease; thus, the linear relationship is

$$l_M = ky$$

The turbulent shearing stress is then

$$(\bar{\tau}_{xy})_{\text{turb}} = \rho k^2 y^2 \left(\frac{\partial \bar{u}}{\partial y} \right)^2$$

To solve this equation we must make a further assumption relating the turbulent shearing stress to y. We know that the turbulent shear must be zero at the wall and at the pipe centerline; hence, it must pass through a region in which it reaches a maximum value. In this region it may be assumed that $(\tau_{xy})_{\text{turb}} \cong \text{constant}$.

Then

$$\rho k^2 y^2 \left(\frac{\partial \bar{u}}{\partial y} \right)^2 = \text{constant}$$

or

$$\frac{\partial \bar{u}}{\partial y} = \frac{C_1}{y}$$

*Hermann Schlichting, *Boundary Layer Theory* (6th ed.) (New York: McGraw-Hill Book Co.), 1968.

This leads to
$$\bar{u} = C_1 \ln y + C_2$$
where C_1 and C_2 are constants, determined from experimental data.

The predicted velocity profile is expected to be applicable only in the region where the turbulent shear is approximately constant. It cannot satisfy the wall condition $\bar{u}(0) = 0$, or the centerline condition, $\partial \bar{u}/\partial y = 0$. Only experimental data can confirm the existence of a region in which the velocity profile is logarithmic.

Extension 7.2.1. The mixing length in a turbulent flow near a wall in an otherwise unbounded stream (a boundary layer) is assumed to have the form given in Fig. E7.2. What is the form for the velocity distribution in the two regions? Develop an expression for $\bar{u}(y)$ in the region removed from the wall.

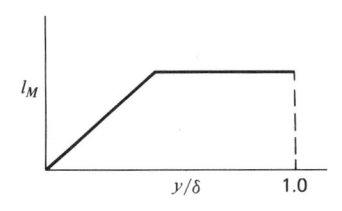

Fig. E7.2.

2. Turbulence Research and Empirical Data

There are two major goals for turbulence research; any specific research project may pursue them singly or in some degree of combination. Basic turbulence research seeks to determine the nature of and the interaction between the fundamental phenomena in the turbulent motion of fluids. Technologically motivated research accepts a given flow field (of technological importance) and seeks to determine its character. A given flow field may be governed by several fundamental phenomena; therefore, technological research deals with more complicated, less instructive flows. From its nature, basic research deals with flow fields which preserve, and which are primarily influenced by, the phenomenon of interest; they often lack direct relationship to technologically important flows.

The theoretical bases for deeper examination of turbulent flows are equations which are derived from the instantaneous Navier-Stokes equations. An example of the quantities to be dealt with is shown in Fig. 7.4. The *correlation functions* (R_{uu}, etc.) defined in Fig. 7.4 have considerable value in defining the character of a turbulent flow. As in the case of the Reynolds stress, if the mechanism causing u'_A is related,

Art. 7.3   APPROACHES TO GENERAL TURBULENCE PROBLEM   321

$R_{AB}(\mathbf{r}_A, \delta\mathbf{r}) = \overline{V_A' V_B'}$

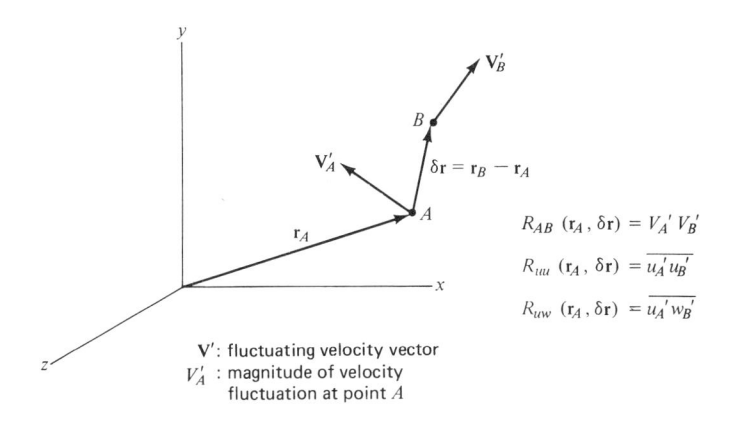

$R_{uu}(\mathbf{r}_A, \delta\mathbf{r}) = \overline{u_A' u_B'}$

$R_{uw}(\mathbf{r}_A, \delta\mathbf{r}) = \overline{u_A' w_B'}$

$\mathbf{V}'$: fluctuating velocity vector
V_A': magnitude of velocity fluctuation at point A

Fig. 7.4. Defining sketch for the correlation function $R_{AB}(\mathbf{r}_A, \delta\mathbf{r})$.

on the average, to the mechanism causing u_B', the correlation function is nonzero. An equation for R_{uu} may be formulated by multiplying the equation for u_A' by u_B' and, similarly, multiplying the equation for u_B' by u_A', adding, and then extracting the time average. (These very complicated expressions will not be given in this text.) From these general relationships, the equations for the Reynolds stress may be obtained by letting $\mathbf{r}_B \to \mathbf{r}_A$.

The equation for the *turbulence kinetic energy* $\overline{q^2}$ may be derived by summing the three component equations for $\overline{u'^2}, \overline{v'^2}, \overline{w'^2}$ at A to obtain the equation for $\overline{q^2}$, defined by

$$\overline{q^2} = \tfrac{1}{2}[\,\overline{u'^2} + \overline{v'^2} + \overline{w'^2}\,] \tag{7.22}$$

From the turbulence kinetic energy equation* we may gain insight into the nature of the turbulence motion by examining the character of the various terms. The following equation illustrates the dominant terms of the turbulence energy equation in a steady two-dimensional turbulent flow:

$$\underbrace{\frac{\partial \overline{q^2}}{\partial t}}_{①} = -\bar{u}\underbrace{\frac{\partial \overline{q^2}}{\partial x}}_{②} - \bar{v}\underbrace{\frac{\partial \overline{q^2}}{\partial y}}_{} \underbrace{}_{\text{AVERAGE TO ZERO}} \qquad \underbrace{}_{\text{THERMAL DISSIPATION}}$$

$$-\underbrace{\frac{\partial \overline{v'q'}}{\partial y}}_{③} - \underbrace{\frac{1}{\rho}\frac{\partial \overline{p'v'}}{\partial y}}_{④} - \underbrace{\overline{u'v'}\frac{\partial \bar{u}}{\partial y}}_{⑤} + \underbrace{(\approx \text{dissipation})}_{⑥} \tag{7.23}$$

(① → zero, steady turbulent flow; ④ flow work; ⑤ average to zero)

*The term "energy budget" is sometimes used to avoid confusion with the energy equation, which expresses the first law of thermodynamics. The present equation deals with kinetic energy only insofar as it is gained by multiplying the momentum/unit mass (i.e., velocity) equation by velocity.

Terms 1 and 2 are the *convective transport* of the turbulence kinetic energy by the mean flow, term 3 is the *diffusive transport* of q^2 by turbulent motion, 4 is the *flow work* of the pressure fluctuations, and term 5 is the *production* of turbulence energy. Term 6 is, to an approximation, the *dissipation*.

Certain characteristics of the terms in the foregoing equation are particularly instructive about the mechanistic processes in turbulent flows. They will be briefly described. Also, it is of some interest to identify the experimental procedures required to assess the magnitude of the terms; this is presented where appropriate.

Terms 1 and 2 account for the net transport of $\overline{q^2}$ into a region surrounding a point of interest. Term 3, which requires the simultaneous measurements of v', u', and w', is rather different from its counterpart for the mean flow, term 2. The $\overline{v'^3}$ part of term 3 involves the skewness of the v'-distribution. If v' were symmetrically distributed about its mean then this term would be zero. Note that $\overline{v'q^2}$ may be positive, negative, or zero, depending upon whether the turbulence velocity tends to convect the turbulence kinetic energy out of or into the volume of interest through its upper ($+y$) surface or whether there is no net convection.

Term 4 represents, for the turbulence kinetic energy, the flow-work effect. However, unlike the similar term for the mean flow, the product of the fluctuating velocity and pressure represents an impossible measurement in terms of conventional techniques. The measurement of p', itself a most difficult quantity to determine accurately, disallows the simultaneous measurement of v' since the pressure probe and the hot-wire probe cannot occupy the same spatial location. Consequently this term, and sometimes term 3, are used as closure terms. That is, all other terms are measured and the value of term 4 is inferred since the sum of all terms is known to be zero. This procedure, although necessary, has its obvious drawbacks.

Term 5 is perhaps the easiest term in the equation to measure; also, it is often one of the quantitatively dominant terms. The identification of this term as the production of turbulence kinetic energy is neither intuitively obvious nor easily shown.* The following descriptive argu-

---

*The complete discussion of the production term 5 is presented in a text by Hinze and another text by Tennekes and Lumley (see Selected References at the end of the chapter). Both presentations use tensor notation; the mathematical manipulations are quite tedious without this shorthand notation. Consequently, the presentation in this text will be restricted to qualitative discussion.

ments will necessarily suffice to establish the interpretation of term 5 as the production term. Since the product of a stress and a velocity can be interpreted as a rate of doing work per unit area, and since the net work done on a given spatial region could result in some combination of the effects of increase in the kinetic energy or dissipation to thermal energy, it is reasonable, and it is possible to show mathematically, that a stress times a velocity gradient is related to the dissipation of kinetic energy. Note that $-\rho \overline{u'v'} \, \partial \bar{u}/\partial y$ is of this form: a stress (the Reynolds stress) times a velocity gradient. However, the mean flow "dissipation" related to the turbulence stresses will not go directly into the thermal form; rather, it goes into turbulence kinetic energy. That is, it represents the loss of mean flow energy which, equivalently, is the production of turbulence kinetic energy.

The foregoing considerations are immediately useful in the interpretation of term 6. Specifically, dissipation of turbulence involves a term like $\overline{\nu(\partial u'/\partial y + \partial v'/\partial x)\partial u'/\partial y}$, which is seen to be a stress τ'_{xy} times a velocity gradient. It is significant that these are stresses and velocity gradients of the fluctuation field. These are the terms which irreversibly convert the kinetic energy of the fluctuations into thermal energy. A term such as term 6 would be almost impossible to measure since, for example, an instantaneous evaluation of $u'(y)$ and a simultaneous evaluation of $v'(x)$ would be required to form the derivatives. Experimentally, these difficulties are "sidestepped" by utilizing a time derivative to approximate the x-derivative (viz., the Taylor hypothesis where $\partial/\partial x = -(1/\bar{u})(\partial/\partial t)$ and special relationships that can be deduced from the mathematical assumption of isotropy which are used to replace the y-derivatives by x-derivatives.

Typical utilizations of Eq. 7.23 should be briefly noted. The equation provides the basis for identifying certain mechanistic features about a flow. For example, at the centerline in a fully developed channel flow terms 1, 2, and 5 are zero. However, term 6 is not zero and thus terms 3 and/or 4 must also be nonzero. The presence of $\overline{q^2}$ at the centerline indicates that it is transported there by terms 3 or 4 and its magnitude compared with $\overline{q^2}$ at other locations suggests the strength of these terms.

Terms 5 and 6 are local effects and, if they are in balance, then the turbulence itself is defined by local characteristics. This situation is often considered necessary for the success of the eddy viscosity models since they suggest a dominantly local character of the transport mechanisms.

A head loss results from the extraction of energy from the mean flow, represented by term 5. Hence, the combination of strong velocity gradients with large turbulence stresses is associated with large losses. Consequently, if such losses are undesirable as in a ducting system, then turning vanes or diffuser passages should be used to limit the extent of separated flow regions where these effects are most prominant. Conversely, if large losses are required for throttling a flow, as in a carburetor or a drinking fountain, then large-scale separation zones are useful.

Analytically, this equation plays an important role in the development of schemes by which the equations of motion may be structured as a complete set. Recall that the Reynolds stress represents an unknown quantity and its presence leads to more unknowns than equations. Since it appears in the turbulence energy equation as well, and if it is presumed that $\overline{q^2} = K\overline{u'v'}$, then the continuity, momentum, and turbulence kinetic energy equations form a closed set if additional assumptions are made regarding terms 3, 4, and 6. The computation of the mean velocity is insensitive to such assumptions; hence, this is a viable technique.

With appropriate restrictions and assumptions, it is possible to relate the correlation coefficient to a description using the concept of Fourier components in the frequency domain. This technique is common for acoustic or electromagnetic waves, in which a total periodic wave form is thought of as the superposition of various sine waves of different amplitudes or strengths. Figure 7.5 illustrates this.

To approximate a turbulent signal in such a manner introduces special features which do not occur with the acoustic or electromagnetic signals. First, the turbulent signal is not periodic, and we must therefore argue that we can pick a long enough time record such that the Fourier components are "averaged" and are therefore effectively steady-state values. Second, the turbulence signal contains "all" frequencies up to the limiting values which correspond to the smallest scales of motion, and its spectral distribution is therefore quite broad. They are contrasted in Figure 7.6; note that the periodic signal has a discrete spectral distributions since it is equivalent to the sum of a finite number of sine waves. The motivation for a description involving the frequency of the fluctuations is easily appreciated in viewing a hot-wire anemometer signal, $u(t)$, on an oscilloscope. In this situation, it is easy to visualize numerous frequencies contributing to the signal. Although we will not consider the details of the derivation here, the relationship between the correlation function (and its associated notion

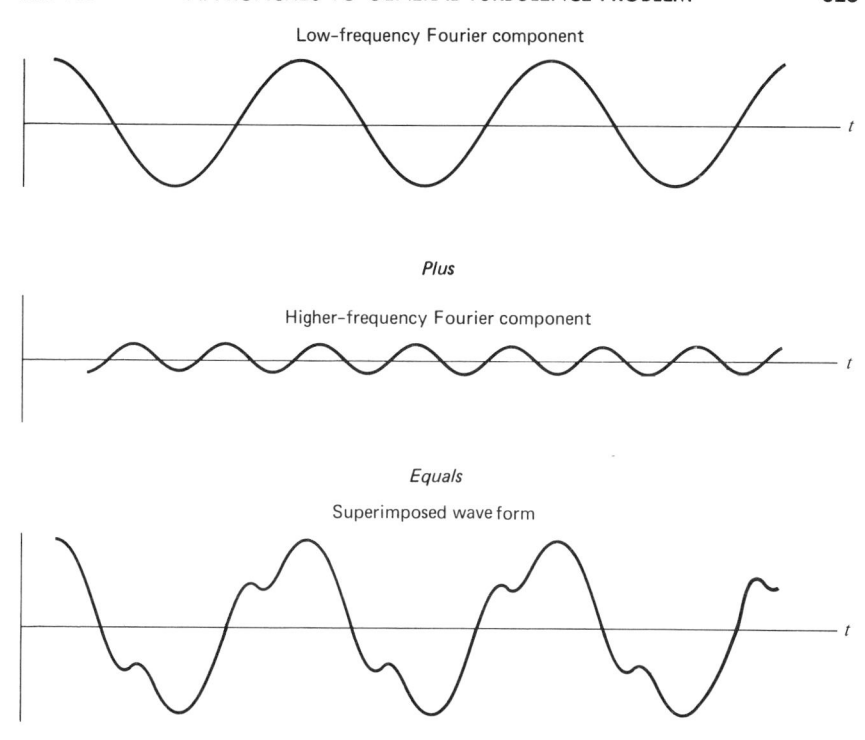

Fig. 7.5. Superposition of sine waves to produce a periodic wave.

of a spatial distribution of waves which can be constructed such that they are equal to the velocity) is related to the frequency distribution. The connection between the two is given by the *Taylor hypothesis*, in which a variation with respect to time is related to a spatial variation as

$$\frac{\partial}{\partial t} = -U \frac{\partial}{\partial x} \qquad (7.24)$$

This relationship implies a "frozen" pattern which is swept past the measuring point at the rate U. For a given velocity, a small characteristic time which is associated with a large frequency implies a small characteristic length. As pointed out with regard to the turbulence kinetic energy equation (7.23), the dissipation is related to the smallest scales of the motion and hence to the highest frequencies. However, in making this observation we must recognize the influence of U on the frequency. For example, the smallest-scale motions in a turbulent boundary layer occur quite close to the wall. This fact is partly obscured, when looking at the frequency content of the signal, by the reduced mean velocity in this region.

326 TURBULENCE Ch. 7

(a) Turbulent-energy fluctuation

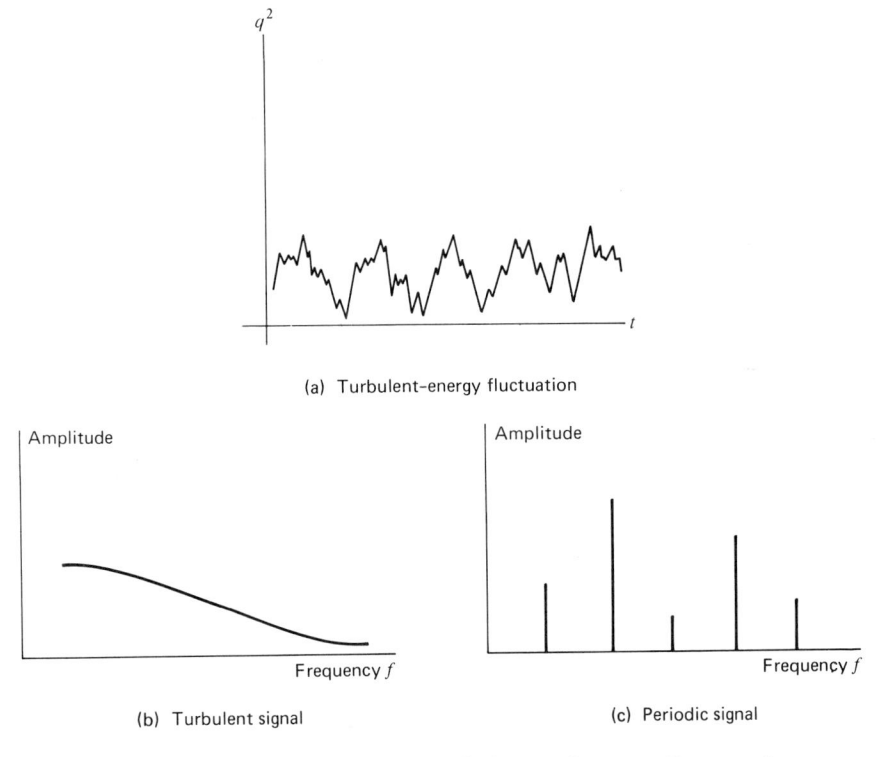

(b) Turbulent signal

(c) Periodic signal

Fig. 7.6. Frequency content of a turbulent and a periodic wave form.

The continuous distribution of amplitude with respect to frequency is a result of an important aspect of turbulent motions, namely, that the energy is transmitted from the large eddies (low-frequency fluctuations) to the smallest eddies (high-frequency fluctuations). This transfer process is related to the nonlinear nature of the governing equations; it plays the important role of limiting the rate at which energy can be dissipated. The turbulence behaves (in some respects) like a device which accepts an input (say, an organized wave motion) and operates on it with a nonlinear transfer function (like $u\, \partial u/\partial x$). Unlike the linear device shown in part a of Fig 7.7 a nonlinear device creates energy at higher frequencies, as shown in part b.

We now have the background to appreciate the following limerick,[*] which (however inadequate its stanza form or rhyme scheme) is ex-

---

[*]Attributed to L. F. Richardson, a researcher in atmospheric turbulence.

Art. 7.3 APPROACHES TO GENERAL TURBULENCE PROBLEM   327

emplary as to fluid mechanics:

> Big whirls make little whirls*
> Which feed on their velocity†
> Little whirls make lesser whirls*
> And so on to viscosity‡

We also now have some of the background to appreciate some classical research studies on turbulence. *Isotropic* turbulence is that for which all average turbulence quantities are independent of direction. This type disallows production of turbulence energy and allows us to concentrate on the behavior of the correlation function R_{AB} and on the spectral transfer and dissipation processes. A less restrictive class, in which the turbulence characteristics are independent of position, is

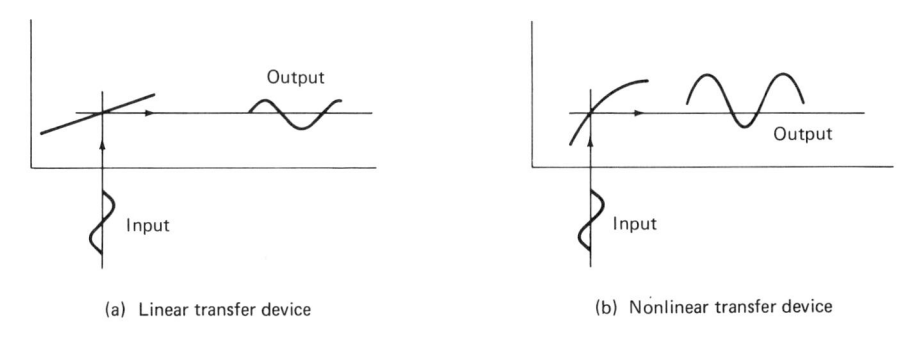

Fig. 7.7. Linear- and nonlinear-device effects in the creation of new spectral-energy content.

called *homogeneous*. These two special kinds of flow are important primarily because they highlight certain phenomena; however, they have also made practical contributions to the study of technologically important flows in that the dissipation process in many "practical" flows is isotropic in character and the spreading of contaminants (i.e., pollution) is controlled by the scale effects of turbulent motion, which are best studied in a homogeneous flow. The fully developed flow in a circular conduit is clearly of technological importance; it is also one of the simplest flows in which production of turbulence energy occurs. Examples of data obtained for these several kinds of flow are presented

---
*See nonlinear effects of Fig. 7.7.
†Let big whirl be represented by $\bar{u}$ and little whirl by q^2 in Eq. 7.23.
‡See the dissipation term (6) in Eq. 7.23.

in Figs. 7.8 and 7.9. The data for the smooth pipe have universal validity (i.e., they are not a function of the laboratory of their origin) because the pipe flow is geometrically unique. These and other basic turbulence data may be found in Hinze* or, of course, their original publication source.

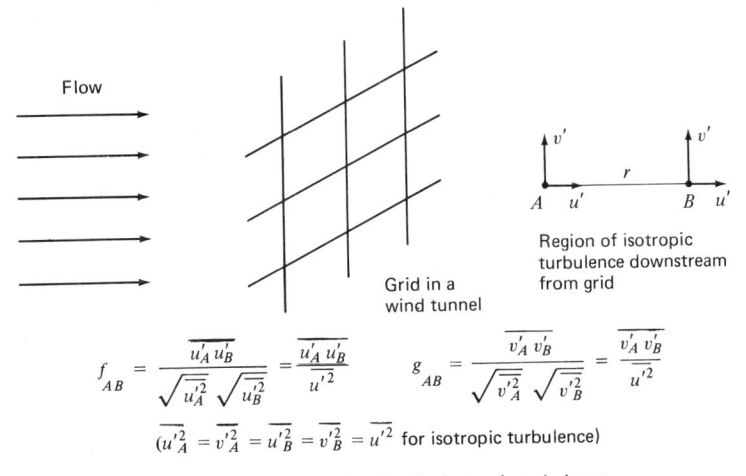

$$f_{AB} = \frac{\overline{u'_A u'_B}}{\sqrt{\overline{u'^2_A}} \sqrt{\overline{u'^2_B}}} = \frac{\overline{u'_A u'_B}}{\overline{u'^2}} \qquad g_{AB} = \frac{\overline{v'_A v'_B}}{\sqrt{\overline{v'^2_A}} \sqrt{\overline{v'^2_B}}} = \frac{\overline{v'_A v'_B}}{\overline{u'^2}}$$

$$(\overline{u'^2_A} = \overline{v'^2_A} = \overline{u'^2_B} = \overline{v'^2_B} = \overline{u'^2} \text{ for isotropic turbulence})$$

(a) Experimental approximation for isotropic turbulence

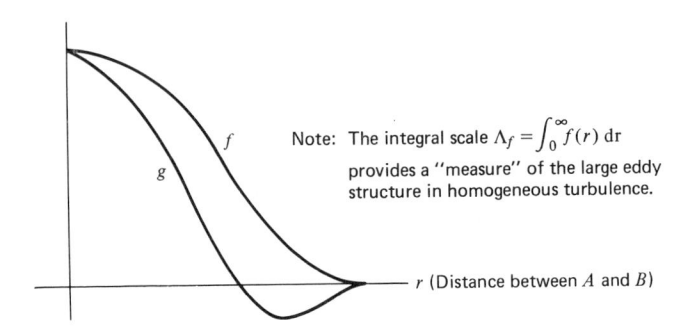

Note: The integral scale $\Lambda_f = \int_0^\infty f(r)\,dr$ provides a "measure" of the large eddy structure in homogeneous turbulence.

r (Distance between A and B)

(b) Correlation functions (normalized), isotropic turbulence. (These are the only two independent correlations for isotropic turbulence.)

Fig. 7.8. Schematic data from an isotropic turbulent flow.

The $f(r)$ and $g(r)$ curves are dependent upon the stage of decay of the isotropic turbulence; the smallest structures will be rapidly

*J. O. Hinze, *Turbulence* (New York: McGraw-Hill Book Co.), 1959.

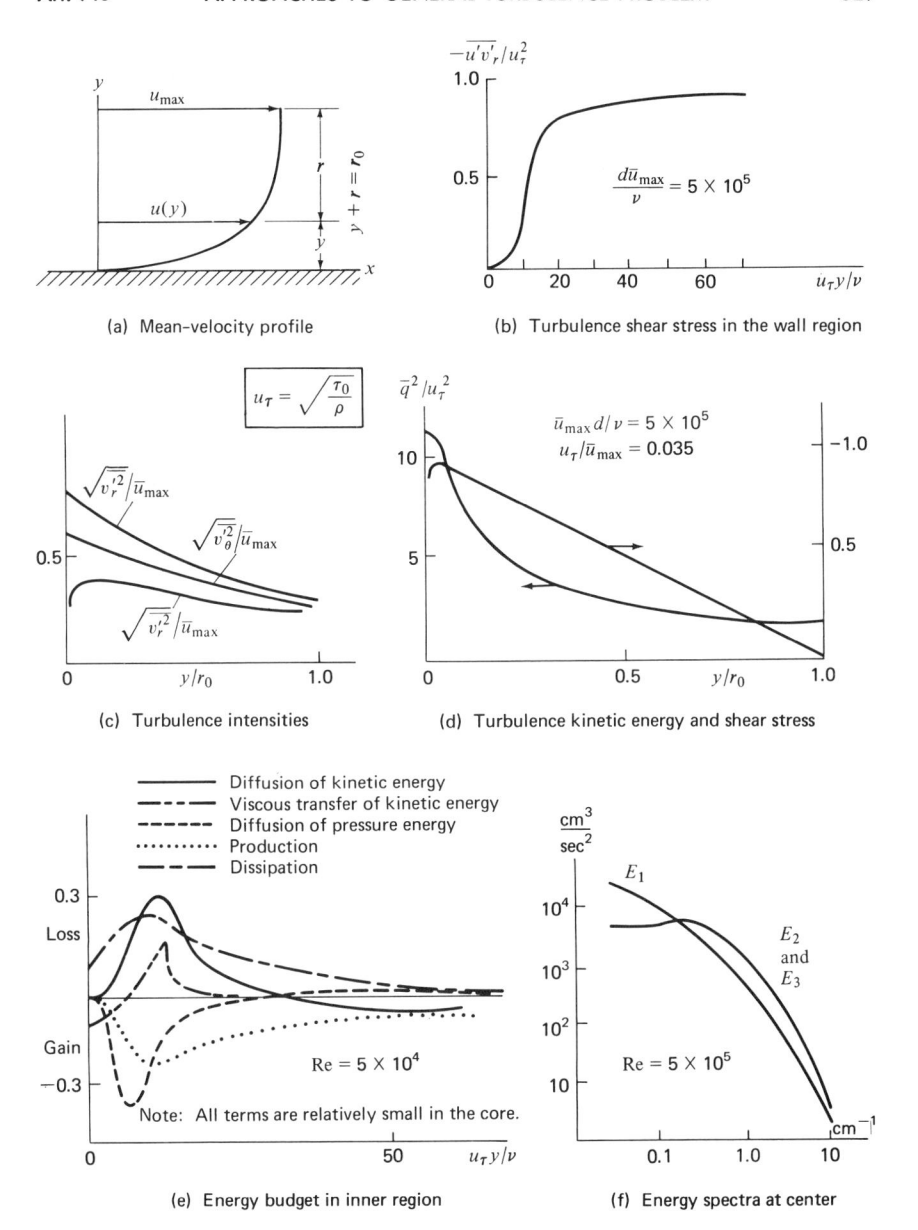

Fig. 7.9. Turbulence data from a fully developed flow in a smooth pipe. (Based on data from J. Laufer, "The Structure of Turbulence in Fully Developed Pipe Flow," NACA Rep. 1174, 1954.)

eliminated by viscous dissipation which will make the curves less steep at $r = 0$. The large structures die more slowly. Because of the disparity between the large and small scale rates, Λ_f increases with respect to the elapsed time from the initiation of the flow. For the pipe flow, the necessary use of the *friction velocity* u_τ ($u_\tau = \sqrt{\tau_0/\rho}$) to correlate the data is an excellent example of the considerations given in the similitude chapter; that is, u_τ is a velocity which characterizes the behavior of the flow, unlike $\bar{u}_{max}$, for example. Note that most of the "action" occurs near the boundary; that is one reason why heat- and mass-transfer rates for turbulent flow are much greater than for laminar flow since the turbulent motions greatly enhance the mixing of the fluid and allow the transport processes to become more effective. The Reynolds shear stress is seen to dominate the total stress, except very near the wall. The pipe will appear to be "smooth" if the roughness elements are in the region where $\overline{u'v'}$ is small and cannot generate turbulence energy in the wakes of the roughness protrusions. The turbulence fluctuations (and $\overline{u'v'}$) extend closer to the wall as the Reynolds number is increased. This subject will be explored more completely in the discussion of the Moody diagram for pipe friction factors, included in the next article.

Example 7.3

Calculate the friction velocity in a 6-in.-high channel if water is flowing. The flow is developed and the pressure drop is measured to be 200 psf over a 10-ft horizontal length.

Solution. The momentum equation applied to a control volume between the two horizontal planes of the channel and 10 ft in length results in

$$2 \times 10 w \tau_0 = (p_1 - p_2) h w$$

where the 2 accounts for both top and bottom surfaces separated a distance h, with w the width of the control volume. This equation yields

$$\tau_0 = \frac{200 \times 6/12}{20} = 5 \text{ psf}$$

Using $\rho = 1.94$ slugs/ft^3, we have

$$u_\tau = \sqrt{\frac{\tau_0}{\rho}} = \sqrt{\frac{5}{1.94}} = 1.61 \text{ fps}$$

7.4 TURBULENT FLOW IN A CIRCULAR PIPE

The specific case of turbulent flow in a circular pipe is of sufficient engineering interest to consider it in some detail. It will also allow us to examine the contributions of research to a flow important in engineering practice. In addition, the discussion will prepare the student for detailed consideration of other important flows.

1. The Inlet Region

A turbulent pipe flow which originates from a reservoir (a region of large volume containing essentially stagnant fluid) and enters a pipe through a smooth inlet will exhibit the entrance and fully developed regions described for a laminar motion (see Fig. 6.3). Starting at the inlet, the shear effects will grow outward from the wall as the streamwise distance is increased. An ever increasing portion of the flow will be sheared until the entire pipe is under the influence of stress effects. Downstream of the location at which the shear effects have propagated to the center of the flow, the turbulence stresses come into an equilibrium condition such that the mean velocity is invariant with respect to streamwise position; that is, the flow is fully developed at the position x_d. The turbulence structure requires an additional development time to become established. This additional time results in a turbulence development length x_T, larger than x_d. The pressure is a linear function of x in the fully developed region; hence, the quantity dp/dx can serve to indicate where a fully developed condition is approached in terms of the pressure gradient. This development length will be termed x_p. We should note that, because the wall-region flow fixes the wall shear stress, which is the dominant effect influencing the pressure drop, and because the wall-region condition is established before the central region of the flow is fully developed, the following relationship describes these three development lengths:

$$x_p < x_d < x_T \tag{7.25}$$

The development length x_d for a turbulent flow is much shorter than for a laminar flow at the same Reynolds number. The actual mechanisms that control the development length are very complex. The entrance geometry, the entrance-velocity profile, the turbulence structure, and the wall roughness all influence the development lengths; consequently, precise relationships are not available. The various development lengths are sketched in Fig. 7.10a. Parts b and c of the figure present experimental data with artifically stimulated transition at the inlet which demonstrate the difference between x_p ($\cong 10d$) and x_d ($\cong 40d$).

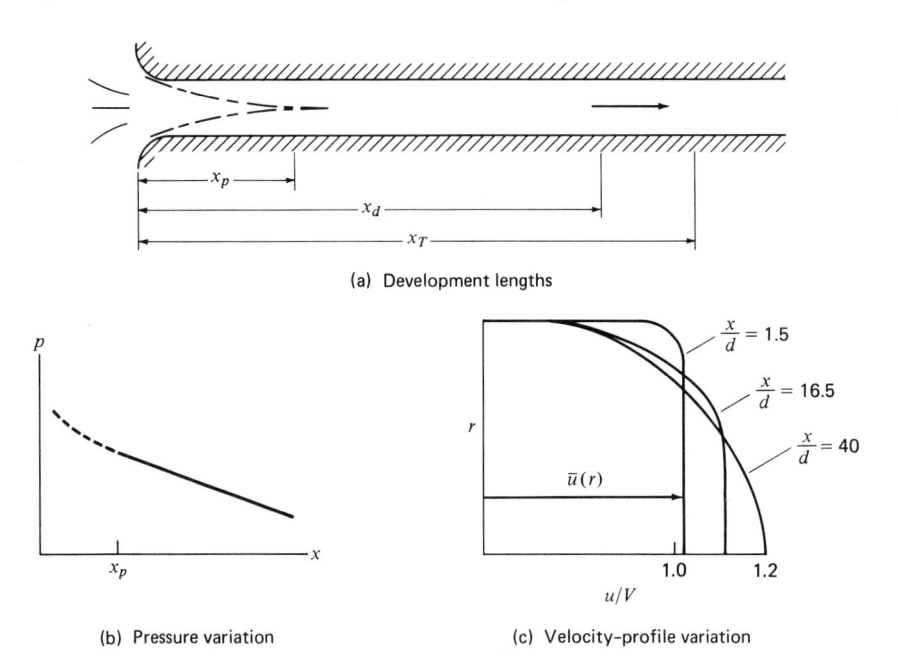

Fig. 7.10. Inlet flow in a pipe. (Part b is based on data from A. R. Barbin and J. B. Jones, "Turbulent Flow in the Inlet Region of a Smooth Pipe," Trans. ASME, 85D, 1963.)

(a) Development lengths

(b) Pressure variation

(c) Velocity-profile variation

2. Shear-Stress Distribution—Fully Developed Flow

The shear stress in a fully developed pipe flow varies linearly with the radius. This variation may be easily shown with the use of a cylindrical control volume, as in Fig. 7.11a. The net momentum flux through the control volume is zero; the pressure and shear forces must be in balance. Consequently

$$\bar{\tau}(r) = -\frac{r}{2}\frac{\bar{p}_2 - \bar{p}_1}{L} \qquad (7.26)$$

For $\bar{\tau} = \tau_0$ at $r = r_0$, the shear stress at the wall is

$$\tau_0 = -\frac{r_0}{2}\frac{\bar{p}_2 - \bar{p}_1}{L} \qquad (7.27)$$

Art. 7.4   TURBULENT FLOW IN A CIRCULAR PIPE   333

(a) Force balance on control volume

(b) Shear-stress distribution in pipe (exaggerated laminar contribution)

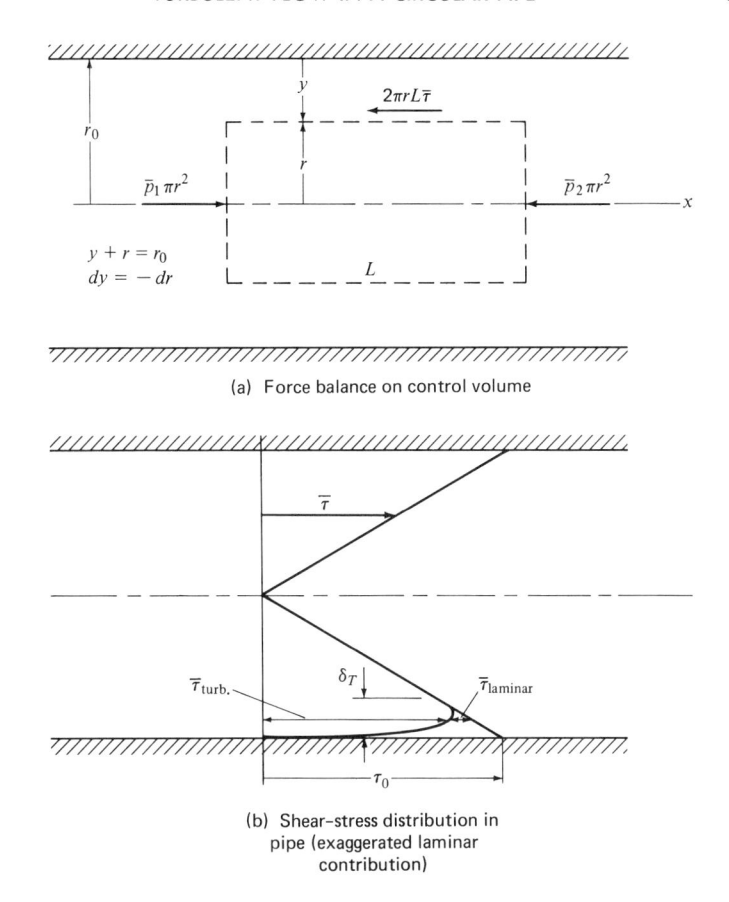

Fig. 7.11. Fully developed pipe flow.

As shown in Fig. 7.11b, the shear stress is composed of a laminar and a turbulent portion. The sketch shows a highly exaggerated laminar contribution; Table 7.1 indicates the distance δ_T from the pipe wall for which the turbulent stress becomes dominant as a function of Reynolds number.* As shown in the table, the extent of the turbulent-dominated motion is quite dependent upon the Reynolds number. In terms of the universal coordinates defined by the turbulence structure near the wall, this quantity is given by the relationship $(Re)_{\delta_\tau} = 30$, that is, $u_\tau \delta_T / \nu = 30$, where $u_\tau = \sqrt{\tau_0/\rho}$. From Fig. 7.9e it can be seen that the dominant production and dissipation of turbulence kinetic energy occur within this region. The value of $u_\tau \delta_T / \nu = 30$ was selected on the basis of the Reynolds stress distribution, as shown in Fig. 7.9b.

*J. O. Hinze, *Turbulence* (New York: McGraw-Hill Book Co.), 1959, Chapter 7.

TABLE 7.1.
Distance from Wall To Establish a Fully Turbulent Condition

$Re = Vd/\nu$	5×10^3	10^4	10^5	10^6
δ_T/d	0.1	0.05	0.006	0.0008

3. Velocity Distribution—Fully Developed Flow

The mean velocity distribution in the region immediately adjacent to the pipe behaves as if it were a viscous flow driven by the shear at its outer boundary as in a Couette motion (see Example 6.5). The resulting velocity distribution is linear and is of the form

$$\frac{\bar{u}}{u_\tau} = \frac{u_\tau y}{\nu} \qquad (7.28)$$

This form of the velocity profile is valid up to a $u_\tau y/\nu$ value of approximately 6. Beyond this position, the experimental data demonstrate a logarithmic form as

$$\frac{\bar{u}}{u_\tau} = A \ln \frac{u_\tau y}{\nu} + B_1 \qquad (7.29)$$

where $A = 2.44$ and $B_1 = 4.9$ are satisfactory values* for a smooth pipe (although there is some ambiguity in the data).

For the remaining core region of the pipe, a velocity-defect expression in which the normalized y-coordinate is given in a purely geometric form (i.e., not dependent upon u_τ) correlates the data and is consequently considered to be indicative of the physical nature of the flow. It is stated as

$$\frac{\bar{u}_{max} - \bar{u}}{u_\tau} = -A \ln \frac{y}{r_0} + B_2 \qquad (7.30)$$

where $A = 2.44$ and $B_2 = 2.5$. The region of overlap between these latter two descriptions lies in the region $u_\tau y/\nu = 500$ to 1000; this "boundary" is dependent upon Reynolds number. This description only approximates the velocity distribution in the central region of the pipe; a more general relationship is written as

$$\frac{\bar{u}_{max} - \bar{u}}{u_\tau} = -A \ln \frac{y}{r_0} + B_2 - h\left(\frac{y}{r_0}\right) \qquad (7.31)$$

*Other data have suggested the use of other numerical values for A and B_1. Values on A range from 2.2 to 2.8, and on B from 4.9 to 6.5.

where $h(y/r_0)$ is a correction factor. Figure 7.12 presents experimental data and the preceding relationships for the inner and outer regions of the pipe flow.

The information contained in these plots is rather subtle, but quite instructive regarding turbulent flows. The form of the velocity profile is Reynolds-number-independent although the region where the outer and inner descriptions are joined approaches the wall for increasing Reynolds number. This independence of the magnitude of the Reynolds number may be appreciated if one recalls the normalized form of the x-component Navier-Stokes equation:

$$\frac{\overline{D}^* \bar{u}^*}{Dt^*} = -\frac{\partial \bar{p}_k^*}{\partial x^*} + \frac{1}{\text{Re}} \nabla^{*2} \bar{u}^* - \left[\frac{\partial \overline{u^{*\prime 2}}}{\partial x^*} + \frac{\partial \overline{u^{*\prime} v^{*\prime}}}{\partial y^*} + \frac{\partial \overline{u^{*\prime} w^{*\prime}}}{\partial z^*} \right] \tag{7.32}$$

The solution to this equation would be a function of the Reynolds number if a purely geometric length scale were used in the region where the viscous stress is important, that is, near the wall. However, the characteristic length and velocity are chosen as ν/u_τ and u_τ, respectively, so that in the normalizing process the Reynolds number does not appear in the differential equation. It is absorbed into the normalized lateral length and a universal curve results. For the region away from the wall, the viscous stress effects, and hence the Reynolds-number influence,* become negligible. Consequently, a purely geometric length scale (y/r_0) may be used for presentation of the data, which are independent of Reynolds number. Note, however, that the velocity must be referred to two different velocity scales as $\bar{u}^* = (u_{\max} - \bar{u})/u_\tau$ since the ratio of $u_{\max}$ to u_τ is dependent upon Reynolds number. (There is an implicit, but not explicit, dependence of $\bar{u}$ on $u_{\max}$ in the region near the wall.) The presence of the correction factor, $h(y/r_0)$, is related only to the inability of a logarithmic form to completely describe the profile. For example, a logarithmic form could not provide for $\partial \bar{u}/\partial y = 0$ at $y = r_0$, as required by symmetry.

The significant feature of the velocity distribution in these three regions is that the descriptions are independent of Reynolds number although the domain of validity of Eqs. 7.29 and 7.30 decreases and increases, respectively, with an increase in Reynolds number.

An alternate form for the description of the mean velocity is given by a relatively simple power-law relationship which is Reynolds-number-

---

*Recall that the reciprocal of the Reynolds number is the coefficient of the viscous term $\left(\frac{1}{\text{Re}} \nabla^{*2} \bar{u}^* \right)$. If Re is large this term may be neglected unless $\nabla^{*2} \bar{u}^*$ is also equally large. The latter is only true very near the wall.

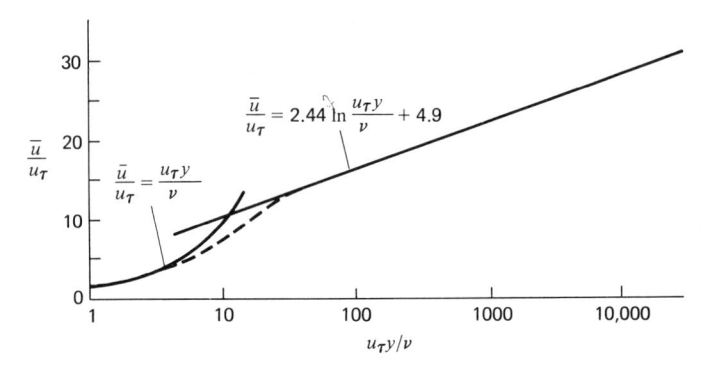

(a) Velocity profile in inner wall region

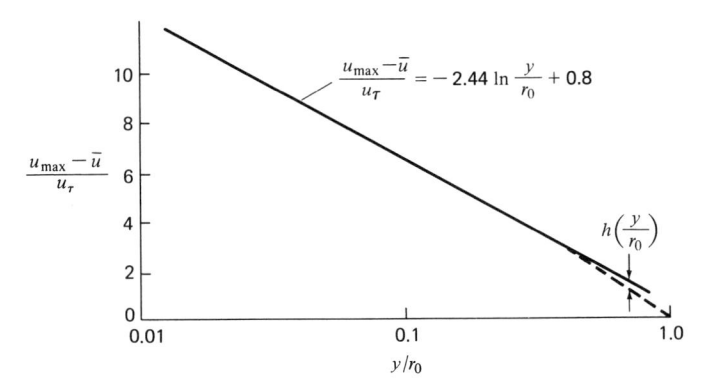

$$\frac{\bar{u}}{u_\tau} = 2.44 \ln \frac{u_\tau y}{\nu} + 4.9$$

$$\frac{\bar{u}}{u_\tau} = \frac{u_\tau y}{\nu}$$

(b) Velocity-defect profile in outer region

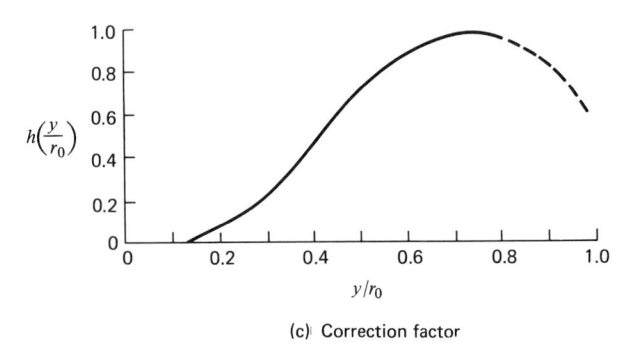

$$\frac{u_{max} - \bar{u}}{u_\tau} = -2.44 \ln \frac{y}{r_0} + 0.8$$

$h\left(\frac{y}{r_0}\right)$

(c) Correction factor

Fig. 7.12. Experimental data for pipe flow. (Based on data from J. Laufer, "The Structure of Turbulence in Fully Developed Pipe Flow," NACA Rep. 1174, 1954.)

dependent but which describes the velocity distribution quite adequately in the central region of the pipe flow. The power-law expression is

$$\frac{\bar{u}}{u_{max}} = \left(\frac{y}{r_0}\right)^{\frac{1}{n}} \tag{7.33}$$

where y is the distance from the pipe wall. The exponent n is related to the friction factor f by the relationship

$$\frac{1}{n} = \sqrt{f} \qquad f \approx 0.1 \tag{7.34}$$

There is no theoretical basis for Eqs. 7.33 and 7.34 but they agree quite well with the experimental data. It should also be noted that the power-law form cannot be valid at the wall since an infinite value of $\partial u / \partial y$ is implied. This form of the velocity distribution would be of use in the calculation of the momentum flux or energy flux of a fully developed pipe flow.

4. Pressure Drop and Losses in a Fully Developed Pipe Flow

The previous sections have demonstrated some of the turbulence quantities, the behavior of the shear stress, and the nature of the mean velocity profile in a fully developed pipe flow. These features of this technologically important flow are instructive about the nature of turbulence and help to identify the important mechanisms which govern the response of the pipe flow to changes in the Reynolds number or geometric (e.g., roughness) conditions. The use of pipe flows in myriad application problems means that compilation of empirical information for direct engineering use is also important. This empirical information is presented in this section. A complete presentation and interpretation of this empirical information rely upon, and consequently further justify, the theoretical considerations which were presented in the control volume portion of this text.

The condition of fully developed flow is sufficient to show that the pressure decreases linearly in fully developed turbulent (as well as laminar) pipe flow.* The Reynolds stress in the equation of motion and the lack of a relationship between this stress and the mean velocity are sufficient to prohibit an analytical prediction of the pressure drop.

---

*This result is not "obvious" but it is easily demonstrated. Eq. 7.27 results from the use of the control volume in Fig. 7.11. For a fully developed flow τ_0 is constant along the pipe. Hence, Δp is linearly related to L. Because of this linear relationship, $dp/dx = (p_2 - p_1)/L$.

However, the similitude considerations of Chapter 4 may be used to show that the normalized pressure drop $\Delta p/\rho V^2$ in a given length of pipe L/d can only be a function of the Reynolds number Vd/ν and the wall conditions (the boundary conditions) for the pipe. The boundary conditions can be satisfactorily accounted for by a roughness parameter e (e/d in dimensionless form). The *roughness* e is a root-mean-square value for the roughness elements of the pipe; that is,

$$e = \frac{1}{l} \left\{ \int_0^l [e'(x)]^2 \, dx \right\}^{1/2} \tag{7.35}$$

where e' is the deviation of the surface from the smooth pipe condition. In general, the head loss, as defined in Chapter 2, accounts for the increase in the thermal terms of the energy equation ($\Delta \tilde{u}$ and $\dot{Q}$) at the expense of the kinetic and potential energies and the flow work. It may be written, in terms of the pressure drop in a constant-area, horizontal pipe and in terms of the friction factor, as

$$-\frac{\Delta p}{\rho g} = h_L = f \frac{L}{d} \frac{V^2}{2g} \tag{7.36}$$

The friction factor f is therefore related to the Reynolds number and the relative roughness; specifically,

$$f = f(\text{Re}, e/d) \tag{7.37}$$

This functional relationship is known from empirical data and is shown schematically in Fig. 7.13. It is known as the *Moody diagram*, after the American engineer L. F. Moody, who assembled the necessary data to develop the plot. This is perhaps the single most useful set of experimental data for engineering applications.

On the Moody diagram, several regimes of flow are recognized. The *laminar-flow regime*, which occurs at relatively low Reynolds numbers, is characterized by the relationship $f = 64/\text{Re}$. A *critical zone*, encountered at higher Reynolds numbers, represents transition from a laminar to a turbulent flow. This transition usually occurs at a Reynolds number of approximately 2000; however, if the pipe has a smooth entrance, if the entering flow has a low fluctuation level ($\overline{u'^2}/V^2$), and if the pipe wall is smooth, the transition process may be delayed to a much higher Reynolds number, possibly 40,000. An *intermediate zone* is then encountered in which the turbulent pipe flow is undergoing a transition from Reynolds-number dependence to Reynolds-number independence. This process is controlled by the wall-roughness ele-

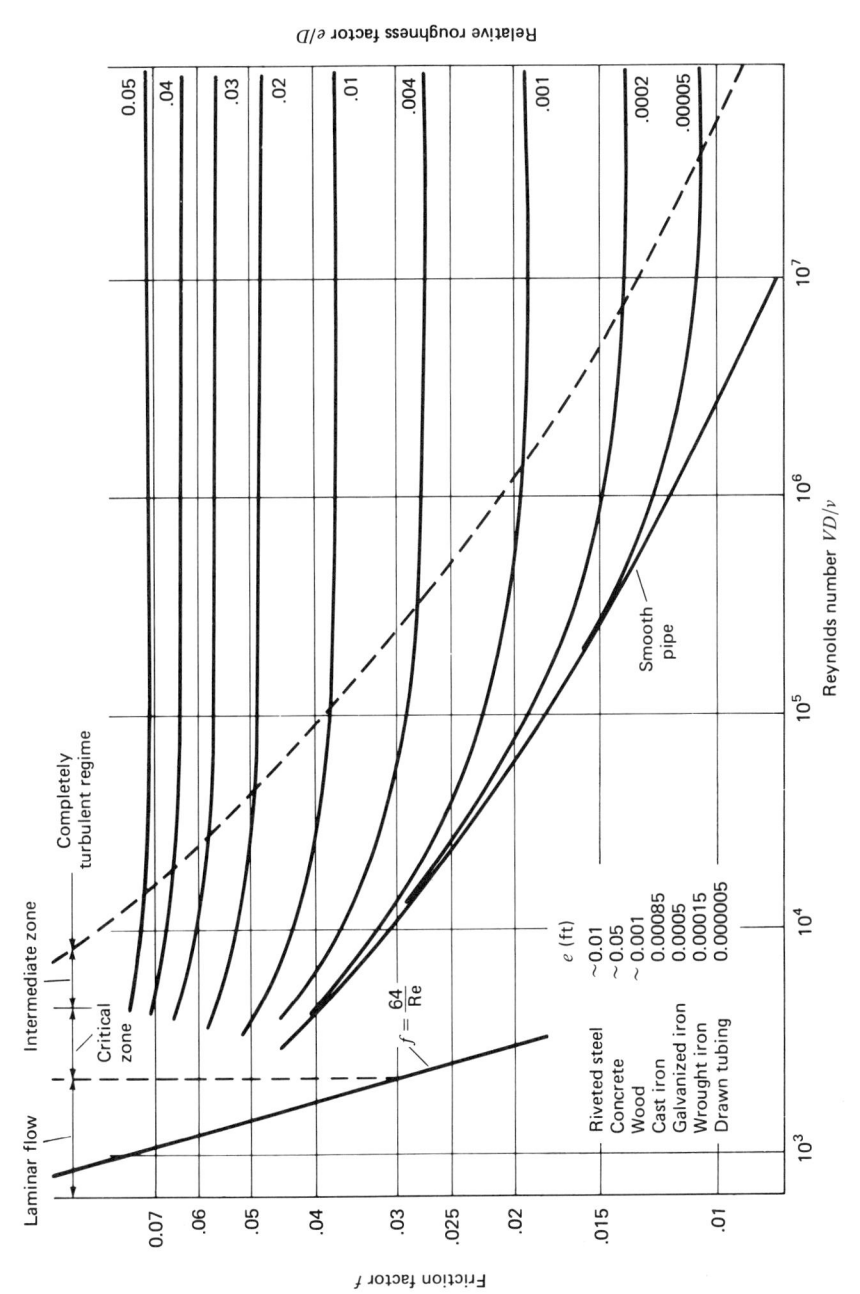

Fig. 7.13. The Moody diagram. (From L. F. Moody, "Friction Factors for Pipe Flow," Trans. ASME, 66, 8 [1944]).

ments. If the elements are submerged in the wall layer, $e \ll \delta_T$, the viscous effects, and hence the Reynolds number, will influence the flow; on the other hand, if the roughness elements protrude through the wall layer, $e > \delta_T$, the viscous effects become negligible and the *completely turbulent regime* is encountered. The wall shear stress, which is directly related to the pressure drop, is directly dependent upon the turbulent stresses for the completely turbulent regime. The discussion in Section 3 of this article may be used to demonstrate the Reynolds-number independence when the turbulence stresses dominate the entire flow field.

5. Evaluation of Minor Losses

The term "minor loss" is used to describe the pressure drop associated with the flow through a geometric passage other than a length of pipe. The flow through a constricted area such as a partially open valve provides a good example (see Fig. 7.14).

The pressure drop across this valve can be considered in the context of the turbulence kinetic energy equation (7.23). The turbulent motion

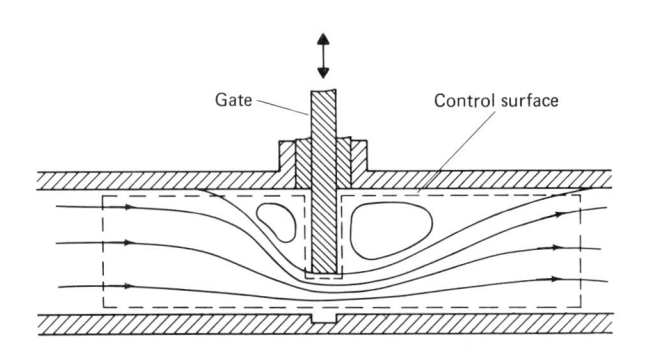

Fig. 7.14. Partially open gate valve.

in the separated flow region is driven by the streaming flow; that is, the streaming flow serves as the energy source to maintain the separated turbulent motion. However, the turbulent motion is constantly dissipating energy to the thermal form (term 6 of Eq. 7.23). In terms of the indicated control volume, the energy necessary to (indirectly) feed the dissipation is extracted from the mean flow in the form of a flow-work decrement, the pressure drop across the valve. When viewed in this context, it is not difficult to see that any geometric passage which involves a region of separated flow will lead to a pressure drop and hence a head loss. This also shows that the minor loss mechanisms are often associated with turbulence energy considerations. Hence, they can be expected to depend upon the square of the velocity scale in the flow; the form

$$h_L = K \frac{V^2}{2g} \qquad (7.38)$$

is suggested. K is a constant based upon the geometry. Table 7.2 provides characteristic K values. Additional values may be found in engineering handbooks.

TABLE 7.2.
Head-Loss Coefficients K

Geometry	K
Globe valve (fully open)	10.0
Angle valve (fully open)	5.0
Gate valve (fully open)	0.19
Close return bend	2.2
Standard tee	1.8
Standard elbow	0.9
Long sweep elbow	0.6
Square-edged entrance	0.5
Well-rounded entrance	0.03
Reentrant entrance	0.8
*Sudden contraction (2 to 1)	0.25†
*Sudden contraction (5 to 1)	0.35†
Sudden enlargement	$(1 - A_1/A_2)^2$‡

*Area ratio.
†Based on V_2.
‡Based on V_1.

The control-volume equations must be used when considering conduit systems. Typically, the flow is steady and the control volumes are fixed in space (nondeformable). The continuity equation then becomes, for an incompressible flow,

$$A_1 V_1 = A_2 V_2 \qquad (7.39)$$

where V_1 and V_2 are the average velocities at sections ① and ②, respectively.

The energy equation between two sections may be written as

$$\dot{Q} - \dot{W}_S = \frac{p_2 - p_1}{\rho}\dot{m} + (\tilde{u}_1 - \tilde{u}_1)\dot{m} + (gh_2 - gh_1)\dot{m}$$

$$+ \int_{A_2} \frac{\bar{u}^3}{2}\rho\, dA - \int_{A_1} \frac{\bar{u}^3}{2}\rho\, dA \quad (7.40)$$

The internal-energy difference minus the heat-transfer rate represents the losses, so we write this equation as

$$-\frac{\dot{W}_S}{\dot{m}} = \alpha_2 \frac{V_2^2}{2} + \frac{p_2}{\rho} + gh_2 - \alpha_1 \frac{V_1^2}{2} - \frac{p_1}{\rho} - gh_1 + gh_L \quad (7.41)$$

where α is the kinetic-energy correction coefficient, defined by

$$\alpha = \frac{\int \bar{u}^3\, dA}{AV^3} \quad (7.42)$$

For turbulent conduit flows, $\alpha \cong 1$. The losses include all minor losses and frictional losses between sections ① and ②.

If a branch occurred in a piping system, these equations would obviously have to be modified to include a third section. Several pipes often branch from a supply pipe so that a number of exiting sections are encountered.

The losses in noncircular conduits can be obtained from circular pipe data. The basis for this is that the frictional effects are rather insensitive to the geometry of the conduit and are fixed by the velocity gradient near the wall. The Reynolds number based upon the hydraulic diameter d_H allows the two flows to be properly related. The hydraulic diameter is defined as

$$d_H = \frac{4A}{P} \quad (7.43)$$

where A is the cross-sectional area and P is the wetted perimeter. Consequently Eq. 7.36 may be reformulated as

$$h_L = f \frac{L}{d_H} \frac{V^2}{2g} \quad (7.44)$$

and f is again given by the Moody diagram.

Example 7.4

Flow through a sudden contraction separates at the contracted entrance (Fig. E7.4) and reattaches downstream with velocity V_2. It reaches a minimum area A_c where the velocity is V_c, as shown. The flow from V_c to V_2 may be considered to be a sudden enlargement flow. Experimentally, the minimum area A_c is found to be $A_c = C_c A_2$, with $C_c = 0.62 + 0.38 \, (A_2/A_1)^3$. Assuming the losses from A_1 to A_c to be negligible, and using the losses for a sudden enlargement from A_c to A_2, determine the loss coefficient for the contraction if $A_1/A_2 = 2$. Compare with that given in Table 7.2.

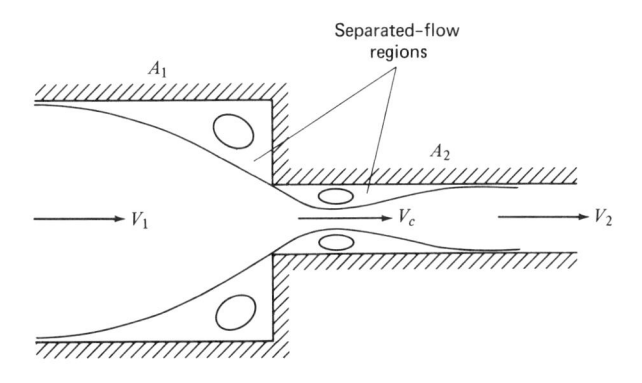

Fig. E7.4.

Solution. The fluid undergoes a gradual contraction from area A_1 to area A_c, the "vena contracta"; this is a relatively efficient process; hence, we will neglect the losses from A_1 to A_c. The loss from A_c to A_2 is given by

$$h_L = \left(1 - \frac{A_c}{A_2}\right)^2 \frac{V_c^2}{2g}$$

where the loss coefficient is found in Table 7.2. We find A_C from

$$A_c = C_c A_2$$
$$= \left[0.62 + 0.38(\tfrac{1}{2})^3\right] A_2 = 0.668 A_2$$

Continuity requires that
$$A_2 V_2 = A_c V_c$$
so that
$$V_c^2 = V_2^2 \left(\frac{A_2}{A_c}\right)^2 = V_2^2/0.668^2$$

344 TURBULENCE Ch. 7

Substituting in the head-loss relationship, we have

$$h_L = \frac{(1 - 0.668)^2}{0.668^2} \frac{V_2^2}{2g}$$

$$= 0.247 \frac{V_2^2}{2g}$$

The loss coefficient is thus 0.247, or approximately 0.25, as given in Table 7.2.

Extension 7.4.1. Calculate the loss coefficient for A_1 extremely large, e.g., an entrance from a reservoir. *Ans.* 0.376

Example 7.5

Water is delivered from a reservoir through a 1000-ft-long, 1-in.-dia. wrought iron pipe to a water tank. The level of the reservoir is 50 ft above the outlet, which empties into the tank 10 ft below the tank water surface. Determine the flow rate. $\nu = 10^{-5}$ ft^2/sec.

Solution. The control volume is drawn from the surface of the reservoir to the surface of the tank. Using a datum 50 ft below the surface of the reservoir, the energy equation (7.41) becomes

$$50 = 10 + h_L$$

The major losses from the frictional effects in the pipe are given by

$$h_L = f \frac{L}{D} \frac{V^2}{2g}$$

$$= 12,000 f \frac{V^2}{2g}$$

where V is the average velocity in the pipe. Since $12,000f$ is quite large (f is usually around 0.02), we may neglect minor losses which occur at the entrance and exit. Then

$$f V^2 = \frac{40 \times 2g}{12,000}$$

This equation relates f and V, as does the Moody diagram. With the equation and the Moody diagram, a trial-and-error solution is possible. To use the Moody diagram we must know e/D. For the wrought iron pipe of this example, using $D = 1/12$ ft, $e/D = 0.0018$. Assume the flow to be in the completely turbulent regime; the Moody diagram then gives $f = 0.021$. From the equation,

$$V = \sqrt{\frac{40 \times 2g}{12,000 \times 0.021}} = 3.19 \text{ fps}$$

Now we must check the Moody diagram again. The Reynolds number is

$$\text{Re} = \frac{VD}{\nu} = \frac{3.19 \times 1/12}{10^{-5}} = 2.26 \times 10^4$$

From the Moody diagram this Reynolds number, with $e/D = 0.0018$, gives $f = 0.028$, higher than our first guess.

As a second guess let $f = 0.03$. The equation gives $V = 2.68$ fps. The Reynolds number is then $\text{Re} = 22{,}300$. From the Moody diagram, $f \cong .03$. An acceptable value of $f = 0.03$ is used. This gives

$$V = \sqrt{\frac{40 \times 2g}{12{,}000 \times 0.03}} = 2.68 \text{ fps}$$

The flow rate is

$$q = AV = \frac{\pi}{4 \times 144} \times 2.68 = 0.0146 \text{ cfs}$$

A trial-and-error solution is typical of a piping system in which the flow rate is not known. If the diameter is unknown, a trial-and-error solution is also required.

Extension 7.5.1. In the example the minor losses were neglected. Assume a reentrant entrance, one elbow, and exit losses of $V^2/2g$. Determine the flow rate, including the minor losses. *Ans.* 0.0146 cfs

7.5 CLOSURE

In this chapter the mathematical and physical complexities of turbulent motion have been considered. The general approach to these motions has been briefly described, and one of the important standard flows, fully developed pipe flow, has been considered in detail.

From this discussion, one can understand why the turbulent flow in a pipe or through a constricted passage should have a pressure drop proportional to the velocity squared (these effects are proportional to the turbulence kinetic energy) and not to the velocity to the first power, as in laminar flow, in which viscous shear effects are dominant. One can also understand the general result that high-Reynolds-number flows will have a character independent of Reynolds number unless a region of viscous shear is important. That is, the velocity field $\bar{u}(x, y)$ in a jet or a wake is governed by the Reynolds stress $-\rho \overline{u'v'}$, which is not dependent upon viscous effects. The friction factor in a pipe is dependent upon Reynolds number if viscous effects are able to fix the wall stress; at sufficiently high Reynolds numbers the roughness elements dominate the viscous wall layer and establish the Reynolds stress throughout the flow. In this case, the friction factor would be Reynolds-number-independent, as shown by the Moody diagram.

Minor-loss coefficients are introduced and shown to be essentially constant for high-Reynolds-number flow situations. When confronted with a new and untabulated geometry involving a minor-loss effect, we might estimate the magnitude of the minor-loss coefficient by estimating the size and intensity of the separated-flow region and then approximating the loss coefficient in terms of other, known values. Example 7.4 demonstrates this technique.

The Moody diagram and the minor-loss tabulations allow the prediction of the pressure-flow characteristics of piping systems. This evaluation is important in the selection of the proper sized pump or fan, or the appropriate pipe dimension. The energy equation in control-volume form is the describing equation for these considerations.

The inclusion of the detailed discussion for the flow in a pipe is prompted by the importance of this flow and by the use of these specific details to demonstrate the important characteristics of turbulent flows, such as the dominance of the Reynolds stress away from a solid surface. Considerations could have been included for other flows. (Similar results for the turbulent boundary layer are included in the discussion of boundary layers in Chapter 9.) Some of the other flows are considered in the context of the problems at the end of this chapter. There is in existence a large literature in which considerable data on many turbulent flow configurations have been recorded. Similarly, encouraging progress has been achieved on specific calculations of turbulent flow fields and on the development of calculation techniques. The material of this chapter is the key to these more advanced and more complete aspects of turbulent flow.

PROBLEMS

7.1 List several examples of fluid flows, not mentioned in the text, that are turbulent flows, and several that are not turbulent. Remember, turbulent flows are high-Reynolds-number flows and must include velocity gradients.

7.2 From the definition of a time-averaged quantity verify that $\overline{u'} = 0$, $\overline{\bar{u}\frac{\partial u'}{\partial x}} = 0$, and $\overline{u'\frac{\partial \bar{u}}{\partial x}} = 0$. Also argue that $\overline{u'\,\partial u'}/\partial x \neq 0$ and $\overline{v'\,\partial u'}/\partial y$ may not be zero.

7.3 Show that $\overline{u'\frac{\partial u'}{\partial x}} + \overline{v'\frac{\partial u'}{\partial y}} + \overline{w'\frac{\partial u'}{\partial z}} = \frac{\partial \overline{u'^2}}{\partial x} + \frac{\partial \overline{u'v'}}{\partial y} + \frac{\partial \overline{u'w'}}{\partial z}$.

PROBLEMS

7.4 Determine an expression for the difference between the time-averaged acceleration

$$\overline{\frac{Du}{Dt}}\hat{i} + \overline{\frac{Dv}{Dt}}\hat{j} + \overline{\frac{Dw}{Dt}}\hat{k}$$

of a fluid particle and the quantity

$$\frac{D\bar{u}}{Dt}\hat{i} + \frac{D\bar{u}}{Dt}\hat{j} + \frac{D\bar{w}}{Dt}\hat{k}$$

Use the results of Problem 7.3. Note that this exercise is the basis for the development of Eqs. 7.10.

7.5 Let u' and v' be given by two sine waves of equal amplitude and frequency. Determine the bounds on K_{uv}.

7.6 Consider a class of problems for which $l_M = ky^2$. Determine the form of the time-averaged turbulent velocity profile in a region in which the turbulent shearing stress is assumed to be constant. Could this profile satisfy a wall boundary condition?

7.7 Assume the eddy viscosity to be constant in a two-dimensional channel flow. Determine the time-averaged velocity distribution. Compare this with a laminar flow.

7.8 For a turbulent flow in a channel, determine the pressure variation $p_k(y)$ in the direction normal to the flow by considering the momentum equation perpendicular to the wall. Show that p_k is maximum at the wall and that the minimum p_k value is related to the maximum value of $\overline{v'^2}$.

7.9 The data given* were obtained in a free air jet 2.5 in. from the center line. Long-term averages of $\bar{u} = 17.4$, $\bar{v} = 1.65$, $\sqrt{\overline{u'^2}} = 7.9$, $\sqrt{\overline{v'^2}} = 2.04$, and $\overline{u'v'} = -10.0$ were determined with the use of an on-line computer capability. Determine $\bar{u}$, $\bar{v}$, $\overline{u'^2}$, $\overline{v'^2}$, and $\overline{u'v'}$ from the data given and compare with the more accurate values. Do these data represent isotropic turbulence?

t (sec)	u (fps)	v (fps)	t (sec)	u (fps)	v (fps)
0.000	26.0	−4.03	0.250	25.0	0.55
0.025	12.1	3.22	0.275	9.4	2.41
0.050	13.8	7.33	0.300	28.7	−6.71
0.075	13.5	6.38	0.325	13.6	1.82
0.100	14.1	4.53	0.350	21.1	−2.94
0.125	17.6	−1.88	0.375	17.0	0.27
0.150	8.6	2.02	0.400	13.0	4.88
0.175	15.9	−4.19	0.425	5.4	0.54
0.200	16.3	2.49	0.450	24.8	1.69
0.225	28.0	1.56	0.475	9.8	1.66

*These axisymmetric jet data were obtained by Dr. S. J. Kleis from a flow system in a mechanical engineering laboratory at Michigan State University.

7.10 Express the z-component and r-component time-averaged momentum equations in cylindrical coordinates for developed flow in a pipe. Assume $\bar{v}_\theta = \bar{v}_r = 0$. (It will be necessary to express the continuity equation in cylindrical coordinates.) Identify the assumption necessary to show that the shear stress distribution in the pipe varies linearly with r, verifying Eq. 7.26.

7.11 The pressure drop in a 50-ft section of 2-in.-dia. water pipe is measured at 20 psi. A velocity probe of the flow yields the following velocity distribution.

r	0	0.1	0.2	0.3	0.4	0.5	0.6	0.7	0.8	0.9	inches
$\bar{u}$	22	22	21.9	21.8	21.6	21.2	20.4	18.8	15.0	8.4	ft/sec

Determine the shearing stress distribution, the velocity gradient at the wall, and the eddy viscosity for each r-location. The kinematic viscosity ν is 10^{-5} ft²/sec.

7.12 Water flows at a Reynolds number of 10,000 in a 2-in.-dia. cast iron pipe. Find the friction velocity and total dragging force over 1000 ft of the pipe. Use $\nu = 10^{-5}$ ft²/sec.

7.13 Gasoline is transported in a 4-in.-dia. wrought iron pipe line. A pressure drop of 20 psi is experienced over a 1000-ft section of the pipe. Determine the flow rate and the horsepower lost in the 1000-ft section. The density of gasoline is 1.3 slug/ft³ and its kinematic viscosity 4×10^{-6} ft²/sec.

7.14 Water at a flow rate of 10.0 cfm is pumped from a reservoir through a 1000-ft section of 2-in.-dia. cast iron pipe. It is to be supplied to a device at 50 psi at the same elevation as the reservoir surface. What horsepower pump must be used at the beginning of the 1000-ft horizontal section? Kinematic viscosity is 10^{-5} ft²/sec.

7.15 You have a choice between a 3-ft-dia. or a 2-ft-dia. riveted steel pipe to transport 1000 gpm of water to a power plant. The larger-diameter-pipe costs more but the losses are greater in a smaller diameter pipe. What percentage decrease in head loss would be realized with the larger diameter pipe? The viscosity is 3×10^{-5} lb-sec/ft².

7.16 Flow through a sudden enlargement is a rather chaotic, time-dependent flow. It may be idealized as shown in the sketch (*facing page*). Applying the integral momentum equation over the control volume shown, show that the head loss coefficient is as given in Table 7.2. State carefully the assumptions required to obtain this result.

7.17 Flow occurs through an orifice plate as shown (*facing page*). Sketch the streamline pattern, and calculate the loss coefficient for the orifice plate by assuming no losses for the contracting region. The minimum area (vena contracta) for the flow jet is $C_c A_2$, with $C_c = 0.62 + 0.38(A_2/A_1)^3$. The area of the orifice is A_2. Solve the problem for $A_1/A_2 = 2$.

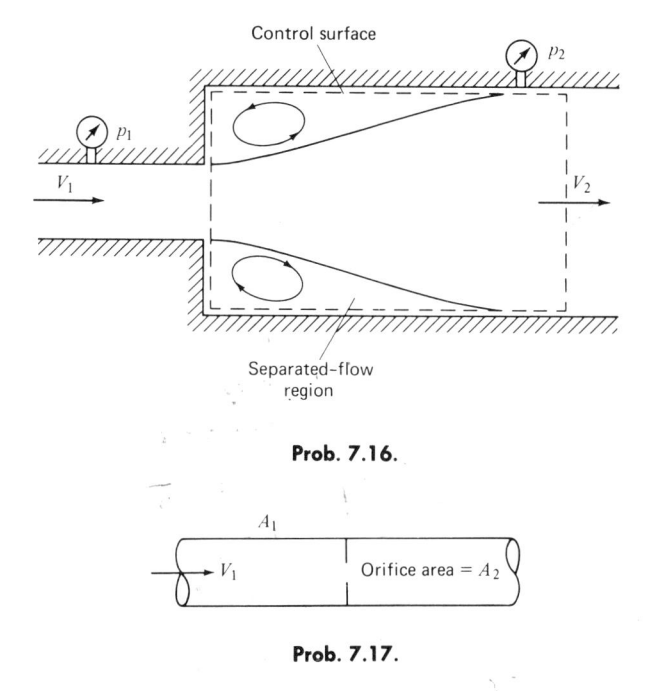

```
                    Control surface
                          │                    ⊙ p₂
    ////////////////////////////////////////
         ⊙ p₁     │    ⎛‾‾⎞              │
    /////////     │    ⎝__⎠              │
    ──→           │                       │  V₂
     V₁           │                       │ ──→
    /////////     │    ⎛‾‾⎞              │
                  │    ⎝__⎠              │
    ////////////////////////////////////////
                       /
              Separated-flow
                 region
```

Prob. 7.16.

```
              A₁
         ┌─────────┬──────────────────┐
    ──→  │         │                  │
     V₁  │         │ Orifice area = A₂│
         └─────────┴──────────────────┘
```

Prob. 7.17.

7.18 Water of velocity V exits from a pipe into a tank of water. Determine the loss coefficient associated with the exit. Place the control surface at the exit of the pipe and around the boundary of the tank.

7.19 Water at 60°F flows from a 4-in.-dia., 50-ft length of galvanized iron pipe which is attached to a reservoir with a reentrant entrance. The head of water in the reservoir is 500 ft above the exit of the pipe. A gate valve is used to control the flow. Determine the flow rate of water from the reservoir with the valve wide open.

7.20 A farmer must pump at least 100 gallons of water per minute from a lake to his field, located 25 ft above the lake. He has a 10-hp pump which is approximately 80 percent efficient. The distance from the lake to the field is 2000 ft. Determine the minimum-sized plastic piping that he must buy. The piping is sized by the half-inch. Use $\nu = 10^{-5}$ ft²/sec.

7.21 A tanker truck is filled from a storage tank holding oil with $\nu = 5 \times 10^{-4}$ ft²/sec. The truck tank is considerably smaller than the storage tank so that the surface can be assumed to be at a constant height above the exit as shown (*next page*). The pipe length from the tank to the truck is 50 ft. Estimate the time required to fill the 5000-gallon truck.

7.22 Water flows from a reservoir through the piping system shown (*next page*). The larger pipe is 100 ft long and the smaller one is 30 ft long. Neglect minor losses and determine the flow rate. Use $\nu = 10^{-5}$ ft²/sec.

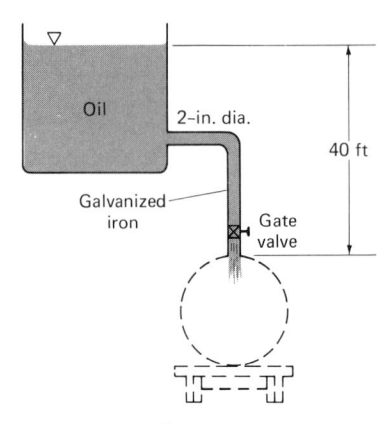

350   TURBULENCE   Ch. 7

Prob. 7.21.

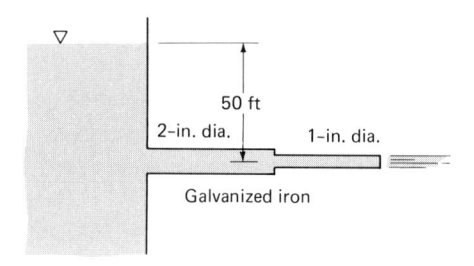

(Figure: Oil reservoir with 2-in. dia. galvanized iron pipe, 40 ft drop, gate valve)

Prob. 7.22.

(Figure: Reservoir with 2-in. dia. galvanized iron pipe, 50 ft, reducing to 1-in. dia.)

7.23 Water flows from a reservoir through a 2-in.-dia., 100-ft-long pipe. The larger cast iron pipe splits into two smaller 1-in.-dia. cast iron pipes. One of the smaller pipes is 50 ft long and exits 80 ft below the reservoir surface; the other is 25 ft long and exits 90 ft below the reservoir surface. Estimate the flow rate from the reservoir. Neglect minor losses. Use $\nu = 10^{-5}$ ft^2/sec.

SELECTED REFERENCES

The motion pictures *Turbulence* (No. 21626; R. W. Stewart, film principal) and *Turbulent Flow* (S. Corrsin, film principal; produced by Milner-Fenwick, Inc., Baltimore, Md.) are recommended to supplement this chapter.

Other discussions of turbulent flow at the level of the present text are found in these works: J. W. Daily and D. R. F. Harleman, *Fluid Dynamics*, Addison-Wesley Publishing Company, Reading, Mass., 1968; Y. S. Yuan, *Foundations of Fluid Mechanics*, Prentice-Hall, Englewood Cliffs, N.J., 1967; J. A. Owczarek, *Introduction to Fluid Mechanics*, International Textbook Co., Scranton, Pa., 1968. For a discussion which provides more information regarding transition, see H. Schlichting, *Boundary Layer Theory*, 6th ed., McGraw-Hill Book Co., New York, 1968. For one which covers the present topics extensively, with a considerable array of experimental data from some of the classical experiments, see J. O. Hinze, *Turbulence*, McGraw-Hill Book Co., New York, 1959. A text essentially predicated upon characteristic scales in turbulent motions and therefore sophisticated and closely associated with experimental observation is H. Tennekes, and J. L. Lumley, *A First Course in Turbulence*, MIT Press, Cambridge, Mass., 1972. A quite readable text devoted to the basic elements of turbulence and its measurement, for which the present text is a suitable preparation, is P. Bradshaw, *An Introduction to Turbulence and Its Measurement*, Pergamon Press, Oxford, 1971.

Two articles suitable for the scientifically mature reader interested in turbulence are these by S. Corrsin: "Outline of Some Topics in Homogeneous Turbulent Flow," *Journal of Geophysical Research*, Vol. 64, pp. 2134–50, Dec. 1959; and "Turbulent Flow," *American Scientist*, Vol. 49, pp. 395–404, Sept., 1961.

Turbulence studies ranging from basic research to applied investigations are reported on in the literature; standard literature-search procedures should be used to recover this information. Documentation for numerous flow fields (e.g., wakes, jets, etc.) may be obtained from such journals as the *Journal of Fluid Mechanics* and the *Journal of Basic Engineering/Journal of Fluids Engineering* (ASME), and those of the American Institute of Aeronautics and Astronautics.

The utilization of the conditionally sampled techniques (briefly mentioned in the text) is reviewed in an article by E. Mollo-Christenson, "Intermittency in Large-Scale Turbulent Flows," to be found in *Annual Review of Fluid Mechanics*, edited by M. Van Dyke, W. G. Vincenti, and J. V. Wehausen, and published by Annual Reviews, Palo Alto, Calif. 1973. The reader interested in pursuing this subject will find the more detailed research articles identified in the foregoing review article available in the journals listed above.

Empirically determined loss coefficients are cataloged in the following publications, among others: "Flow of Fluids," Technical Paper 409, Crane Co., Chicago, 1942; "Flow of Fluids Through Valves, Fittings and Pipe," Technical Paper 410, Crane Co., Chicago, 1957; *Fan Engineering*, 6th ed., Buffalo Forge Co., Buffalo, N. Y., 1961.

Numerous special investigations to document specific loss characteristics have been carried out by engineering experiment stations at various universities. Characteristic examples are: A. P. Kratz and J. R. Fellows, "Pressure Losses Resulting from Changes in Cross-Sectional Area in Air Ducts," University of Illinois, Engineering Experiment Station Bulletin 300, Urbana, Ill., 1938; and J. B. Hamilton, "Suppression of Pipe Intake Losses by Various Degrees of Rounding," University of Washington, Engineering Experiment Station Bulletin 51, Seattle, 1928.

The technical societies whose published literature and special publications should be checked for cataloged information on head losses are:
American Society of Mechanical Engineers (ASME)
American Society of Civil Engineers (ASCE)
American Society of Heating and Refrigeration Engineers (ASHRE);
Society of Automotive Engineers (SAE)

8

Inviscid Flows

8.1 INTRODUCTION

The subject of inviscid flows has typically been associated with flow around bodies where the viscous effects are confined to a thin layer near the body and to the wake region, if one exists, behind the body. The flow outside this thin viscous boundary layer and the wake region is insensitive to viscosity and is thus considered an *inviscid flow*. In addition to the inviscid flow around bodies there are several examples of internal flows which can be considered inviscid. The most familiar is the inviscid core in an entrance flow; however, flow around a turbine blade and flow through a short contraction are also internal flows with inviscid regions.

The designation inviscid flow is appropriate because we are interested in describing flow fields for which the effects of viscous shear (and Reynolds stresses for the time-averaged flow) are negligible. This can be most succinctly expressed by reconsidering the Navier-Stokes equation,

$$\frac{D\mathbf{V}}{Dt} = -\frac{1}{\rho}\nabla p_k + \nu \nabla^2 \mathbf{V} \tag{8.1}$$

and noting that the neglect of the viscous term would result from a relatively small value for the product of ν and $\nabla^2 \mathbf{V}$. If a fluid were employed for which $\nu = 0$ then this term would be identically zero. Since no fluid satisfies this condition we must search for flow situations in which the product is *relatively* small; these situations are associated with relatively small values for the fluid property ν (e.g., fluids like air and water) and for relatively large-scale lengths which characterize the velocity derivatives. The flow around a porpoise could be approximated by an inviscid flow but flow around a microorganism would be viscous-dominated and the inviscid approximation would not be acceptable.

From the considerations of similitude presented in Chapter 4, we note that the reciprocal of the Reynolds number is the coefficient of the viscous term in the normalized equation,

$$\frac{D^*\mathbf{V}^*}{Dt^*} = -\nabla^* p_k^* + \frac{1}{\mathrm{Re}} \nabla^{*2} \mathbf{V}^* \tag{8.2}$$

For a sufficiently large Reynolds number, inviscid flows are satisfactory approximations to the real flow only outside the thin viscous wall layer and the wake region, that is, outside the regions in which the viscous terms or the Reynolds stress terms are important.

If the condition $\mu \nabla^2 \mathbf{V} \ll \nabla p_k$ is met, then the momentum equation becomes

$$\frac{D\mathbf{V}}{Dt} = -\frac{1}{\rho} \nabla p_k \tag{8.3}$$

and the equation for an inviscid flow results. Since this is the same equation that would govern the behavior of an *ideal fluid*, i.e., one with $\nu = 0$, the term ideal or inviscid fluid is sometimes used. We will not use such terminology here since it is important that the student recognize that the character of the flow—not that of the fluid—justifies the use of Eq. 8.3.

Equation 8.3 is termed the Euler equation; historically it predates the Navier-Stokes equation. An important difference between the Navier-Stokes equation and the Euler equation is the order of the equations. The first-order Euler equation requires one less boundary condition. The no-slip condition for the tangential velocity at a surface is not required for the Euler equation; however, since a solid boundary is also a streamline, the normal component of velocity is zero at the boundary.

In the first part of this chapter we will consider inviscid flows which do not possess vorticity, usually referred to as *potential flows* or *irrotational flows*. Art. 8.7 will then be devoted to inviscid flows which may have vorticity.

Potential flows are of primary importance to the aerodynamicist. The potential flow solution for an airfoil provides an accurate approximation for the lift. It is therefore not surprising that engineers in the aircraft industry and those concerned with aerodynamic problems have made the greatest contributions to potential flow theory in the past few decades. Potential flow also exists in the flow outside of the wake regions in flow around an automobile; however, since aerodynamic forces are of secondary importance in the design of an automobile, little effort has been made in determining its external flow characteristics. The complexity of the automobile's geometry and the three-dimensional nature of the flow field obviously make it an extremely difficult problem. On the other hand, the streamlined aircraft and nearly two-dimensional airfoil have proven to be quite susceptible to potential-flow analysis.

Establishing the behavior of inviscid flows before studying the boundary-layer growth on a body is well advised since the inviscid-flow results must be used as an input to the boundary-layer problem. The velocity at the edge of the boundary layer matches the velocity from inviscid flow theory (see Fig. 8.1), and the pressure gradient in the boundary layer is determined from the inviscid-flow solution.

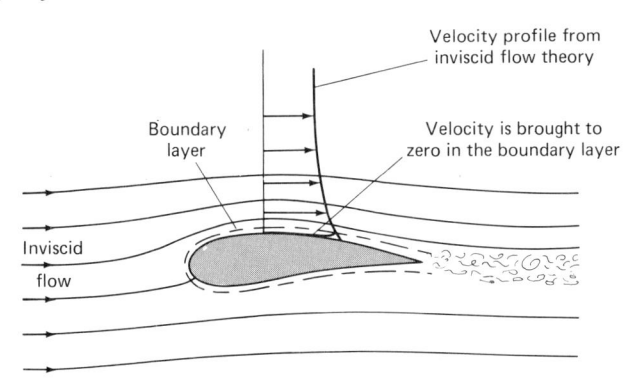

Fig. 8.1. Flow around an airfoil, showing inviscid-flow region.

A second motivation for the examination of potential and other inviscid flow solutions is to be able to predict the flow behavior in a proposed configuration. The viscous and/or turbulent stresses which always accompany fluid flows are confined to the boundary layers which are often sufficiently thin to allow the bulk flow pattern to be approximated by that of an inviscid flow. Therefore, a first approximation to many flow fields, especially flow around objects, may be provided by an inviscid solution. The engineer must be careful, however, since the inviscid approximation completely fails for certain geometries. For example, in the high Reynolds number flow around a blunt object, such as the sphere in Fig. 8.2, the inviscid-flow approximation is definitely in error for the separated flow over the rear portion of the sphere. It is, however, a reasonable approximation to the flow over the front of the sphere. Consider also the flow of fluid through a gradual enlargement in a channel, sometimes called a diffuser. The rise in pressure in the enlarging region may force the slow-moving fluid in the boundary layer to separate from the wall on one side or the other, thus destroying a possible inviscid approximation. For a large-angle diffuser the flow would separate from both sides with a jet of fluid flowing through the center region. For small diffuser angles, and for sufficiently short lengths, the flow would remain attached and an inviscid approximation could be useful. This is shown in the "no stall" region of Fig. 8.3a. For relatively large angles or large L/w ratios the flow stalls, as indicated on the figure. The "transitory stall" regime is characterized by a highly pulsating flow. In Fig. 8.3b the "effectiveness" of the diffuser in increasing the pressure is presented for a particular diffuser with $L/w = 8$. In general, the inviscid approximation is useful for high-Reynolds-number flows in regions where the flow remains attached to the boundary.

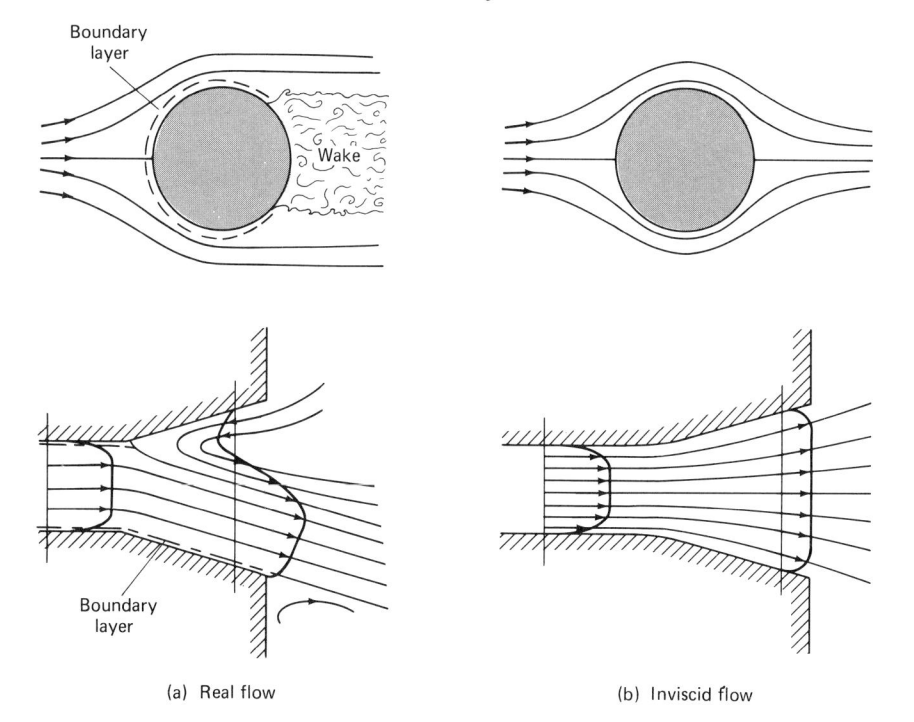

(a) Real flow

(b) Inviscid flow

Fig. 8.2. Flow around a sphere and in a diffuser.

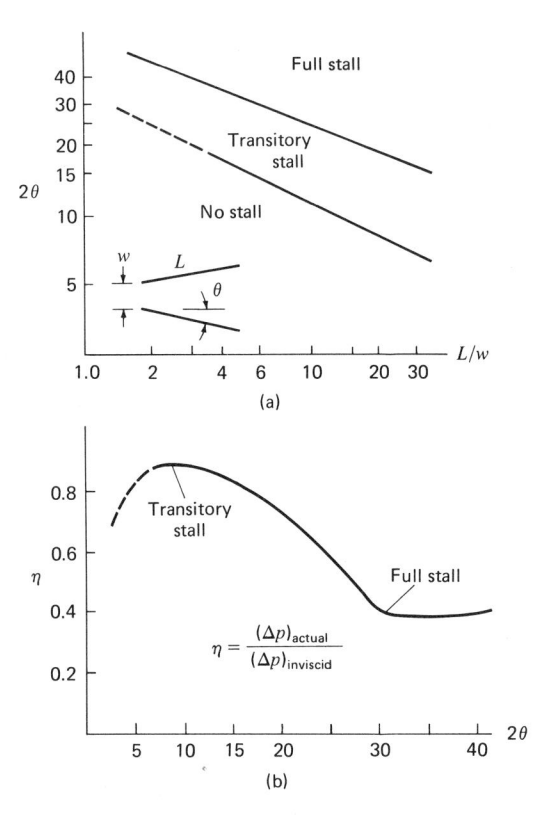

Fig. 8.3. Flow in a plane-wall subsonic diffuser. (From "Optimum Design of Straight-Walled Diffusers" by D. E. Abbott, S. J. Kline, and R. W. Fox, J. Basic Eng., Sept. 1959, p. 321.)

8.2 THE EULER-s AND EULER-n EQUATIONS

The Euler equation is particularly instructive for the expression of certain phenomena and flow behavior when it is formulated in streamline coordinates. We will consider only two-dimensional flows involving the streamwise distance s, with the unit vector $\hat{s}$, and the normal coordinate n with unit vector $\hat{n}$. This coordinate system is shown in Fig. 8.4. The substantial derivative can then be written in the form.

$$\frac{D\mathbf{V}}{Dt} = \frac{\partial}{\partial t}(V\hat{s}) + V\frac{\partial}{\partial s}(\hat{s}V)$$

$$= \hat{s}\frac{\partial V}{\partial t} + V\frac{\partial \hat{s}}{\partial t} + \hat{s}V\frac{\partial V}{\partial s} + V^2\frac{\partial \hat{s}}{\partial s} \qquad (8.4)$$

Note that $V_n = 0$, so that $V_s = V$. By using

Art. 8.2   THE EULER-s AND EULER-n EQUATIONS

$$\frac{\partial \hat{s}}{\partial s} = \lim_{\Delta s \to 0} \frac{\Delta \hat{s}}{\Delta s} = \lim_{\Delta s \to 0} \frac{-\hat{n}\,\Delta\alpha}{\Delta s} = \lim_{\Delta s \to 0} \frac{-\hat{n}\,\Delta\alpha}{R\Delta\alpha} \tag{8.5}$$

we find that

$$\frac{\partial \hat{s}}{\partial s} = -\frac{\hat{n}}{R} \tag{8.6}$$

Similarly,

$$\frac{\partial \hat{s}}{\partial t} = -\hat{n}\,\frac{\partial \theta}{\partial t} \tag{8.7}$$

where θ is the angle between the tangent to the instantaneous streamline and the arbitrary but fixed reference line. Hence,

$$\frac{D\mathbf{V}}{Dt} = \hat{s}\left(\frac{\partial V}{\partial t} + V\frac{\partial V}{\partial s}\right) - \hat{n}\left(V\frac{\partial \theta}{\partial t} + \frac{V^2}{R}\right) \tag{8.8}$$

We may also express the pressure gradient in terms of s and n:

$$\nabla p_k = \frac{\partial p_k}{\partial s}\hat{s} + \frac{\partial p_k}{\partial n}\hat{n} \tag{8.9}$$

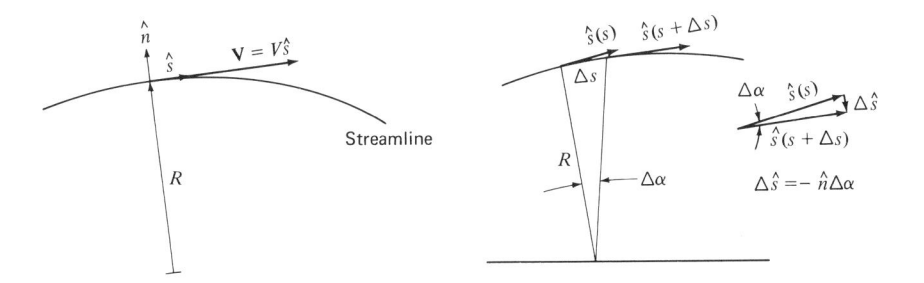

Fig. 8.4. Streamwise coordinates.

The Euler-s equation, expressing the momentum equation along a streamline, is found by substituting the foregoing expressions in Eq. 8.4 and equating the coefficients of s. There results

$$\frac{\partial V}{\partial t} + V\frac{\partial V}{\partial s} = -\frac{1}{\rho}\frac{\partial p_k}{\partial s} \tag{8.10}$$

The Euler-n equation, expressing the momentum normal to the streamline, is

$$V\frac{\partial \theta}{\partial t} + \frac{V^2}{R} = \frac{1}{\rho}\frac{\partial p_k}{\partial n} \tag{8.11}$$

The integration of the Euler-s equation provides a relationship between velocity and pressure along a streamline for *steady flow* (neglecting body forces, so that $p_k = p$):

$$\int_{①}^{②} V \frac{\partial V}{\partial s} ds + \int_{①}^{②} \frac{1}{\rho} \frac{\partial p}{\partial s} ds = 0 \quad \text{or} \quad \frac{V_2^2 - V_1^2}{2} + \int_{①}^{②} \frac{dp}{\rho} = 0$$
(8.12)

where ① and ② are two points on a given streamline. This form accounts for the possibility of a compressible flow, that is, $\rho = \rho(p)$. If the flow is incompressible, then the result may be expressed as

$$\frac{V^2}{2} + \frac{p}{\rho} = \text{const} \tag{8.13}$$

which is the often-utilized *Bernoulli equation*. In this form, the Bernoulli equation is seen to be valid for (i) steady, (ii) incompressible, (iii) inviscid flow, (iv) along a streamline, (v) referred to an inertial reference frame, (vi) with negligible body forces. If body forces were included the quantity gh would be added to the left side of Eq. 8.13.

A pressure-gradient-driven secondary flow is shown in Fig. 8.5. The bulk flow in the pipe establishes a radial pressure gradient such that the pressure at 1 is less than at 2. This pressure gradient results from $\partial p/\partial n = \rho V^2/R$. The velocity near the wall of the pipe is less than that of the bulk flow; hence, the velocity magnitude of this fluid is not sufficient to balance the pressure gradient. Consequently, a secondary flow is established in which the fluid near the wall responds to the radial pressure gradient by moving from 2 to 1 along the circumference of the pipe.

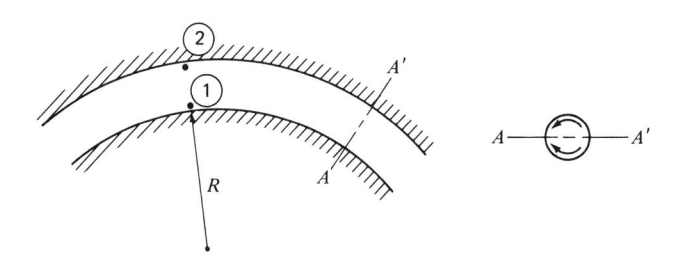

Fig. 8.5. Secondary flow in a curved conduit.

The primary utilization of the Euler-s equation is in terms of the Bernoulli integral whereas the n equation finds considerable utility in the description of flow phenomena. That is, the qualitative knowledge that the pressure increases outwardly across curved streamlines is quite

Art. 8.2   THE EULER-s AND EULER-n EQUATIONS

often valuable in the interpretation of a flow field even when quantitative values are difficult to estimate because of the unknown radius of curvature R and the unknown dependence of the velocity V with respect to the normal coordinate n.

Example 8.1

A pitot tube in a turbulent flow gives a time-varying reading even though the time-averaged velocity at a point is constant. Analyze the response of the probe in terms of the Euler-s equation and determine the source for the time-varying reading.

Solution.   The appropriate form of the Euler-s equation is (assuming incompressible flow)

$$\frac{\partial V}{\partial t} + V\frac{\partial V}{\partial s} = -\frac{1}{\rho}\frac{\partial p}{\partial s}$$

The integral of this equation along a streamline, which originates in the flow field at some position unaffected by the presence of the probe and terminates at the face of the pitot tube, is

$$\int_{(1)}^{(2)} \frac{\partial V}{\partial t}\,ds + \frac{V_2^2 - V_1^2}{2} + \frac{p_2 - p_1}{\rho} = 0$$

or

$$p_0 = p_1 + \rho\frac{V_1^2}{2} + \rho\int_{(1)}^{(2)} \frac{\partial V}{\partial t}\,ds.$$

where p_0 is the pressure measured by the pitot tube. Characteristic streamlines are shown in Fig. E8.1.

From this it is seen that a major contribution to the unsteadiness of the pitot tube response is the changing location of the stagnation streamline origin. That is, the streamline closer to the boundary will, in general, have a lower stagnation pressure since viscous effects have acted upon these fluid elements

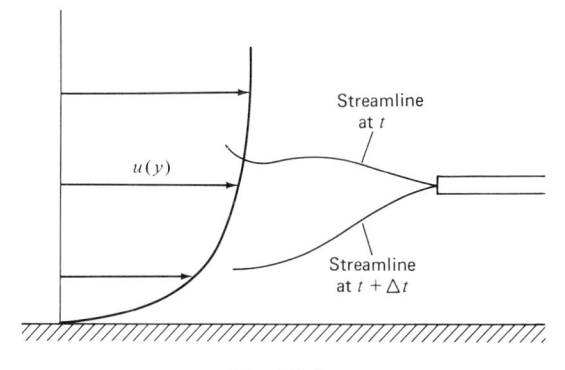

Fig. E8.1

360 INVISCID FLOWS Ch. 8

for a longer time, resulting in a lower velocity. In a turbulent flow large-amplitude fluctuations do occur so that this phenomenon is often observed.

A second source of the time variation of p_0 is the integral quantity. The fluctuations in V are a result of the turbulent eddy structures in the flow, and these structures have characteristic length $(s_2 - s_1)$; this leads to a nonzero value for the integral.

Example 8.2

Although the measurement of the pressure at a solid boundary is one of the most reliable measurements in fluid mechanics, it is subject to errors resulting from the presence of burrs at the lip of the hole. Explain the influence of a burr on the upstream and the downstream edges of a hole used for a static tap measurement.

Solution. The presence of a burr will cause the fluid to deflect vertically upward with a maximum point of the arch slightly downstream of the burr. This will cause a low pressure region in the separated wake behind the burr and a high pressure region in front of the burr. A characteristic streamline pattern is shown by the sketch. In Fig. E8.2(a) the streamline curvature indicates an outward-pointing normal vector so that from Eq. 8.11, for steady flow,

$$\frac{\partial p}{\partial n} = \rho \frac{V^2}{R} \quad \text{or} \quad \Delta p \cong \rho \frac{V^2}{R} \Delta n$$

From this we note that $p_A < p_B$. The reading would be low. From Fig. E8.2(b) the streamline curvature indicates a normal vector pointing toward point C so that $p_C > p_D$. A high reading would be recorded.

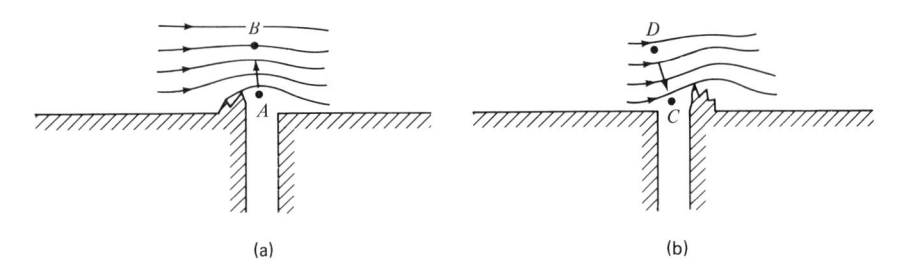

Fig. E8.2

8.3 EQUATIONS OF POTENTIAL FLOW

Initially, nonvortical or irrotational fluid may become rotational under the direct action of viscous diffusion or noninertial acceleration effects. Consequently, there are entire flows, which are driven by

Art. 8.3 EQUATIONS OF POTENTIAL FLOW

pressure or gravitational forces, in which the bulk of the flow is irrotational, that is, a flow in which each fluid element may accelerate or deform but not rotate. In general, the fluid near a solid boundary, where the viscous effects cause a no-slip condition, will not be approximated by an irrotational flow. A necessary and sufficient condition for identifying a flow as irrotational is

$$\nabla \times \mathbf{V} = 0 \tag{8.14}$$

This means that the velocity field $\mathbf{V}$ is a conservative vector field given by the gradient of a *scalar potential function* ϕ (see Art. 1.3); that is,

$$\mathbf{V} = \nabla \phi \tag{8.15}$$

Note that the *vector* velocity is obtained from a knowledge of the *scalar* function ϕ. In scalar form, Eq. 8.15 includes the three equations

$$u = \frac{\partial \phi}{\partial x}$$

$$v = \frac{\partial \phi}{\partial y} \tag{8.16}$$

$$w = \frac{\partial \phi}{\partial z}$$

The continuity equation for an incompressible flow is

$$\nabla \cdot \mathbf{V} = \frac{\partial u}{\partial x} + \frac{\partial v}{\partial y} + \frac{\partial w}{\partial z} = 0 \tag{8.17}$$

Using Eqs. 8.16 the continuity equation, in terms of the velocity potential, becomes

$$\frac{\partial^2 \phi}{\partial x^2} + \frac{\partial^2 \phi}{\partial y^2} + \frac{\partial^2 \phi}{\partial z^2} = 0 \tag{8.18}$$

which is *Laplace's equation*.

The momentum equation (3.52), without the viscous term, which is negligible in the inviscid flow, and with the use of Eq. (1.61) is

$$\frac{\partial \mathbf{V}}{\partial t} + \nabla \left(\frac{V^2}{2} \right) = -\frac{\nabla p}{\rho} - g \nabla h \tag{8.19}$$

where h is the vertical dimension and we have used $\nabla \times \mathbf{V} = 0$. Using

$\mathbf{V} = \nabla\phi$, this becomes

$$\nabla\left[\frac{\partial\phi}{\partial t} + \frac{V^2}{2} + \frac{p}{\rho} + gh\right] = 0 \qquad (8.20)$$

at every point in the potential flow. This means that

$$\frac{\partial\phi}{\partial t} + \frac{V^2}{2} + \frac{p}{\rho} + gh = \text{const} \qquad (8.21)$$

For steady flows the Bernoulli equation results,

$$\frac{V^2}{2} + \frac{p}{\rho} + gh = \text{const} \qquad (8.22)$$

This equation is valid everywhere, not just along a streamline.

To solve for a potential flow around a body we must first determine ϕ such that Laplace's equation (8.18) is satisfied, then find the velocity field from Eqs. 8.16 and the pressure field from Eq. 8.22. The pressure field can then be integrated over the area of interest to give a force. This would be the technique followed to determine the lift on an airfoil.

The immediate problem is the determination of the potential function ϕ for a particular problem of interest. The general three-dimensional problem will not be studied in this text because conventional methods are restricted to either plane, two-dimensional flows or axisymmetric flows. For both of these special classes of flows it is possible to define a function $\psi(x, y)$ called a *stream function* which is constant along a streamline. Since the flow is tangential to a solid surface, ψ is constant along a body. For the two-dimensional plane problem, the continuity equation is

$$\frac{\partial u}{\partial x} + \frac{\partial v}{\partial y} = 0 \qquad (8.23)$$

If we define

$$u = \frac{\partial\psi}{\partial y}, \qquad v = -\frac{\partial\psi}{\partial x} \qquad (8.24)$$

then continuity is automatically satisfied for an inviscid or a viscous flow. Using the equation for a streamline (Eq. 1.50) with $\mathbf{V} = u\,\hat{i} + v\,\hat{j}$ and $d\mathbf{R} = dx\,\hat{i} + dy\,\hat{j}$, we see that $v\,dx - u\,dy = 0$ along a streamline. This is exactly $d\psi = \frac{\partial\psi}{\partial x}dx + \frac{\partial\psi}{\partial y}dy = 0$, using Eqs. 8.24. Hence, we conclude that ψ is constant along a streamline. For a plane irrotational flow the first two components of vorticity, ξ and η, are identically zero; the third component gives

$$\zeta = \frac{\partial v}{\partial x} - \frac{\partial u}{\partial y} = 0$$

or
$$\frac{\partial^2 \psi}{\partial x^2} + \frac{\partial^2 \psi}{\partial y^2} = 0 \tag{8.25}$$

Consequently, the stream function also satisfies Laplace's equation in an irrotational flow; therefore, if we can determine the stream function, the velocity can be obtained from Eqs. 8.24 and the pressure from Eq. 8.22.

Two techniques may be employed to determine the potential function ϕ or the stream function ψ. The first technique is to solve directly Laplace's equation with the appropriate boundary conditions, using either a numerical technique or, possibly, the separation of variables method. The second and often utilized technique is to investigate some simple functions that satisfy Laplace's equation and then to superimpose these simple functions, which is allowable because Laplace's equation is linear, to provide the flow around the body of interest. This second method will be emphasized since it is the most commonly used procedure for potential-flow considerations.

It should be emphasized that we need only determine the ψ function or the ϕ function to within a constant, since a constant can be added to either of these functions and it will not effect the velocity field or the pressure distribution.

A note on boundary conditions for potential flows is in order. Laplace's equation is second-order and requires boundary conditions on the complete boundary enclosing a particular region of interest; that is, the stream function (or the velocity potential) or its derivative must be known over the *entire* surface. Consider a flow around a body shown in Fig. 8.6. The dotted surface *and* the body form the surface surrounding the region of interest. The condition at large distances from the body would be

$$u = U, v = 0 \quad \text{or} \quad \psi = Uy \tag{8.26}$$

The no-slip condition is no longer required on the body's surface since the effects of viscosity are neglected. Hence, we need not require the tangential component on the surface to be zero. The body is a streamline and along a streamline the stream function is constant. Thus, on the body we can choose the constant to be zero (it is arbitrary) so that $\psi = 0$ on the body and the stream function is specified on the entire surface. The condition at the body for the velocity potential is more difficult to specify so the stream function is generally used.

Another observation for plane irrotational flows is that

$$\frac{\partial \psi}{\partial y} = \frac{\partial \phi}{\partial x}, \quad \frac{\partial \psi}{\partial x} = -\frac{\partial \phi}{\partial y} \tag{8.27}$$

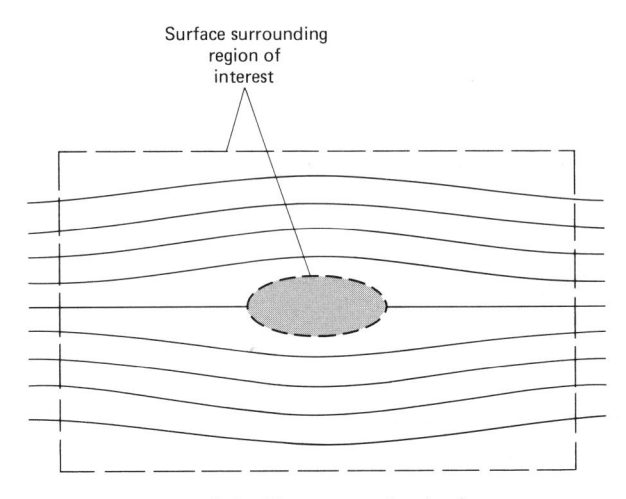

Fig. 8.6. Flow around a body.

These follow from Eqs. 8.16 and 8.24. They are the famous *Cauchy-Riemann equations* and enable us to use the theory of complex variables in our two-dimensional, plane problems. The functions ϕ and ψ are *harmonic functions* and form an *analytic complex function* ($\phi + i\psi$) called the *complex velocity potential*. Conformal transformations, along with all the complex variable theory, can thus be used for this class of problems. We will not use complex variables in this text but it is interesting to note the restricted class of problems for which it is useful in fluid mechanics, namely, plane, incompressible, irrotational flows.

Example 8.3

Show that in a two-dimensional incompressible flow the difference in the stream function between any two streamlines represents the flow rate per unit of depth between the two streamlines.

Solution. The infinitesimal flow rate per foot of depth flowing past the elemental distance dl (Fig. E8.3) is

$$dq = u\,dy - v\,dx$$

where the negative sign results since to go from ψ to $\psi + d\psi$ we must move in the negative x-direction. Substituting for u and v from eqs. (8.24) gives

$$dq = \frac{\partial \psi}{\partial y}dy + \frac{\partial \psi}{\partial x}dx$$

$$= d\psi$$

Art. 8.3   EQUATIONS OF POTENTIAL FLOW   365

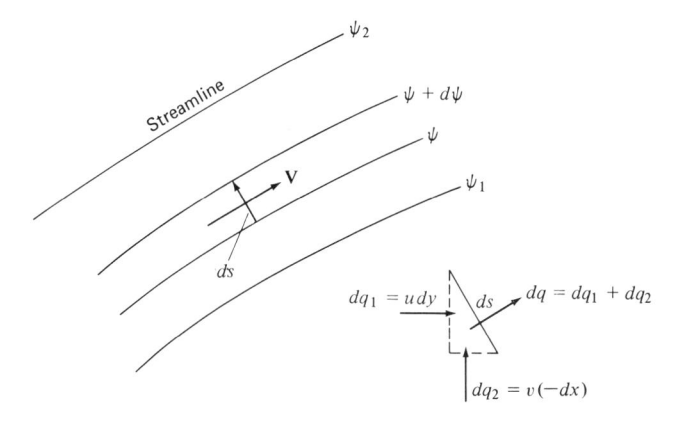

Fig. E8.3

If we integrate this from streamline ① to streamline ② we obtain
$$q = \psi_2 - \psi_1$$
for the incompressible, two-dimensional flow.

Example 8.4

The streamfunction for a particular flow is given as $\psi(x, y) = x^2 - y^2$. Is this flow irrotational? If so, calculate the velocity potential.

Solution. The velocity components are
$$u = \frac{\partial \psi}{\partial y} = -2y$$
and
$$v = -\frac{\partial \psi}{\partial x} = -2x$$

The vorticity components are then
$$\xi = 0, \quad \eta = 0, \quad \zeta = -2 + 2 = 0$$

The flow is irrotational, since all vorticity components are zero. A particle would not rotate, it would only deform.

The velocity potential is found as follows. From the first equation,
$$u = \frac{\partial \phi}{\partial x} = -2y$$
so that
$$\phi = -2xy + f(y)$$

Differentiating the above with respect to y gives
$$\frac{\partial \phi}{\partial y} = -2x + \frac{\partial f}{\partial y}$$

Equating this to $v = -2x$ gives $f = C$ where C is a constant. The velocity potential is then

$$\phi = -2xy + C$$

The constant C is not important since it does not affect the velocity or pressure fields. Hence, it is often set equal to zero.

Example 8.5

Show that the potential lines and the streamlines for a two-dimensional, incompressible, inviscid flow intersect one another at right angles, with the result that a curvilinear grid is formed.

Solution. Two contours intersect at right angles if their slopes form negative reciprocals at each point in the flow field. The local slope of a constant-ψ line can be expressed in terms of the ratio of the velocity components; that is

$$\text{Slope of constant-}\psi \text{ line} = v/u$$

Along a line of constant ϕ, $d\phi = 0$, so that

$$d\phi = \frac{\partial \phi}{\partial x} dx + \frac{\partial \phi}{\partial y} dy = 0$$

Hence, the slope $\dfrac{dy}{dx}$ of a constant-ϕ line is given as

$$\left.\frac{dy}{dx}\right|_{\phi=\text{const}} = \frac{-\partial\phi/\partial x}{\partial\phi/\partial y} = -u/v$$

This slope of a constant-ϕ line is seen to be the negative reciprocal of the slope of a constant-ψ line. Hence, the two lines are orthogonal everywhere their slope is defined. This feature of the ϕ- and ψ-lines results in the formation of an "orthogonal grid" of curvilinear squares. Such a grid is shown in Fig. E8.5.

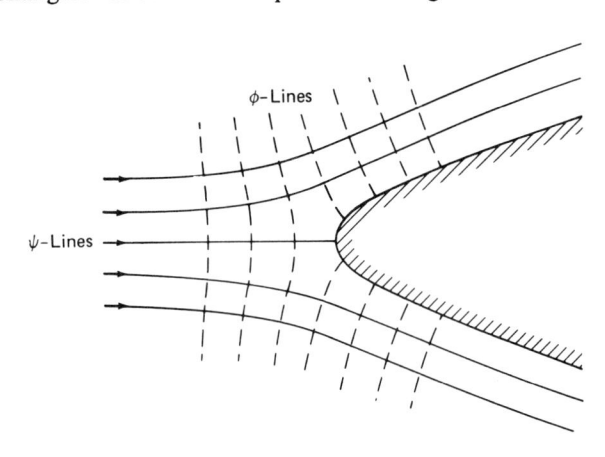

Fig. E8.5

Example 8.6

Write the viscous term for a Newtonian incompressible, homogeneous flow in terms of the vorticity and show that vorticity always accompanies viscous effects. Then show that for a potential flow the viscous term vanishes.

Solution. The viscous term for an incompressible, homogeneous fluid is $\mu \nabla^2 \mathbf{V}$. We can use a vector identity which states that

$$\nabla \times (\nabla \times \mathbf{V}) = \nabla(\nabla \cdot \mathbf{V}) - (\nabla \cdot \nabla)\mathbf{V}$$

This can be verified by expanding both sides in cartesian coordinates.

The quantity $\nabla \cdot \mathbf{V}$ is zero because of continuity; hence,

$$\nabla^2 \mathbf{V} = -\nabla \times (\nabla \times \mathbf{V})$$

Thus, the viscous term can alternately be written as $-\mu \Delta \times \omega$, so that if this term is not zero then the vorticity cannot be zero. We see then, that viscous effects are non-zero only in regions of vorticity. The converse is usually, but not necessarily, true; that is if vorticity is non-zero the viscous term is non-zero (unless $\Delta \times \omega \cong 0$). Using

$$\mu \nabla^2 \mathbf{V} = -\mu \nabla \times \omega$$

we see that if ω is everywhere zero then the viscous term vanishes.

Extension 8.6.1. Assume that a potential-flow solution exists for flow around a body. Viscous effects are confined to a thin boundary layer surrounding the body. Does the potential flow solution satisfy the complete Navier-Stokes equations for the flow external to the boundary layer?

Ans. Yes

Extension 8.6.2. Identify whether the bulk flow in the following problems would or would not be adequately discribed as a potential flow: the flow exiting from a bellows, the flow inside a journal bearing, the flow over an airplane, the flow which escapes from a closet as the door is closed.

8.4 SOME SIMPLE PLANE POTENTIAL FLOWS

We will now investigate some rather simple functions which satisfy Laplace's equation, $\nabla^2 \psi = 0$. Any function satisfying this equation represents a potential flow. Whether it is of particular interest to the engineer depends on the streamline pattern represented by $\psi(x, y)$ constant; that is, whether such a function includes a form of an object of interest. Some functions which give streamline patterns of interest are considered in the following sections of this article.

1. Uniform flow

Since Laplace's equation is second-order, a first-order dependence of ψ on x and y represents a possible stream function. Specifically,

$$\psi = Ax + By \tag{8.28}$$

The velocity components are

$$u = \frac{\partial \psi}{\partial y} = B \tag{8.29}$$

and

$$v = -\frac{\partial \psi}{\partial x} = -A \tag{8.30}$$

This represents a uniform flow as shown in Fig. 8.7. If $A = 0$, the uniform flow is only in the x-direction.

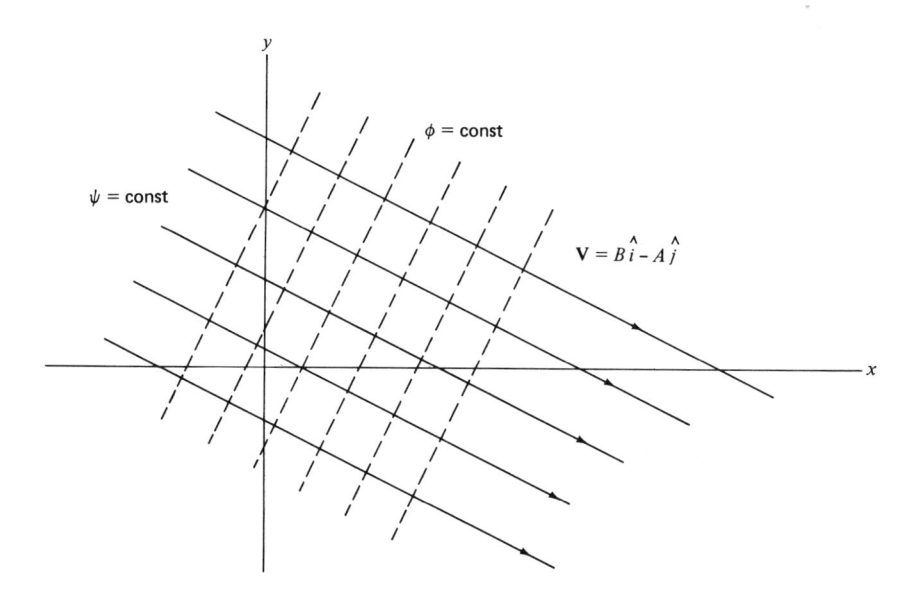

Fig. 8.7. Uniform flow.

The velocity potential for this uniform flow can be shown to be

$$\phi = Bx - Ay \tag{8.31}$$

2. Stagnation Flow

Another simple function which satisfies Laplace's equation is

$$\psi = Axy \tag{8.32}$$

The velocity components are

$$u = \frac{\partial \psi}{\partial y} = Ax \tag{8.33}$$

and

$$v = -\frac{\partial \psi}{\partial x} = -Ay \tag{8.34}$$

The streamlines represented by $\psi = 0$ are the x-axis and the y-axis. Since any streamline can be replaced by a solid boundary we can consider this to be flow in a corner or flow against a wall, as shown in Fig. 8.8. Both velocity components are zero at the origin; hence this is

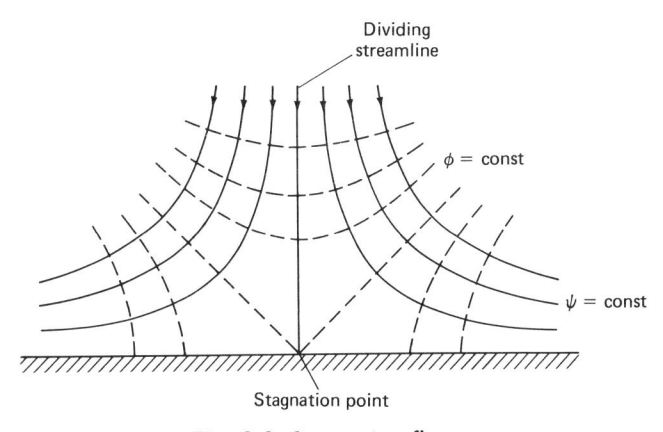

Fig. 8.8 Stagnation flow.

often referred to as *stagnation flow*. The velocity potential can be found, using Eqs. 8.16, to be

$$\phi = \frac{A}{2}(x^2 - y^2) \tag{8.35}$$

An interesting feature of a stagnation flow is that the streamline passing through the stagnation point divides the flow so that part of the flow proceeds in one direction and part in another. For flow around an airfoil the dividing streamline separates the flow that proceeds over the top of the airfoil from that which travels underneath the airfoil.

3. Sources and Sinks

For many applications it is more convenient to use polar coordinates, shown as r and θ in Fig. 8.9. Laplace's equation, in polar coordinates, is

$$\nabla^2 \psi = \frac{1}{r}\frac{\partial}{\partial r}\left(r\frac{\partial \psi}{\partial r}\right) + \frac{1}{r^2}\frac{\partial^2 \psi}{\partial \theta^2} = 0 \tag{8.36}$$

The velocity components are

$$v_r = \frac{1}{r}\frac{\partial \psi}{\partial \theta} \qquad v_\theta = -\frac{\partial \psi}{\partial r} \qquad (8.37)$$

which follow from the continuity equation,

$$\frac{1}{r}\frac{\partial}{\partial r}(rv_r) + \frac{1}{r}\frac{\partial v_\theta}{\partial \theta} = 0 \qquad (8.38)$$

Consider the simple harmonic function

$$\psi = A\theta \qquad (8.39)$$

Using Eqs. 8.37, the velocity components are

$$v_r = \frac{A}{r} \qquad v_\theta = 0 \qquad (8.40)$$

Since v_θ is everywhere zero, the streamline pattern must be represented by radial lines emanating from the origin, as shown in Fig. 8.9. If A is positive a source is represented; if A is negative, a sink results.

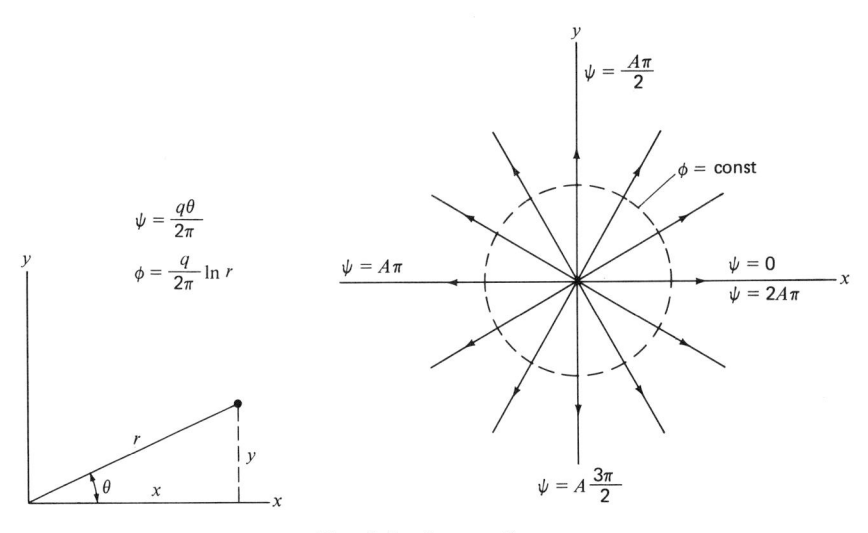

Fig. 8.9. Source flow.

The velocity at a particular r is constant for all θ. Hence we can integrate around a circle enclosing the origin to obtain the flow rate q as

$$q = \int_0^{2\pi} v_r r\, d\theta$$

$$= 2\pi A \qquad (8.41)$$

In terms of the *source strength q*, the streamfunction is

$$\psi = \frac{q}{2\pi} \theta \qquad (8.42)$$

where q is measured in ft^3/sec/ft of depth.

In cartesian coordinates the streamfunction is

$$\psi = A \tan^{-1} \frac{y}{x} \qquad (8.43)$$

and the velocity components are

$$u = \frac{Ax}{x^2 + y^2}, \qquad v = \frac{Ay}{x^2 + y^2} \qquad (8.44)$$

The associated velocity potential would be

$$\phi = A \ln \sqrt{x^2 + y^2} \qquad (8.45)$$

in cartesian coordinates. In polar coordinates, ϕ would be

$$\phi = A \ln r \qquad (8.46)$$

4. An Irrotational Vortex

Another simple function of interest, because of its resulting streamline pattern, is

$$\psi = A \ln r \qquad (8.47)$$

This satisfies Laplace's equation everywhere but at the origin, where $r = 0$. Hence the flow must be irrotational everywhere, except possibly at the origin. The velocity components are

$$v_r = 0, \qquad v_\theta = -\frac{A}{r} \qquad (8.48)$$

and obviously represent circular streamlines about the origin, shown in Fig. 8.10. The velocity increases as the origin is approached; a tornado is a good example of this type of motion.

The *circulation* Γ is defined as (counterclockwise is positive)

$$\Gamma = \oint \mathbf{V} \cdot d\mathbf{s} \qquad (8.49)$$

For the specific case of the irrotational vortex and for the contour

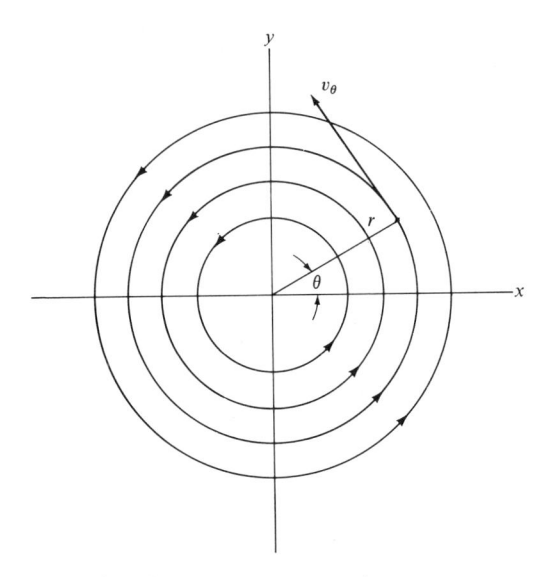

Fig. 8.10. An irrotational vortex.

formed by a circle around the origin, Γ may be expressed as

$$\Gamma = \int_0^{2\pi} v_\theta(r \, d\theta) = -2\pi A \qquad (8.50)$$

The stream function, in terms of the *vortex strength* Γ is

$$\psi = -\frac{\Gamma}{2\pi} \ln r \qquad (8.51)$$

The reason why circulation exists in an irrotational flow is that we have integrated around a singularity. If any path had been chosen which did not enclose the origin, Γ would have been zero. At the origin there exists an infinite vorticity, with the vorticity zero everywhere else.

In cartesian coordinates the streamfunction is

$$\psi = A \ln \sqrt{x^2 + y^2} \qquad (8.52)$$

and the velocity components are

$$u = \frac{Ay}{x^2 + y^2} \qquad v = -\frac{Ax}{x^2 + y^2} \qquad (8.53)$$

The corresponding potential function would be

$$\phi = -A \tan^{-1} \frac{y}{x} \qquad (8.54)$$

or
$$\phi = -A\theta \tag{8.55}$$

5. A Doublet

We can create another simple function which satisfies Laplace's equation by the method of superposition. Place a source at $x = -\epsilon$ and a sink at $x = +\epsilon$, where ϵ is a small quantity (see Fig. 8.11). The stream function is

$$\psi = A \tan^{-1} \frac{y}{x+\epsilon} - A \tan^{-1} \frac{y}{x-\epsilon} \tag{8.56}$$

Remembering that

$$\partial f/\partial x = \lim_{\epsilon \to 0} \frac{f(x+\epsilon, y) - f(x-\epsilon, y)}{2\epsilon}$$

we may put Eq. 8.56 in the form

$$\psi = 2\epsilon A \frac{\partial}{\partial x}\left(\tan^{-1}\frac{y}{x}\right)$$

$$= 2\epsilon A \left[-\frac{y}{x^2+y^2}\right] \tag{8.57}$$

where $\epsilon \to 0$ and $A \to \infty$ so that ϵA remains constant. The resultant flow is called a *doublet*. Defining the *doublet strength* μ to be $\mu = 2\epsilon A$, the streamfunction is

$$\psi = -\frac{\mu y}{x^2+y^2} \tag{8.58}$$

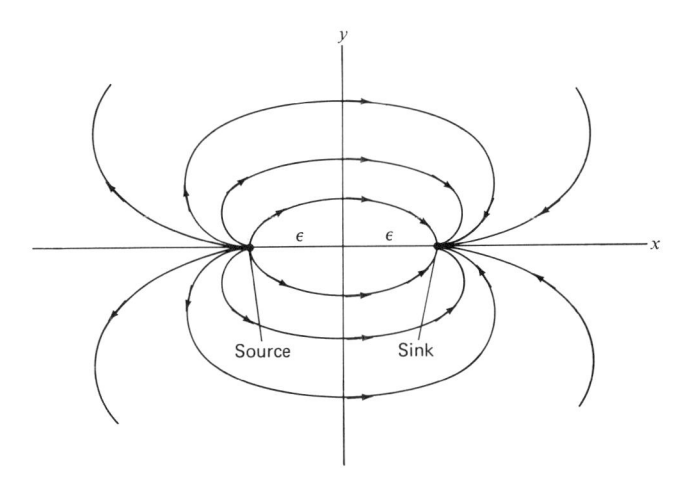

Fig. 8.11. Superposition of a source and sink. If $\epsilon = 0$, a doublet is formed.

This doublet is oriented in the negative x-direction. It could have been directed differently if the original source and sink had been placed along a different line; however, it is most common to orient the doublet as represented in Eq. 8.58.

In polar coordinates, the streamfunction is

$$\psi = -\frac{\mu \sin \theta}{r} \tag{8.59}$$

with the velocity components

$$v_r = -\frac{\mu \cos \theta}{r^2} \qquad v_\theta = -\frac{\mu \sin \theta}{r^2} \tag{8.60}$$

The cartesian velocity components are

$$u = -\frac{\mu(x^2 - y^2)}{(x^2 + y^2)^2} \qquad v = -\frac{2\mu xy}{(x^2 + y^2)^2} \tag{8.61}$$

The corresponding velocity potential for the doublet is

$$\phi = \frac{\mu x}{x^2 + y^2} \tag{8.62}$$

or, in polar coordinates,

$$\phi = \frac{\mu \cos \theta}{r} \tag{8.63}$$

The resulting doublet flow is shown in Fig. 8.12 with both streamlines and potential lines shown. For ϕ and ψ constant, Eqs. 8.58 and 8.62

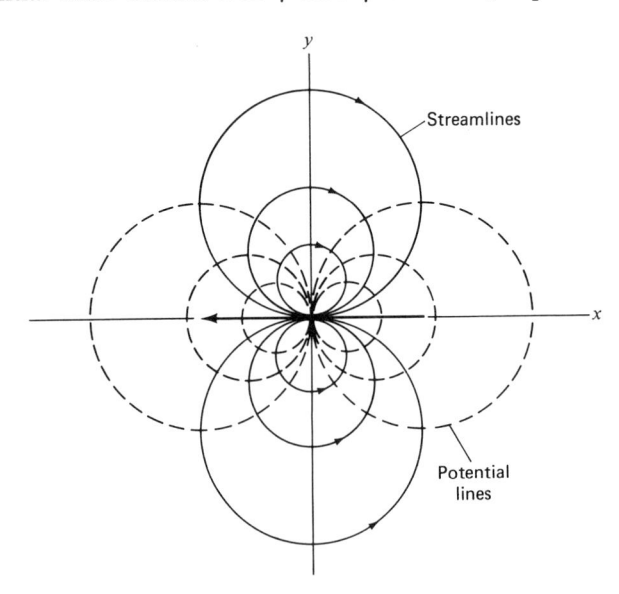

Fig. 8.12. The doublet oriented along the x-axis.

show that two families of circles result, all passing through the origin. Note the orthogonality between the streamlines and potential lines.

Example 8.7

Determine the velocity potential ϕ for a source flow.

Solution. Using polar coordinates the velocity components are related to the velocity potential with

$$v_r = \frac{\partial \phi}{\partial r} \qquad v_\theta = \frac{1}{r}\frac{\partial \phi}{\partial \theta}$$

Hence, from eq. (8.40),

$$\frac{\partial \phi}{\partial r} = \frac{A}{r}$$

giving

$$\phi = A \ln r + f(\theta)$$

Then, since $v_\theta = 0$ for a source,

$$\frac{1}{r}\frac{\partial \phi}{\partial \theta} = \frac{1}{r}\frac{\partial f}{\partial \theta} = 0$$

Thus, $f(\theta)$ is at most a constant; but, since we only wish to determine ϕ to within a constant we simply let the constant be zero. This will not affect the velocity field or the pressure field. Finally,

$$\phi = A \ln r$$

The constant-ϕ lines are circles about the origin. They are always normal to the streamlines.

Extension 8.7.1. The function ϕ, in general, should include a constant so that

$$\phi = A \ln r + C$$

Why can we choose $C = 0$? If $C = 10$, how would the pressure, velocity components, and stresses change?

8.5 SUPERPOSITION

The flows that are presented in the previous article are referred to as "simple flows"; their flow patterns are easily perceived and their mathematical descriptions are uncomplicated. These simple flows may define a flow field of engineering interest but their principal use is in the construction of more complicated flow fields by the process of *superposition*, that is, by adding two or more flows. In this sense, the simple flows are the basic building blocks for two-dimensional plane flows. We can now superimpose the ψ-functions to create the flows of interest. This superposition is allowable because the governing equations for the ϕ- and ψ-functions are linear; specifically, $\nabla^2 \phi = \nabla^2 \psi = 0$. We simply add any combination of the stream functions together and we are assured that the new function satisfies all the basic equations. The purpose is to create a ψ or ϕ-function which represents a flow field of interest; numerous examples are given below. It will become clear that if one streamline can be identified which has a desired geometric shape, then this streamline will be designated as the "body" and, in general, the streamlines beyond this region taken as the flow around the body. Note that, since no flow crosses a streamline, its role is identical to that of a solid surface.

1. Flow Past a Half-body

If a uniform flow in the x-direction is combined with a source flow, the stream function and velocity potential are

$$\psi = Uy + \frac{q}{2\pi} \tan^{-1} \frac{y}{x} \tag{8.64}$$

$$\phi = Ux + \frac{q}{2\pi} \ln \sqrt{x^2 + y^2} \tag{8.65}$$

The streamline which divides the source flow from the external flow forms a half-body shown in Fig. 8.13. The value of the stream function on the body would be $\psi = q/2$ since $\psi = 0$ on the positive x-axis. On

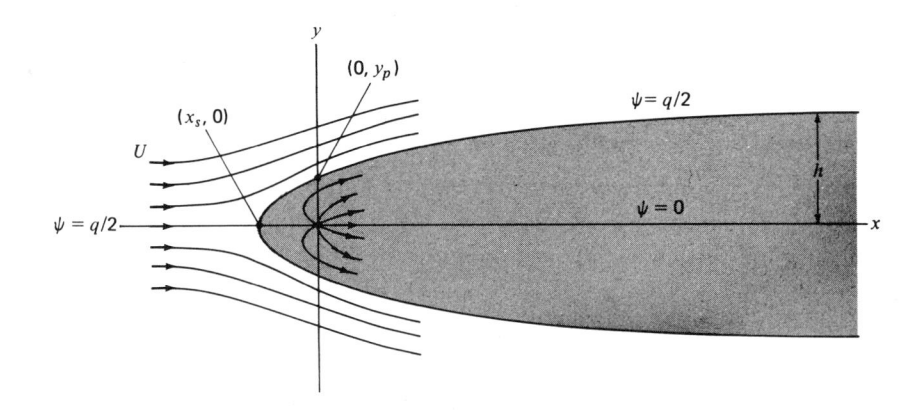

Fig. 8.13. Flow past a half-body.

the negative x-axis, $\tan^{-1} y/x = \pi$ and Eq. 8.64 gives the stream function as $\psi = q/2$.

It will be instructive to find the stagnation point, the y-intercept, and the asymptotic dimension of the half-body. The stagnation point occurs on the x-axis at the point where $u = 0$ and $y = 0$. By differentiating Eq. 8.64 with respect to y, we have, along the x-axis,

$$u = U + \frac{qx}{2\pi(x^2 + \cancel{y^2}^0)} \tag{8.66}$$

Setting $u = 0$ the x-coordinate of the stagnation point is given by

$$x_s = -\frac{q}{2\pi U} \tag{8.67}$$

The y-intercept occurs where the $\psi = q/2$ streamline intersects the y-axis. The value $q/2$ is chosen for ψ since we know that the quantity of fluid flowing between the x-axis and the curve designating the body surface is half of the fluid emitting from the source, namely, $q/2$. See Example 8.3. At the y-intercept point $x = 0$, so that

$$\psi = Uy_P + \frac{q}{2\pi} \tan^{-1} {\left. y_P \middle/ 0 \right.}^{\pi/2} = q/2 \tag{8.68}$$

giving

$$y_P = q/4U \tag{8.69}$$

The asymptotic dimension h would be found from letting $x \to \infty$ and $y = h$, namely,

$$\psi = Uh + \frac{q}{2\pi} \tan^{-1} {\left. h \middle/ \infty \right.}^0 = q/2 \tag{8.70}$$

so that

$$h = \frac{q}{2U} \tag{8.71}$$

2. Flow Past a Cylinder

The combination of a uniform flow in the x-direction and a doublet oriented in the negative x direction will result in flow past a cylinder, shown in Fig. 8.14. The stream function and velocity potential are

$$\psi = Ur \sin \theta - \frac{\mu \sin \theta}{r} \tag{8.72}$$

$$\phi = Ur \cos \theta + \frac{\mu \cos \theta}{r} \tag{8.73}$$

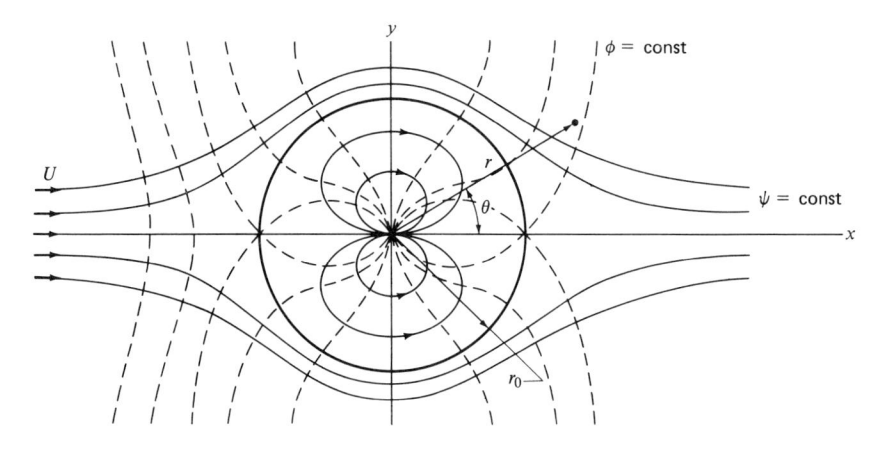

Fig. 8.14. Flow past a cylinder, showing orthogonality between streamlines and potential lines.

The radius of the cylinder is determined by locating the radius at which the radial component of velocity $v_r = 0$. The r-component of velocity, using Eq. 8.37, is

$$v_r = \left(U - \frac{\mu}{r^2}\right) \cos \theta \tag{8.74}$$

Setting $v_r = 0$ gives the radius of the cylinder, as

$$r_0 = \sqrt{\frac{\mu}{U}} \tag{8.75}$$

The stagnation points are located by setting $v_\theta = 0$ with $r = r_0$. Making use of Eq. 8.37, the θ-component of velocity is

$$v_\theta = -\left(\frac{\mu}{r^2} + U\right) \sin \theta \tag{8.76}$$

The stagnation points on the cylinder are thus located at $\theta_s = 0°$ and $\theta_s = 180°$. At these angles and $r = r_0$ both v_θ and v_r are zero.

Art. 8.5 SUPERPOSITION

The streamfunction, in terms of the cylinder radius, is then

$$\psi = Ur \sin \theta - \frac{r_0^2 U \sin \theta}{r} \qquad (8.77)$$

The velocity on the cylinder where $r = r_0$ is found by substituting Eq. 8.75 in 8.76, namely,

$$v_c = -2U \sin \theta \qquad (8.78)$$

with $v_r = 0$. The pressure distribution on the cylinder, neglecting body-force effects, is found by using Bernoulli's equation and is

$$p_c = p_0 - 2\rho U^2 \sin^2 \theta \qquad (8.79)$$

where p_0 is the pressure at the stagnation point where the velocity is zero.

We observe that the pressure is maximum at the stagnation point, decreases to a minimum value on the top and bottom of the cylinder, and reaches the same maximum value at the opposite stagnation point. Because of this symmetrical distribution no drag would result. This simple result has been obtained using a circular cylinder; however, it is applicable to all non-circular cylinders in potential flows. *In an irrotational flow around a body, the drag is always zero.* For the circular cylinder, the pressure distribution is shown in Fig. 8.15 for an irrotational flow, (part a) and for a real flow with separation (part b). The

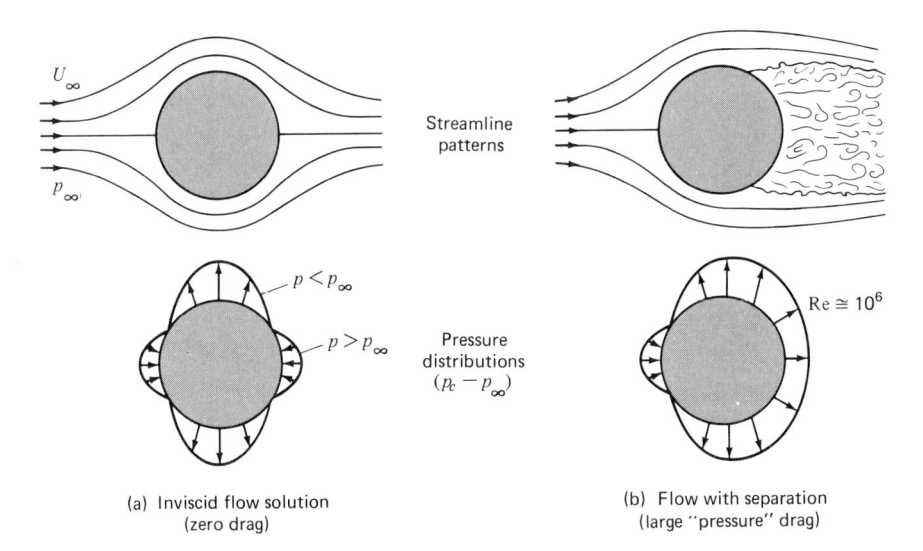

Fig. 8.15. Pressure distribution on a cylinder.

pressure distribution predicted from the potential flow analysis is a good approximation to the real pressure distribution close to the point of separation. In the rotational wake region the irrotational solution is no longer an acceptable approximation to the actual flow. If the Reynolds number associated with the flow is low enough, no separation will occur; however, then viscous effects cannot be neglected for this situation and again the potential flow theory is not an acceptable approximation to the flow. The fact that the potential flow solution is an acceptable approximation to the flow up to the separation point makes it important to engineers solving for flows around bodies. In fact, for streamlined bodies the flow may not separate even at large Reynolds numbers; for these aerodynamic bodies the potential flow approximates the flow over the whole body. The low-pressure region of a separated flow is, in general, the dominant contributor to the aerodynamic drag of bluff bodies. Flow around bodies is considered in Chapter 10.

Flow around a circular cylinder is often used to illustrate a graphical method which gives an approximate solution to an irrotational plane flow. The method is based on the orthogonality which exists between the streamlines and potential lines. To utilize the method, streamlines are sketched so that they are equally spaced in regions of uniform flow; then potential lines are sketched in, equally spaced in regions of uniform flow and with the same spacing as the streamlines. Near the body the potential lines are sketched in with the same approximate spacing as the streamlines. The resulting pattern of streamlines and potential lines is a *flow net*, shown in Fig. 8.16. Potential flow theory states that streamlines and potential lines intersect at right angles,

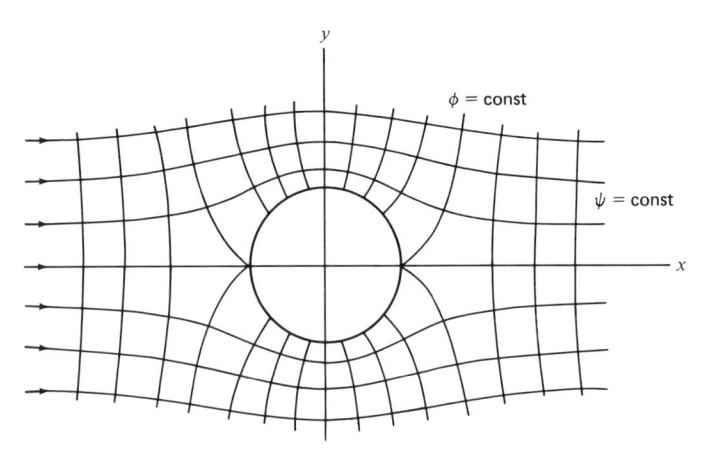

Fig. 8.16. Flow net for flow around a cylinder.

hence following the foregoing procedure all spaces in the flow net should approximate squares, and in regions of uniform flow, the spaces should be squares. An initial sketch of the flow net usually results in regions in which the spaces are elongated or rectangular. This indicates that either the streamlines or the potential lines are sketched too close together. A second attempt at the sketch should produce spaces that are more nearly squares, and a flow pattern that is more nearly correct. Iterations are continued until all the spaces approximate squares as closely as possible. Usually a coarse grid is used to start the process. After several iterations a finer grid is introduced to give more accurate results. After the flow net is completed, the velocity at a point can be approximated by observing the distance between the streamlines in the vicinity of the point. The velocity is inversely proportional to the distance between the streamlines. For example, if the distance between two streamlines in the vicinity of a point is one half of the distance between the same two streamlines in the region where the uniform velocity is U, then continuity can be used to show that the velocity at the point would be approximately $2U$.* The pressure change can then be approximated with the use of Bernoulli's equation.

3. Flow Around a Cylinder with Circulation

If an irrotational vortex is added to the streamfunction of Eq. (8.77) there results

$$\psi = Ur \sin \theta - \frac{r_0^2 U \sin \theta}{r} - \frac{\Gamma}{2\pi} \ln r \qquad (8.80)$$

The velocity distribution on the cylinder with radius $r_0 = \sqrt{\mu/U}$ is given by

$$v_c = -2U \sin \theta + \frac{\Gamma}{2\pi} \sqrt{\frac{U}{\mu}} \qquad (8.81)$$

The pressure distribution on the cylinder is

$$p_c = p_0 - \rho \frac{U^2}{2} \left[2 \sin \theta - \frac{\Gamma}{2\pi r_0 U} \right]^2 \qquad (8.82)$$

where p_0 is the stagnation point pressure. The lift is found by integrating the vertical component of the pressure force, shown in Fig. 8.17,

---

*Alternatively, the velocity can be obtained from the approximation $\Delta\phi/\Delta s$, where Δs is the streamwise spacing of the curvilinear square.

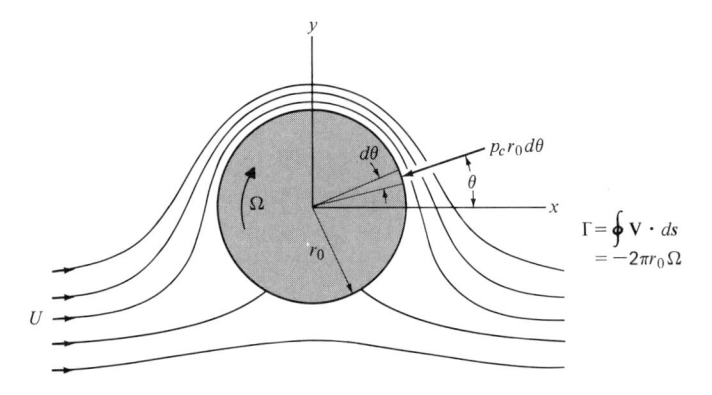

Fig. 8.17. Flow around a cylinder with circulation.

and is
$$L = -\int_0^{2\pi} p_c \sin\theta \, r_0 d\theta \tag{8.83}$$

With the expression for the pressure p_c from Eq. 8.82, this may be integrated to give
$$L = -\rho U \Gamma \tag{8.84}$$

This simple expression for the lift is also applicable to all non-circular cylinders. It and the zero drag conclusion form the *Kutta-Joukowsky theorem*.

4. Series of Sources and Sinks

A series of sources and sinks could be situated on the x-axis and superimposed on a uniform flow to create flow around a streamlined body, shown in Fig. 8.18a. A series of sources and sinks of various strengths would be combined with a uniform flow as

$$\psi = Uy + \sum_{i=1}^{N} \frac{q_i}{2\pi} \tan^{-1} \frac{y}{x - x_i} \tag{8.85}$$

So that no net flow emanates from the body, we require that

$$\sum_{i=1}^{N} q_i = 0 \tag{8.86}$$

By letting $q_i \to 0$ at either end of the symmetrical body we can force the body to be pointed; if q_i is finite at an end it will be blunt.

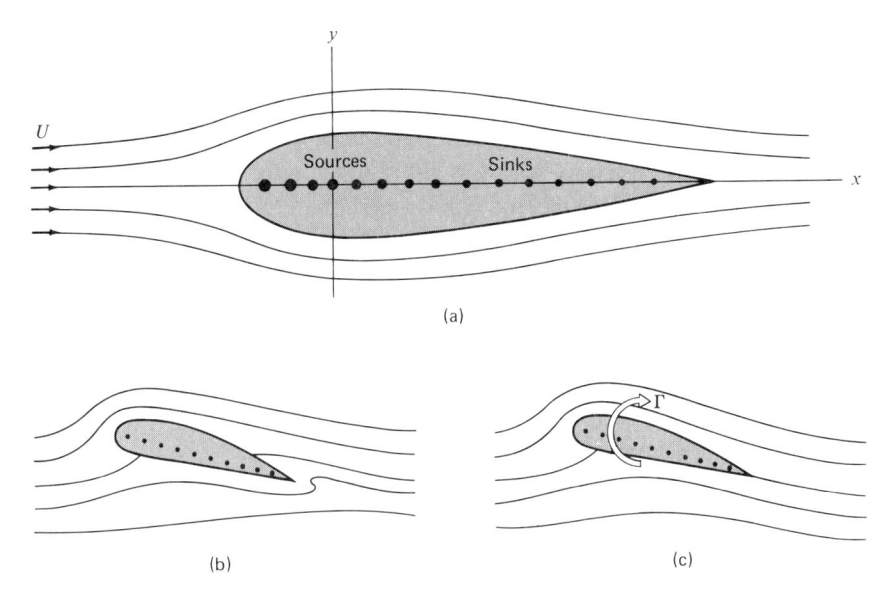

Fig. 8.18. Flow around a streamlined body.

The sources and sinks could be distributed along a line, as shown in Fig. 8.16b, to give the airfoil an angle of attack. This would, however, result in a flow as shown. The flow would turn the corner at the trailing edge with a stagnation point located on the top surface. Of course, this implies a tremendously large normal pressure gradient at the trailing edge since $V \neq 0$ as $R \to 0$ in $\partial p/\partial n = \rho V^2/R$; hence the fluid would not turn the corner at the trailing edge but would be as shown in Fig. 8.16c. In order that our potential flow be a good approximation to the real flow a vortex is superimposed as shown. The strength Γ of the vortex is chosen such that the dividing streamline leaves the rear of the airfoil at the trailing edge. The condition that the dividing streamline be located at the trailing edge is the *Kutta condition*. The resulting lift on the airfoil is $-\rho U \Gamma$, and is a good approximation to the lift on the actual airfoil.

The technique of superimposing sources and sinks is difficult if we wish to form a predetermined body. The more common technique used today is to distribute sources and sinks along the surface of a known body, for example an airfoil, and then to determine the source and sink strengths by requiring the normal component of velocity on the body to vanish. The digital computer is obviously very handy in problems where a large number of sources and sinks are involved.

5. Image Flow

Some interesting flows can be generated by the *method of images*. For example, if a source flow next to a plane wall a distance d away were desired we could generate this flow by placing a source at the position $(d, 0)$ and its image, a source of equal strength, at the position $(-d, 0)$. This is shown in Fig. 8.19. Because of symmetry the y-axis becomes a streamline and hence can be replaced with a solid boundary. The combined velocity potential is

$$\phi = \frac{q}{2\pi} \ln \sqrt{(x-d)^2 + y^2} + \frac{q}{2\pi} \ln \sqrt{(x+d)^2 + y^2} \quad (8.87)$$

resulting in velocity components

$$u = \frac{q}{2\pi} \left[\frac{x-d}{(x-d)^2 + y^2} + \frac{x+d}{(x+d)^2 + y^2} \right]$$

$$v = \frac{q}{2\pi} \left[\frac{y}{(x-d)^2 + y^2} + \frac{y}{(x+d)^2 + y^2} \right]$$

(8.88)

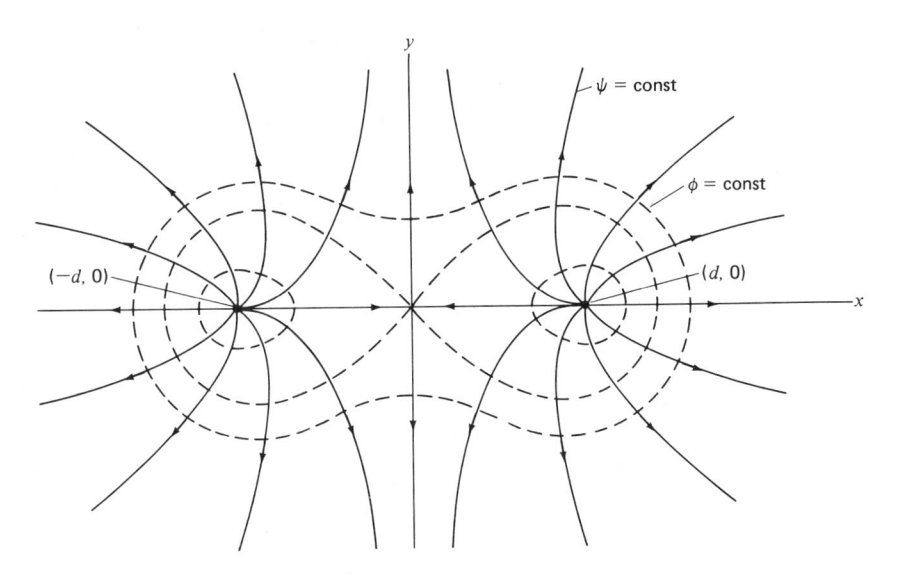

Fig. 8.19. Source flow near a plane wall.

Art. 8.5 SUPERPOSITION 385

On the y-axis, $x = 0$ and from the above expression for u we see that $u = 0$. Only v is non-zero along the y-axis. Thus, the velocity vector is tangent to the y-axis, and the y-axis must be a streamline. The flow on the right of the y-axis or the left then represents a source flow near a wall. The origin is a stagnation point and the flow in the near vicinity of the origin would be quite similar to that near the stagnation point of the first section of Art. 8.4.

Various other flows of importance can be formed by the method of images. A source or sink flow in the end of a channel can be generated by superimposing an infinite series of sources or sinks along the y-axis. An infinite series of doublets distributed along the y-axis superimposed with a uniform flow could result in flow around a cylinder in a channel.

Example 8.8

Determine the angles which locate the two stagnation points on a rotating cylinder. In particular, if an 8-in. dia. cylinder rotates at 200 rpm in a 10-fps free stream, locate the two stagnation points.

Solution. The velocities, calculated from the streamfunction in eq. (8.80) are

$$v_\theta = -\frac{\partial \psi}{\partial r} = -U \sin \theta - \frac{r_0^2}{r^2} U \sin \theta + \frac{\Gamma}{2\pi r}$$

$$v_r = \frac{1}{r} \frac{\partial \psi}{\partial \theta} = U \cos \theta - \frac{r_0^2}{r^2} U \cos \theta$$

On the cylinder $r = r_0$ and v_r is obviously zero. We wish to find the point where $v_\theta = 0$, the stagnation point. Setting $v_\theta = 0$ gives

$$\sin \theta_s = \frac{\Gamma}{4\pi r_0 U}$$

The angles which locate the two stagnation points are thus given from the expression

$$\theta_s = \sin^{-1}\left(\frac{\Gamma}{4\pi r_0 U}\right)$$

The circulation is calculated by considering the streamline of the vortex (see Fig. 8.10) at $r = r_0$ to have a velocity of $r_0 \Omega$, where Ω is the angular velocity of the cylinder. The circulation is then (counterclockwise is positive)

$$\Gamma = \oint \mathbf{V} \cdot d\mathbf{s} = -r_0 \Omega (2\pi r_0) = -\frac{200 \times 2\pi}{60} \times \frac{2\pi}{9} = -14.6 \text{ ft}^2/\text{sec}$$

so that

$$\theta_s = \sin^{-1}\left(-\frac{14.6}{40\pi/3}\right) = \sin^{-1}(-0.348)$$

Hence,

$$\theta_s = 200.4°, 339.6°$$

Extension 8.8.1. Determine the Ω in rpm which would cause only one stagnation point to exist on the cylinder. *Ans.* 60 rad/sec

Extension 8.8.2. Determine the pressure distribution on the cylinder of Example 8.8 and calculate the maximum and minimum values of $(p_c - p_\infty)$. Water is flowing. *Ans.* 97 psf, −609 psf

Extension 8.8.3. An approximation to the flow described in this example can be created by the physical situation shown in Fig. E8.8. A flat ribbon, wrapped tightly around the cylinder from a roll of kitchen towels works well. Do the experiment. Explain why viscosity is *necessary* to approximate this inviscid flow problem. Explain why the cylinder executes a looping motion.

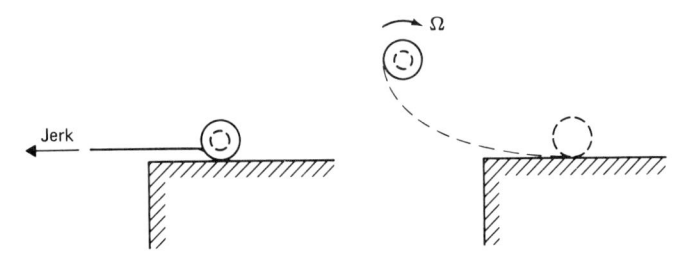

Fig. E8.8

8.6 AXISYMMETRIC FLOWS

Laplace's equation in spherical coordinates for the axisymmetric case is

$$\nabla^2\phi = \frac{1}{r^2}\frac{\partial}{\partial r}\left(r^2\frac{\partial\phi}{\partial r}\right) + \frac{1}{r^2\sin\theta}\frac{\partial}{\partial\theta}\left(\sin\theta\frac{\partial\phi}{\partial\theta}\right) = 0 \quad (8.89)$$

The simple potential function

$$\phi = \frac{A}{r} \quad (8.90)$$

satisfies Laplace's equation and represents a *point source* with streamlines emanating along radial lines in all directions from the origin of r.

This point source with fluid emanating from a single point is quite different from the plane source with fluid emanating from a line. An integration of the velocity around a spherical surface allows us to introduce the source strength q and write

$$\phi = -\frac{q}{4\pi r} \tag{8.91}$$

for a source.

By combining a source at $x = -\epsilon$ with a sink of equal strength at $x = +\epsilon$ the axisymmetric doublet results as $\epsilon \to 0$ with velocity potential

$$\phi = \frac{\mu \cos \theta}{r^2} \tag{8.92}$$

If we combine a uniform flow and this doublet flow we would have flow around a sphere. The velocity potential would be

$$\phi = \frac{\mu \cos \theta}{r^2} + Ur \cos \theta \tag{8.93}$$

This flow is shown in Fig. 8.20 along with a graphical presentation of the coordinates r, θ, and x.

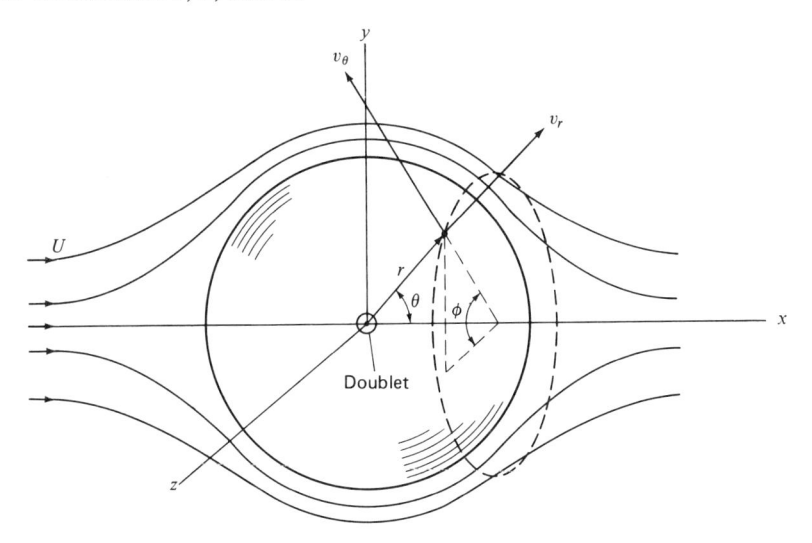

Fig. 8.20. Flow around a sphere.

The velocity vector would be found from

$$\mathbf{V} = \frac{\partial \phi}{\partial r} \hat{e}_r + \frac{1}{r} \frac{\partial \phi}{\partial \theta} \hat{e}_\theta \tag{8.94}$$

where θ is measured from the positive x-axis. For flow around a sphere

$$v_r = \frac{\partial \phi}{\partial r} = U \cos \theta - \frac{2\mu \cos \theta}{r^3} \tag{8.95}$$

$$v_\theta = \frac{1}{r} \frac{\partial \phi}{\partial \theta} = -U \sin \theta - \frac{\mu \sin \theta}{r^3} \tag{8.96}$$

The radius of the sphere is found to be

$$r_0 = \left(\frac{2\mu}{U}\right)^{1/3} \tag{8.97}$$

A uniform flow could be superimposed with a series of sources and sinks or a series of doublets placed along a straight line segment to create flow around an axisymmetric body, as was shown in Fig. 8.18, for the plane flow.

It should be pointed out here that a stream function for the axisymmetric flow could have been defined with the aid of the continuity equation in spherical coordinates. But the velocity potential and streamfunction no longer satisfy the Cauchy-Riemann equations so that complex variables is not of use in axisymmetric flows. It may be useful, however, to determine the streamfunction since it is constant along a solid boundary. The continuity equation, in spherical coordinates, for axisymmetric flows is

$$\frac{1}{r^2} \frac{\partial}{\partial r}(r^2 v_r) + \frac{1}{r \sin \theta} \frac{\partial}{\partial \theta}(v_\theta \sin \theta) = 0 \tag{8.98}$$

The axisymmetric stream function would then be defined from

$$v_r = \frac{1}{r^2 \sin \theta} \frac{\partial \psi}{\partial \theta}, \quad v_\theta = -\frac{1}{r \sin \theta} \frac{\partial \psi}{\partial r} \tag{8.99}$$

Using Eqs. 8.95 and 8.96 the streamfunction for flow around a sphere would be

$$\psi = \tfrac{1}{2} U r^2 \sin^2 \theta - \frac{\mu \sin^2 \theta}{r} \tag{8.100}$$

Example 8.9

A point source is placed at $x = -a$ and an equal strength sink at $x = +a$ (Fig. E8.9). Determine the maximum radius r_0 of the axisymmetric body formed if the source and sink are placed in a uniform flow.

Art. 8.6 AXISYMMETRIC FLOWS 389

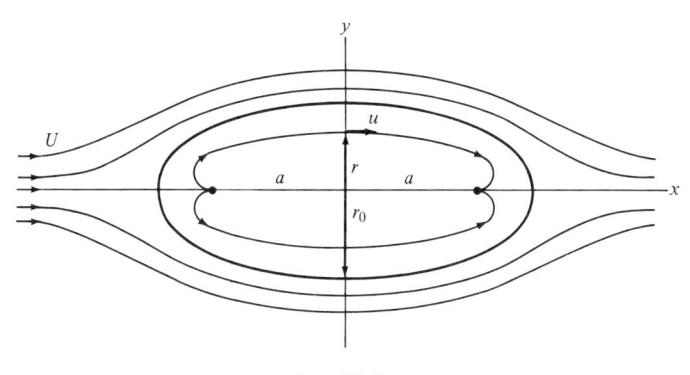

Fig. E8.9

Solution. Using Eq. 8.91 in cartesian coordinates, the velocity potential is

$$\phi = Ux - \frac{q}{4\pi\sqrt{(x+a)^2 + y^2 + z^2}} + \frac{q}{4\pi\sqrt{(x-a)^2 + y^2 + z^2}}$$

Because of symmetry, the body will have its maximum thickness at $x = 0$, as shown. One may find the body radius by integrating from $r = 0$ to $r = r_0$ so that the total mass flux across the area of integration is q. The x-component of velocity is

$$u = \frac{\partial \phi}{\partial x} = U + \frac{(x+a)q}{4\pi[(x+a)^2 + y^2 + z^2]^{3/2}} - \frac{(x-a)q}{4\pi[(x-a)^2 + y^2 + z^2]^{3/2}}$$

Along the $x = 0$ plane

$$u = U + \frac{2qa}{4\pi(a^2 + y^2 + z^2)^{3/2}}$$

$$= U + \frac{qa}{2\pi(a^2 + r^2)^{3/2}}$$

The flow rate q through a circle of radius r_0 is

$$q = \int_0^{r_0} \left[U + \frac{qa}{2\pi(a^2 + r^2)^{3/2}} \right] 2\pi r \, dr$$

$$= U\pi r_0^2 - qa \left[\frac{1}{\sqrt{a^2 + r_0^2}} - \frac{1}{a} \right]$$

For a particular set of flow parameters r_0 could be determined.

Extension 8.9.1. Determine the maximum body radius r_0 if $U = 10$ fps, $a = 4$ ft and $q = 100$ ft^3/sec. *Ans.* 1.71 ft

Extension 8.9.2. Determine the length of the body for these parameters. *Ans.* 9.82 ft

8.7 ROTATIONAL INVISCID FLOWS

In the preceding articles the vorticity was considered zero. It is appropriate to discuss briefly the governing equations for inviscid flows which have vorticity. The differential momentum equation, in vector form, for an inviscid flow, is

$$\frac{\partial \mathbf{V}}{\partial t} + \tfrac{1}{2}\nabla V^2 + \boldsymbol{\omega} \times \mathbf{V} = -\frac{\nabla p}{\rho} - g\nabla h \qquad (8.101)$$

By taking the curl of both sides of this equation we obtain, for a constant-density flow,

$$\frac{\partial}{\partial t}(\nabla \times \mathbf{V}) + \nabla \times (\boldsymbol{\omega} \times \mathbf{V}) = 0 \qquad (8.102)$$

recalling that the curl of the gradient of a scalar quantity is zero. It can be shown that

$$\nabla \times (\boldsymbol{\omega} \times \mathbf{V}) = (\mathbf{V} \cdot \nabla)\boldsymbol{\omega} - (\boldsymbol{\omega} \cdot \nabla)\mathbf{V} \qquad (8.103)$$

(This can be verified by expanding both sides.) The momentum equation, in terms of vorticity, is then

$$\frac{\partial \boldsymbol{\omega}}{\partial t} + (\mathbf{V} \cdot \nabla)\boldsymbol{\omega} = (\boldsymbol{\omega} \cdot \nabla)\mathbf{V} \qquad (8.104)$$

or

$$\frac{D\boldsymbol{\omega}}{Dt} = (\boldsymbol{\omega} \cdot \nabla)\mathbf{V} \qquad (8.105)$$

This equation is the inviscid counterpart of the vorticity transport equation (5.14).

Equation 8.105 allows us to state that *if an inviscid flow is irrotational*, that is ω is everywhere zero, *then it must remain irrotational*; for if $D\boldsymbol{\omega}/Dt = 0$ everywhere in the flow, then ω at the next instant must be zero and hence ω must remain at the constant zero value. This may be referred to as the *persistence of irrotationality*.

For a two-dimensional plane flow, two vorticity components are zero ($\xi = \eta = 0$), and one velocity component is zero ($w = 0$). Equation 8.105 reduces to

$$\frac{D\zeta}{Dt} = 0 \tag{8.106}$$

allowing us to state that the vorticity of a fluid particle in a plane inviscid flow cannot change as the particle moves along.

Of course, viscosity can create vorticity in an otherwise irrotational flow, and can cause vorticity to change in a plane flow; however, in the foregoing, viscous effects have been neglected.

Returning to Eq. 8.101, if a constant-density flow is restricted to be steady,

$$\nabla\left[\frac{V^2}{2} + \frac{p}{\rho} + gh\right] = \mathbf{V} \times \boldsymbol{\omega} \tag{8.107}$$

Let $V^2/2 + p/\rho + gh = \Phi$; then, from Fig. 8.21, we see that $\nabla\Phi$ is normal to both a streamline and a vortex line. But, we recall that $\nabla\Phi$ is normal to constant-Φ lines; hence we conclude that

$$\frac{V^2}{2} + \frac{p}{\rho} + gh = \text{const} \tag{8.108}$$

along a streamline *or a vortex line* and, more generally, to the plane defined by the family of vortex lines that intersects a family of

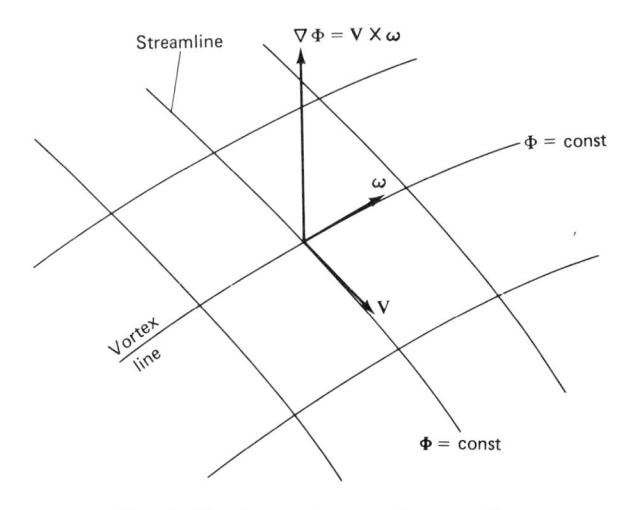

Fig. 8.21. Streamlines and vortex lines.

streamlines. Note that Bernoulli's equation for rotational flows is therefore more restrictive than for irrotational flows where Eq. 8.108 may be applied to all points in the flow.

8.8 HELE-SHAW FLOW

The Hele-Shaw apparatus uses a viscous-dominated flow between parallel plates to provide a visualization technique for the study of inviscid flows. This seeming paradox is at least potentially resolved when one realizes that the viscous effects are dominant for the flow between the plates (with respect to z) whereas the "irrotational" motion is in a direction parallel to the plates. The relationship of the two motions will be shown by the Navier-Stokes equations.

Consider a flow system as shown in Fig. 8.22. The velocity has only two components because of the small and constant spacing between the plates. Hence, the Navier-Stokes equations in the x- and y-directions are

$$u \frac{\partial u}{\partial x} + v \frac{\partial u}{\partial y} = -\frac{1}{\rho}\frac{\partial p}{\partial x} + \nu\left(\frac{\partial^2 u}{\partial x^2} + \frac{\partial^2 u}{\partial y^2} + \frac{\partial^2 u}{\partial z^2}\right) \quad (8.109)$$

$$u \frac{\partial v}{\partial x} + v \frac{\partial v}{\partial y} = -\frac{1}{\rho}\frac{\partial p}{\partial y} + \nu\left(\frac{\partial^2 v}{\partial x^2} + \frac{\partial^2 v}{\partial y^2} + \frac{\partial^2 v}{\partial z^2}\right) \quad (8.110)$$

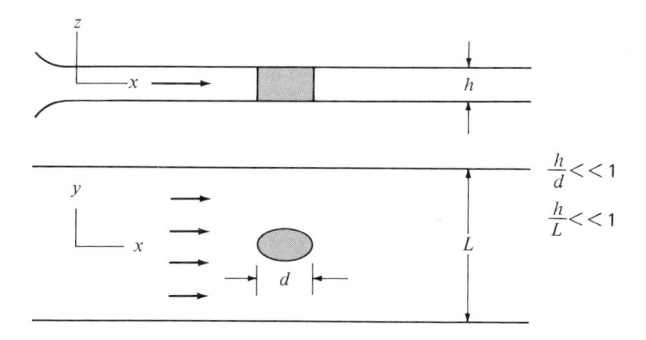

Fig. 8.22. Hele-Shaw apparatus.

For a viscous dominated flow, involving very slow motion, the obvious restrictions and assumptions lead to

$$\frac{\partial p}{\partial x} = \mu \frac{\partial^2 u}{\partial z^2} \quad \text{and} \quad \frac{\partial p}{\partial y} = \mu \frac{\partial^2 v}{\partial z^2} \quad (8.111)$$

Because the z-spacing is small, the second derivative with respect to z

dominates the x- and y-derivatives. These equations may be integrated, giving

$$u = \frac{1}{2\mu} \frac{\partial p}{\partial x} \left[z^2 - \frac{h^2}{4} \right] \tag{8.112}$$

and

$$v = \frac{1}{2\mu} \frac{\partial p}{\partial y} \left[z^2 - \frac{h^2}{4} \right] \tag{8.113}$$

Let $u = U(x, y)$ and $v = V(x, y)$ at $z = 0$; then

$$-\frac{\partial p}{\partial x} = \frac{8\mu}{h^2} U(x, y) \tag{8.114}$$

and

$$-\frac{\partial p}{\partial y} = \frac{8\mu}{h^2} V(x, y) \tag{8.115}$$

Differentiate Eq. 8.114 with respect to y and Eq. 8.115 with respect to x; subtract, and there results

$$\frac{\partial V}{\partial x} - \frac{\partial U}{\partial y} = 0 \tag{8.116}$$

But this is precisely the condition that the lateral flow has zero z-component vorticity since

$$(\nabla \times \mathbf{V})_z = \frac{\partial v}{\partial x} - \frac{\partial u}{\partial y}$$

and since the flow between the plates is in the direction imposed by the U- and V-components, the motion appears to be irrotational in this plane!

Hence, the Hele-Shaw apparatus provides a viscous dominated motion which can be used to visualize an irrotational flow. The apparatus is often simply a large, flat horizontal surface over which a thin layer of water flows at very low speeds. The position where $z = 0$ is then the free surface. Various objects are placed on the table and pellets of potassium permanganate are dropped into the water in front of the object. Potential flow streamlines are then visible.

PROBLEMS

8.1 Give several examples of flows or portions of flows that can be considered inviscid. Also give examples of flows or flow regions for which the inviscid-flow approximation would not be acceptable.

8.2 Include the body-force term due to gravity in the Euler-s equation and integrate along a streamline for an incompressible, steady flow. State some specific problems in which the body-force term would be significant and some in which it would be insignificant.

8.3 A vena contracta, a minimum area, forms at the outlet of a sharp-edged orifice; such an orifice is shown in the sketch. (a) Based on the trajectory which a particle must follow starting from point A, describe why such a vena contracta must occur. (b) Will an air jet exhausting into the atmosphere show a vena contracta?

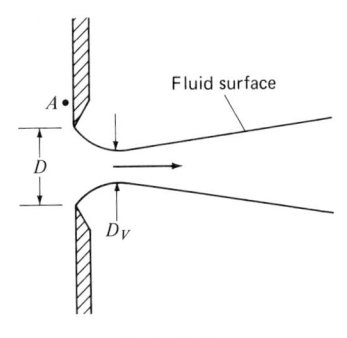

Prob. 8.3

8.4 Describe how a perfume bottle atomizer works, using an appropriate form of Bernoulli's equation.

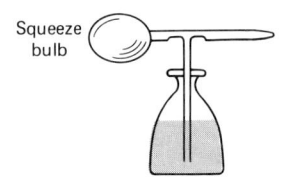

Prob. 8.4

8.5 Explain which way (and why) a right-handed pitcher can most easily throw a curve ball.

8.6 A curved flow passage is shown which involves the pressure distributions indicated in the graph. Using the appropriate forms of the Euler equa-

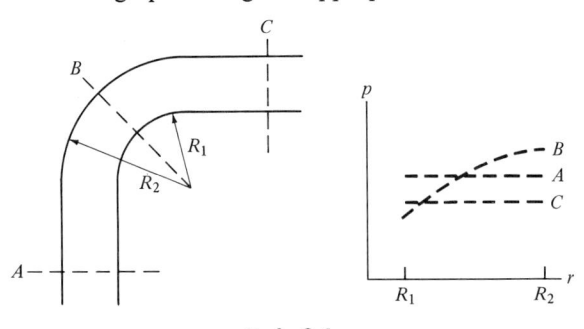

Prob. 8.6

tions, explain the observed form of the pressure distributions. Is the velocity around the bend higher on the inside or the outside?

8.7 Given: the velocity potential $\phi = x^2 - 2xy - y^2$. Can this represent an imcompressible-fluid flow? If so, find the stream function ψ.

8.8 An aircraft is moving at a speed of 300 mph at an elevation where the pressure is 10 psi. Determine the expected pressure at the stagnation point and at a point on the upper surface of the wing where the velocity, from potential-flow theory, is calculated to be 400 mph. Assume incompressible flow for a first approximation.

8.9 Sketch regions of vorticity and regions of zero vorticity for the following: (a) entrance flow in a pipe; (b) flow over two-dimensional airfoil; (c) flow over a sphere, with a separated region; and (d) flow around a Greyhound bus.

8.10 It is proposed that, to find the potential flow of fluid around a cylinder of radius R located in the center of a channel of depth h, the governing equations be solved using a finite difference scheme. State the equation to be solved and the necessary boundary conditions. The flow rate is q ft³/sec/ft. What is the velocity profile upstream from the cylinder?

8.11 Can the stream function $\psi = A(x^2 - y^2)$ be used for a potential flow? If so, sketch the streamline pattern. Also, determine the potential function and sketch several potential lines.

8.12 Show that $\psi = Ar^{\pi/\alpha} \sin(\pi\theta/\alpha)$ represents flow in a corner, as shown. Show also the following: (a) if $\alpha = \pi/2$, stagnation flow results; (b) if $\alpha = \pi$, uniform flow results; and (c) if $\alpha = 2\pi$, flow around a semi-infinite flat plate results. Sketch some streamlines for the flow of part c.

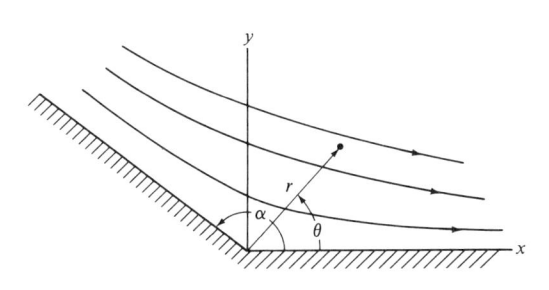

Prob. 8.12

8.13 A velocity potential function is given by $\phi = 10x + 40x/(x^2 + y^2)$. (a) Verify that ϕ satisfies Laplace's equation, $\nabla^2\phi = 0$. (b) Determine the stream function and sketch the streamline corresponding to $\psi = 0$. (c) Find the pressure distribution in the water along the x-axis, assuming the pressure at $x = -\infty$ is 10 psi. (d) Locate any stagnation points.

8.14 Derive the expression for the velocity potential for (a) a plane vortex and (b) a plane doublet.

8.15 Using Eqs. 8.53, show that, for a vortex flow, the velocity vector is normal to a radius vector and hence in the θ-direction. Determine the magnitude of the velocity vector and compare with Eq. 8.48.

8.16 Show that the vorticity of an irrotational vortex is everywhere zero except at the origin, where it is infinite.

8.17 Superimpose a source of strength $q = 8$ at $x = -2$, a sink of equal strength at $x = 2$, and a uniform flow in the x-direction of $U = 10$. Does this represent flow around a closed body? If so, determine the maximum thickness of the body. What is the value of the stream function all along the x-axis and on the body?

8.18 Place a source at $(0, -\epsilon)$ and a sink of equal strength at $(0, \epsilon)$; let ϵ approach 0 and the source and sink strengths approach infinity. Derive an expression for the stream function of a doublet oriented along the y-axis.

8.19 A uniform flow of 20 fps is combined with a doublet of strength $\mu = 80$ ft^3/sec situated at the origin to give flow around a cylinder. (a) Sketch the velocity along the y-axis. (b) Determine the deceleration at the point $(-4, 0)$. (c) Find the force of the water on the front half of the 20-ft-long cylinder if the stagnation-point pressure is 5 psi.

8.20 Air flows around a 2-ft-dia., 100-ft-high pole at 50 fps. Assuming a potential flow over the front half and a separated flow at constant pressure over the rear half, approximate the pressure drag on the pole. See Fig. 8.15b.

8.21 Air flows over the symmetrical plane body shown. Using a flow net, approximate the streamline pattern, assuming no separation. Estimate the minimum pressure on the body if the pressure in the free stream is 15 psi.

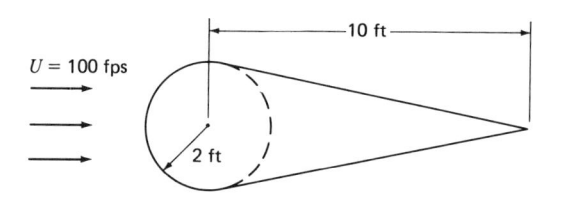

Prob. 8.21

8.22 Sketch the streamline pattern with the use of a flow net for the plane contraction shown. Assume the streamlines are equally spaced downstream of the contraction. Estimate the pressure at the point $(1, 1)$ if the pressure upstream is 20 psi and the flow rate is 100 ft^3/sec/ft of water.

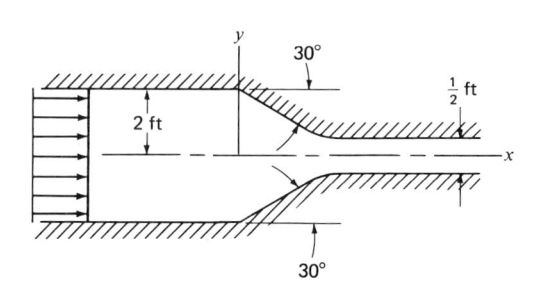

Prob. 8.22

PROBLEMS

8.23 Determine the rotational speed Ω at which a 2-in.-dia. cylinder would have to rotate in a 20-fps free-stream flow so that only one stagnation point would exist on the cylinder. Sketch the streamline pattern.

8.24 A 2-ft.-dia. cylinder is rotated at 400 rpm in the flow moving at 10 fps. Locate the stagnation point (or points), and sketch the streamline pattern. *Hint*: Be careful; the stagnation point may not be on the cylinder.

8.25 Gas is stored in an underground storage and pumped out as needed. The streamline pattern in porous media can be approximated by a potential flow. A well is placed in the plane layer containing the gas, next to two impervious rock layers which intersect at right angles. The well extracts the gas uniformly throughout the layer. For an extraction rate of 5 ft^3/sec/ft, determine the velocity that would be expected along the x-axis. (*Hint*: Use the method images.)

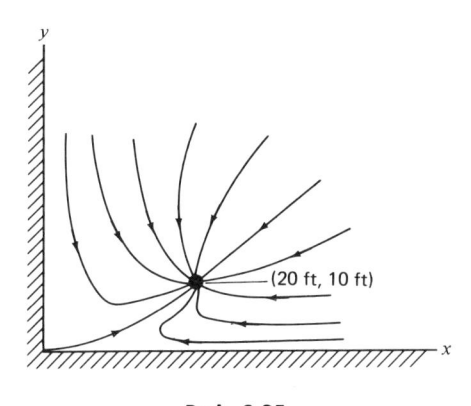

Prob. 8.25

8.26 We wish to determine the potential-flow field around a cylinder in a channel, with the channel height large compared with the diameter of the cylinder. Superimpose a uniform flow and a large number of doublets and show with the use of a digital computer that such a flow results.

8.27 A potential flow into a sink in the end of a channel is desired. Superimpose a large number of sinks to give the desired flow. Determine $u(x)$ along the x-axis.

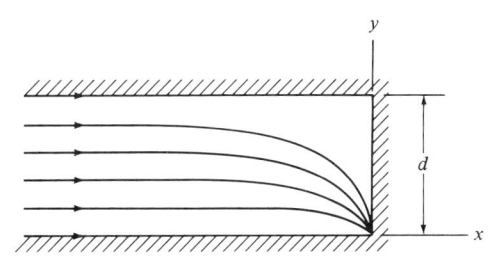

Prob. 8.27

INVISCID FLOWS

8.28 Derive an expression for the stream function for a point source and sketch a streamtube containing two neighboring streamlines. Also sketch a potential surface for the point source.

8.29 Determine the velocity at distances of 2 ft and 2 inches from a point sink if fluid is being withdrawn from a reservoir at a rate of 10 ft^3/sec. The sink is removed from any solid boundaries.

8.30 Determine the cartesian components u, v, w of the velocity vector for a point source. Show that they satisfy the continutiy equation $\partial u/\partial x + \partial v/\partial y + \partial w/\partial z = 0$.

8.31 A point sink is placed 2 ft from an impervious boundary. Determine the velocity distribution along the stagnation line from the sink to the boundary. The extraction rate is 6 ft^3/sec.

8.32 Verify the expression (8.100) for the streamfunction ψ for axisymmetric potential flow around a sphere. Determine the maximum velocity on a 8-in.-dia. sphere placed in a 20-fps uniform flow.

8.33 A point source of strength q is placed in a uniform flow. Determine the y-intercept of the half-body formed, the location of the stagnation point, and the asymptotic dimension of the axisymmetric body. Use $U = 10$ fps and $q = 200$ cfs.

8.34 Air flows over a large spherical weather balloon. Estimate the pressure drag if the flow can be approximated by a potential flow on the front half. The flow separates at the position of lowest pressure, and this low pressure is assumed to exist over the entire rear area of the sphere. See Fig. 8.15b for a similar type of flow. Use $U = 50$ mph, $r_0 = 10$ ft, and $\rho = 0.0024$ slug/ft^3.

8.35 Start with the differential momentum equation written in component form and, using rectangular cartesian coordinates, show that for a two-dimensional plane flow $D\zeta/Dt = 0$. (Differentiate the x-equation with respect to y and the y-equation with respect to x and subtract the resulting equations).

8.36 Using Eq. 8.106, argue that inviscid flow after a two-dimensional contraction cannot exist as shown. What's wrong with it? Assume that the parabolic profile is generated by viscosity but that through the short contraction viscous effects are negligible. Sketch the probable velocity distribution at section 2.

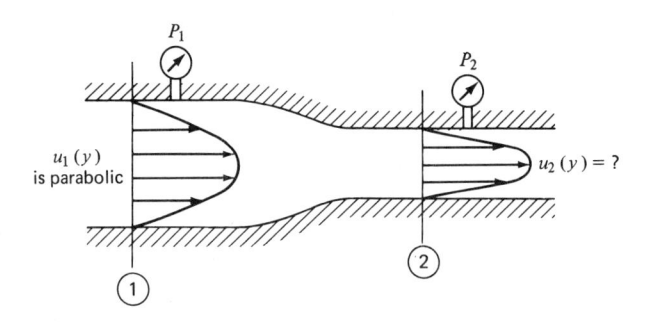

Prob. 8.36

8.37 For the contraction shown in Prob. 8.36, approximate the incoming profile with a linear distribution. At the wall the velocity is zero and at the centerline it is 20 fps. The dimensions before and after the contraction are 4 in. and 2 in., respectively. Determine the velocity distribution at section 2. Also, find the downstream location of the streamline that was 1 in. from the wall at section 1.

SELECTED REFERENCES

The motion picture *Pressure Fields and Fluid Acceleration* (No. 21609; A. H. Shapiro, film principal) is suggested as a supplement to the material in this chapter.

The classical subject of potential flow is considered in most texts. Alternate discussions at about the level of this book are provided by two works: S. Eskinazi, *Principles of Fluid Mechanics*, 2nd ed., Allyn & Bacon, Boston, 1968; and R. H. Sabersky, A. J. Acosta, and E. G. Hauptman, *Fluid Flow*, 2nd ed., Macmillan Co., New York, 1971.

Complex-variable theory may be used to describe the potential motion of a fluid; for a readable discussion of such considerations, see H. R. Vallentine, *Applied Hydrodynamics*, Butterworths Scientific Publications, London, 1959. A text which separates, but sufficiently covers, both the introductory viewpoint and the complex-variable considerations—as well as many applications—is J. M. Robertson, *Hydrodynamics in Theory and Application*, Prentice-Hall, Englewood Cliffs, N.J., 1965. A more mathematically oriented account of this subject is well presented by G. K. Batchelor, *An Introduction to Fluid Dynamics*, Cambridge University Press, London, 1967. A classic in this subject area is L. M. Milne-Thomson, *Theoretical Hydrodynamics*, 4th ed., Macmillan Co., New York, 1960.

9

Boundary-Layer Flows

9.1 INTRODUCTION

Prior to the twentieth century the fluid mechanics community was quite adept at formulating analyses of inviscid flows; it was also able to derive quite satisfactory solutions for creeping flows or conduit flows completely dominated by viscosity. However, a viscous flow around a body, especially a well-streamlined body like an airfoil, could not be modeled satisfactorily. The inviscid, potential flow around an airfoil gives a very good approximation to the real flow pattern and a surprisingly good approximation to the lift, but it predicts a zero drag —a very poor prediction, of course. (The prediction of acceptable lift but zero drag is commonly known as *D'Alembert's paradox*.) Because of this shortcoming, inviscid flows were commonly considered of only academic interest. This was especially true of flows around blunt bodies, where separation occurred. The engineer was forced to rely on experimental work accompanied by empirical formulas.

In 1904, Ludwig Prandtl, a young German engineer, presented a brief paper at a meeting of mathematicians. This paper (which was not accorded widespread recognition or acceptance until the 1920's) not only resolved the D'Alembert paradox, but inaugurated a series of investigations and further developments which have been among the most important contributions in all of fluid dynamics. The calculations of flow over airfoils, around turbine and compressor blades, and in the inlet to conduits are all made feasible by the insights derived from Prandtl's theory. In this theory it is proposed that all the viscous effects are concentrated in a thin layer near the boundary and that outside this layer the fluid behaves as though it is inviscid. *This inviscid external flow is closely approximated by the potential flow around the body, if we completely neglect the boundary layer.* It is only because this viscous layer is thin that this approximation may be made. If the flow separates from the boundary, the potential-flow solution is a good approximation to the actual flow only up to the separated region. Since the viscous layer is thin, the governing equations can be simplified enough to enable us to obtain an acceptable solution. The inviscid-flow solution is used as boundary conditions for the viscous-boundary-layer problem. With the viscous-layer flow properly solved, we can predict the point of separation and calculate the drag on bodies. In this context, boundary-layer theory provides an important ingredient for the solution to the problem of high-Reynolds-number flow around bodies.

Boundary-layer theory is one of the most important areas of study in fluid mechanics. The subject is of recent origin and, hence—not surprisingly—an active focus of considerable current research. Some simple two-dimensional layers are shown in Fig. 9.1. Three-dimensional layers are found on most complicated geometries, such as turbine blades, pump impellers, automobile fenders, and the like. In this chapter, we will investigate only a few simple two-dimensional layers, both laminar and turbulent. However, the methods to be employed are similar to, and serve as the basis for, the methods required to consider more complicated boundary-layer flows. Hence, our limited treatment is appropriate for this introduction to the vast subject area of boundary layers within fluid mechanics. We further categorize a boundary layer as either laminar or turbulent depending upon the absence, or presence, respectively, of significant Reynolds stress effects in the boundary layer.

The boundary layer and its important characteristics are defined and discussed in the following as an introduction to an analysis of this flow field. Since there is no obvious division between the inviscid flow and the viscous layer, an arbitrary definition is employed: *The edge of the boundary layer, with thickness designated by δ, is the locus of the points at which the velocity is 99 percent of the potential-flow wall velocity***; see

---

* This percentage is arbitrary; values such as 0.95 or 0.90 are sometimes used. The value 0.99 is often difficult to distinguish when viewing experimental data because of scatter in the measurements, presence of a nonuniform free-stream velocity, or both. A physically more satisfactory thickness could be based upon a sufficiently small value of the vorticity; this is not a common practice and it would also be difficult to measure.

(a) Plane flows (b) Axisymmetric flows

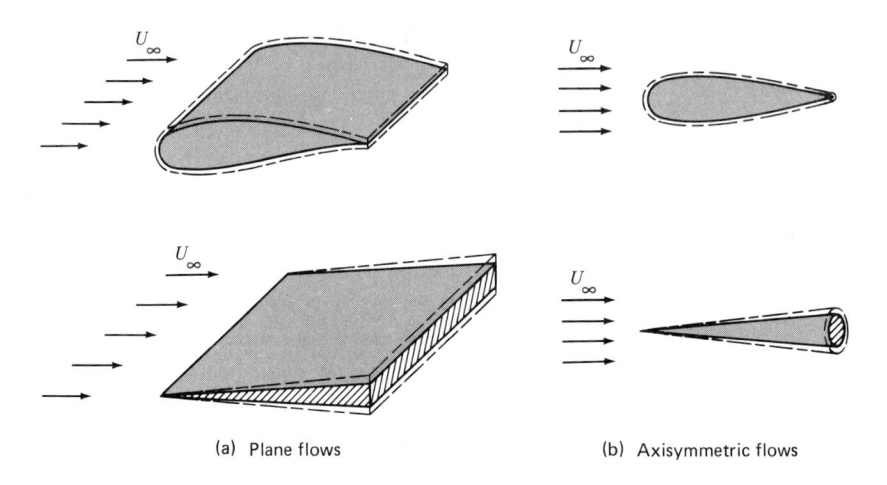

Fig. 9.1. Two-dimensional boundary layers.

Fig. 5.7. The wall potential-flow velocity is usually called the *free-stream velocity*.

At a given x-location at the wall, the boundary-layer velocity distribution $u(y)$ has a slope $\partial u/\partial y$, which is positive. If the pressure is increasing in the x-direction—an *adverse pressure gradient*—the slope $(\partial u/\partial y)_{\text{wall}}$ will decrease with x. If the adverse pressure gradient persists, at some point—the *point of separation*—$(\partial u/\partial y)_{\text{wall}}$ will be zero and the flow will separate from the boundary; see Fig. 9.2. The *separation streamline*, indicated by the solid line, separates the main flow from the separated flow; there is no net flow into or out of the separated region, although turbulent mixing—and, consequently, turbulent stresses—are often quite important in this region of the flow field. (An airfoil is designed so that separation, usually referred to as stall, does not exist. Slots in the wing itself are often used to prevent separation by bleeding off the low-velocity fluid near the surface, or injecting high-velocity air near the surface.)

The boundary layer usually begins as a laminar flow; with sufficient development length it undergoes transition to turbulent flow, as shown diagramatically by Fig. 9.3. This transition to turbulence occurs at a nominal value of x_T, determined from $U_\infty x_T/\nu = 3 \times 10^5$ to 6×10^5 for flow over a smooth flat plate with zero pressure gradient. This value is rather strongly dependent upon the amplitude of fluctuations in the free stream, with the lower limit corresponding to high-amplitude fluctuations (or rough surfaces). Laminar boundary layers have been observed up to Reynolds numbers in excess of $U_\infty x_T/\nu = 10^6$ for extremely small fluctuation levels. The quantity $U_\infty x/\nu$ defines the

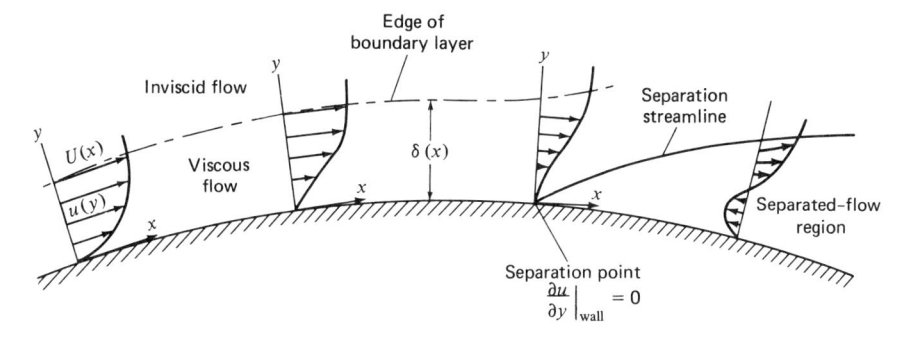

Fig. 9.2. The boundary layer with separation (separated region exaggerated).

local Reynolds number, Re_x, and $U_\infty x_T/\nu$ the critical Reynolds number Re_{crit}. For a smooth plate and a relatively quiescent free stream, the beginning of the transition region is characterized by small-amplitude plane waves becoming unstable and growing; these then break down three-dimensionally and appear as turbulent "bursts" of fluid projected out into the boundary layer from near the wall. As the fluid travels farther downstream the burst rate increases and the whole boundary layer becomes turbulent except for a sublayer adjacent to the wall, which, because of its low speed, remains free of significant turbulent stresses although it does have a strongly fluctuating character. This unsteady viscous sublayer continues to "feed" the turbulent boundary layer with the bursts.

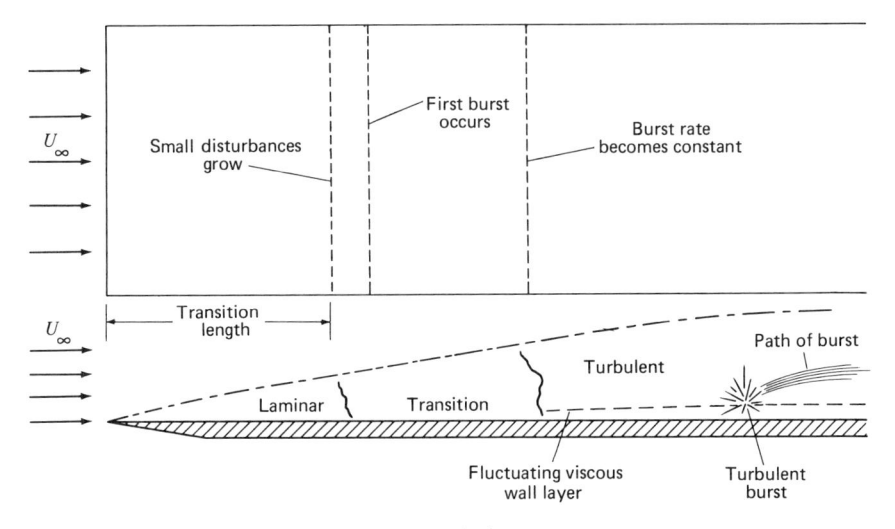

Fig. 9.3. The laminar-to-turbulent transition process.

The turbulent flow is characterized by a wall shearing stress τ_0 much larger than that of a laminar flow. Its time-averaged velocity distribution is shown in Fig. 9.4; If the flow were laminar with the same boundary-layer thickness the distribution would be as shown by the dashed profile. The time-averaged boundary-layer edge for the turbulent flow is identified by δ; the instantaneous edge would be as shown. The turbulent boundary layer thickens much more rapidly than the laminar layer, so that at a particular x-location we may expect a substantially larger value of the boundary-layer thickness δ for the turbulent layer.

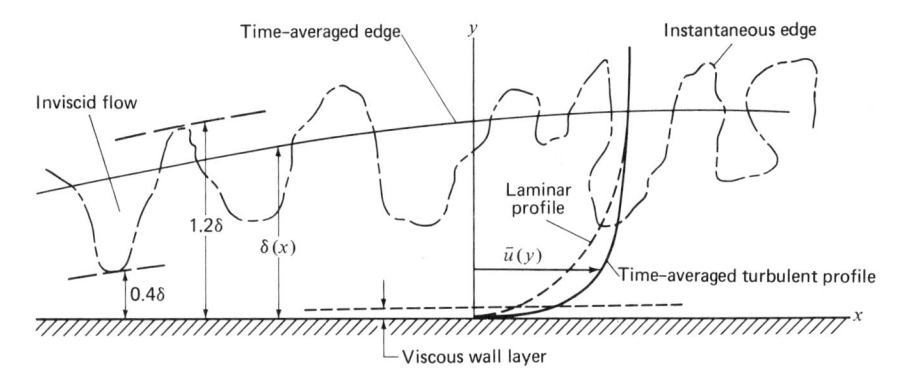

Fig. 9.4. A turbulent boundary layer.

For a given flow, the boundary-layer growth $\delta(x)$, velocity profile $\bar{u}(x, y)$, wall shearing stress $\tau_0(x)$, point of transition, and separation point each depend on several characteristics, in varying combinations. These include the pressure gradient, the amplitude and frequency of the free-stream disturbances, the wall roughness, the wall heat transfer, the wall mass transfer for the case of porous walls or walls with slots, and other features such as magnetic fields, non-Newtonian properties, wall vibrations, and so forth.

A final qualitative consideration in this introductory article is to emphasize the relationship of potential-flow theory to boundary-layer theory. The boundary layer in which the viscous effects are concentrated is very thin when compared with the dimensions of the body over which the fluid is flowing. Hence, we assume the thin boundary layer does not substantially alter the external flow from the potential flow calculated by completely neglecting viscous effects, as was done in Chapter 8. The potential flow provides a necessary velocity boundary condition at the outer edge of the boundary layer; it is the velocity at the boundary of the potential flow. The pressure gradient dp/dx in the boundary layer is also very important. Because the boundary layer is thin the pressure does not vary significantly over the depth of the layer, that is, $\partial p/\partial y \cong 0$ in the boundary layer; hence, the boundary-layer pressure gradient dp/dx is approximately the same as that of potential flow at the boundary.

9.2 THE LAMINAR-BOUNDARY-LAYER EQUATIONS

We will begin the consideration of boundary-layer flows by examining the simplest possible case: a flat plate immersed in an unbounded stream at a zero angle of attack, shown in Fig. 9.3. This flow is often used to approximate real flow situations and thus has been the subject of considerable study. The construction of an appropriately simplified model is first presented. The solution to the resulting equations is presented in the following article.

The information of fluid-mechanical interest is readily provided by the solution to the governing equations. For example, one might be interested in the drag on the flat plate or the velocity at a given distance from the surface. These are obtained by solving the appropriate equations and then obtaining the desired information from the particular solutions. The problems will demonstrate this procedure.

The zero-pressure-gradient solution is the only exact solution that we will consider. Other solutions are available for special pressure-gradient conditions; these are discussed in the references listed at the end of this chapter. For turbulent flows and general pressure-gradient flows, it is necessary to resort to numerical or approximate solution techniques. The approximate technique based upon an integral equation formulation is presented in a later article; both laminar and turbulent flows are considered. It is, of course, possible to confine one's initial study of boundary-layer solutions to the integral formulation.

A boundary layer is considered to be "thin" even though the boundary-layer thickness may be quite large; for example, the boundary-layer thickness associated with the wind blowing inland over New York City may be greater than the height of the Empire State Building. The word "thin" means that the thickness of the boundary layer δ is small compared to the distance L over which it is being developed. This distance might be the length of an airfoil or a submarine or the many miles out in the Atlantic Ocean where the boundary-layer growth starts as the wind blows toward New York City. Figure 9.5, which shows a boundary layer with appropriate lateral and longitudinal scales, demonstrates this property of being thin. We usually sketch a velocity profile, as in Fig. 9.3, in an exaggerated manner; but we must always keep in mind that the boundary layer is very thin compared to the dimensions of the body over which the fluid is flowing. If the free-stream velocity in Fig. 9.5 were increased to 100 fps, the boundary-layer thickness at any location where the boundary layer was still laminar would be decreased by a factor of 10. Drawn to scale in Fig. 9.5, the boundary layer would be hardly noticed! Yet, in that thin viscous layer the velocity must be brought to zero. The velocity gradients must be extremely large to accomplish this. In fact, the gradients are so large on reentering satellites that the viscous effects cause temperatures sufficient to "burn up" all unshielded vehicles. (See the viscous dissipation function Φ of the differential energy equation in Chapter 3.)

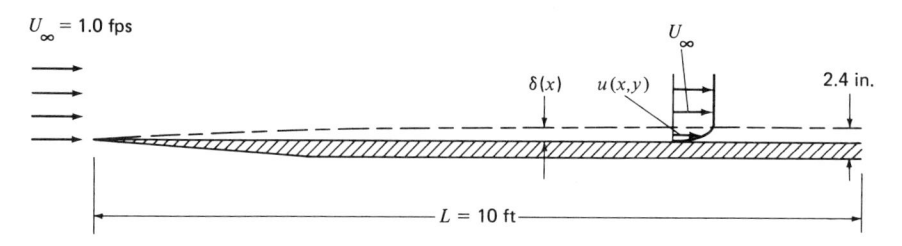

Fig. 9.5. Actual dimensions of boundary growth of air moving over a flat plate.

We normalize, or "nondimensionalize," the equations, using the length L and the uniform flow velocity U_∞. The dimensionless boundary-layer thickness $\delta^* = \delta/L$ is then indeed small. The dimensionless variables are

$$x^* = \frac{x}{L} \quad y^* = \frac{y}{L} \quad u^* = \frac{u}{U_\infty}$$

$$v^* = \frac{v}{U_\infty} \quad p^* = \frac{p}{\rho U_\infty^2} \quad U^* = \frac{U}{U_\infty} \tag{9.1}$$

The Navier-Stokes equations for this two-dimensional, steady flow are then

$$u^* \frac{\partial u^*}{\partial x^*} + v^* \frac{\partial u^*}{\partial y^*} = -\frac{\partial p^*}{\partial x^*} + \frac{1}{\text{Re}} \left(\frac{\partial^2 u^*}{\partial x^{*2}} + \frac{\partial^2 u^*}{\partial y^{*2}} \right) \tag{9.2}$$

$$u^* \frac{\partial v^*}{\partial x^*} + v^* \frac{\partial v^*}{\partial y^*} = -\frac{\partial p^*}{\partial y^*} + \frac{1}{\text{Re}} \left(\frac{\partial^2 v^*}{\partial x^{*2}} + \frac{\partial^2 v^*}{\partial y^{*2}} \right) \quad (9.3)$$

where

$$\text{Re} = \frac{U_\infty L}{\nu} \quad (9.4)$$

These equations may be simplified by an *order-of-magnitude analysis*, the name given to an orderly approximation to the magnitude of the terms in the differential equations. The orders of magnitude, increasing from a small quantity to a large quantity, are

$$\delta^{*2} \ll \delta^* \ll 1 \ll \frac{1}{\delta^*} \ll \frac{1}{\delta^{*2}} \quad (9.5)$$

Think of δ^* as a small number less than 1, say 10^{-2}; then the orders above follow. Figure 9.5 shows the δ-value associated with an air flow over a flat plate. For this condition it is not difficult to accept δ^* as being very small.

The x^*-component of the dimensionless velocity is approximately of order one, so we write

$$u^* = 0(1) \quad (9.6)$$

In the boundary layer (see Fig. 9.6c), u^* changes from 0 to U^* over the thickness δ^*; hence

$$\frac{\partial u^*}{\partial y^*} \cong \frac{\Delta u^*}{\Delta y^*} \cong \frac{1}{\delta^*} \quad \text{or} \quad \frac{\partial u^*}{\partial y^*} = 0\left(\frac{1}{\delta^*}\right) \quad (9.7)$$

since $U^* = 0(1)$. Likewise we see that

$$\frac{\partial^2 u^*}{\partial y^{*2}} \cong \frac{\Delta(\partial u^*/\partial y^*)}{\Delta y^*} \cong \left(\frac{1}{\delta^*}\right)\left(\frac{1}{\delta^*}\right) \quad \text{or} \quad \frac{\partial^2 u^*}{\partial y^{*2}} = 0\left(\frac{1}{\delta^{*2}}\right)$$

$$(9.8)$$

Because u^*, at a fixed y^*, decreases at most to zero over the dimensionless length of unity, we write

$$\frac{\partial u^*}{\partial x^*} \cong \frac{\Delta u^*}{\Delta x^*} \cong \frac{1}{1} \quad \text{or} \quad \frac{\partial u^*}{\partial x^*} = 0(1) \quad (9.9)$$

Likewise

$$\frac{\partial^2 u^*}{\partial x^{*2}} = 0(1) \quad (9.10)$$

408 BOUNDARY-LAYER FLOWS Ch. 9

(a) Dimensional airfoil

(b) Dimensionless airfoil

(c) Dimensionless boundary layer. x^* is always measured along the boundary and y^* is always normal to the boundary

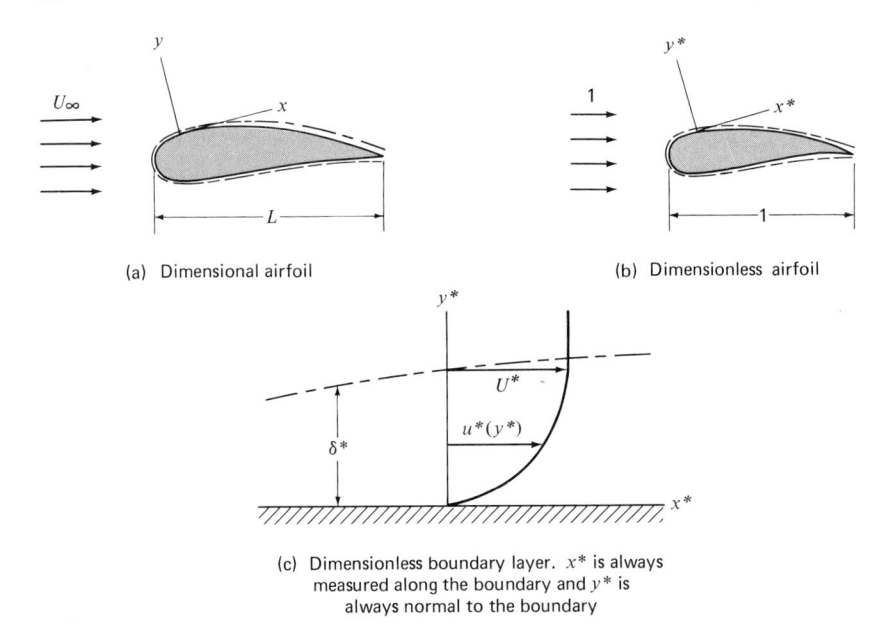

Fig. 9.6. Dimensionless boundary layer.

From continuity,

$$\frac{\partial u^*}{\partial x^*} = -\frac{\partial v^*}{\partial y^*} \qquad (9.11)$$

Integrating to obtain v^* in the boundary layer, there results

$$v^* = \int_0^{y^*} -\frac{\partial u^*}{\partial x^*}\, dy^* \cong 0(1)\,0(\delta^*) = 0(\delta^*) \qquad (9.12)$$

since the limit y^* on the integral must be less than or equal to δ^*. We may now approximate the magnitude of the terms in Eq. 9.2 by

$$\underset{(1)}{u^*}\,\underset{(1)}{\frac{\partial u^*}{\partial x^*}} + \underset{(\delta^*)}{v^*}\,\underset{\left(\frac{1}{\delta^*}\right)}{\frac{\partial u^*}{\partial y^*}} = -\underset{(1)}{\frac{\partial p^*}{\partial x^*}} + \frac{1}{\text{Re}}\left(\underset{(1)}{\overset{\text{neglect}}{\cancel{\frac{\partial^2 u^*}{\partial x^{*2}}}}} + \underset{\left(\frac{1}{\delta^{*2}}\right)}{\frac{\partial^2 u^*}{\partial y^{*2}}}\right) \qquad (9.13)$$

where the order of magnitude is given below the term. The term $\partial p^*/\partial x^*$ is at most of order one; it is found by differentiating Bernoulli's equation, which in dimensionless form is

$$\frac{U^{*2}}{2} + p^* = \text{const} \qquad (9.14)$$

For a flat plate $\partial p^*/\partial x^*$ is zero and thus would not influence the analysis.

In Eq. 9.13 the nonviscous terms are all of order one; hence, if viscous effects are important, the viscous term must be at most of order one (an equation may not have a single term of larger order than all the others). Thus

$$\frac{1}{\text{Re}} \frac{\partial^2 u^*}{\partial y^{*2}} = 0(1) \tag{9.15}$$

which requires that

$$\text{Re} = 0\left(\frac{1}{\delta^{*2}}\right) \gg 1 \tag{9.16}$$

(See Eq. 9.8.) This shows that *a boundary layer exists only at a large Reynolds number*.

Proceeding to the y-component momentum equation, repeating the foregoing analysis, and writing the orders of magnitude below the terms, we have

$$u^* \frac{\partial v^*}{\partial x^*} + v^* \frac{\partial v^*}{\partial y^*} = -\frac{\partial p^*}{\partial y^*} + \frac{1}{\text{Re}}\left(\frac{\partial^2 v^*}{\partial x^{*2}} + \frac{\partial^2 v^*}{\partial y^{*2}}\right) \tag{9.17}$$

(1)   (δ^*)   (δ^*)   (1)   (δ^{*2})   (δ^*) $\left(\frac{1}{\delta^*}\right)$

which shows that $\partial p^*/\partial y^* = 0(\delta^*)$, that is, that the pressure variation across the boundary layer is small. Hence, a good approximation is to neglect $\partial p^*/\partial y^*$, so that $p^* = p^*(x^*)$. This means that the pressure at the outer edge of the boundary layer is the same as the pressure at the wall; thus, we can use the pressure in the boundary layer to be that of inviscid-flow theory at the wall. The inviscid-flow solution is obtained or physical data are used for the pressure distribution; hence we know $U^*(x^*)$, and $\partial p^*/\partial x^*$ (this is usually written as dp^*/dx^* since p^* is only a function of x^*). *These quantities are only functions of x^* since they are evaluated at the solid boundary.*

We thus have two unknowns, u^* and v^*, to be determined from the laminar boundary-layer equations. The equations are often called the *Prandtl boundary-layer equations*; for a two-dimensional boundary layer, they are written as

$$u^* \frac{\partial u^*}{\partial x^*} + v^* \frac{\partial u^*}{\partial y^*} = -\frac{dp^*}{dx^*} + \frac{1}{\text{Re}} \frac{\partial^2 u^*}{\partial y^{*2}} \tag{9.18}$$

$$\frac{\partial u^*}{\partial x^*} + \frac{\partial v^*}{\partial y^*} = 0 \tag{9.19}$$

The boundary-layer equations are still nonlinear and are, in general, very difficult to solve. However, they are much easier to solve than the complete set of equations, which has three unknowns, u, v, and p.

For steady plane flow, the Prandtl boundary-layer equations, in dimensional form, are

$$u\frac{\partial u}{\partial x} + v\frac{\partial u}{\partial y} = -\frac{1}{\rho}\frac{dp}{dx} + \nu\frac{\partial^2 u}{\partial y^2} \tag{9.20}$$

$$\frac{\partial u}{\partial x} + \frac{\partial v}{\partial y} = 0 \tag{9.21}$$

with boundary conditions

$$u = v = 0 \quad \text{at } y = 0$$

$$u = U(x) \quad \text{at } y = \delta \tag{9.22}$$

For the general case, it is often useful to resort to a numerical solution of these equations if the problem at hand is sufficiently important to warrant the considerable effort and expense represented by such a procedure. Note, however, that this could represent considerably less effort and expense than physically constructing and testing numerous airfoil or compressor blade designs. For a flat plate and certain other flows, it is not necessary to resort to a purely numerical solution; rather, a transformation can be found that allows the partial differential equations to be written as ordinary differential equations, which are, of course, much easier to solve. Transformations which accomplish this trick are called *similarity transformations*. We will illustrate a similarity transformation for flow over a flat plate in the next article.

The stream function ψ may be introduced to facilitate the solution. Note that it allows the information of the continuity equation to be introduced into the momentum equation. Using $u = \partial\psi/\partial y$ and $v = -\partial\psi/\partial x$, Eq. 9.21 is identically satisfied, and Eq. 9.20 is, for our steady, incompressible flow,

$$\frac{\partial\psi}{\partial y}\frac{\partial^2\psi}{\partial x\partial y} - \frac{\partial\psi}{\partial x}\frac{\partial^2\psi}{\partial y^2} = -\frac{1}{\rho}\frac{dp}{dx} + \nu\frac{\partial^3\psi}{\partial y^3} \tag{9.23}$$

This is a partial differential equation and requires the following boundary conditions:

$$\frac{\partial\psi}{\partial y} = -\frac{\partial\psi}{\partial x} = 0 \quad \text{at } y = 0$$

$$\frac{\partial\psi}{\partial y} = U(x) \quad \text{at } y \text{ large} \tag{9.24}$$

This equation may be solved for ψ instead of solving the set of Eqs. 9.20 and 9.21 for u and v.

9.3 LAMINAR-BOUNDARY-LAYER FLOW OVER A FLAT PLATE

The boundary-layer equation in terms of the stream function (Eq. 9.23) along with its boundary conditions (Eq. 9.24) may be simplified for flow over a flat plate with zero pressure gradient. Many practical applications can be approximated as such a flow: wind blowing over the earth, flow over a thin airfoil, and flow along the edge of a ship, for example. It also is the simplest boundary-layer flow; hence, many important aspects of the boundary-layer phenomenon may be understood from a thorough grasp of flow over a flat plate.

The boundary-layer equations of Prandtl are a much simpler mathematical description than the full Navier-Stokes equations; there are two equations instead of three, and the pressure is no longer an unknown. However, the momentum equation is still a nonlinear, second-order partial differential equation, with the consequence that general solutions are not usually available. Blasius, a student under Prandtl, was able to manipulate the equations such that the separation of variables resulted in two ordinary differential equations. Blasius then developed a series as an approximate analytical solution to describe the boundary-layer flow. The agreement of the experimental data with Blasius' solution confirmed the assumptions made by Prandtl in the development of the boundary-layer equations. A major contribution of Blasius was the transformation of coordinates which allowed the development of the ordinary differential equations; these considerations are presented below.

The general strategy employed may be described as follows:

1. Introduce two new spatial coordinates of the form

$$\xi = \xi(x, y) \quad \text{and} \quad \eta = \eta(x, y) \qquad (9.25)$$

2. Express the stream function ψ and its derivatives in terms of ξ and η, using

$$\frac{\partial \psi}{\partial x} = \frac{\partial \psi}{\partial \xi}\frac{\partial \xi}{\partial x} + \frac{\partial \psi}{\partial \eta}\frac{\partial \eta}{\partial x} \quad \text{and} \quad \frac{\partial \psi}{\partial y} = \frac{\partial \psi}{\partial \xi}\frac{\partial \xi}{\partial y} + \frac{\partial \psi}{\partial \eta}\frac{\partial \eta}{\partial y} \qquad (9.26)$$

3. Substitute back into Eq. 9.23 and attempt to separate variables as

$$\psi(\xi, \eta) = G(\xi)F(\eta) \qquad (9.27)$$

If the variables do separate and the new boundary conditions can be satisfied, then ordinary differential equations result and a solution is much simpler than would have been necessary for the partial differential equation. This scheme will be demonstrated for flow over a flat plate with zero pressure gradient.

Let the coordinate transformations take the specific form

$$\xi = x \quad \text{and} \quad \eta = y\sqrt{\frac{U_\infty}{\nu x}} \tag{9.28}$$

This choice is based on experience; there is no way to predict with certainty what, if any, transformations will allow an ordinary differential equation to be obtained. The derivatives of the stream function in terms of ξ and η are

$$\frac{\partial \psi}{\partial y} = \sqrt{\frac{U_\infty}{\nu \xi}} \frac{\partial \psi}{\partial \eta} \quad \text{and} \quad \frac{\partial \psi}{\partial x} = -\frac{\eta}{2\xi}\frac{\partial \psi}{\partial \eta} + \frac{\partial \psi}{\partial \xi} \tag{9.29}$$

with similar, more complicated expressions for the higher derivatives. After substituting in Eq. 9.23 and simplifying, there results, with $dp/dx = 0$,

$$-\frac{1}{2\xi}\left(\frac{\partial \psi}{\partial \eta}\right)^2 + \frac{\partial \psi}{\partial \eta}\frac{\partial^2 \psi}{\partial \eta \partial \xi} - \frac{\partial \psi}{\partial \xi}\frac{\partial^2 \psi}{\partial \eta^2} = \nu \frac{\partial^3 \psi}{\partial \eta^3}\sqrt{\frac{U_\infty}{\nu \xi}} \tag{9.30}$$

Assuming the η-dependence and the ξ-dependence separate, as shown in Eq. 9.27, there results

$$-\frac{1}{2\xi}GF'^2 + F'(F'G') - G'FF'' = F'''\sqrt{\frac{U_\infty \nu}{\xi}} \tag{9.31}$$

We observe that if we let

$$G = \sqrt{U_\infty \nu \xi} \tag{9.32}$$

then Eq. 9.31 indeed separates, that is, Eq. 9.32 expresses G as a function only of ξ and Eq. 9.31 reduces to

$$FF'' + 2F''' = 0 \tag{9.33}$$

which expresses F as only a function of η.

The boundary conditions (Eq. 9.24) become

$$\frac{\partial \psi}{\partial \eta} = \frac{\partial \psi}{\partial \xi} = 0 \quad \text{at } \eta = 0 \quad \text{and} \quad \frac{\partial \psi}{\partial \eta} = \sqrt{\nu U_\infty \xi} \quad \text{for } \eta \text{ large} \tag{9.34}$$

Art. 9.3 LAMINAR-BOUNDARY-LAYER FLOW OVER A FLAT PLATE 413

or, in terms of F,

$$F' = F = 0 \quad \text{at } \eta = 0 \quad \text{and} \quad F' = 1 \quad \text{for } \eta \text{ large} \quad (9.35)$$

The value for "η large" will emerge in the solution to Eq. 9.33. The stream function, finally, is *(INTEGRATING 9.34)*

$$\psi = \sqrt{U_\infty \nu \xi}\, F(\eta) \quad (9.36)$$

FUNCTION OF η AND ξ

resulting in velocity components

$$u = \frac{\partial \psi}{\partial y} = U_\infty F' \quad (9.37)$$

and

$$v = -\frac{\partial \psi}{\partial x} = \frac{1}{2}\sqrt{\frac{\nu U_\infty}{x}}\,(\eta F' - F) \quad (9.38)$$

We see that u is a function only of η. With the foregoing transformation of coordinates and separation of variables, a *similarity* solution for u has been found: that is, it depends on only *one* variable. Only special types of boundary-layer flows allow such a solution. It is of interest to note that the velocity component v depends on both x and η.

In 1908, using a series expansion, Blasius found the solution to Eq. 9.33. A more accurate numerical solution is presented in Table 9.1. We see from the table that the point where u is 99 percent of the free-stream velocity is approximately $\eta = 5$. This is, by definition, the edge

TABLE 9.1
Solution of the Boundary-Layer Equations for Flow over a Flat Plate with Zero Pressure Gradient *HELPS YOU CALCULATE*

CHANGE IN u WITH RESPECT TO y

$\eta = y\sqrt{\dfrac{U_\infty}{\nu x}}$	F	$F' = u/U_\infty$	$\tfrac{1}{2}(\eta F' - F)$	F''
0	0	0	0	0.3321
1	0.1656	0.3298	0.0821	0.3230
2	0.6500	0.6298	0.3005	0.2668
3	1.397	0.8461	0.5708	0.1614
4	2.306	0.9555	0.7581	0.0642
5	3.283	0.9916	0.8379	0.0159
6	4.280	0.9990	0.8572	0.0024
7	5.279	0.9999	0.8604	0.0002
8	6.279	1.0000	0.8605	0.0000

of the boundary layer, $y = \delta$. Hence, from Eq. 9.28,

$$\delta = 5\sqrt{\frac{\nu x}{U_\infty}} \tag{9.39}$$

Using $\frac{\partial u}{\partial y} = \frac{\partial u}{\partial \eta}\frac{\partial \eta}{\partial y} = U_\infty F'' \underbrace{\frac{\partial \eta}{\partial y}}_{\text{GIVES SLOPE OF THE CURVE}}$, the shearing stress at the wall is

$$\tau_0 = \mu \frac{\partial u}{\partial y}\bigg|_{\text{wall}} = \underbrace{0.332\mu U_\infty}_{F''(0)}\sqrt{\frac{U_\infty}{\nu x}} = 0.332\rho U_\infty\sqrt{\frac{\nu U_\infty}{x}} \tag{9.40}$$

The shearing stress may be non-dimensionalized by dividing τ_0 by $\frac{1}{2}\rho U_\infty^2$. This dimensionless shearing stress is the *local skin friction coefficient*, given by

$$c_f = \frac{\tau_0}{\frac{1}{2}\rho U_\infty^2} = \frac{0.664}{\sqrt{U_\infty x/\nu}} \longrightarrow \text{NONDIMENSIONAL } x \tag{9.41}$$

The drag on one side of the flat plate of length L, per unit width, is

$$\text{Drag} = \int_0^L \tau_0\, dx = \frac{1.33\rho U_\infty^2 L}{2\sqrt{U_\infty L/\nu}} \tag{9.42}$$

The dimensionless drag, the *skin friction coefficient*, is then

$$C_f = \frac{\text{Drag}}{\frac{1}{2}\rho u_\infty^2 L} = \frac{1.33}{\sqrt{U_\infty L/\nu}} \tag{9.43}$$

The *displacement thickness* δ_d and *momentum thickness* θ, quantities which are often used with the description of a boundary layer, may also be extracted from the results presented in Table 9.1. By definition

WELL DEFINED
$$\begin{cases} \delta_d = \int_0^\delta \left(1 - \frac{u}{U_\infty}\right) dy \\ \theta = \int_0^\delta \frac{u}{U_\infty}\left(1 - \frac{u}{U_\infty}\right) dy \end{cases} \tag{9.44}$$

In terms of the new variables these may be written as

$$\delta_d = \int_0^\delta (1 - F')\sqrt{\frac{\nu\xi}{U_\infty}}\, d\eta$$

$$\theta = \int_0^\delta F'(1 - F')\sqrt{\frac{\nu\xi}{U_\infty}}\, d\eta \tag{9.45}$$

Integration yields the results

$$\delta_d = \sqrt{\frac{\nu x}{U_\infty}} \, [\eta - F]_0^5 = 1.72\sqrt{\frac{\nu x}{U_\infty}}$$

$$\theta = 0.664\sqrt{\frac{\nu x}{U_\infty}} \tag{9.46}$$

The displacement thickness is observed to be approximately one third the boundary-layer thickness.

The preceding development is restricted to laminar flow, but a boundary-layer flow will undergo transition to turbulence if L is sufficiently large: hence, in the drag, shearing-stress, and boundary-layer-thickness relationships the magnitude of L and x must be less than that of the transition length. The transition length for flow over a smooth flat plate with zero pressure gradient is 600,000 $\nu/U_\infty \gtrsim x_T \gtrsim 300{,}000\nu/U_\infty$ with the smaller value corresponding to a high free stream disturbance level.

Another comment should be made here. The relationship for shearing stress indicates that $\tau_0 \to \infty$ as $x \to 0$. We know that the shearing stress will not actually be infinite at $x = 0$ so the relationship (Eq. 9.40) must not be valid for very small x; and indeed, it is not. The boundary-layer assumptions are not applicable at small x, for example, $\partial u/\partial x \neq O(1)$; there are noticeable deviations with experimental results for $\mathrm{Re}_x = U_\infty x/\nu < 10^4$. However, the relationship for the drag is very close to that measured experimentally provided that $\mathrm{Re}_L = U_\infty L/\nu > 10^4$.

9.4 CONTROL-VOLUME EQUATIONS FOR THE BOUNDARY LAYER

The integral equation for the boundary layer, which allows for an approximate solution for both laminar and turbulent flows, will now be presented. The control volume is shown in Fig. 9.7 along with the mass and momentum balances. Applying the conservation of mass for steady flow, we find from Fig. 9.7b that the mass entering the top of the control volume ($\dot{m}$ represents the mass flux) is

$$\begin{aligned}\dot{m}_{\mathrm{top}} &= \dot{m}_{\mathrm{out}} - \dot{m}_{\mathrm{in}} \\ &= \frac{\partial}{\partial x}\left[\int_0^\delta \rho u\, dy\right] dx\end{aligned} \tag{9.47}$$

considering a unit of depth into the paper. From parts c and d of Fig.

BOUNDARY-LAYER FLOWS

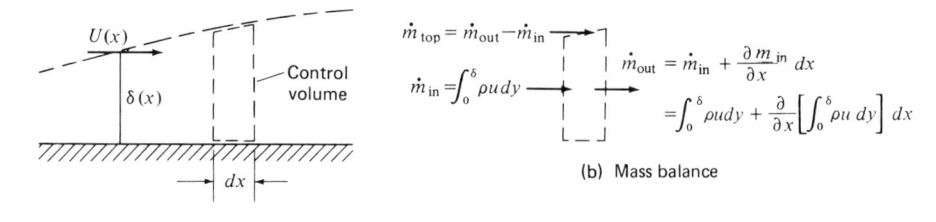

Fig. 9.7. Control volume for a plane boundary layer.

(a) Control volume

(b) Mass balance
$$\dot{m}_{top} = \dot{m}_{out} - \dot{m}_{in}$$
$$\dot{m}_{in} = \int_0^\delta \rho u \, dy$$
$$\dot{m}_{out} = \dot{m}_{in} + \frac{\partial \dot{m}_{in}}{\partial x} dx = \int_0^\delta \rho u \, dy + \frac{\partial}{\partial x}\left[\int_0^\delta \rho u \, dy\right] dx$$

(c) Forces

(d) Momentum flux
$$\dot{mom}_{top} = \dot{m}_{top} U(x)$$
$$\dot{mom}_{in} = \int_0^\delta \rho u^2 \, dy$$
$$\dot{mom}_{out} = \dot{mom}_{in} + \frac{\partial \dot{mom}_{in}}{\partial x} dx = \int_0^\delta \rho u^2 \, dy + \frac{\partial}{\partial x}\left[\int_0^\delta \rho u^2 \, dy\right] dx$$

9.7 we see that the x-component of the momentum integral equation yields

$$\Sigma F_x = \dot{mom}_{out} - \dot{mom}_{in} - \dot{mom}_{top} \qquad (9.48)$$

where móm represents the momentum flux in the x-direction.

Neglecting higher-order terms, this may be put in the form

$$-\delta \frac{\partial p}{\partial x} dx - \tau_0 dx = \frac{\partial}{\partial x}\left[\int_0^\delta \rho u^2 \, dy\right] dx - U(x) \frac{\partial}{\partial x}\left[\int_0^\delta \rho u \, dy\right] dx \qquad (9.49)$$

(PRESSURE GRADIENT FORCE) (SHEAR FORCE)

where the last term accounts for the x-component momentum of the fluid which enters through the top of the control volume. $U(x)$ is the velocity at the edge of the boundary layer; τ_0 is the shearing stress at the wall; and p is the pressure, which changes with x but not across the thin layer. Dividing both sides by dx, and noting that the integrals on the right-hand side and the pressure are functions of x only (the y-dependence integrates out), Eq. 9.49 becomes

$$-\delta \frac{dp}{dx} - \tau_0 = \frac{d}{dx}\int_0^\delta \rho u^2 \, dy - U \frac{d}{dx}\int_0^\delta \rho u \, dy \qquad (9.50)$$

Art. 9.4   CONTROL-VOLUME EQUATIONS FOR BOUNDARY LAYER

Assuming constant density, this may be written as [FREE STREAM VELOCITY]

$$\tau_0 = -\delta \frac{dp}{dx} + \rho \frac{d}{dx}\int_0^\delta u(U-u)\,dy - \rho\left(\frac{dU}{dx}\right)\int_0^\delta u\,dy \quad (9.51)$$

[TIED WITH]

usually referred to as the *von Kármán integral equation*. It simply relates the shearing stress at the wall to the velocity field and the wall pressure distribution.

The momentum thickness and displacement thickness are respectively [MOMENTUM THICKNESS]

$$\theta = \frac{1}{U^2}\int_0^\delta u(U-u)\,dy \quad (9.52)$$

and [DISPLACEMENT THICKNESS]

$$\delta_d = \frac{1}{U}\int_0^\delta (U-u)\,dy \quad (9.53)$$

Then, von Kármán's integral equation can be manipulated to yield

$$\tau_0 = \rho\frac{d}{dx}(U^2\theta) + \rho\delta_d U\frac{dU}{dx} \quad (9.54)$$

[$\tau_0 = \rho U^2 \frac{d\theta}{dx}$ FOR 2.P.G.]   [=0 FOR A 2.P.G]

where we have used Bernoulli's equation, $U^2/2 + p/\rho = $ constant, to write

$$-\delta\frac{dp}{dx} = \rho\frac{dU}{dx}\int_0^\delta U\,dy \quad (9.55)$$

(See Problem 9.11.) It is possible to give a physical description to the displacement thickness δ_d. See Prob. 9.9 for such a description.

It should be emphasized here that the $U(x)$ at the outer edge of the boundary layer and $p(x)$ in the boundary layer are, as a first approximation, the wall velocity and pressure distributions found from potential flow.

If $U = $ const then $dp/dx = 0$, and Eq. 9.54 would reduce to

$$\tau_0 = \rho U^2 \frac{d\theta}{dx} \quad (9.56)$$

This would be the situation for a uniform flow over a flat plate.

An assessment of the foregoing assumptions indicates that the equations are valid for laminar *or* turbulent flow. Also, at the wall the flow is always viscous (even if the boundary layer is turbulent), so that

$$\tau_0 = \mu\left.\frac{\partial u}{\partial y}\right|_{\text{wall}} \quad (9.57)$$

where τ_0 and $\partial u/\partial y$ are functions of time if the wall layer is fluctuating. A *local skin friction coefficient* is often introduced, defined by

$$c_f = \frac{\tau_0}{\frac{1}{2}\rho U_\infty^2} \tag{9.58}$$

The von Kármán integral equation (9.51) may be considered to be a first-order equation for $\delta(x)$. The velocity field $u(x\,y)$ must be known to evaluate the integrals, and a suitable expression for τ_0 must be obtained to solve for $\delta(x)$. Conversely, if $\delta(x)$ were known along with $p(x)$ and a reasonable approximation for $u(x,y)$, the shearing stress τ_0 could be determined. Although, this equation may be considered to be the functional relationship between these several flow variables, its primary use is to calculate $\delta(x)$ or $\tau_0(x)$. The requirement of the velocity field to evaluate the terms of Eq. 9.51 is, upon consideration, rather strange since one would expect that, if the velocity field $u(x,y)$ could be calculated, $\delta(x)$ would be defined and the pressure and shear-force values would be available from $\tau_0 = \mu(\partial u/\partial y)_{\text{wall}}$ and $p(x)$ from the Bernoulli equation, using the velocity values outside the boundary layer. The contribution of Eq. 9.51 is that it allows $\delta(x)$ to be calculated from a postulated $u(x,y)$. This procedure is effective because the $\delta(x)$, so calculated, is relatively insensitive to the assumed $u(x,y)$. This will be shown by an example. An approximate solution for a zero-pressure-gradient flow (see Fig. 9.8) may be compared with an analytical solution. The agreement achieved for this case encourages the extension of the technique to other cases for which an analytical solution is difficult, laminar flows with arbitrary pressure gradients or turbulent boundary-layer flows.

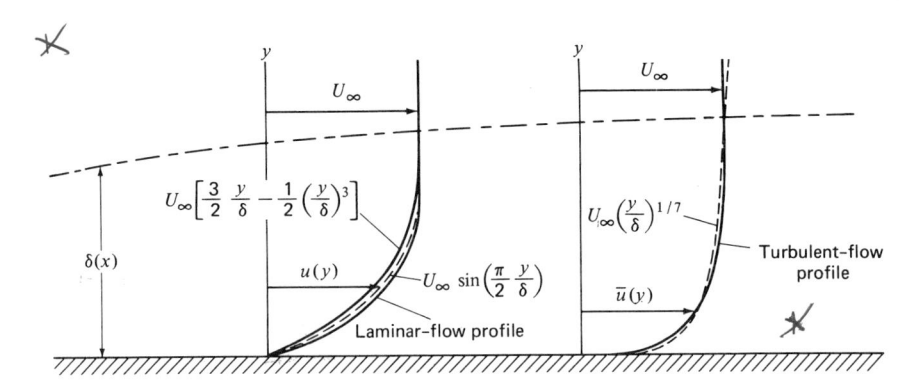

Fig. 9.8. Some approximate velocity distributions for flow over a flat plate.

Art. 9.4 CONTROL-VOLUME EQUATIONS FOR BOUNDARY LAYER

Example 9.1

Consider an incompressible boundary-layer flow with U_∞ = constant. Assuming a velocity distribution $u = U_\infty(2y/\delta - y^2/\delta^2)$, determine the wall shearing stress $\tau_0(x)$ and the boundary-layer thickness $\delta(x)$. (A parabolic velocity profile is an approximation to a laminar boundary-layer profile.)

Solution. If U = constant, Bernoulli's equation yields $dp/dx = 0$. Then, from Eq. 9.51,

$$\tau_0 = \rho \frac{d}{dx} \int_0^\delta u(U_\infty - u)\, dy$$

$$= \rho \frac{d}{dx} \int_0^\delta U_\infty^2 \left(\frac{2y}{\delta} - \frac{y^2}{\delta^2} \right)\left(1 - \frac{2y}{\delta} + \frac{y^2}{\delta^2} \right) dy$$

$$= \rho U_\infty^2 \frac{d}{dx}\left(\frac{2\delta}{15} \right)$$

$$= 0.133\, \rho U_\infty^2 \frac{d\delta}{dx} \qquad (E9.1)$$

But Eq. 9.57 allows us to equate this to $\mu(\partial u/\partial y)_{\text{wall}}$ giving

$$\mu U_\infty \left(\frac{2}{\delta} - \cancel{\frac{2\cancel{y}}{\delta}}^{0} \right) = 0.133\rho U_\infty^2 \frac{d\delta}{dx}$$

or

$$\frac{\mu\, dx}{0.0665\rho U_\infty} = \delta\, d\delta$$

Using $\delta = 0$ at $x = 0$, this integrates to

$$\frac{\delta^2}{2} = \frac{\mu x}{0.0665\rho U_\infty}$$

or

$$\boxed{\delta = 5.48 \sqrt{\frac{\mu x}{\rho U_\infty}}}$$

→ SIMILAR TO BLASIUS BUT
BLASIUS COEFF. = 5

Substitute this back into Eq. E9.1 and obtain

$$\tau_0 = 0.133\rho U_\infty^2\, 2.74 \sqrt{\frac{\mu}{\rho U_\infty x}}$$

$$= 0.365\rho U_\infty^2 \sqrt{\frac{\nu}{U_\infty x}}$$

420 BOUNDARY-LAYER FLOWS Ch. 9

These compare favorably with the more exact solutions for a laminar boundary layer (as determined in Art. 9.3), which are

$$\delta = 5.0\sqrt{\frac{\nu x}{U_\infty}} \quad \text{and} \quad \tau_0 = 0.332\rho U_\infty^2 \sqrt{\frac{\nu}{U_\infty x}}$$

Extension 9.1.1. Starting with a second-order polynomial $u(y) = A + By + Cy^2$, state the three conditions on the velocity field which are sufficient to evaluate A, B, and C. Then show that the form given in Example 9.1 results.

Extension 9.1.2. Assume that the boundary layer over a section of the tail of a commercial jetliner is laminar, and may be approximated by flow over a flat plate. Determine the maximum thickness of the layer if the jet is traveling at 30,000-ft altitude at 480 mph. If the section were 2 ft long, would the assumption of laminar flow be tenable?   *Ans.*   0.0596 in.; no

Example 9.2

Find an expression for the shear force exerted on the two parallel walls of a wide channel of length L (Fig. E9.2) in terms of θ and δ_d. The fluid enters the channel with uniform velocity U_0 and the length of the channel is less than the development length.

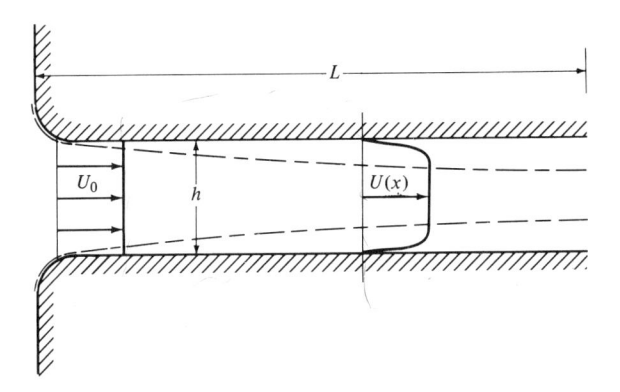

Fig. E9.2

Solution. The integral continuity equation gives

$$hU_0 = 2\int_0^\delta u \, dy + U(x)[h - 2\delta]$$

We may write

$$\delta U(x) = \int_0^\delta U(x) \, dy$$

since in the integration process $U(x)$ is independent of y. The continuity

Art. 9.4 CONTROL-VOLUME EQUATIONS FOR BOUNDARY LAYER

equation may then be written as

$$hU_0 = Uh - 2\int_0^\delta (U - u)\, dy$$
$$= Uh - 2\delta_d U$$

The velocity outside the boundary layer at any section is then

$$U(x) = \frac{U_0}{1 - 2\delta_d/h}$$

The shear force is found by integrating the shearing stress; it is, per unit of width,

$$F = 2\int_0^L \tau_0\, dx$$

From Eq. 9.54 this becomes

$$F = 2\int_0^L \rho \frac{d}{dx}(U^2\theta)\, dx + 2\int_0^L \rho \delta_d U \frac{dU}{dx}\, dx$$

For the last integral we need dU/dx; it is found by differentiating the expression for $U(x)$:

$$\frac{dU}{dx} = \frac{2U_0}{h}\left(1 - \frac{2\delta_d}{h}\right)^{-2} \frac{d\delta_d}{dx}$$

Hence,

$$F = 2\rho U_e^2 \theta_e + \frac{4\rho U_0^2}{h}\int_0^{(\delta_d)_e} \delta_d \left(1 - \frac{2\delta_d}{h}\right)^{-3} d\delta_d$$

where the subscript e refers to the exit quantities. The integral above may be integrated by parts, so that

$$F = 2\rho \left\{ U_e^2 \theta_e + U_0^2 \frac{(\delta_d)_e^2}{h}\left[1 - \frac{2(\delta_d)_e}{h}\right]^{-2} \right\}$$

For a short channel, $(\delta_d)_e/h \ll 1$, and

$$F \cong 2\rho U_e^2 \theta_e$$

Extension 9.2.1. Relate the pressure gradient in the channel to the displacement thickness $\delta_d(x)$.   *Ans.* $\dfrac{2\rho U_0^2 h^2}{(2\delta_d - h)^3}\dfrac{d\delta_d}{dx}$

Extension 9.2.2. To solve the example problem numerically a velocity distribution of the form $u(x, y) = U(x)(A + By + Cy^2 + Dy^3)$ could be assumed. Evaluate the quantities A, B, C, and D by using these conditions: (1)

$u = 0$ at y = 0; (2) $\frac{\partial u}{\partial y} = 0$ at $y = \delta$; (3) $u = U(x)$ at $y = \delta$; and (4) at $y = 0$, the pressure gradient balances the viscous stress, that is, $dp/dx = \mu(\partial^2 u/\partial y^2)_{y=0}$. Use $dp/dx = -\rho U dU/dx$.

9.5 TURBULENT-BOUNDARY-LAYER FLOW OVER A FLAT PLATE

The turbulent boundary layer is, perhaps, the most important of the generalized technologically important flows; it has certainly received the most attention. Turbulent boundary layers occur in slicing and hooking golf balls, on airfoils and turbine blades, and in your trachea (windpipe). They are more complicated than the fully developed pipe flow because the nonlinear convective acceleration terms are present; in general, there is also a pressure gradient. The zero-pressure-gradient case has been extensively investigated in terms of its turbulence structure, and data like the data previously presented for the pipe are available for the zero-pressure-gradient boundary layer.*

Several aspects of the turbulent-boundary-layer structure are presented here. They serve as examples of the characteristics revealed by experimental research as well as to define the important physical characteristics of a boundary layer. The mean velocity profile $\bar{u}(x, y)$ in the boundary layer may be divided into three zones, each of which exhibits a *self-similar* or *self-preserving* character. The terms "self-similar" and "self-preserving" both mean that a dimensionless quantity depends on only one lateral or transverse dimensionless variable. Hence, the distribution preserves its normalized shape as the flow proceeds downstream. An example of this is the $u(\eta)$ velocity profile, as discussed for the laminar boundary layer of Art. 9.3.

The self-similar velocity distribution in the inner region, shown in Fig. 9.9a, can be represented by

$$\frac{\bar{u}}{u_\tau} = f(u_\tau y/\nu) \qquad (9.59)$$

where the friction velocity $u_\tau = \sqrt{\tau_0/\rho}$ is again introduced. The relationship (9.59) is often referred to as the *law of the wall*. The velocity profile is self-similar in that $\bar{u}^* = \bar{u}^*(y^*)$. Note that the characteristic

---

* There is a wealth of data and interpretations available in the volumes from a conference devoted to the calculation procedures for turbulent boundary layers. See "Computation of Turbulent Boundary Layers" (1968 AFOSR IFP–Stanford Conference), Thermosciences Division, Stanford University.

Art. 9.5　TURBULENT-BOUNDARY-LAYER FLOW OVER A FLAT PLATE

velocity u_τ and the characteristic length ν/u_τ used in this turbulent-boundary-layer normalization depend upon the velocity gradient at the wall, since $\tau_0 = \mu(\partial u/\partial y)_{y=0}$. The self-similar zone which extends from the wall to $y^* \cong 5$ is the *viscous wall layer*, in which

$$\frac{\bar{u}}{u_\tau} = \frac{u_\tau y}{\nu} \tag{9.60}$$

The viscous stresses dominate the Reynolds stresses in the wall layer and the time-averaged linear velocity profile results in a constant average shearing stress; however, the time-dependent velocity profile is a randomly fluctuating profile. It is not turbulent even though random fluctuations do occur since the Reynolds stress, $\overline{\rho u' v'}$, is a negligible fraction of the net shear stress. The wall layer is a time-dependent laminar layer.

The *turbulent zone*, which extends from $y^* = 50$ to $y^* \cong 1000$, is identified with the average profile

$$\frac{\bar{u}}{u_\tau} = 2.44 \ln \frac{u_\tau y}{\nu} + 4.9 \tag{9.61}$$

In this self-similar zone, if the average shearing stress is assumed to be constant and the mixing length (see Example 7.2) assumed to be given by $l_M = ky$, then

$$\bar{\tau} = \rho l_M^2 \left(\frac{\partial \bar{u}}{\partial y} \right)^2$$

$$= \rho k^2 y^2 \left(\frac{\partial \bar{u}}{\partial y} \right)^2 = \text{const} \tag{9.62}$$

This can be solved to give

$$\bar{u} = c_1 \ln y + c_2 \tag{9.63}$$

or, introducing the normalizing quantities,

$$\frac{\bar{u}}{u_\tau} = c_3 \ln \frac{u_\tau y}{\nu} + c_4 \tag{9.64}$$

Experimental data agree with this expression if $c_3 = 2.44$ and $c_4 = 4.9$.

The viscous wall layer and turbulent zone are connected by a *buffer zone*, in which the experimental data do not agree well with either Eq. 9.61 or 9.60.

The *outer region* is characterized by the self-similar relationship

$$\frac{U_\infty - \bar{u}}{u_\tau} = f\left(\frac{y}{\delta}\right) \tag{9.65}$$

This is often referred to as the *velocity-defect law*. The part of this region nearer the wall is the turbulent zone; in terms of the velocity defect the velocity profile in the turbulent zone is

VELOCITY DIFFERENCE

$$\frac{U_\infty - \bar{u}}{u_\tau} = -2.44 \ln \frac{y}{\delta} + 2.5 \tag{9.66}$$

The third self-similar zone occurs in the outer portion of the boundary layer. It is identified by the profile[†]

$$\frac{U_\infty - \bar{u}}{u_\tau} = 9.6\left(1 - \frac{y}{\delta}\right)^2 \tag{9.67}$$

These relationships for the outer region are applicable for smooth or rough walls.

It is noted that for large $(u_\tau y/\nu)$ values the data in Fig. 9.9a are not self-similar (they are dependent on y^* and x^*); however, when the appropriate characteristic values are used a self-similar form results, as illustrated in Fig. 9.9b.

Although it is a useful relation for certain calculations, as will be shown in the following article, the simple power law relation of Fig. 9.9c has no theoretical basis; it does not fit the inner profile at all well (e.g., $\partial \bar{u}/\partial y|_{y=0} = \infty$) and the exponent n is a function of the local Reynolds number.

Various expressions for the local skin friction coefficient ($c_f = \tau_0/\frac{1}{2}\rho U_\infty^2$) exist for flow over a smooth flat plate with zero pressure gradient. All of the velocity-profile expressions have included τ_0 through the friction velocity $u_\tau = \sqrt{\tau_0/\rho}$; however, it is impossible to extract an explicit relationship for τ_0. Using $\tau_0 = \mu(\partial \bar{u}/\partial y)_{\text{wall}}$, for example, the viscous-wall-layer profile produces the identity $\tau_0 \equiv \tau_0$. Using $\bar{u} = U_\infty$ at $y = \delta$, relationship 9.67 produces $0 \equiv 0$. We are left with the velocity-profile equation (Eq. 9.61) for the turbulent zone, which is not applicable at the wall or the outer edge of the boundary

---

[†] Some authors prefer $(U_\infty - \bar{u})/u_\tau = -3.8 \ln (y/\delta)$ instead of Eq. 9.67.

Art. 9.5   TURBULENT-BOUNDARY-LAYER FLOW OVER A FLAT PLATE   425

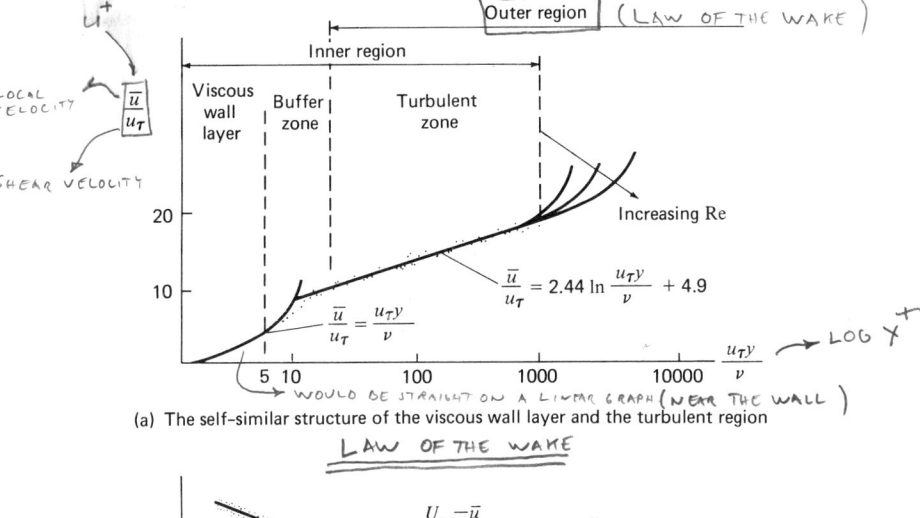

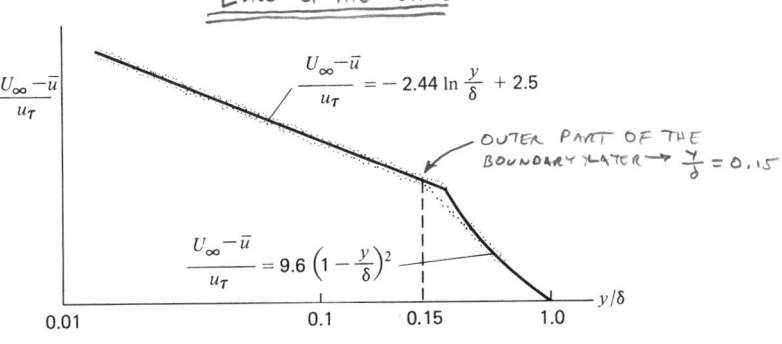

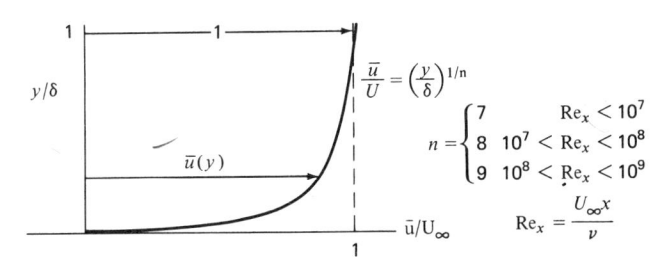

(a) The self-similar structure of the viscous wall layer and the turbulent region

$$\frac{\bar{u}}{u_\tau} = 2.44 \ln \frac{u_\tau y}{\nu} + 4.9$$

$$\frac{\bar{u}}{u_\tau} = \frac{u_\tau y}{\nu}$$

(b) The self-similar structure of the outer region

$$\frac{U_\infty - \bar{u}}{u_\tau} = -2.44 \ln \frac{y}{\delta} + 2.5$$

$$\frac{U_\infty - \bar{u}}{u_\tau} = 9.6 \left(1 - \frac{y}{\delta}\right)^2$$

(c) Common mode of presenting the mean velocity in the boundary layer

$$\frac{\bar{u}}{U} = \left(\frac{y}{\delta}\right)^{1/n}$$

$$n = \begin{cases} 7 & Re_x < 10^7 \\ 8 & 10^7 < Re_x < 10^8 \\ 9 & 10^8 < Re_x < 10^9 \end{cases}$$

$$Re_x = \frac{U_\infty x}{\nu}$$

Fig. 9.9. Velocity profiles in the zero-pressure-gradient turbulent boundary layer. (The curves in parts a and b are based on data taken from F. H. Clauser, *Advances in Applied Mechanics,* Vol. 4, Academic Press, N.Y., 1956.)

layer. However, overlooking these shortcomings, we put Eq. 9.61 in the form

$$\sqrt{2/c_f} = 2.44 \ln\left[\sqrt{\frac{c_f}{2}}\frac{\delta U_\infty}{\nu}\right] + 4.9 \qquad (9.68)$$

where we have imposed the condition $\bar{u} = U_\infty$ at $y = \delta$. This equation implicitly relating c_f to $\delta(x)$, yields results which do not agree well with experimental data. It is difficult to measure the boundary-layer thickness δ accurately, so Clauser* introduced the displacement thickness δ_d and arrived at the more accurate relation

$$\sqrt{2/c_f} = 2.44 \ln \frac{\delta_d U_\infty}{\nu} + 4.3 \qquad (9.69)$$

This relationship allows us to determine the local skin friction, or local wall shear, by measuring the displacement thickness. This is quite useful since it is difficult to directly measure the wall shearing stress.

If it is assumed that the laminar portion of the boundary layer is very short, an approximate and empirical relationship[†] between the local skin friction coefficient and Re_x is

$$c_f = \frac{0.370}{(\log_{10} \text{Re}_x)^{2.58}} \qquad (9.70)$$

where $\text{Re}_x = xU_\infty/\nu$.

Assuming turbulent flow from the leading edge, the dimensionless drag force, or skin friction drag coefficient, is found to be

$$C_f = \frac{0.427}{(\log_{10} \text{Re}_L - 0.407)^{2.64}} \qquad (9.71)$$

where $\text{Re}_L = LU_\infty/\nu$. Eq. 9.71 agrees with experimental data on a zero-pressure-gradient flat plate for $10^6 < \text{Re}_L < 10^9$.

Example 9.3

Determine the viscous wall-layer thickness and the boundary-layer thickness on the roof of a bus, 6 ft behind the windshield. Assume the boundary layer has zero thickness at the leading edge of the roof. The bus is traveling at 60 mph in otherwise still air at 68°F. Neglect any pressure-gradient effects.

---

[*] F. H. Clauser, "The Turbulent Boundary Layer," *Advances in Applied Mechanics*, Vol. 4 (New York, Academic Press), 1956, pp. 2–51.

[†] F. Schultz-Grunow, National Advisory Committee on Aeronautics, Tech. Memo. 986, 1941

Art. 9.5 TURBULENT-BOUNDARY-LAYER FLOW OVER A FLAT PLATE

Solution. The boundary layer undergoes transition at approximately $Re_{crit} = 3 \times 10^5 = 88 x_T / 1.6 \times 10^{-4}$. This gives a laminar region of less than $\frac{1}{2}$ ft. Let us neglect this laminar portion and assume the boundary layer to be turbulent from the leading edge. This allows us to determine the wall shearing stress from expression 9.70. It is

$$\frac{\tau_0}{\frac{1}{2}\rho U_\infty^2} = \frac{0.370}{\left(\log_{10} \frac{88 \times 6}{1.6 \times 10^{-4}}\right)^{2.58}} = 0.00292$$

or

$$\frac{\tau_0}{\rho} = 0.00292 \times \frac{88^2}{2} = 11.3 \text{ ft}^2/\text{sec}^2$$

Hence, at $x = 6$ ft,

$$u_\tau = \sqrt{\tau_0/\rho} = 3.36 \text{ fps}$$

The viscous wall layer has a thickness δ_v, given by

$$\frac{u_\tau \delta_v}{\nu} = 5$$

or

$$\delta_v = \frac{5 \times 1.6 \times 10^{-4}}{3.36} = 2.38 \times 10^{-4} \text{ ft}$$

To determine the turbulent boundary-layer thickness at $x = 6$ ft, we find the velocity at the outer edge of the turbulent zone. From Eq. 9.66, it is, using $y/\delta = 0.15$ from Fig. 9.9b,

$$\frac{88 - \bar{u}}{3.36} = -2.44 \ln 0.15 + 2.5 = 7.1$$

This yields

$$\bar{u} = 64.2 \text{ fps}$$

Then from Eq. 9.61, which is also valid in the turbulent zone, we have

$$\frac{64.2}{3.36} = 2.44 \ln \frac{3.36 y}{1.6 \times 10^{-4}} + 4.9$$

Solving for y, we find the outer edge of the turbulent zone to be located at the distance

$$y = 0.0162 \text{ ft}$$

The edge of the turbulent zone occurs at $y/\delta = 0.15$. Hence,

$$\delta = \frac{y}{0.15} = \frac{0.0162}{0.15} = 0.108 \text{ ft}$$

Extension 9.3.1. Calculate the velocity at edge of the viscous wall layer.
Ans. 16.8 fps

Extension 9.3.2. Calculate the thickness of the buffer zone.
Ans. 0.00208 ft

9.6 POWER-LAW FORM OF THE TURBULENT BOUNDARY LAYER

We have shown in Art. 9.4 how the von Kármán integral equation (9.51) can be used to determine $\delta(x)$ for a laminar boundary layer. For this solution, it was necessary to assume a form for the expression $u/U_\infty = f(y/\delta)$ and to evaluate the wall shear stress in terms of $f'(0)$. Equation 9.51 is not dependent upon the laminar-flow condition; it applies equally well to a turbulent boundary layer. However, the details of the calculation procedure are dependent on the state of the flow. They are considered below. We will perform the calculation for a zero-pressure-gradient boundary layer; the difficulties encountered for a nonzero pressure gradient will be noted.

The integrals involving $\bar{u}(x, y)$ can be evaluated from an assumed form for $\bar{u}(x, y)$. We will use the simple power-law expression for the assumed form because of considerable reduction in complexity as compared with the self-similar wall and outer-region profiles, as shown in Fig. 9.9. The power-law form is given by

$$\frac{\bar{u}}{U_\infty} = \left(\frac{y}{\delta}\right)^{\frac{1}{n}} \qquad n = \begin{cases} 7 & \text{Re}_x < 10^7 \\ 8 & 10^7 < \text{Re}_x < 10^8 \\ 9 & 10^8 < \text{Re}_x < 10^9 \end{cases} \qquad (9.72)$$

where $\text{Re}_x = U_\infty x / \nu$. Note that this form of the velocity profile cannot be used to establish τ_0 using the expression

$$\tau_0 = \mu \left.\frac{\partial u}{\partial y}\right|_{\text{wall}} \qquad (9.73)$$

since the derivative is infinite at $y = 0$ for the power-law form. However, we recall that the law-of-the-wall characteristic quantities were u_τ and ν/u_τ for the velocity and length, respectively. Introducing these, we can write Eq. 9.72 as

$$\frac{\bar{u}/u_\tau}{U_\infty/u_\tau} = \frac{\left(u_\tau y/\nu\right)^{1/n}}{\left(u_\tau \delta/\nu\right)^{1/n}} \qquad (9.74)$$

Art. 9.6 POWER-LAW FORM OF TURBULENT BOUNDARY LAYER

The denominators of the expression above are simply the numerators evaluated at $y = \delta$. Assuming the relationship

$$\frac{\bar{u}}{u_\tau} = C\left(\frac{u_\tau y}{\nu}\right)^{1/n} \tag{9.75}$$

to be valid for all y we may write

$$\frac{U_\infty}{u_\tau} = C\left(\frac{u_\tau \delta}{\nu}\right)^{1/n} \tag{9.76}$$

Solving for u_τ^2, this may be expressed as

$$u_\tau^2 = \left(\frac{U_\infty}{C}\right)^{\frac{2n}{n+1}} \left(\frac{\nu}{\delta}\right)^{\frac{2}{n+1}} \tag{9.77}$$

Introducing the wall shear, $u_\tau^2 = \tau_0/\rho$, the local skin friction coefficient becomes

$$c_f = \frac{\tau_0}{\tfrac{1}{2}\rho U_\infty^2} = C_1 \left(\frac{\nu}{U_\infty \delta}\right)^{\frac{2}{1+n}} \tag{9.78}$$

where the constant C_1 is evaluated by comparison with experimental data. It has been evaluated for $n = 7$ to be 0.046. The local skin friction coefficient is then

$$c_f = 0.046 \left(\frac{\nu}{U_\infty \delta}\right)^{1/4} \quad (\mathrm{Re}_x < 10^7) \tag{9.79}$$

This relationship was originally established by Blasius and is often referred to as *Blasius' formula*.

The power-law form results in the following value for the momentum thickness:

$$\theta = \delta \int_0^\delta \frac{u}{U_\infty}\left(1 - \frac{u}{U_\infty}\right)\frac{dy}{\delta} = \delta \int_0^1 \left(\frac{y}{\delta}\right)^{\frac{1}{n}} \left[1 - \left(\frac{y}{\delta}\right)^{\frac{1}{n}}\right] d\left(\frac{y}{\delta}\right)$$

$$= \delta \int_0^1 \eta^{1/n}[1 - \eta^{1/n}]\, d\eta = \frac{\delta n}{2 + 3n + n^2} \tag{9.80}$$

Equation 9.51, for the case of a zero-pressure-gradient flow, may be written as Eq. 9.56, which is

$$\tau_0 = \rho U_\infty^2 \frac{d\theta}{dx} \tag{9.81}$$

Using the empirical relationship 9.79 and the results of Eq. 9.80, we then have

$$\tau_0/\rho U_\infty^2 = 0.023\left(\frac{\nu}{U_\infty \delta}\right)^{1/4}$$

$$= \frac{n}{2 + 3n + n^2} d\delta/dx \qquad (9.82)$$

or

$$\frac{d\delta}{dx} = \frac{2 + 3n + n^2}{n} 0.023\left(\frac{\nu}{U_\infty \delta}\right)^{1/4} \qquad (9.83)$$

Relationship 9.79 for c_f is valid only for length Reynolds numbers less than 10^7; consequently, we will use the value $n = 7$. The differential equation for $d\delta/dx$ therefore becomes

$$\frac{d\delta}{dx} = \frac{72}{7} 0.023 \left(\frac{\nu}{U_\infty \delta}\right)^{1/4}$$

or

$$\int_0^\delta \delta^{1/4} d\delta = \frac{1}{4.23}\left(\frac{\nu}{U_\infty}\right)^{1/4} \int_0^x dx \qquad (9.84)$$

where we assume the laminar portion of the boundary layer to be short so that at $x = 0$, $\delta = 0$.

Integrating gives

$$\delta^{5/4}(x) = \left(\frac{\nu}{U_\infty}\right)^{1/4}\left(\frac{1}{4.23}\right)\left(\frac{5}{4}\right)x \qquad (9.85)$$

or

$$\frac{\delta}{x} = 0.380\left(\frac{\nu}{U_\infty x}\right)^{1/5} \qquad \text{Re}_x < 10^7 \qquad (9.86)$$

By substituting back into Eq. 9.79 we find that the local skin friction coefficient is

$$c_f = 0.059/\text{Re}_x^{1/5} \qquad \text{Re}_x < 10^7 \qquad (9.87)$$

and the skin friction drag coefficient is

$$C_f = 0.074/\text{Re}_L^{1/5} \qquad \text{Re}_L < 10^7 \qquad (9.88)$$

These specific results are limited by the assumption of the power-law form and the Reynolds-number limitation of the Blasius skin friction

Art. 9.6   POWER-LAW FORM OF TURBULENT BOUNDARY LAYER

relationship (9.79). The technique indicated is general and serves to show what the necessary steps are in the formulation of the von Kármán equation to calculate $\delta(x)$ for a turbulent boundary layer.

Example 9.4

Determine the boundary-layer thickness at the end of, and the shear force on the surface of, a flat roof which is 16 ft wide and 10 ft long in the streamwise direction of a 15-fps wind. Use $\nu = 1.5 \times 10^{-4}$ ft^2/sec and assume a zero-pressure-gradient flow.

Solution. The transition length x_T for this flow is found from $U_\infty x_T/\nu \cong 3 \times 10^5$. This gives

$$x_T = \frac{3 \times 10^5 \times 1.5 \times 10^{-4}}{15} = 3 \text{ ft}$$

Since this is a significant fraction of the length of the roof, the laminar-boundary-layer portion will have to be accounted for.

The boundary-layer thickness $\delta(x)$ at the end of the plate is found from Eq. 9.86 for a turbulent layer; but, since the turbulent layer did not start at the origin of the plate, it is necessary to consider a fictitious plate for which the turbulent layer starts at $x' = 0$. The length coordinates of the two plates are related by a "matching condition"; specifically, $\theta_{\text{laminar}} = \theta_{\text{turb}}$ at the transition point of the real flow will be used to calculate this fictitious origin. By this choice the shear force is preserved in the fictitious problem since Eq. 9.56 allows us to write $\int_0^{x'} \tau_0 \, dx = \rho U_\infty^2 \theta$.

From Eq. 9.46 the laminar momentum thickness is

$$\theta_l = 0.664\sqrt{\frac{\nu x_T}{U_\infty}} = 0.664\sqrt{\frac{1.5 \times 10^{-4} \times 3}{15}} = 3.64 \times 10^{-3} \text{ ft}$$

Since Re $= 15 \times 10/(1.5 \times 10^{-4}) = 10^6$, which is less than 10^7, the power-law form of the turbulent layer with $n = 7$ is used. For this, Eq. 9.80 gives (using $\theta_T = \theta_l$),

$$\delta(x') = \frac{72}{7} \theta_T(x') = \frac{72}{7} \times 3.64 \times 10^{-3} = 3.75 \times 10^{-2} \text{ ft}$$

The fictitious plate length x' is then found from Eq. 9.86:

$$\delta(x') = 0.380\left(\frac{\nu}{U_\infty}\right)^{1/5}(x')^{4/5}$$

or

$$x' = \left[\frac{3.75 \times 10^{-2}}{0.380}\left(\frac{15}{1.5 \times 10^{-4}}\right)^{1/5}\right]^{5/4} = 0.983 \text{ ft}$$

The boundary-layer thickness at the end of the 10 ft length, or at the end of the fictitious plate of length $10 - x_T + x' = 7.983$ ft, is

$$\delta = 0.380\left(\frac{\nu}{U_\infty}\right)^{1/5} 7.98^{4/5} = 0.38\left(\frac{1.5 \times 10^{-4}}{15}\right)^{1/5} 7.98^{4/5} = 0.200 \text{ ft}$$

The shear force per unit width is given in terms of θ as

$$F_s/w = \int_0^x \tau_0 \, dx = \rho U_\infty^2 \theta$$

Using Eq. 9.80 for θ, we have

$$F_s = \rho U_\infty^2 \theta w = 0.00238 \times 15^2 \times \frac{7}{72} \times 0.2 \times 16 = 0.167 \text{ lb}$$

Example 9.5

The one-seventh power-law relationship for a turbulent boundary layer is a simpler form to use for calculations than the empirical relationships of the previous article. However, it is not as accurate. Compare the following boundary layer quantities for a water flow at 40 fps over a 10 ft long flat plate using the relationships for both descriptions: (a) the displacement thickness, (b) the local skin friction coefficient at $x = 10$ ft, and (c) the skin friction drag coefficient. Use $\nu = 10^{-5}$ ft^2/sec and assume turbulent flow from the leading edge.

Solution. The boundary-layer quantities will first be calculated, using the empirical relationships. To determine the local skin friction coefficient c_f, using Eq. 9.69, we must first determine the displacement thickness δ_d. It is given by Eq. 9.53 as

$$\delta_d = \frac{1}{U_\infty} \int_0^\delta (U_\infty - \bar{u}) \, dy$$

For the turbulent boundary layer (see Fig. 9.9b), we neglect the small contribution to this integral of the viscous wall layer giving

$$\delta_d = \frac{1}{U_\infty} \int_{\delta_v}^{.15\delta} u_\tau\left(-2.44 \ln \frac{y}{\delta} + 2.5\right) dy + \frac{1}{U_\infty} \int_{.15\delta}^\delta u_\tau 9.6\left(1 - \frac{y}{\delta}\right)^2 dy$$

where δ_v is the viscous wall layer thickness. To determine a numerical value for δ_d we must evaluate u_τ, δ_v, and δ. Following the steps of Example 9.3, these three quantities are evaluated as follows.

Equation 9.70 gives

$$c_f = \frac{\tau_0}{\frac{1}{2}\rho U_\infty^2} = \frac{0.370}{\left(\log_{10} \frac{40 \times 10}{10^{-5}}\right)^{2.58}} = 1.97 \times 10^{-3}$$

and
$$u_\tau = \sqrt{\tau_0/\rho} = \sqrt{1.97 \times 10^{-3} \times 40^2/2} = 1.26 \text{ fps}$$
Hence
$$\delta_v = \frac{5\nu}{u_\tau} = \frac{5 \times 10^{-5}}{1.26} = 3.97 \times 10^{-5} \text{ ft}$$

To find δ we first determine the velocity at the edge of the turbulent zone, $y = 0.15\delta$, using Eq. 9.66. We have
$$\frac{40 - \bar{u}}{1.26} = -2.44 \ln .15 + 2.5$$
giving
$$\bar{u} = 31.0 \text{ fps}$$
Then from Eq. 9.61 we have
$$\frac{31.0}{1.26} = 2.44 \ln \frac{1.26y}{10^{-5}} + 4.9$$
or, at the edge of the turbulent zone,
$$y = 0.0255 \text{ ft}$$
so that
$$\delta = \frac{y}{0.15} = \frac{0.0255}{0.15} = 0.170 \text{ ft}$$

Returning to the integral equation given above, there results
$$\delta_d = \frac{u_\tau}{U_\infty} \int_{3.97 \times 10^{-5}}^{0.15\delta} \left(-2.44 \ln \frac{y}{\delta} + 2.5\right) dy + \frac{u_\tau}{U_\infty} \int_{0.15\delta}^{\delta} 9.6\left(1 - \frac{y}{\delta}\right)^2 dy$$
$$= \frac{u_\tau}{U_\infty} \times 3.74\delta$$
$$= \frac{1.26}{40} \times 3.74 \times .170 = 0.020 \text{ ft}$$

Finally, from Eq. 9.69 there results
$$c_f = \frac{2}{\left[2.44 \ln \dfrac{40 \times .02}{10^{-5}} + 4.3\right]^2} = 1.97 \times 10^{-3}$$

The drag coefficient is found from Eq. 9.71 to be
$$C_f = \frac{0.427}{\left[\log_{10} \dfrac{40 \times 10}{10^{-5}} - 0.407\right]^{2.64}} = 2.36 \times 10^{-3}$$

Similar quantities may be calculated with the one-seventh power-law form. The displacement thickness is

$$\delta_d = \int_0^\delta \left(1 - \frac{u}{U_\infty}\right) dy = \int_0^\delta \left[1 - \left(\frac{y}{\delta}\right)^{1/7}\right] dy = \frac{\delta}{8}$$

Using Eq. 9.86 for δ, we find

$$\delta_d = 10 \times 0.380 \left(\frac{10^{-5}}{40 \times 10}\right)^{1/5} / 8 = 0.0143 \text{ ft}$$

Equation 9.87 can be used to compute c_f as

$$c_f = \frac{0.059}{\left(\dfrac{40 \times 10}{10^{-5}}\right)^{1/5}} = 1.78 \times 10^{-3}$$

The drag coefficient is found from Eq. 9.88 to be

$$C_f = \frac{0.074}{\left(\dfrac{40 \times 10}{10^{-5}}\right)^{1/5}} = 2.23 \times 10^{-3}$$

The results of these two calculation techniques are summarized in the table below. The percentage of error is based upon the data from the empirical relationships.

	Empirical Relationships	One-Seventh Power Law	Percentage of Error
c_f	1.97×10^{-3}	1.78×10^{-3}	-9.6
C_f	2.36×10^{-3}	2.23×10^{-3}	-5.5
δ_d	2.00×10^{-2} ft	1.43×10^{-2} ft	-28.5

9.7 ADDITIONAL CONSIDERATIONS CONCERNING THE TURBULENT BOUNDARY LAYER

An important characteristic of the turbulent boundary layer is shared with the free shear flows of jets and wakes: its boundary with the external fluid is a very thin sheet, (of the order of 1 mm for typical laboratory flows) called the *viscous superlayer*, which is plastered over the interior turbulent fluid. Since only viscous effects can transmit vorticity and since the boundary layer, wake, or jet "grows" by capturing the external, nonvortical fluid, it must do so via a region in which the length scale is sufficiently small that the viscous effects are important; hence the viscous superlayer. Figure 9.10 is an instantaneous sketch of an actual boundary layer. A measuring probe at a fixed y-value sees the passage of these "hills" of turbulence and the valleys of nonvortical fluid. The result is an output as shown on the sketch.

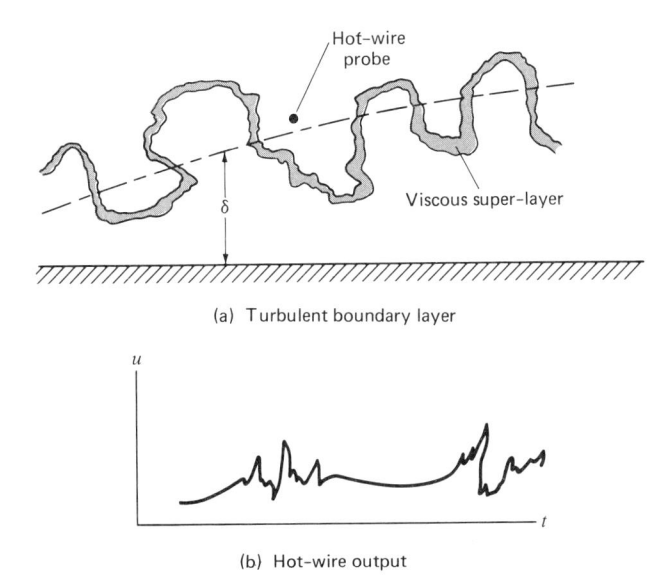

Fig. 9.10. An instantaneous picture of the turbulent boundary layer.

The substitution $\bar{u} + u'$ for the instantaneous-velocity component u is certainly not incorrect for the situation shown in Fig. 9.10b; however, neither is it the most helpful in order to understand the physical nature of the flow. The average and the mean square of the velocity interior to the superlayer are obviously different from the same quantities in the flow exterior to it. Their separate documentation would lead to a more instructive representation of the complete flow field. Indeed, modern research is currently involved in extending our basic descriptions of turbulent motions to include more extensive correlations and "conditional" measurements such as the $\bar{u}$ and u' values both interior and exterior to the superlayer.

9.8 EFFECTS OF A PRESSURE GRADIENT

Nearly all of the examples and results presented in the introduction to boundary-layer theory in this chapter have been for flow on a flat plate with zero pressure gradient. Obviously, the pressure gradient very strongly influences the boundary-layer growth, the transition length, the wall shear, and the other boundary-layer parameters. Although many real flows may be approximated by a zero-pressure-gradient condition, there are numerous other flows for which pressure-gradient effects are important. The pressure gradient plays an important role if the curvature of the body in the flow direction is significant. Body curvature results in both $\partial p/\partial x$ and $\partial p/\partial y$ in the vicinity of the boundary. The change in pressure across the thin boundary layer is generally small, so that $\partial p/\partial y \simeq 0$ in the boundary layer. The streamwise pressure gradient $\partial p/\partial x$ may be quite small and still influence the boundary layer significantly, if it acts over a sufficient length.

To observe qualitatively the role of the pressure gradient, consider the boundary-layer equation (9.20) for a steady flow. At the boundary where $y = 0$ the no-slip condition requires that $u = v = 0$. The result, [FROM NAVIER STOKES]

$$\mu \frac{\partial^2 u}{\partial y^2}\bigg|_{y=0} = \frac{dp}{dx} \qquad (9.89)$$

shows that the curvature of the velocity profile at the wall is directly dependent on the pressure gradient. For a zero pressure gradient the curvature of the velocity profile is zero at the wall and, of course, must be zero at the outer edge of the boundary layer where viscous effects become negligible. The curvature must be negative in the boundary layer since the slope $\partial u/\partial y$ decreases from the wall where it is maximum to zero at the edge of the layer. This is shown in Fig. 9.11a.

For a negative (favorable) pressure gradient, the curvature is negative at the wall and decreases to zero at the outer edge of the layer, as shown in Fig. 9.11b. The velocity profile is similar to that of the zero-pressure-gradient flow except that it is fuller throughout.

For a positive (adverse) pressure gradient, the curvature is positive at the wall, and must approach zero from the negative side at the outer edge of the layer. This requires an inflection point ($\partial^2 u/\partial y^2 = 0$) in the boundary-layer velocity profile where the slope $\partial u/\partial y$ reaches a maximum. This is shown qualitatively in Fig. 9.11c. For the adverse-pressure-gradient flow, the fluid is attempting to flow into a region of higher pressure. This retards the flow and, if the positive pressure

Art. 9.8 EFFECTS OF A PRESSURE GRADIENT

(a) Zero pressure gradient

(b) Negative pressure gradient — FAVORABLE PRESSURE GRADIENT — ACCELERATING THE FLUID

(c) Positive pressure gradient — ADVERSE

BOUNDARY LAYER ON ITS WAY TO SEPARATION

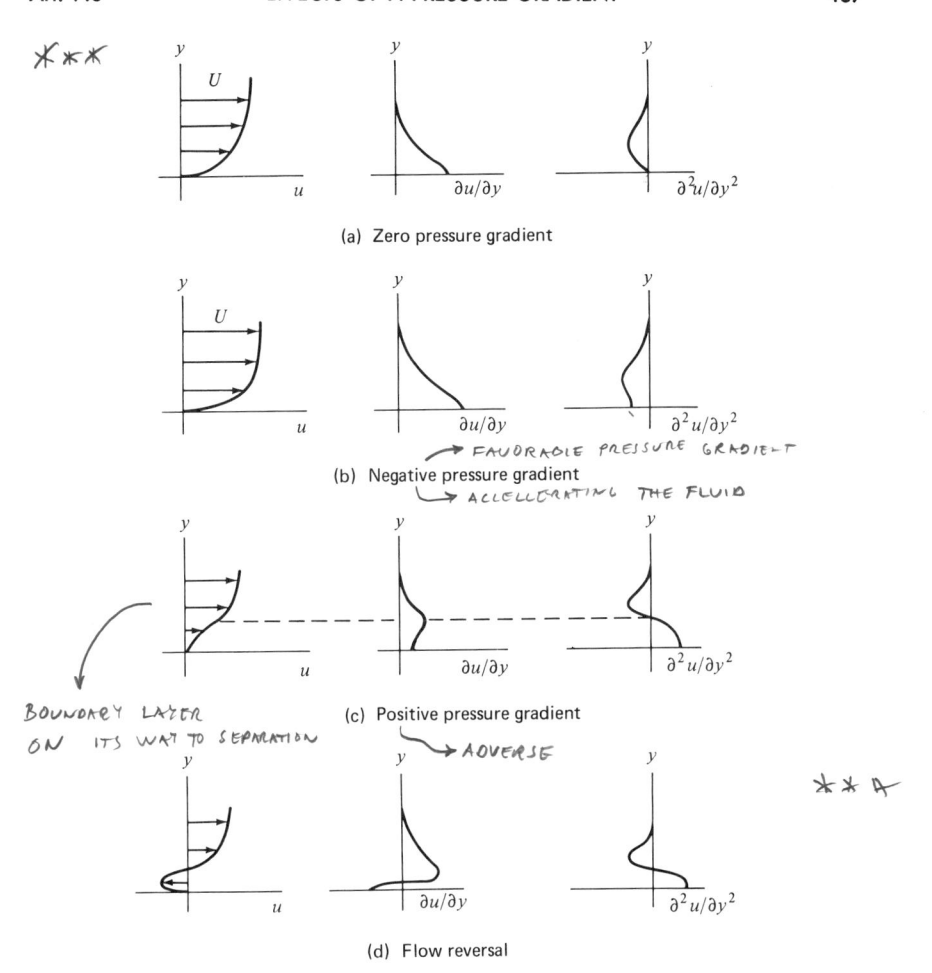

(d) Flow reversal

Fig. 9.11. Effect of pressure gradient on the velocity profile.

gradient acts over a sufficient length, the flow near the wall may actually be reversed as shown in Fig. 9.11d. The x-location at which the velocity gradient $\partial u/\partial y$ is zero at the wall locates the separation point, shown in Fig. 9.2.

The boundary layer thickens at a greater rate for adverse pressure gradients and at a lesser rate for favorable pressure gradients. The shearing stress at the wall decreases as the pressure gradient increases. The position at which transition occurs is also sensitive to the pressure gradient. The laminar portion of the boundary layer is shortened as the pressure gradient increases and lengthened for favorable pressure gradients.

It is very difficult, even for flow on a flat plate, to solve the boundary-layer equations with a nonzero pressure gradient. However, using numerical techniques the boundary-layer equations can be solved with various pressure gradients on flat plates or on curved bodies.

9.9 CLOSURE

Boundary-layer flows have been studied for zero-pressure-gradient flow over a flat plate. The influence of the pressure gradient has been discussed. There are a number of parameters, other than the pressure gradient, that affect the boundary layer. These include the wall roughness, the fluctuations in the free stream, and the wall heat transfer. Not only is the magnitude of the wall roughness significant, but the shape and positioning of the roughness elements also affect the boundary layer. A rough wall results in early transition to turbulence and higher wall shearing stress. A wall may appear to be "rough" in one region of the boundary layer and "smooth" in another region. In a laminar boundary layer the wall always appears to be smooth. In a turbulent layer the wall is termed "rough" if the height of the roughness elements is approximately equal to or greater than the thickness of the viscous wall layer. Since the viscous wall layer is thinnest immediately after transition, the wall becomes smoother as the flow develops downstream and the viscous wall layer thickens.

The free-stream fluctuations, sometimes referred to as free stream turbulence, also affect the transition process. For extremely small fluctuations a laminar flow has been observed at local Reynolds numbers as high as 2.8×10^6; in fact, this appears to be an upper limit for the critical Reynolds number. As the fluctuation amplitude is increased, the transition length x_T decreases. The frequencies associated with the fluctuations are also significant.

Heat transfer influences the boundary layer primarily through the dependence of the viscosity on the temperature. For a liquid flow, the viscosity decreases near a heated wall and this variation must be accounted for in the differential equations. A heated wall for a liquid flow would result in a fuller velocity profile and a decrease in the boundary-layer thickness. Since the viscosity increases with temperature in a gas flow, the opposite effects are observed with a heated wall.

The fundamental ideas which form boundary-layer theory have been presented in this chapter. Quantitatively, we have considered flow over a smooth flat plate with zero pressure gradient. A more advanced course in boundary-layer theory would present techniques which could be used in more complex flow situations.

PROBLEMS

9.1 Follow a fluid particle which enters the laminar boundary layer and proceeds along the plate shown. Trace out a typical pathline. Also sketch a time-averaged streamline. For the pathline, consider a situation in which the particle encounters a burst in the middle of the transition region.

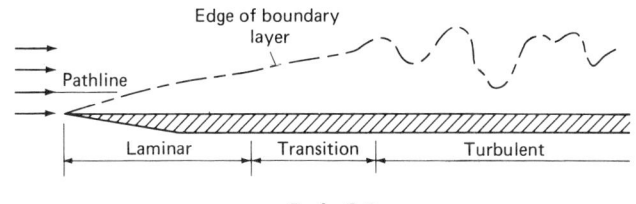

Prob. 9.1

9.2 A flat plate is inserted in a two-dimensional converging channel as shown. Assuming the inviscid flow in the contraction to be uniform, i.e., $U = U(x)$, determine the free-stream velocity which exists on the flat plate. Also determine the pressure gradient dp/dx for the boundary layer. $\rho = 0.002$ slug/ft^3.

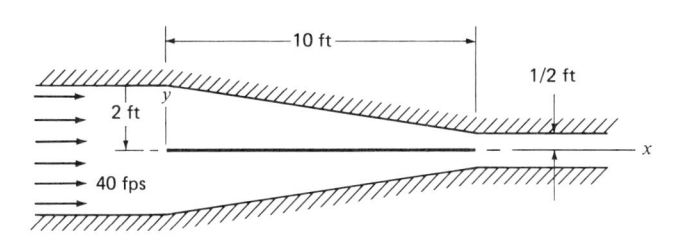

Prob. 9.2

9.3 An airfoil on a commercial jetliner can be crudely approximated by a flat plate. Determine the distance from the leading edge at which transition to turbulence can be expected if the critical Reynolds number is 600,000. The jet travels at 500 mph at an altitude of 30,000 ft.

9.4 Water flows over the step as shown (*next page*) and separates, reattaching somewhere downstream. At section two the velocity distribution is approximated as shown. Locate the position of the separation streamline at section two.

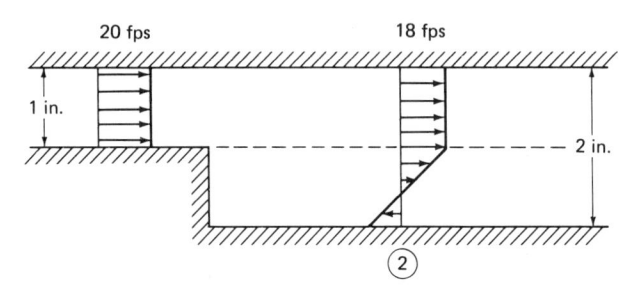

Prob. 9.4

9.5 Verify Eq. 9.30 by obtaining expressions for the necessary higher-order derivatives in Eq. 9.23, substituting in Eq. 9.23, and simplifying assuming $dp/dx = 0$. Then assume that $\psi = F(\eta)G(\xi)$ and show that Eq. 9.31 results. Using the relationship 9.32 finally, verify Eq. 9.33.

9.6 Assume a cubic profile in a boundary-layer flow

$$u = U_\infty \left[\frac{3}{2} \frac{y}{\delta} - \frac{1}{2} \left(\frac{y}{\delta} \right)^3 \right]$$

Determine the displacement thickness and compare with the exact solution given in Eq. 9.46.

9.7 Show that the integral momentum equation can be simplified to result in Eq. 9.49. Explain why the height of the downstream face of the control volume in Fig. 9.7c increases by the amount $(\partial \delta / \partial x) \, dx$.

9.8 Using an infinitesimal element of height y as shown, determine a relationship for the velocity component v of the fluid leaving the top of the element by considering the integral continuity equation. Also, show that for $y > \delta$, v is constant and equal to $-\int_0^\delta \frac{\partial u}{\partial x} \, dy$.

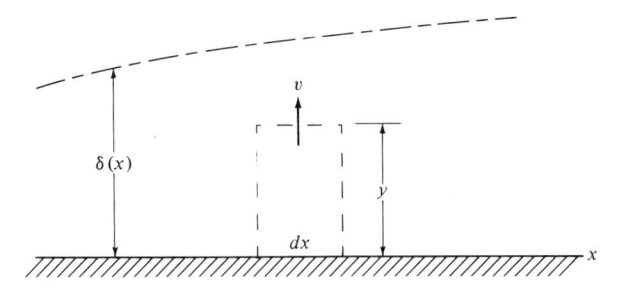

Prob. 9.8

9.9 If a fluid flows between parallel walls in a channel in which boundary layers are growing on the walls, the inviscid free stream flow will increase in magnitude as the flow proceeds down the channel. Show that in order to maintain a constant free stream value for the velocity it is necessary to deflect each wall a distance $\delta_d(x)$, the displacement thickness. (This is done in wind tunnels when a precise zero pressure gradient condition is required.)

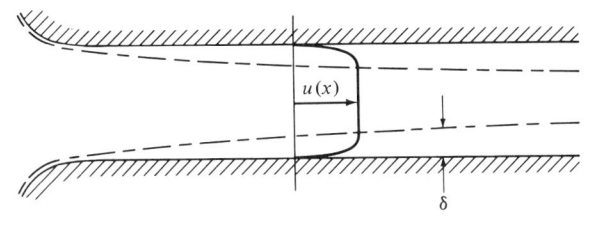

Prob. 9.9

9.10 Show that Eq. 9.51 follows from Eq. 9.50. Explain why the $\dfrac{d}{dx}$ operator cannot be moved inside the integral and why we write dp/dx instead of $\partial p/\partial x$.

9.11 (a) Show that Eq. 9.55 follows from Bernoulli's equation. *Hint*: $\int_0^\delta U\,dy = U\delta$ since $U = U(x)$. (b) We have neglected body forces in our development. Show that Eq. 9.54 is acceptable even if body forces are important; assume flow on a vertical flat plate and correct Eqs. 9.49, 9.50, and 9.51.

9.12 Assume a linear approximation to the velocity distribution in a laminar boundary layer on a flat plate in a uniform flow of water and calculate the drag force on a 20-ft-long plate. The plate is 10 ft wide; $U_\infty = 20$ fps and $\mu = 2 \times 10^{-5}$ lb-sec/ft².

9.13 Sketch a typical velocity profile on a flat plate with zero pressure gradient. Identify the quantity $(U_\infty - u)$. Also identify the area on the velocity profile sketch which is equivalent to $\delta_d U_\infty$. Show the relative magnitudes of the quantities δ, δ_d, and θ.

9.14 The ratio δ_d/θ is termed the shape factor H. Calculate H for a linear profile $u = U_\infty y/\delta$, and a parabolic profile $u = U_\infty(2y/\delta - y^2/\delta^2)$. Then infer, that $H_{\text{favorable}} < H_{\text{zero}} < H_{\text{adverse}}$.

9.15 For the physical setup of Prob. 9.2, find an approximate expression for the total shear force acting on the 12-ft-long, 10-ft-wide plate if a zero-pressure-gradient boundary layer grows on the plate from $x = 0$ to $x = 2.0$ ft, the point at which the contraction starts. The approximation is made that the contraction keeps the boundary layer from further growth; that is, $d\delta/dx = 0$ for $x > 2$ ft. The boundary-layer velocity profile is approximated by $u = U_\infty[2y/\delta - y^2/\delta^2]$. (This would be an approximation for a laminar boundary layer.)

9.16 A parabolic velocity distribution is assumed for a boundary layer on a flat plate with zero pressure gradient. Determine how the displacement and momentum thicknesses vary with the x-coordinate.

9.17 A laminar boundary-layer flow occurs over a flat plate with a pressure gradient imposed so that $U = 3 + x/3$. If $\delta_d = 1.74\sqrt{\nu x/U}$ and $\theta = 0.657\sqrt{\nu x/U}$, find dp/dx and τ_0 at the location $x = 3$ ft. Compare this value of τ_0 with that obtained for a zero-pressure-gradient flow with $U_\infty = 4$ fps. See the results in Example 9.1.

9.18 Assuming a velocity distribution of $u = U_\infty \sin[(y/\delta)(\pi/2)]$ in a boundary layer, calculate the velocity component v at $y = 0, = \delta/2, = \delta, = 10\delta$, for $x = 10$ ft. Over the flat plate, water flows with $U_\infty = 10$ fps and $\nu = 10^{-5}$ ft²/sec.

9.19 Air at 59°F flows parallel to a 12-ft-long, 6-ft-wide flat plate at 20 fps. Determine the drag, and the displacement and boundary-layer thicknesses at $x = 12$ ft, assuming (a) laminar flow, and (b) turbulent flow from the leading edge and extending over the entire plate. (See Example 9.5)

9.20 Air at 68°F flows parallel to a flat plate 12 ft long at 40 fps. On a single graph, sketch both the velocity profile in the boundary layer at $x = 12$ ft for a laminar layer and for a layer which is assumed turbulent from the leading edge.

9.21 Assume that the function $\bar{u} = U_\infty(y/\delta)^{1/7}$ approximates the velocity distribution in a turbulent boundary-layer flow on a flat plate. Can this velocity distribution be used to give the shearing stress at the wall? Does it satisfy the necessary conditions at the edge of the boundary layer? For $U_\infty = 80$ fps and air at standard conditions, determine δ for the turbulent boundary layer at $x = 6$ ft. Assume the turbulent boundary layer to exist from the leading edge.

9.22 If the wind blowing toward the land from the ocean is estimated to start 40 miles away, and if a wind speed of 40 fps at a temperature of 60°F is assumed, (a) estimate the boundary-layer thickness and (b) the thickness of the viscous wall layer at the land; (c) also estimate the displacement thickness. Assume Eq. 9.66 to be valid over the outer portion of the boundary layer.

9.23 Determine the relationship for the momentum thickness for laminar flow over a flat plate with zero pressure gradient. Use the von Kármán integral equation (Eq. 9.56) along with the exact solution for the wall shear stress given by Eq. 9.40. Write an integral relationship for θ in terms of F and η; express this as θ/δ.

9.24 A jet aircraft is taking off at 180 mph with the air at 68°F. At a distance of 6 ft from the leading edge of the airfoil, approximate the following: (a) the viscous wall layer thickness; (b) the velocity at the edge of the viscous wall layer; (c) the value of y at the outer edge of the turbulent zone; and (d) the boundary-layer thickness. Assume the boundary layer is turbulent from the leading edge.

9.25 Estimate the total viscous drag on one side of a ship traveling at 20 knots. A side is approximately 20 ft by 200 ft. Also determine the boundary-layer thickness at the end of the ship and the shearing stress at the midpoint.

9.26 Estimate the horsepower necessary to overcome the viscous drag on a dirigible balloon that is moving in still air at 40 fps at an elevation of 5000 ft. It is 15 ft in diameter and 100 ft long. Neglect the drag over the nose of the dirigible. Would this be a good approximation to the required horsepower?

9.27 As a result of a separation bubble at the upstream end of a flat wall, the $x = 0$ location for the subsequent boundary-layer growth is not clearly defined. The velocity profile at a location 1 ft downstream where the flow is straightened out following the separation is approximated by a power-law form, $n = 7$, and shows a boundary-layer height of 0.60 in. The free-stream velocity is 80 fps; the fluid is air at standard conditions. Determine the boundary-layer thickness at a point 2.5 ft farther downstream; also, find the shear force which exists on the wall between the two locations.

9.28 A diffuser is a device with diverging walls which seeks to slow a fluid stream in such a manner that p increases as U decreases. Show that for the laminar-flow diffuser shown in the sketch (a) $\partial^2 u / \partial y^2 > 0$ at $y = 0$, and (b) the shear stress $\mu(\partial u / \partial y)$ reaches a maximum at some distance from the wall. (This may be interpreted as the effect of the main flow in "dragging the fluid along the wall in order to overcome the retarding effect of the increasing pressure.")

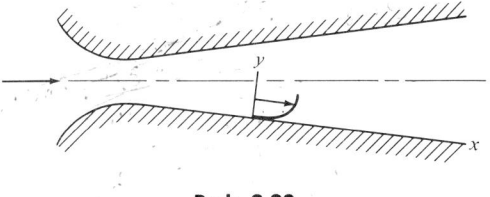

Prob. 9.28

9.29 Consider the following questions about the diffuser flow of Prob. 9.28 and justify your responses by a reasonable discussion. (a) Would it aid the diffuser performance if the wall boundary layers were turbulent? (b) Would it be advisable to have rough walls if the flow were always laminar? If the flow were near transition at the separation point? (c) Would the diffuser behave differently if a disturbed free stream, such as the outlet of a fan, were present as compared with a uniform and low-turbulence free-stream flow?

SELECTED REFERENCES

The motion pictures *Fundamentals of Boundary Layers* (No. 21623; F. H. Abernathy, film principal) and *Boundary Layer Control* (No. 21614) are recommended as useful supplements to this chapter.

Introductory discussions of boundary layers are provided in most texts. The two following are suggested for discussions at a level similar to that of the present text: J. A. Owczarek, *Introduction to Fluid Mechanics*, International

Textbook Co., Scranton, Pa., 1968; J. W. Daily, and D. R. F. Harleman, *Fluid Dynamics*, Addison-Wesley Publishing Co., Reading, Mass., 1968. A more comprehensive discussion of the laminar boundary layer and boundary-layer transition is to be found in H. Schlichting, *Boundary Layer Theory*, 6th ed., McGraw-Hill Book Co., New York, 1968. Turbulent-boundary-layer data are available in J. O. Hinze, *Turbulence*, McGraw-Hill Book Co., New York, 1959.

Two comprehensive articles, each summarizing the information available up to its publication date, are: F. H. Clauser, "The Turbulent Boundary Layer," *Advances in Applied Mechanics*, Vol. 4, H. L. Dryden and Theodore von Kármán, eds., Academic Press, New York, 1956; and L. S. G. Kovasznay, "The Turbulent Boundary Layer," *Annual Review of Fluid Mechanics*, Vol. 2, M. Van Dyke, W. G. Vincenti, and J. V. Wehausen, eds., Annual Reviews, Palo Alto, Calif., 1970.

The proceedings of the Stanford Conference constitute an excellent documentation of boundary-layer data as well as "current" prediction schemes. See *Computation of Turbulent Boundary Layers*, Vol. 1, S. J. Kline, M. V. Morkovin, G. Sovran, and D. J. Cockrell, eds., and Vol. 2, D. C. Coles and E. A. Hirst, eds., Thermosciences Division, Department of Mechanical Engineering, Stanford University, Stanford, Calif., 1969.

10

External Flows

10.1 INTRODUCTION

The study of flows around bodies—external flows—is of primary importance to the aerodynamicist in his study of flow around the various components of an aircraft. Many of the developments in the fluid dynamics of external flows have been motivated by such aerodynamic problems. However, the motion of a ship, torpedo, automobile, hovercraft, sediment in a stream, raindrops, and red blood cells in blood plasma, as well as the flow of air past buildings, smokestacks, bridges, and trees and of water past bridge abutments and piers, are also examples which can benefit from a general understanding of external flows.

It is extremely difficult to completely solve for the velocity and pressure distributions in the flow field external to a body. The flow fields external to the two simplest geometries, the infinite circular cylinder and the sphere, have been determined for a very limited range of Reynolds numbers (Re < 5) and are not of much engineering interest. For high-Reynolds-number flows, the approximate technique of considering all viscous effects to be concentrated in the boundary layer and thus solving the boundary layer problem has yielded acceptable results for flows around streamlined bodies without separation. This technique requires advanced numerical methods and is beyond the scope of this text. The external flows involving separated regions have not, in general, been treated analytically; it is simply too difficult, but it is this type of flow which is encountered most often in engineering applications. There has been some effort in predicting the point of separation, especially on the more simple geometries. For these reasons, external flows have been, essentially, a subject for which empirical documentation has been of primary importance. For many bodies, the experimental data necessary to calculate the quantities of interest to the engineer has been obtained. This is the type of information to be presented in this chapter.

We will discuss some of the phenomena associated with external flows in an effort to understand qualitatively the flow of fluid around bodies. The influence of compressibility and cavitation will also be discussed.

10.2 SOME PHENOMENA ASSOCIATED WITH EXTERNAL FLOWS

In this article we will consider the various phenomena associated with flow around bodies; one of the most important of these phenomena is the resultant force acting on the body due to the fluid. The *drag*, that component of the force parallel to the direction of and resisting the motion, is of concern in most external flows. The component of force normal to the direction of motion, the *lift*, is also of obvious importance in many flows. The types of flows considered may be subdivided into three categories: (1) incompressible immersed flows involving such objects as submarines, fish and birds, automobiles, trains, parachutes, helicopters, airplanes at low speeds,, frisbees, buildings, and smoke stacks; (2) compressible flows involving high-speed aircraft ($V > 300$ mph), reentering satellites, and bullets; and (3) flows which involve a free surface or an interface between two fluids as in the case of surface vessels. Of primary concern to us will be incompressible immersed flows not involving a free surface.* Some results will, however, be given to show the influence of compressibility.

Separation is of particular interest when studying external flows. It occurs when the main stream "separates" from the surface of the body (see Fig. 10.1). This may result in a substantial alteration of the flow field from the case of no separation. It accounts for "stall" and subsequent loss of control of aircraft, for inefficient operation of turbines, for poor performance of diffusers, and for the large drag associated with blunt bodies.

A phenomenon of interest occurs over a certain range of Reynolds number when fluid is flowing past a circular, or nearly circular,

---

*Free-surface effects become significant whenever the depth of a slender body is below the surface less than five times the length of the body.

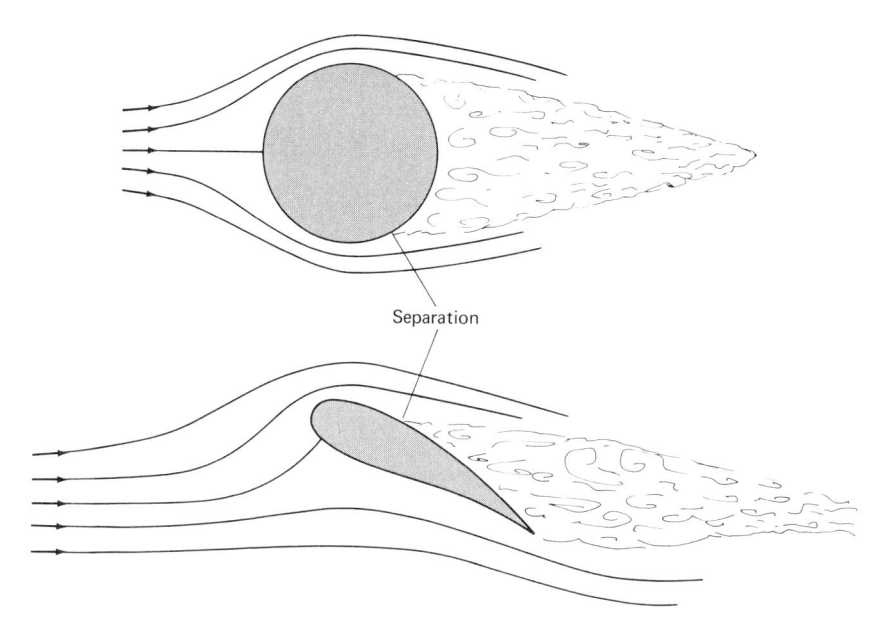

Fig. 10.1. Examples of separation of fluid on a body.

cylinder. The separation point on the cylinder is not stationary but oscillates around its average location, such as between A and B in Fig. 10.2. A vortex is generated as the separation point moves from B to A, is then "shed" from the cylinder and the separation suddenly moves back to point B. This process continues with a shedding frequency ω and results in an oscillatory force acting on the cylinder. If ω is near the natural frequency of the cylinder, the resulting condition of resonance* may be sufficient to cause structural damage. A dramatic case of this phenomenon was the wind-driven oscillations which caused the collapse of a suspension bridge near Tacoma, Washington. Power lines, TV antennas, and other structures have also been damaged by this effect. Care must be taken in the design of structures which exhibit this phenomenon so that their natural frequency is quite different from that of the shedding vortices.

The wake downstream of the cylinder maintains the vortex structure and is referred to as a *Kármán vortex street*. This vortex structure occurs in the Reynolds number range $40 < \text{Re} < 10{,}000$. Figure 10.2b

---

*Resonance is an interesting phenomenon that may occur when an input frequency of a force on a body is equal to (or very close to) the natural frequency of the body. The natural frequency is that frequency with which a body would vibrate if it were given a "knock."

(a) Kármán vortex street

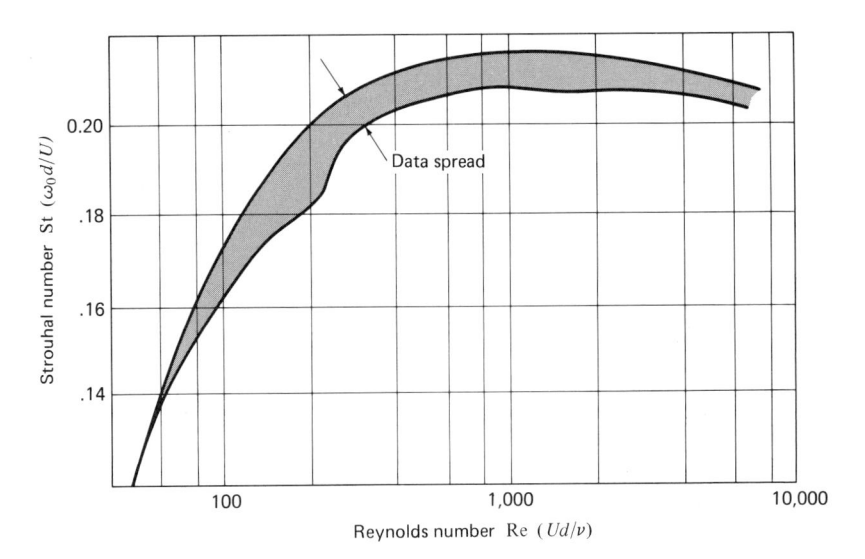

(b) Strouhal number vs. Reynolds number

Fig. 10.2. Oscillatory flow in the wake of a circular cylinder. (Based on the data of A. Roshko, "On the Development of Turbulent Wakes from Vortex Streets," NACA Rep. 1191, 1954).

is taken from a comprehensive investigation of these effects by Roshko.* Two of the conclusions from his study are:

(1) Periodic wake phenomena behind bluff cylinders may be classified into two distinct Reynolds number ranges (joined by a transition range). For a circular cylinder these are:

> Stable range $\quad 40 < \mathrm{Re} < 150$
> Transition range $\quad 150 < \mathrm{Re} < 300$
> Irregular range $\quad 300 < \mathrm{Re} < 10{,}000+$

*A. Roshko, "On the Development of Turbulent Wakes from Vortex Streets," NACA Rep. 1191, 1954.

In the stable range the classical, stable Kármán streets are formed; in the irregular range the periodic shedding is accomplished by irregular, or turbulent, velocity fluctuations. (2) A velocity meter based upon the relation between velocity and shedding frequency is practical.

The essential idea advanced in conclusion (2) is that the velocity of a flow may be determined by relating the Strouhal number St (a dimensionless shedding frequency) to the Reynolds number by using the data of Fig. 10.2b. Example 10.1 illustrates the velocity meter.

Another phenomenon of occasional interest is that of aerodynamic heating of an object moving at high speed, an excellent example being spacecraft reentry. Included in this area of study would be the influence of a heated body on the various flow parameters such as boundary layer thickness and drag.

Cavitation is a phenomenon encountered when a body moves through a liquid at speeds sufficient to create local pressures at or below the vapor pressure. The drag characteristics of a body are affected by these conditions and thus must be considered in the analysis of, for example, a marine propeller.

Example 10.1

The velocity of an airstream is to be measured. The velocity range expected is from 1 fps to 10 fps. Evaluate the possibility of determining the velocity of the airstream by (a) a pitot probe with an attached manometer, and (b) vortex shedding.

Solution. From the simplified relationship between the reading of a manometer attached to a pitot probe and the velocity (see Example 1.18), assuming $T = 70°F$ (530°R) and $H = 29.5$ in. Hg,

$$V = 15.9\sqrt{\frac{530h}{29.5}} = 67\sqrt{h}$$

Consequently, for $V = 1$ and 10 fps, $h = 0.00022$ and 0.022 in. H$_2$O, respectively. Since the typical accuracy for a carefully constructed micromanometer is of the order of 0.001 in. H$_2$O the minimum velocity cannot be successfully measured with a pitot probe.

If the shedding technique were used, its application would be somewhat dependent upon the available instrumentation. A hot-wire anemometer, an electronic device which is sensitive to velocity perturbations, may be used to determine the frequency by displaying the output signal on an oscilloscope or a recorder. Only the frequency is desired, thus it is not necessary to obtain a voltage output–velocity correlation curve, a rather difficult task. (If a wave analyzer is available the frequency can be measured directly.)

We select a cylinder size to make the hot-wire sensing most simple, for example, $d = 0.5$ in. For this condition the lower Reynolds number of interest (using $\nu = 1.6 \times 10^{-4}$ ft^2/sec) is

$$\text{Re} = \frac{1 \times 0.5/12}{1.6 \times 10^{-4}} = 260$$

and the higher Reynolds number is 2600. The corresponding Strouhal numbers from Fig. 10.2b, are 0.20 and 0.21, respectively. The lower frequency of the oscillations is found from

$$\text{St} = \frac{\omega_0 d}{U} = 0.20 \quad \text{or} \quad \omega_0 = \frac{0.20 \times 1}{0.5/12} = 4.8 \text{ rad/sec}$$

with the higher frequency equal to 50.4 rad/sec. These frequencies are easily measured with the suggested equipment.

The vortex shedding method is more accurate than the pitot probe method at both of the velocities of this example. At 10 fps the difference is not as pronounced. Obviously there will be a velocity at which the pitot probe method becomes more accurate.

10.3 SEPARATION

Separation occurs when the main stream flow leaves the body and forms a free-stream surface, a *dividing stream surface*, in the interior of the fluid. This stream surface may exhibit considerable unsteadiness. The *wake* can be identified as that region of the flow behind a body which is vortical (or sheared) if the approach flow is irrotational. (Note that since the shear effects propagate beyond the dividing stream surface, the boundary of the wake and the dividing stream surface are the same only at separation.) The separation phenomenon, the dividing stream surface, and the wake of a body are shown in Fig. 10.3.

For an inviscid flow, the main stream remains attached to the body, as was observed in Chapter 8. For the real flow situation, the flow over the forward portion of the body is similar to the inviscid flow (i.e., the pressure distribution and streamline pattern are similar), but the presence of the separation phenomenon makes the real and inviscid cases for the rest of the body quite different. A quantitative example of this is presented later.

The location of the separation point is strongly dependent upon the geometry of the body. An abrupt change in the geometry, such as a backward facing step or a corner of a building, will cause the flow to separate. The flow cannot follow the geometry of the body since to do

SEPARATION

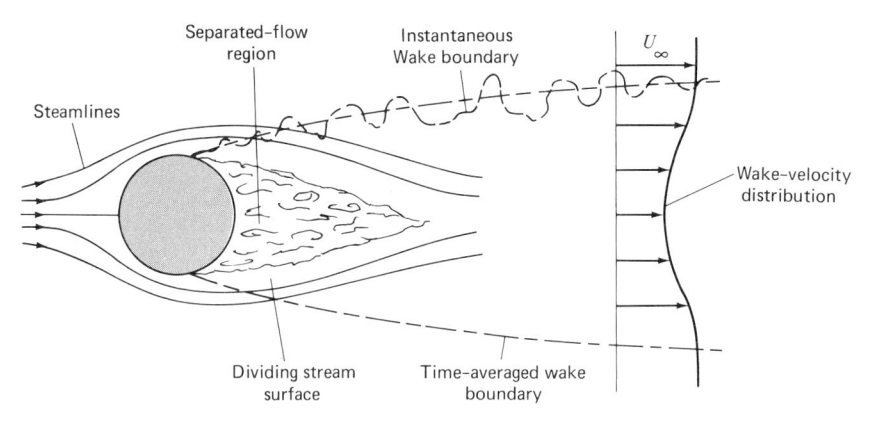

Fig. 10.3. Details of a separated flow.

so would require it to turn a sharp angle, hence a small radius of curvature and, from the Euler-n equation, a correspondingly large pressure gradient normal to the curved streamlines. That is, from

$$\frac{\partial p}{\partial n} = \rho \frac{V^2}{R} \tag{10.1}$$

$\frac{\partial p}{\partial n} \to \infty$ as $R \to 0$ for $V \neq 0$. The flow field responds to this "unreasonable" requirement by forming its own boundary for which the required pressure gradients are available.

The main stream may separate from a body because of an adverse (positive) pressure gradient. Qualitatively, momentum of the fluid near the surface may be insufficient to overcome the effect of the increasing pressure which exerts a net retarding force on each material element. When this happens, the fluid defines a new path for which its momentum is compatible with the pressure gradients of the new flow problem. For example, the inviscid flow over a cylinder involves a forward stagnation point, fluid acceleration to a maximum velocity at the maximum thickness, minimum pressure at the maximum thickness, and a deceleration to the rear stagnation point where the pressure rises to $\rho U_\infty^2/2$ above the ambient value. The boundary layer fluid, whose momentum is reduced by frictional effects, cannot negotiate the region of pressure rise to the rear stagnation point; rather, it separates from the body and forms the dividing stream surface of Fig. 10.4 resulting in a much smaller pressure rise. The following analysis demonstrates that a streamwise increase in pressure is a necessary condition to produce separation in the absence of the radius-of-curvature effects noted above.

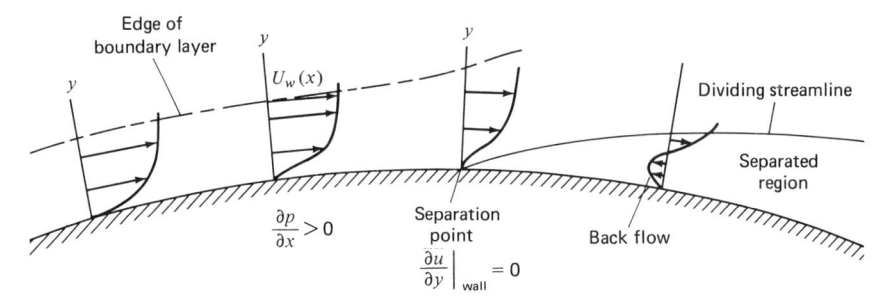

Fig. 10.4. Flow separation caused by an adverse pressure gradient.

The mathematical condition for separation must first be established. Let x be in the streamwise direction and let y be normal to the wall. Upstream of the separation point, $(\partial u/\partial y)_{\text{wall}} > 0$ and downstream of the separation point $(\partial u/\partial y)_{\text{wall}} < 0$. This condition identifies whether the flow near the wall is in the streamwise or backflow direction. The separation point is defined as the point where $(\partial u/\partial y)_{\text{wall}} = 0$. The condition that $\partial u/\partial y = 0$ at the wall for separation and the fact that $\partial u/\partial y$ is small at the edge of the boundary layer indicates that a local maxima exists in $\partial u/\partial y$ for a separated flow. If $\partial u/\partial y = $ maximum for $y \neq 0$, then $\partial^2 u/\partial y^2 = 0$ for the same y value. The decrease in $\partial u/\partial y$ near the edge of the boundary layer indicates that $\partial u^2/\partial y^2 < 0$ in this region; hence, $\partial^2 u/\partial y^2 > 0$ at $y = 0$ for a separated flow. The Navier-Stokes equations evaluated at $y = 0$ (i.e., for $u = v = 0$) show that

$$0 = -\frac{1}{\rho}\frac{\partial p}{\partial x} + \nu \left(\frac{\partial^2 u}{\partial y^2}\right)_{\text{wall}} \quad (10.2)$$

Consequently, $\partial p/\partial x > 0$ if $\partial^2 u/\partial y^2 > 0$. Hence, $\partial p/\partial x$ must be greater than zero, that is, an adverse pressure gradient must exist, if separation is to occur in the absence of curvature effects.

Separation is also influenced by a number of characteristics of the flow. The Reynolds number is the primary parameter influencing separation; of secondary importance are the free-stream fluctuation amplitude, wall roughness, and wall temperature. At sufficiently low Reynolds numbers, separation occurs only at sudden changes in the geometry; for flow around blunt objects with no sudden change in the geometry, for instance, a sphere, it would not occur. At some particular Reynolds number, a separated region would appear near the rear of a blunt object; and, as the Reynolds number increased, the separated region would increase, the boundary layer upstream of the separation point remaining laminar. At a sufficiently large Reynolds number the boundary layer before separation undergoes transition from a laminar to a turbulent layer. The fluid near the wall for the turbulent boundary layer contains substantially more momentum than the laminar layer at the same free stream velocity (see Fig. 10.5). This additional momentum in the boundary layer is more capable of overcoming the adverse effects of the positive pressure gradient; hence, a turbulent boundary larger separates farther downstream than the laminar layer. The resultant aft movement of the separation point is observed in the drag characteristics of spheres and cylinders; this is explored in the next section.

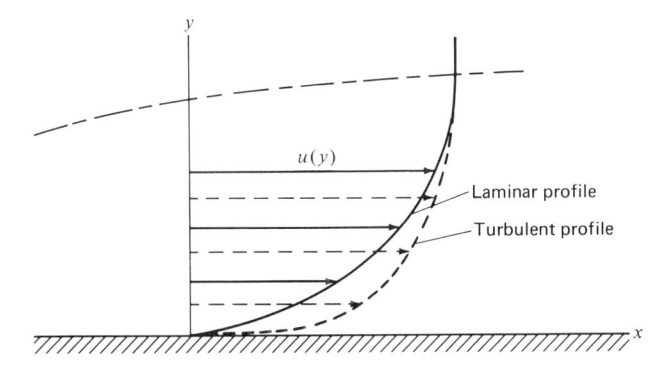

Fig. 10.5. Boundary-layer velocity profiles for laminar and turbulent flow.

For plane, two-dimensional bodies we can predict the point of separation analytically for a steady laminar flow by solving the boundary-layer equations

$$u\frac{\partial u}{\partial x} + v\frac{\partial u}{\partial y} = -\frac{dp}{dx} + \frac{1}{Re}\frac{\partial^2 u}{\partial y^2}$$

$$\frac{\partial u}{\partial x} + \frac{\partial v}{\partial y} = 0 \qquad (10.3)$$

where the pressure gradient dp/dx is that of potential flow at the wall. The boundary conditions are

$$\begin{aligned} u &= 0 & \text{at } y &= 0 \\ u &= U_w(x) & \text{at } y &= \delta \end{aligned} \qquad (10.4)$$

where $U_w(x)$ is the potential flow velocity at the wall. The coordinates x and y are as shown in Fig. 10.4. The boundary-value problem above is solved, and the point at which $(\partial u/\partial y)_{\text{wall}} = 0$ is defined as the point of separation. A numerical solution, using a digital computer, would be very useful in developing such a solution to this problem. However, it is considered beyond the scope of this text.

A similar set of equations would be solved if the boundary layer became turbulent before separation except the usual Reynolds stress terms must be included. Also, if an axisymmetric problem (a sphere, for example) were to be considered, appropriate terms to account for the curvature must be included.

10.4 LIFT AND DRAG

Lift and drag are defined as the force exerted on a body by a flowing fluid in the normal and streamwise directions, respectively. The forces are the net result of the pressure and shear-stress distributions and, consequently, the force is related to the character of the flow field over the body. The lift force L and the drag force D may be defined as

$$L = -\int_A p\hat{n}\cdot\hat{j}\, dA + \int_A \tau_0\cdot\hat{j}\, dA \tag{10.5}$$

and

$$D = -\int p\hat{n}\cdot\hat{i}\, dA + \int_A \tau_0\cdot\hat{i}\, dA \tag{10.6}$$

where τ_0 is the shear stress vector in the plane of the surface, $\hat{i}$ is the unit vector in the streamwise direction and $\hat{j}$ is transverse to it in the positive direction (not parallel and normal to the surface) in which the lift is desired. Lift and drag coefficients may be defined in terms of these forces:

$$C_L = \frac{L}{\frac{1}{2}\rho U_\infty^2 A} \tag{10.7}$$

$$C_D = \frac{D}{\frac{1}{2}\rho U_\infty^2 A} \tag{10.8}$$

where the normalizing factor $\frac{1}{2}\rho U_\infty^2 A$ is that used for incompressible flows in the similitude discussion in Chapter 4. This is seen to be well motivated since the lift and drag depend upon the pressure and the shear stress, which are both normalized by $\frac{1}{2}\rho U_\infty^2$. The area A is a characteristic area, such as the frontal projected area of a cylinder (for the drag coefficient) or the planform area for the lift on an airfoil.

The normalized pressure and the velocity distributions are dependent on the boundary conditions and the Reynolds number for any given

Art. 10.4 **LIFT AND DRAG** 455

flow field (Chapter 4 gives the basic considerations supporting this statement). Hence, it is directly established that the lift and drag coefficients are dependent on the same factors. Detailed considerations of the drag coefficient for several different flow fields will be presented in the next section; general characteristics regarding lift and the particular case of the drag coefficient for flow about a sphere will be presented in this section.

1. Circulation

The lift on a body is usually of interest under the conditions of high Reynolds numbers and motion through a stationary medium. Consequently, the restrictions to inviscid, irrotational flow are realistic constraints for an otherwise general discussion of lifting bodies. From the inviscid-flow chapter, we know that the lift on a two-dimensional cylinder with circulation is given by the simple relationship

$$L = -\rho U_\infty \Gamma \qquad (10.9)$$

It is plausible, but not easily shown, that the lift on any two-dimensional body in inviscid, irrotational motion is given by Eq. 10.9. The use of Eq. 10.9 to describe the lift on a body can be further appreciated in the context of an example.

Consider a flat plate, oriented at a slight angle, in a uniform airstream as shown in Fig. 10.6a. The streamline pattern shown is that of an inviscid, irrotational motion without circulation. Parts a and b are both unrealistic, in that the effects of viscosity are ignored; however, Fig. 10.6a is also unrealistic since it supposes that the flow will come around the trailing edge before it separates from the plate. The flow will not, of course, follow such a path on the forward or aft ends of the plate; the forward position will be altered by a separation bubble and a reattached flow near the leading edge. A much more significant alteration is signified by the flow pattern of Fig. 10.6b. The stagnation

(a) Inviscid, irrotational flow without circulation

(b) Inviscid, irrotational flow with circulation

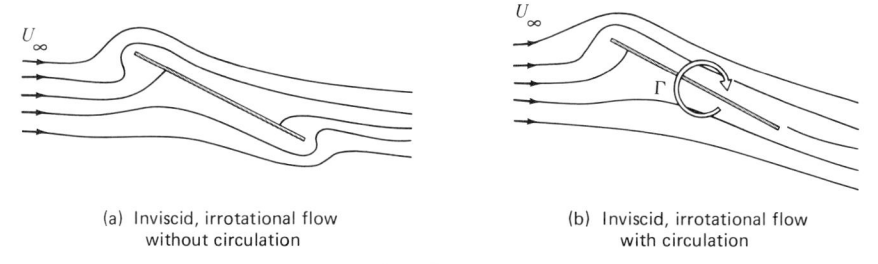

Fig. 10.6. Flow over a flat plate at a given angle of attack.

streamline, which defines the plane of the body, leaves the aft edge of the plate if a circulation Γ is added to the flow pattern; this is termed the *Kutta condition*. This amount of circulation may be considered responsible for the lift of magnitude $-\rho U_\infty \Gamma$, which acts upon the plate of Fig. 10.6b.

The relationship for the rate of change of circulation, Eq. 5.25, allows a different and perhaps a more realistic conceptualization of the existence of lift. The flat plate used above will be replaced by an airfoil shape to aid the realism. Figure 10.7 shows at time t an airfoil which started from rest at $t = 0$; the fluid is stagnant. A simply connected contour C has a segment of its contour in the free portion of the fluid and remainder at the surface of the airfoil. The initial circulation about C is zero; the rate of change of Γ is given by (see Eq. 5.25)

$$\frac{d\Gamma}{dt} = \oint_C \nu \nabla^2 \mathbf{V} \cdot d\mathbf{s} - \int_A [\nabla \times (\boldsymbol{\omega} \times \mathbf{V})] \cdot \hat{n} \, dA \qquad (10.10)$$

As the airfoil accelerates from zero velocity, the surface effects ($\nu \nabla^2 \mathbf{V}$) increase the magnitude of the circulation Γ from its initial zero value. However, as the velocity increases, the area integral, which represents the net convection of ω out of the planar region bounded by the curve, is also increasing. When a steady flow is attained, $d\Gamma/dt = 0$ and the continual production effect of the viscous term is exactly balanced by the vorticity flux.

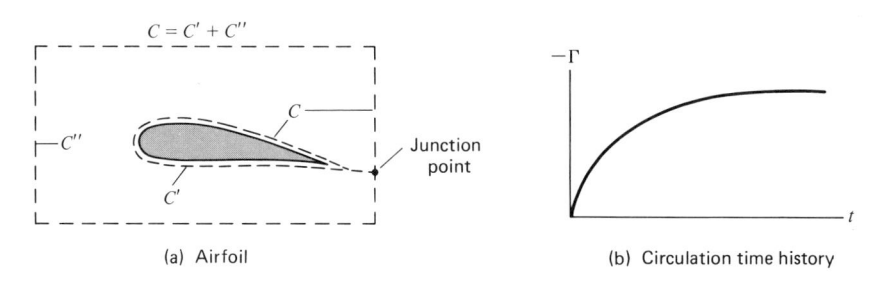

(a) Airfoil

(b) Circulation time history

Fig. 10.7. Two-dimensional airfoil with attached contour and its circulation history for an airfoil motion initiated from rest.

Curve C is composed of two segments, C' and C''. The curves C' and C'' are allowed to meet at a common juncture point, and the curve C'' encloses all of the space inside of the large square; this is not a simply connected domain. Hence, we can consider that the airfoil is removed and a single concentrated vortex provides the steady state circulation Γ. That is, the lift on the vortex filament is the same as the lift on the airfoil, $-\rho U_\infty \Gamma$. A material curve which contains the airfoil and is placed at a large distance from the airfoil will have a circulation which remains constant at its initial value of zero. Consequently, the negative circulation associated with C'' which moves with the airfoil will be balanced by the circulation of the "starting vortex" which is shed as the airfoil begins its motion. This is demonstrated in Fig. 10.8.

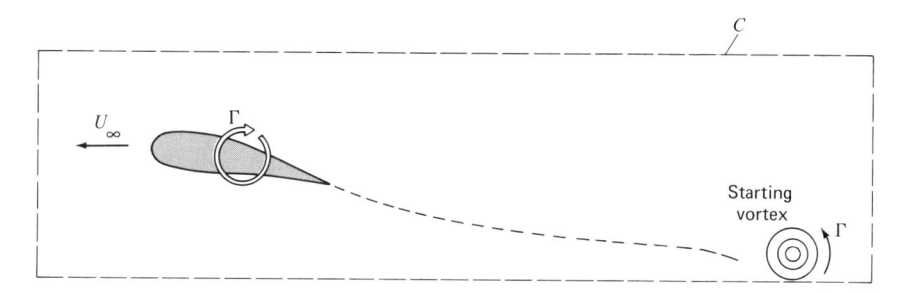

Fig. 10.8. Starting vortex for an airfoil.

2. Drag

In general, the low pressure in the wake region on the downstream part of a body, combined with the high stagnation pressure on the front of a body, results in a contribution to the total drag called the *pressure drag* or *form drag*. The remaining contribution to the drag is the *viscous drag* or *frictional drag* resulting from the shearing stresses acting on the surface on the body. The viscous drag is usually a small part (less than 10%) of the total drag for bodies with relatively large separated regions such as those shown in Fig. 10.1. For streamlined bodies at small angles of attack or for bodies at low Reynolds numbers, such as those shown in Fig. 10.9, the separated region is either small or non-existent and the viscous drag is the dominant contribution to the drag.

These effects are clearly demonstrated in the drag coefficient for a sphere, as shown in Fig. 10.10a. The pressure distribution over the dia-

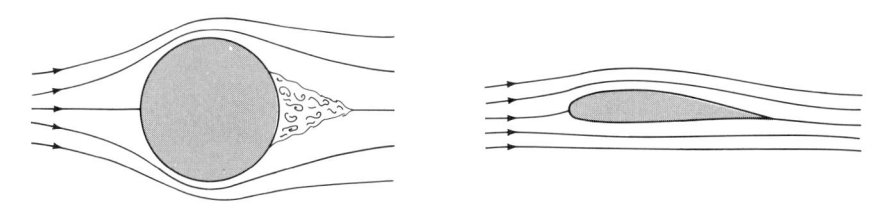

Fig. 10.9. Flows dominated by viscous drag.

metral plane is presented in Fig. 10.10b. For quite small Reynolds numbers, there is a negligible separated region and the viscous stress is the dominant contribution to the drag. A creeping flow exists for Re < 1 and laminar separation occurs at the rear of the cylinder at Re ≅ 50. The drag coefficient decreases with increasing Reynolds number for these Reynolds number values. The influence of the initial stages of separation (the separated region enlarges until Re = 3 × 10^4) and the eventual dominance of the pressure effects in determining the drag are shown by the leveling off of the drag coefficient with increasing Reynolds number. The dramatic drop in the C_D curve is the result of transition from a laminar to a turbulent boundary layer. Recall that the turbulent boundary layer could penetrate farther into the region of increasing pressure because of its increased momentum near the sur-

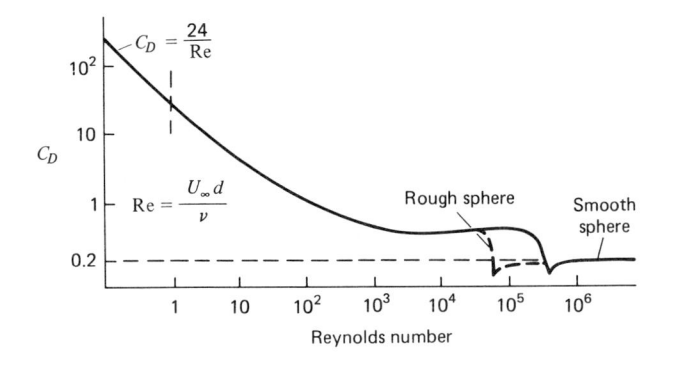

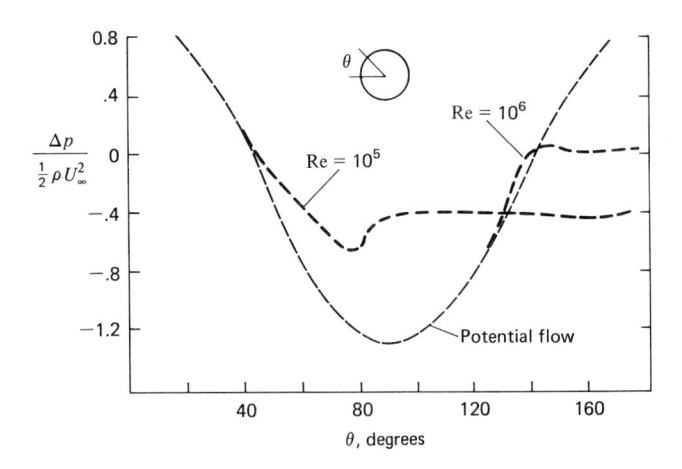

Fig. 10.10. Drag coefficient and pressure distribution for flow around a sphere.

face. The consequence is that the size of the low pressure wake region is substantially reduced, and hence the contribution of the low-pressure wake to the total drag force is similarly reduced. Fig. 10.11 shows the significant alteration in the flow pattern resulting from the transition phenomenon.

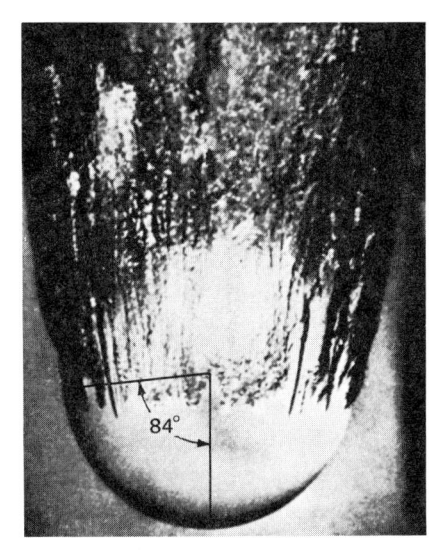

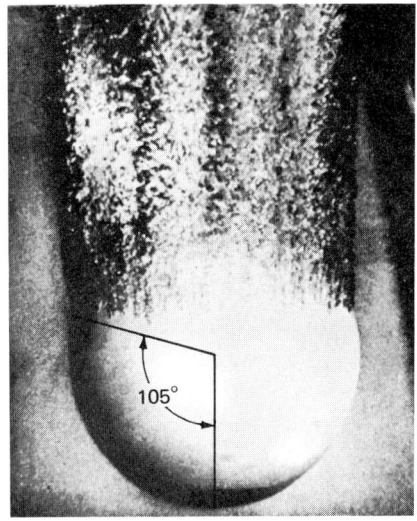

(a) Laminar boundary layer before separation

(b) Turbulent boundary layer before separation

Fig. 10.11. Influence of boundary-layer transition on flow around a sphere, Re = 200,000. (Courtesy of U.S. Naval Ordnance Test Station, Pasadena, Calif.)

The influence of the boundary conditions is manifest in two different ways regarding the transition point and hence the location of the drop in the C_D curve. The free-stream fluctuation intensity influences the Reynolds number $(U_\infty d/\nu)$ required for transition, Re_{crit}, on the smooth sphere. The fluctuation intensity enters as a boundary condition in that it, along with the mean velocity magnitude U_∞, is used to characterize the velocity which approaches the sphere. The second boundary-condition effect is the influence of the surface roughness on the Re_{crit} value. Since the velocity goes to zero at the actual surface location, the roughness, as characterized by e/d, enters in the problem in terms of the no-slip boundary condition. It is possible to provide artificial roughness to cause an early transition of the laminar boundary layer. A trip wire or sand glued to the surface is often effective for laboratory studies. The fact that boundary layer transition delays separation and reduces the drag coefficient has found practical application in the design of golf balls. The dimples on the golf ball enhance transition and hence cause a drop to the lower portion of the C_D curve ($C_D = 0.2$) at a smaller Reynolds number. The turbulent boundary layer is also much less sensitive to rotational (hook-slice) effects in terms of the influence of angular motion on the separation point. The motion of a smooth table tennis ball when thrown with spin attests to the importance of such separation effects with a laminar boundary layer.

10.5 STEADY-FLOW DRAG CHARACTERISTICS FOR IMMERSED BLUNT OBJECTS

From our discussion of similitude in chapter four we know that the dimensionless drag (the drag coefficient) is a function of the Reynolds number, which is the only parameter that exists in the governing differential equations for flows in which compressibility, unsteady, gravitational, and surface tension effects are not important. Drag coefficient versus Reynolds number curves will be presented for some common bodies. The pressure distribution on several of the bodies will also be presented. The bodies will be assumed to be immersed in an infinite stream flowing with an upstream velocity U_∞.

1. Cylinders

The drag coefficient curve and the pressure distribution for flow around a circular cylinder of infinite length are shown in Fig. 10.12. The flow is completely laminar until turbulence is generated in the vortex-wake region at Reynolds numbers greater than 5000. The creeping-flow regime terminates at $\text{Re} \cong 1$, separation occurs at $\text{Re} \cong 5$ and a laminar Kármán vortex street occurs over the range $40 \gtrsim \text{Re} \gtrsim 5000$. The wake then becomes turbulent and the vortex street is not observed at $\text{Re} > 10{,}000$. At $\text{Re} \cong 200{,}000$ for a smooth cylinder and 40,000 for a very rough cylinder, the boundary layer just before the separation point becomes turbulent and a sudden drop in the drag occurs. This drop can also be triggered at the lower Reynolds numbers if the free-stream flow has large amplitude fluctuations. For high-Reynolds-number flows, $\text{Re} > 10^6$, the drag coefficient is constant (0.4) so that the drag is directly proportional to the velocity squared; this is the "completely turbulent" regime of flow.

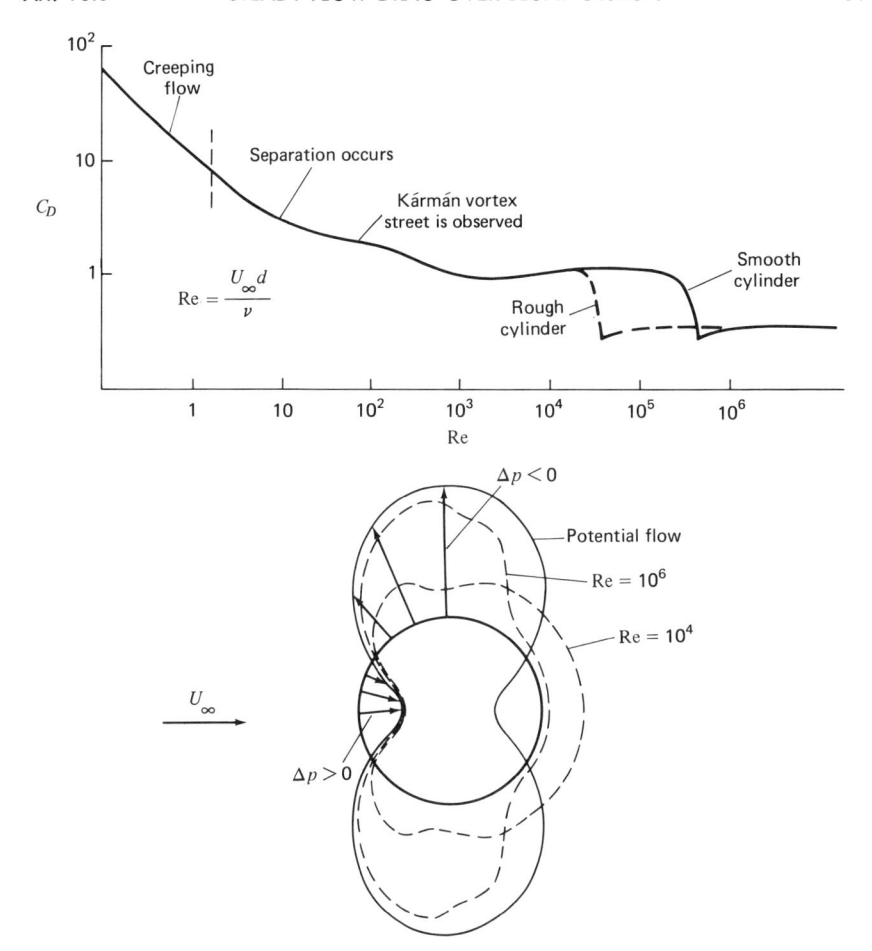

Fig. 10.12. Drag-coefficient curve and pressure distribution for flow around a circular cylinder. (From J. W. Daily and D. R. F. Harleman, *Fluid Dynamics*, Addison-Wesley, 1966.)

Periodic vortex shedding initiates at $Re \simeq 44$ with an associated Strouhal number ($St = \omega d / U_\infty$) of 0.12. The Strouhal number reaches a value of 0.21 at $Re \simeq 1000$ and maintains this value until the vortex shedding phenomenon disappears in the "completely turbulent" flow regime. During the vortex shedding the drag coefficient maintains a time average value of $C_D = 1.0$.

Drag coefficients for circular cylinders of finite length and elliptical cylinders of infinite length are presented in Table 10.1.[2]

TABLE 10.1
Drag Coefficients of Finite-Length Circular Cylinders* and of Infinite-Length Elliptic Cylinders

Circular Cylinder		Elliptic Cylinder		
$\dfrac{\text{Length}}{\text{Diameter}}$	$\dfrac{C_D}{C_{D_\infty}}$	$\dfrac{\text{Major Axis}}{\text{Minor Axis}}$	Re	C_D
∞	1	2	4×10^4	0.6
40	0.82	4	10^5	0.46
20	0.76	4	2.5×10^4 to 10^5	0.32
10	0.68	8	2.5×10^4	0.29
5	0.62	8	2×10^5	0.20
3	0.62			
2	0.57			
1	0.53			

*C_{D_∞} is the drag coefficient for an infinite-length circular cylinder obtained in Fig. 10.12.

TABLE 10.2
Drag Coefficients for Various Blunt Objects

Object		L/w	Re	C_D
Square cylinder	→□ w	∞	3.5×10^4	2.0
	→◇ w	∞	10^4–10^5	1.6
Rectangular plates	→❘ w	∞	$> 10^3$	2.0
		20	$> 10^3$	1.5
		5	$> 10^3$	1.2
		1	$> 10^3$	1.1
Automobile				
1920		—	$> 10^5$	0.9
Modern		—	$> 10^5$	0.3
Circular cylinder	→▭ w	0	$> 10^3$	1.10
		4	$> 10^3$	0.90
		7	$> 10^3$	1.0
Volkswagen bus		—	$> 10^5$	0.42

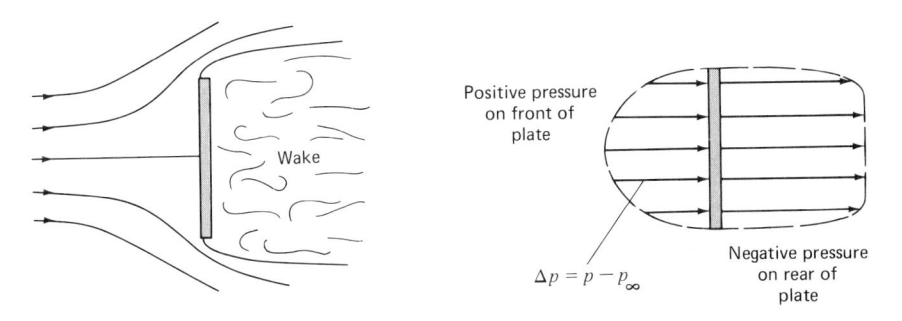

Fig. 10.13. Streamline pattern and pressure distribution for flow around a flat plate and a disc.

2. Flat Plates, Discs, and Other Blunt Objects

For flow around a flat plate or a disc, the drag is due to the pressure drag only; the viscous drag is negligible. The separation region does not change with increasing Reynolds number; thus it is not surprising to find that the drag coefficient is constant for Reynolds numbers above the creeping-flow range. For an infinite flat plate, the drag coefficient is 2.0 and for a disc it is 1.1. The streamline pattern and pressure distribution are shown in Fig. 10.13. Drag coefficients for flat plates with finite width and various other blunt objects are presented in Table 10.2.

Example 10.2

The drag coefficient on a modern automobile is approximately constant for all driving speeds. If the horsepower of the engine is increased by 25 percent approximately what percentage of increase in top speed can be expected?

Solution. The integral energy equation for the automobile shows that

$$-\dot{W}_s = \text{Drag} \times V$$

Assuming that the efficiencies of the two engines are the same, we have

$$\frac{(hp)_2}{(hp)_1} = \frac{(\text{Drag})_2 \times V_2}{(\text{Drag})_1 \times V_1}$$

Let the horsepower of the second engine be 1.25 times the horsepower of the first; then, using drag = $C_D \times \frac{1}{2}\rho V^2 A$, we have

$$1.25 = \frac{V_2^3}{V_1^3} \quad \text{or} \quad V_2 = 1.078 V_1$$

assuming $(C_D)_1 = (C_D)_2$, as suggested. The percentage of increase in velocity is then

$$\text{Percentage of increase} = \frac{V_2 - V_1}{V_1} \times 100 = 7.8\%$$

Extension 10.2.1. If the thermal and mechanical efficiencies of an engine are approximately constant between the speeds of 50 mph and 60 mph, determine the increase in fuel consumption for a 200-mile trip by traveling 60 mph instead of 50 mph. *Ans.* 44%

Example 10.3

A square service station sign, 15 ft on a side, is secured on top of an 80-ft-high pole which is 1 ft in diameter. Approximate the force on the supporting structure due to the wind drag for a 60-mph wind. Would you expect vortex shedding at this wind speed?

Solution. The drag on the sign is found, using a C_D value from Table 10.2, to be

$$\text{Drag} = \tfrac{1}{2}\rho V^2 A C_D$$

$$= \tfrac{1}{2} \times 0.0024 \times 88^2 \times 15^2 \times 1.1$$

$$= 2300 \text{ lb}$$

The contribution to the drag by the pole, assuming a velocity of 88 fps acting over the complete length, is

$$F_P = \tfrac{1}{2}\rho V^2 A C_D$$

$$= \tfrac{1}{2} \times 0.0024 \times 88^2 \times 1 \times 80 \times 0.4$$

$$= 300 \text{ lb}$$

where C_D is found in Fig. 10.12 using a Reynolds number of $\text{Re} = Vd/\nu = 88 \times 1/1.6 \times 10^{-4} = 5.5 \times 10^5$. Assuming that large-amplitude fluctuations occur near the surface of the earth, due to the turbulent ground boundary layer, we use $C_D = 0.4$. The total force is

$$F = 2300 + 300 = 2600 \text{ lb}$$

The Reynolds number of 5.5×10^5 is in the completely turbulent-flow regime, so that vortex shedding would not be occurring.

10.6 FLOW CHARACTERISTICS FOR AIRFOILS

It is the strong adverse pressure gradient that leads to separation on a blunt object such as a cylinder. Airfoils are streamlined bodies designed to reduce the magnitude of the adverse pressure gradient. By eliminating the separated region, or by making it very small, we can reduce the drag coefficient by a factor of 10 or more. It is interesting to note that the drag on a cylinder is 10 times larger than on an airfoil shape with the same frontal (or projected) area. The drag for the airfoil is primarily viscous drag associated with the boundary layer on the airfoil.

As the angle of attack of an airfoil is increased, the adverse-pressure-gradient increases in magnitude and separation is encouraged. For an aircraft, relatively strong adverse pressure gradients occur at takeoff and landing, when the speed is reduced and the lift must be maintained. Separation-control techniques must then be employed to eliminate a large-scale separation, which would negate the increased lift from the larger angle of attack.* The most common technique is to design gaps which open up so that the higher pressure on the bottom of the airfoil forces a stream of high-momentum air into the boundary layer, thereby delaying or eliminating separation. An example of this type of separation control is shown in Fig. 10.14 for the airfoil of a proposed short-take-off-and-landing (STOL) aircraft. Here the high-velocity exhaust from fans is passed through slots for separation control.

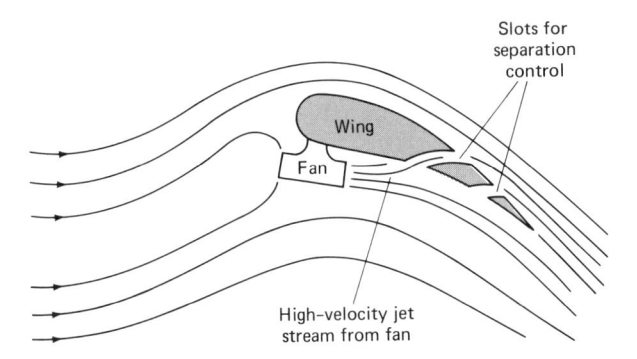

Fig. 10.14. Separation-control technique for a proposed STOL wing-fan system.

The drag of an airfoil is usually presented in its relationship to the lift force, which is of course of primary importance. The drag is

---

*An alternate approach is often used on landing. Large flaps distended from the main wing section serve to provide an upward force as a result of the downward-directed momentum flux. There is a large drag force associated with this configuration which serves as an aerodynamic brake.

typically minimum when the angle of attack is zero, and gradually increases as the angle of attack is changed from zero. The lift is typically positive for zero angle of attack and increases as the angle of attack is increased. (See Fig. 10.15 for the drag and lift coefficients of a representative airfoil.) The plot of C_D vs. C_L is quite pertinent for an airfoil. Aerodynamically, it gives the "cost" as a function of the "benefit." Such plots, and a knowledge of C_D for the other segments of the aircraft, along with its gross weight and the thrust of the engines, allow the cruise condition to be estimated. The lift coefficients reach a high of approximately 1.6 for conventional airfoils; lift coefficients for the above-described STOL airfoils may reach magnitudes of 8 to 10 before separation ("stall") becomes significant.

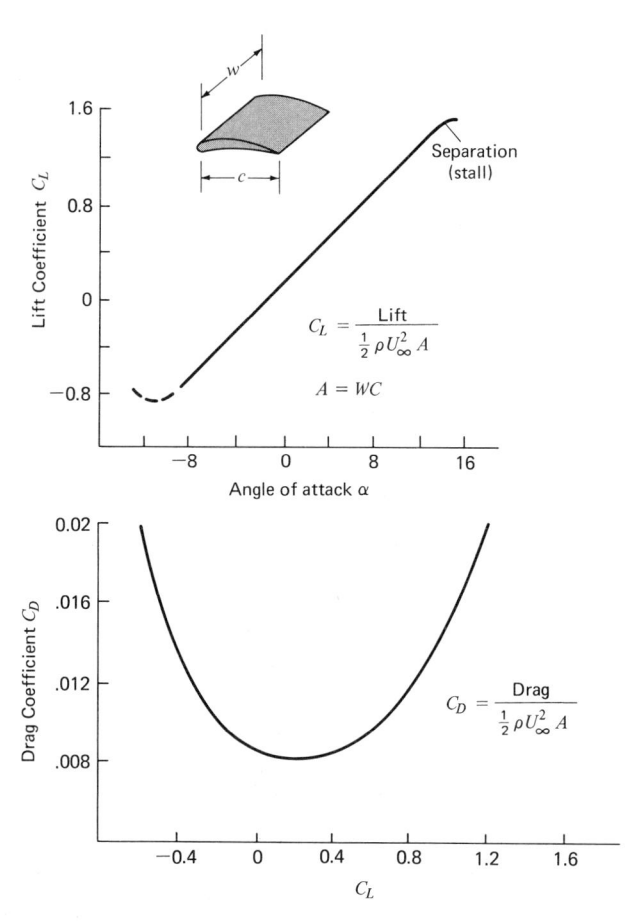

Fig. 10.15. Drag and lift coefficients for a typical airfoil for $Re > 10^6$.

Art. 10.6 FLOW CHARACTERISTICS FOR AIRFOILS

An interesting phenomenon associated with a lifting body such as an airfoil is the requirement that circulation exist in the plane perpendicular to the axis of the body. In Chapter 8 the lift was given by

$$\text{Lift} = -\rho U_\infty \Gamma \qquad (10.11)$$

Thus, as we increase the lift, the magnitude of the circulation is increased. The increase in lift with respect to angle of attack can be directly related to the increase in circulation magnitude about the airfoil. It is not difficult to imagine the dramatic effect that separation could impose on the magnitude of Γ. This stall condition is shown in Fig. 10.15. As the flow separates from the airfoil a dramatic decrease in the aerodynamic lift is encountered. However, large lift for low forward speed is required for landing and takeoff. If a large angle of attack is used to gain the large lift, separation may occur; as already mentioned, active separation-control techniques are employed to eliminate or delay the undesirable separation, as shown in Fig. 10.16. These techniques are (1) slotting the airfoil to feed high-velocity air from the lower part of the airfoil into the low-speed boundary layer on the upper part, (2) suction of the low-speed fluid from near the wall on the upper surface, and (3) tripping the laminar boundary layer to create turbulence. These techniques all serve to increase the momentum in the boundary layer, thus delaying or preventing separation.

Fig. 10.16. Flow over an airfoil, with and without suction.

Fig. 10.17 shows the vortex pattern for a finite airfoil. The requirement that a vortex tube may not be broken may be used to explain the trailing vortex segments which connect to the starting vortex. The three-dimensional field of the trailing vortices, which endanger small aircraft flying behind large aircraft, causes a significant downwash, especially in the vicinity of the tail, and strongly influences the positioning of the tail control surfaces. Often these surfaces are located above the aircraft body, out of the region of large downwash. The tip vortices and starting vortex eventually diffuse, due to the action of viscosity, so that after some period of time they are not noticeable.

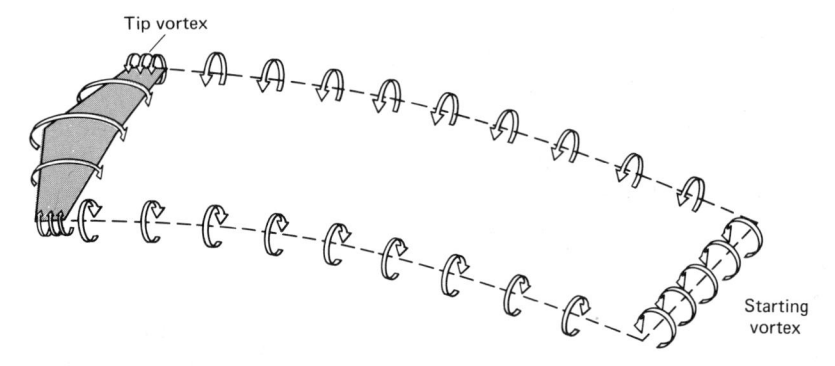

Fig. 10.17. Vortex system associated with an airfoil.

10.7 EFFECT OF COMPRESSIBILITY ON DRAG

Compressibility effects become important at speeds exceeding a Mach number of approximately 0.3. For high-Reynolds-number flows the drag coefficient becomes Reynolds-number-independent and the Mach number is the governing parameter. The drag coefficient is quite constant on conventional airfoils for Mach numbers up to 0.75; then a sudden rise occurs until the Mach number is approximately unity. The drag coefficient then decreases, as shown in Fig. 10.18. By sweeping back the wings, as is done in commercial subsonic aircraft, the sudden rise in the drag coefficient can be delayed to higher Mach numbers, 0.9 possibly. It is the component of velocity normal to the leading edge of the sweptback wing which must have an associated Mach number less than 0.75. Cruise speeds at $M = 0.8$ for subsonic aircraft with sweptback wings and $M = 2.0$ for supersonic aircraft are reasonable, considering the drag characteristics. Compressibility also affects the drag coefficient for flow normal to a flat plate. For an insulated plate, the results are as shown in Fig. 10.19; for a plate with heat transfer, the results would be altered.

10.8 EFFECT OF CAVITATION ON DRAG

Cavitation occurs when the local pressure becomes less than the local vapor pressure of the liquid. Equivalently, cavitation occurs whenever

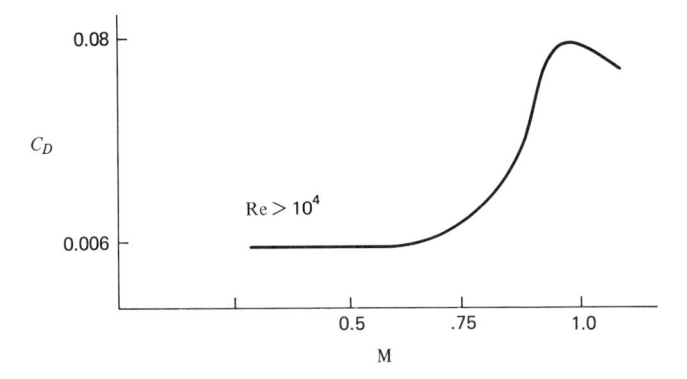

Fig. 10.18. Drag coefficient versus Mach number for a typical unswept airfoil.

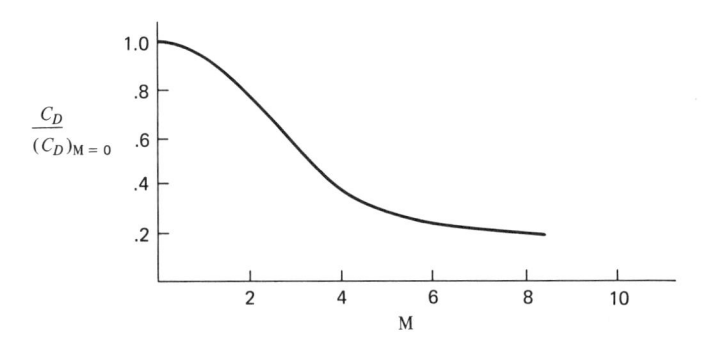

Fig. 10.19. Drag coefficient on a flat—compressibility effect.

the *cavitation number* K, defined as

$$K = \frac{p_\infty - p_v}{\frac{1}{2}\rho U_\infty^2} \tag{10.12}$$

is less than the critical cavitation number K_{crit}. This critical value is dependent on the particular shape of the body. As K decreases, the cavitation increases in intensity; and for sufficiently small K, "super cavitation" results. The drag characteristics of a body are sensitive to cavitation; and, experimentally, it has been found that, for small cavitation number,

$$C_D = C_D(0)(1 + K) \tag{10.13}$$

where $C_D(0)$ is the drag coefficient of the body for $K = 0$. This relationship is acceptable for most geometries up to $K \cong 0.4$. Table 10.3 lists some of the drag coefficients for $K = 0$ for $\text{Re} \cong 10^5$.

The drag and lift coefficients and the cavitation number are shown in Table 10.4 for a typical nonsymmetric hydrofoil for $10^5 < \text{Re} < 10^6$.

TABLE 10.3
Drag Coefficients for Zero Cavitation Number for Blunt Objects

Two-Dimensional Body			Axisymmetric Body		
Geometry	θ	$C_D(0)$	Geometry	θ	$C_D(0)$
Flat plate	—	0.88	Disc	—	0.8
Wedge	120	0.74	Cone	120	0.64
	90	0.64		90	0.52
	60	0.49		60	0.38
	30	0.28		30	0.20
Circular cylinder	—	0.50	Sphere	—	0.30

Note: $Re = 10^5$.

TABLE 10.4
Drag and Lift Coefficients and Cavitation Number for a Typical Hydrofoil

Angle (°)	Lift Coefficient (C_L)	Drag Coefficient (C_D)	Cavitation Number (K)
−2	0.2	0.014	0.5
0	0.4	0.014	0.6
2	0.6	0.015	0.7
4	0.8	0.018	0.8
6	0.95	0.022	1.2
8	1.10	0.03	1.8
10	1.22	0.04	2.5

10.9 ADDED MASS

For a body moving at constant velocity, such as has been studied in the previous articles of this chapter, the applied force must overcome the drag and the weight. However, if a body is accelerating then an additional force is required to accelerate the body and to accelerate the fluid in the vicinity of the body. The resistance to the motion provided by the accelerating fluid is in addition to the drag force which may be present. The accelerating body of mass M appears to be more resistant to the motion than it should be owing to the fluid being accelerated; it appears to have an added mass M_a which is independent of the velocity for an irrotational motion of the fluid around the body. Using Newton's second law

$$F - D = (M + M_a)\frac{dV_B}{dt} \tag{10.14}$$

where V_B is the velocity of the body and D is the drag force. We have assumed a horizontal motion.

The added mass is typically related to the mass m_f of fluid displaced by the body. The relationship is

$$m_a = km_f \tag{10.15}$$

where k is the *added mass coefficient*. The added mass coefficient is actually a second-order tensor. Presenting it as a scalar coefficient results in a different coefficient for each direction of motion relative to a fixed axis on the body. That is, for an ellipsoidal body, the coefficient k changes as the direction of motion changes relative to, say, the major axis. Some of the more common coefficients for irrotational motions are: for an ellipsoid with the major axis twice the minor axis and moving parallel to the major axis, $k = 0.2$; for a sphere, $k = 0.5$; and for a long cylinder moving normal to its axis, $k = 1.0$.

For dense bodies accelerating in the atmosphere it is acceptable to neglect the added mass since the mass of the displaced air would be negligible when compared with the mass of the body. However, to determine the acceleration of bodies in liquids the added mass must be accounted for. As a body is set in motion from rest the motion is initially irrotational so that the added mass coefficients given above may be used. They also find use in the study of oscillating bodies.

PROBLEMS

10.1 Sketch the separated flow on a turbine blade and a diffuser. Also sketch the probable airflow around an automobile, showing the various separated regions.

10.2 Sketch the probable pressure distribution over the top surface and the bottom surface of a stalled airfoil, and of the same airfoil with no separation.

10.3 Indicate how separation may be avoided on the airfoil of an aircraft during take-off. Explain why it is beneficial to avoid separation.

10.4 Would you expect the flow to separate on a submersible, a one-man submarine, traveling at 2 fps? Explain in detail why a sudden drop would occur in the drag vs. velocity curve as the velocity is increased.

10.5 We wish to calculate the point of separation of water for flow around a circular cylinder. Write the equations and boundary conditions explicitly and discuss how one would find a solution.

10.6 Approximate the maximum force to be expected on a flag pole composed of two section, 6 and 9 in. in diameter, respectively; each section is 50 ft long. Assume a maximum wind velocity of 120 mph. Over what range of wind velocities would a transverse oscillatory force be expected on the pole?

10.7 Over what range of velocities would you expect an electrical wire $\frac{1}{4}$ in. in diameter to periodically shed vortices? Determine the range of frequencies associated with the shedding.

10.8 A large 100-ft-high smoke stack is to be designed for wind speeds of up to 150 mph. Approximate the maximum design force on it if its diameter is 6 ft at the top and 20 ft at the bottom.

10.9 Approximate the terminal velocity of a 2-ft-dia. sphere weighing 100 lb/ft^3 if it is dropped (a) from a plane and (b) from a ship.

10.10 The flow around the front of a sphere can be approximated by potential flow so that $U_w = (3/2)U_\infty \sin \theta$. The flow separates at $\theta = 90°$, with the pressure in the separated region assumed to be equal to the pressure at the $\theta = 90°$ position. Determine the drag coefficient.

10.11 Approximate the maximum density of a 0.2 in.-dia. sphere falling in water at 60°F so that a separated flow just exists.

10.12 A 4 × 8 ft rectangular sign weighs 200 lb. At what wind speed will the sign tip over if it is on supports which extend 3 ft from the base? The 8 ft dimension is vertical.

10.13 (a) Estimate the drag force on a car traveling at 100 mph if the drag coefficient is approximately 0.4. The projected area is 25 ft^2. (b) Determine the maximum possible speed of a Volkswagen bus with a 40-*hp* engine (80-percent mechanical efficiency) on the level with no tail wind. Assume $A = 30$ ft^2.

10.14 Estimate the minimum speed necessary for a 50-ton aircraft to take off, using (a) conventional airfoils with no boundary-layer control with a total wing area of 500 ft^2, and (b) a STOL airfoil with the same wing area.

10.15 Explain why the pilot of a small aircraft must be very careful when flying behind a large aircraft, even though the distance between may be several miles.

10.16 The drag coefficient for a jet aircraft is constant at approximately 0.006. Determine the percentage of reduction in drag as the aircraft reduces its speed from 500 mph to 400 mph. It maintains a constant elevation.

10.17 For an elevation of 30,000 ft, determine the maximum velocity at which a subsonic aircraft should travel if the wings are not swept back. For a sweptback angle of 30°, determine the increased cruise speed.

10.18 A submarine travels at a speed of 50 fps. A cylindrical tube 15 in. in diameter and 6 ft long is extended perpendicular to the submarine hull. (a) Estimate the drag on the tube if the submarine is 100 ft below the surface. (b) At what speed would cavitation be significant if $K_{crit} = 1.0$?

10.19 A 75-lb ball, 1 ft in diameter, is held stationary beneath the water and released. Determine the initial acceleration of the ball (a) neglecting the added mass and (b) accounting for the added mass.

10.20 A submersible has a length of 14 ft and a maximum diameter of 7 ft. It is at rest beneath the water and begins to move. Estimate the error in the initial acceleration if the added mass is neglected.

SELECTED REFERENCES

The motion picture *The Fluid Dynamics of Drag* (Nos. 21601, 21602, 21603, and 21604; A. H. Shapiro, film principal) would provide a helpful supplement to the text material in this chapter.

A companion text for the film is available: A. H. Shapiro, *Shape and Flow*, Doubleday & Co., Garden City, New York, 1961.

The subject of lift and drag for real objects relies heavily upon empirical documentation of lift and drag coefficients. A volume devoted to drag coefficients is S. F. Horner, *Fluid Dynamic Drag*, Midland Park, N.J. (published by the author), 1958. For aerodynamic characteristics of airfoils a classic reference (including a classification scheme) is I. H. Abbott, A. E. von Doenhoft, and L. S. Stivers, Jr., "Summary of Airfoil Data," NACA Report. No. 824, 1945. Aeronautical engineers have been the chief contributors to and users of the information regarding lift and drag coefficients. Further exploration of these topics is therefore best pursued in their reference texts; see, for example, A. M. Kuethe and J. D. Schetzer, *Foundations of Aerodynamics*, John Wiley & Sons, New York, 1950.

11

Compressible Flows

11.1 INTRODUCTION

In the previous chapters the discussion has been broadly restricted to incompressible, homogeneous flows. At low speeds, less than a Mach number of 0.3, gas flows are treated as incompressible flows. This is justified because the density variations caused by the flow itself are negligible. Incompressible gas flows are involved in a large number of flows of technical interest, as attested by the numerous examples in the preceding chapters. However, there are many flows of engineering interest in which the density cannot be assumed constant. Included are all supersonic flows, that is, Mach number greater than unity, and subsonic flows in which the Mach number exceeds approximately 0.3. Supersonic flows are encountered around aircraft moving at speeds greater than that of sound, and in rocket nozzles, and supersonic wind tunnels. Subsonic compressible flows are observed around commercial airliners, in fan-jet engines, in high-speed (but subsonic) wind tunnels, in air compressors, and in turbines. There are also examples of compressible effects important in liquid flows; two common examples are water hammer and a blast wave created by an underwater explosion. In fact, the finite speed of propagation of waves in a solid, as in an earthquake, is also due to the compressibility of the solid.

We now wish to introduce the effects of compressibility into some simple uniform-flow situations. By "uniform flow" we mean that the velocity at a given streamwise location in the pipe or channel is a constant and hence does not vary across the conduit. The control-volume formulation of the various laws will be used along with some

Art. 11.1 INTRODUCTION

thermodynamic relationships modified to reflect the variability in density ρ. We will work problems which involve gases such as air, oxygen, and carbon dioxide, for pressure and temperature conditions such that they behave as perfect gases.

The equation of state for a perfect gas is

$$p = \rho RT \tag{11.1}$$

in which R is the gas constant. This introduces another variable, the absolute temperature T; thus an additional relationship is necessary, a process equation. The most common processes are the *isentropic* (adiabatic frictionless) *process*,

$$p/\rho^\gamma = \text{const} \tag{11.2}$$

the *isothermal process*,

$$p/\rho = \text{const} \tag{11.3}$$

and the *isobaric process*,

$$p = \text{const} \tag{11.4}$$

The symbol γ stands for the ratio of specific heats,

$$\gamma = c_p/c_v \tag{11.5}$$

and may be taken as constant over relatively large ranges of temperature.

Enthalpy is a useful thermodynamic property for compressible-flow problems; it is defined by

$$h = \tilde{u} + p/\rho \tag{11.6}$$

Recall that the enthalpy is also given by

$$h = c_p T \tag{11.7}$$

and the internal energy by

$$\tilde{u} = c_v T \tag{11.8}$$

for a perfect gas. The foregoing equations may be combined to provide a relationship between c_p, c_v, and R, namely,

$$c_p = c_v + R \tag{11.9}$$

or, with the use of Eq. 11.5,

$$\gamma = 1 + R/c_v \tag{11.10}$$

The thermodynamic property *entropy* is quite important for the description of compressible flows. Its change is defined by the second law of thermodynamics to be

$$\Delta S \geq \int \frac{\delta Q}{T} \tag{11.11}$$

and for a perfect gas, the specific-entropy change is

$$\Delta s = c_p \ln \frac{T_2}{T_1} - R \ln \frac{p_2}{p_1} \tag{11.12}$$

These are some of the quantities introduced into the equations which describe compressible gas flow. Because of thermodynamic considerations, these flows are necessarily more complicated than incompressible flows. For this reason, we restrict ourselves to ducts with gradually varying cross-sectional areas so that the simplification of the uniform-flow assumption can be made. With this assumption, the continuity, momentum, and energy integral equations for steady flow between two sections may be expressed as

$$\rho_1 A_1 V_1 = \rho_2 A_2 V_2 \tag{11.13}$$

$$\Sigma \mathbf{F} = \rho_1 A_1 V_1 (\mathbf{V}_2 - \mathbf{V}_1) \tag{11.14}$$

$$\frac{\dot{Q} - \dot{W}_S}{\dot{m}} = \frac{V_2^2}{2} + \frac{p_2}{\rho_2} + \tilde{u}_2 - \frac{V_1^2}{2} - \frac{p_1}{\rho_1} - \tilde{u}_1 \tag{11.15}$$

Potential-energy changes, $g(z_2 - z_1)$, are usually neglected in gas flows. The energy equation is often written in terms of enthalpy, with the use of Eq. 11.6, as

$$\frac{\dot{Q} - \dot{W}_S}{\dot{m}} = \frac{V_2^2}{2} + h_2 - \frac{V_1^2}{2} - h_1 \tag{11.16}$$

or, in terms of temperature,

$$\frac{\dot{Q} - \dot{W}_S}{\dot{m}} = \frac{V_2^2}{2} + c_p T_2 - \frac{V_1^2}{2} - c_p T_1 \tag{11.17}$$

Four flow and state variables, p, ρ, T, and V, are involved in these three basic equations. The equation of state for an ideal gas, $p = \rho RT$, may be used to introduce a fourth relationship involving the three state variables. The temperature appears explicitly only in the energy equation, and the equation of state; these two may be combined to reduce the number of equations and unknowns to three. Using $c_p = R\gamma/(\gamma - 1)$ and the equation of state, the energy equation may be written as

$$\frac{\dot{Q} - \dot{W}_S}{\dot{m}} = \frac{V_2^2}{2} + \frac{\gamma}{\gamma - 1}\frac{p_2}{\rho_2} - \frac{V_1^2}{2} - \frac{\gamma}{\gamma - 1}\frac{p_1}{\rho_1} \qquad (11.18)$$

The continuity, momentum, and modified energy equations constitute the basic equations in analyzing compressible flows. We will use them to consider compressible-flow behavior in devices of engineering interest.

11.2 COMPRESSIBILITY AND THE SPEED OF SOUND

In Chapter 5, several fluid-dynamics phenomena were considered with reference to the equation of motion. Specific types of flow were then considered in later chapters: viscous flow, turbulent flow, and inviscid flow. For each of these conditions, we noted the relative magnitude of the viscous shear term as expressed by the Reynolds number.

In a similar manner, it is possible to identify other phenomena via the magnitude of the other terms of the momentum equation. For example, the coefficient of the gravity body-force term represents the effect of a free surface for a liquid motion or the importance of the buoyancy effects. Similarly, the importance of compressibility effects is evidenced in the coefficient of the pressure-gradient term. The isentropic stagnation properties can be defined as suitable normalizing quantities. These normalizing properties could be the pressure and density that would be obtained if the fluid were brought to rest without friction or heat-transfer effects. These quantities could also be the pressure and density upstream of a body or in a pressurized reservoir tank for a conduit flow. Using p_0 and ρ_0 as the upstream pressure and density, then $p^* = p/p_0$, $\rho^* = \rho/\rho_0$, and $V^* = V/V_0$; the momentum equation (see Eq. 4.27) becomes [†]

$$\rho^* \frac{D^*\mathbf{V}^*}{Dt^*} = -\frac{1}{M_0^2 \gamma}\nabla^* p^* + \frac{1}{\text{Re}}\left[\nabla^{*2}\mathbf{V}^* + \frac{1}{3}\nabla^*(\nabla^* \cdot \mathbf{V}^*)\right] \qquad (11.19)$$

where the Mach number M_0 is defined as

$$M_0 = V_0/a_0 \qquad (11.20)$$

[†]See Example 11.1 for a derivation of the coefficient of the pressure term.

If we neglect the viscous terms, which is often possible at large Reynolds numbers, the Mach number M_0 and the specific heat ratio γ, characterize the problem.

A sound wave may be an unfamiliar concept; the concept of a wave on the surface of a liquid is not. A pebble dropped in a body of still water sends a ripple or a gravity wave traveling outward in a radial pattern. The liquid-surface rise and fall, which denote the passage of the gravity wave, may be considered analogous to the rise and fall of pressure during the passage of a sound (or acoustic) wave. The magnitude of the variations in pressure may be considered very small in ratio to the pressure magnitude of the ambient fluid; so acoustic pressure may be conveniently expressed in terms of the millibar unit, defined as one dyne per square centimeter.

The speed of sound is the speed at which a sound wave, a pressure disturbance of small amplitude, travels through a fluid. To determine this speed consider a control volume moving with the sound wave, as shown in Fig. 11.1. Relative to the moving control volume, the fluid speed approaching will be V; as the fluid passes through the region of the small pressure change the velocity and density will undergo incremental changes. A steady-flow situation results for a wave traveling at constant velocity. The continuity integral equation requires that

$$\rho A V = (\rho + \Delta \rho)(V + \Delta V)A \tag{11.21}$$

which, by rearranging, may be put in the form

$$\Delta V = -\frac{V \Delta \rho}{\rho + \Delta \rho} \tag{11.22}$$

The momentum equation gives

$$pA - (p + \Delta p)A = (\rho + \Delta \rho)(V + \Delta V)^2 A - \rho V^2 A \tag{11.23}$$

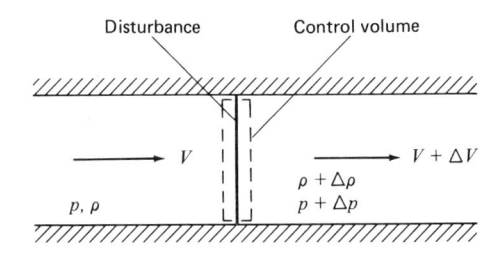

Fig. 11.1. A control volume enclosing a segment of a sound wave moving with a velocity V.

Art. 11.2 COMPRESSIBILITY AND THE SPEED OF SOUND

With the use of Eq. 11.21, this simplifies to

$$-\Delta p = \rho V \Delta V \qquad (11.24)$$

Combining this with Eq. 11.22 gives

$$V^2 = \frac{\Delta p}{\Delta \rho}\left(1 + \frac{\Delta \rho}{\rho}\right) \qquad (11.25)$$

If we assume the amplitude of the wave to be small, then $\Delta\rho/\rho \ll 1$, and $\Delta p/\Delta\rho \simeq \partial p/\partial\rho$. We thus have determined the velocity of propagation of a sound wave, designated by a, to be

$$a = \sqrt{\frac{\partial p}{\partial \rho}} \qquad (11.26)$$

The quantity a is called the *speed of sound* or the *sonic velocity*.

Small-amplitude, moderate-frequency waves (up to approximately 18,000 Hz) travel isentropically, so that p/ρ^γ = constant. Then,

$$a = \sqrt{\frac{\gamma p}{\rho}} = \sqrt{\gamma RT} \qquad (11.27)$$

for a perfect gas. At high frequency, small-amplitude waves generate heat and the process is more nearly isothermal. Then, for a perfect gas, Eq. 11.26 gives

$$a = \sqrt{RT} \qquad (11.28)$$

Waves of small amplitude also propagate through liquids and solids owing to compressibility effects. The quantity $\rho\,\partial p/\partial\rho$ is called the *bulk modulus* and is approximately constant. For water it is equal to 310,000 psi, although it does vary slightly with pressure and temperature. It is also highly dependent on the amount of dissolved gases.

The Mach number at a point in a flow is defined as

$$\mathrm{M} = V/a \qquad (11.29)$$

where the speed of sound and the velocity are measured at the same point. The speed of sound is given by Eq. 11.27. It is used when referring to the Mach number at a point since waves are assumed to travel isentropically in an otherwise undisturbed flow. Thus the Mach number at a point is written as

$$M = \frac{V}{\sqrt{\gamma p/\rho}} = \frac{V}{\sqrt{\gamma RT}} \qquad (11.30)$$

for a perfect gas.

If a source of small disturbances is initiated from a stationary location, waves travel outward in a radial manner with the speed of sound, as shown in Fig. 11.2. The position of the wave is shown after a time Δt and then at multiples of Δt. For a moving source, the waves would travel as shown in parts b and c of Fig. 11.2. If the velocity of the moving source, say an airplane, is less than the speed of sound, disturbances would travel ahead of the plane. If the source is traveling faster than the speed of sound, no warning would be given to an observer until after the source had passed by. The zone outside of the cone is a *zone of silence*. The situation in part b occurs when $M_0 < 1$, and in c when $M_0 > 1$. The cone formed in part c is called the *Mach cone*; the angle α is given by

$$\alpha = \sin^{-1} \frac{a_0}{V_0} = \sin^{-1} \frac{1}{M_0} \qquad (11.31)$$

The discussion above is limited to small-amplitude waves, often referred to as Mach waves. They are formed on the needle-nose of an aircraft or on sharp leading edges of an airfoil. For a blunt-nosed aircraft traveling at supersonic speeds a large-amplitude wave, called a *shock wave*, is formed. The situation in part c of Fig. 11.2 is qualita-

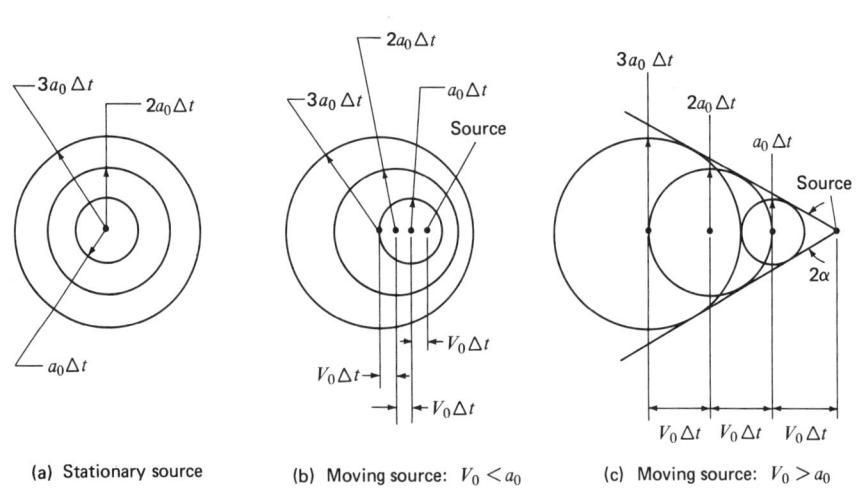

Fig. 11.2. Waves propagating from a noise source.

tively the same, but the shock wave angle is quite different. Oblique shock waves will be considered later; the normal shock wave will be examined in the following article.

Example 11.1

Show that the normalizing factors introduced lead to the term $\dfrac{1}{\gamma M_0^2}\nabla^* p^*$.

Solution. The convective acceleration is normalized by $\rho_0 V_0^2/L$; the pressure-gradient term is normalized by p_0/L. Hence, dividing by $\rho_0 V_0^2/L$ yields $p_0/\rho_0 V_0^2$ as the coefficient of the pressure-gradient term. Since $p/\rho = RT$,

$$\frac{p_0}{\rho_0 V_0^2} = \frac{RT_0}{V_0^2} = \frac{\gamma RT_0}{\gamma V_0^2}$$

and since $\gamma RT_0 = a_0^2$ and $M_0 = V_0/a_0$ this term is

$$\frac{p_0}{\rho_0 V_0^2} = \frac{1}{\gamma M_0^2}$$

11.3 THE NORMAL SHOCK WAVE

The preceding article established the propagation speed of a small-amplitude pressure disturbance, $a = \sqrt{\partial p/\partial \rho}$. This result depends upon the condition that the property changes be small when compared to the base value. It is not difficult to imagine physical problems for which the restriction to small changes is not applicable. For example, the flow in a rifle barrel ahead of the bullet, the exit flow from a rocket or jet engine nozzle, and the air flow around a supersonic aircraft all involve changes in the pressure which are large with respect to the base value. As a specific example, the pressure directly in front of a bullet moving at approximately 3000 fps is nearly 30 times larger than the pressure in the surrounding atmosphere. The velocity, density, and temperature also undergo large changes from their base values for such flow situations.

The property changes which occur across a shock wave take place over an extremely short distance, one that is comparable with the mean free path of the molecules; it is typically of the order of 10^{-5} in. Complicated phenomena such as viscous dissipation and heat conduction occur inside the shock wave and are best described in terms of the material behavior on the molecular scale; these effects are not of interest in this introductory presentation. In our discussion we will treat the shock wave as a simple discontinuity in the flow and use the integral equations to relate the quantities of interest.

A shock wave may occur as a line of discontinuity of the properties oriented on an angle to the flow, or it may occur normal to the flow, depending on the geometry and the direction of flow. These are referred to as *oblique shock waves* and *normal shock waves*, respectively. Normal shocks, which we will consider in this article, are most often encountered in enclosed channels or directly in front of a blunt object. Oblique shock waves will be considered in Art. 11.6.

The velocity, pressure, and density before and after a normal shock wave will be related by use of the governing equations. The control volume used for this analysis is chosen to enclose the shock wave, as shown in Fig. 11.3. This control volume can be used for a normal shock moving with velocity V_1, or it can be used for a stationary shock wave. The integral continuity equation gives, assuming $A_1 = A_2$ since the shock wave is so thin,

$$\rho_1 V_1 = \rho_2 V_2 \tag{11.32}$$

The momentum equation results in

$$p_1 A_1 - p_2 A_2 = \rho_2 V_2^2 A_2 - \rho_1 V_1^2 A_1 \tag{11.33}$$

or, using Eq. 11.32 with $A_1 = A_2$, we have

$$p_1 - p_2 = \rho_1 V_1 (V_2 - V_1) \tag{11.34}$$

The energy equation (Eq. 11.18) provides a third relationship between V, p, and ρ. Specifically, with $\dot{Q} = \dot{W}_S = 0$,

$$\frac{V_2^2 - V_1^2}{2} + \frac{\gamma}{\gamma - 1}\left(\frac{p_2}{\rho_2} - \frac{p_1}{\rho_1}\right) = 0 \tag{11.35}$$

Eqs. 11.32, 11.34, and 11.35 provide three equations to determine any

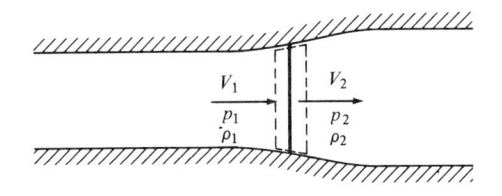

Fig. 11.3. A stationary shock wave in a conduit.

three unknowns in the problem. If the flow quantities, p_1, ρ_1, and V_1, are known in front of the normal shock wave then the necessary information may be determined after the shock wave.

Even though the equations above are sufficient to determine quantities of interest on either side of a shock wave, it is convenient to express the equations in terms of the Mach numbers, M_1 and M_2. The momentum equation (Eq. 11.34) may be written as

$$p_1\left(1 + \frac{\rho_1 V_1^2}{p_1}\right) = p_2\left(1 + \frac{\rho_2 V_2^2}{p_2}\right) \tag{11.36}$$

Introducing $M = V/\sqrt{p\gamma/\rho}$, the momentum equation becomes

$$p_1(1 + \gamma M_1^2) = p_2(1 + \gamma M_2^2) \tag{11.37}$$

The energy equation (Eq. 11.35) may be rearranged, using $p/\rho = RT$, as

$$T_1\left(1 + \frac{\gamma - 1}{\gamma R T_1}\frac{V_1^2}{2}\right) = T_2\left(1 + \frac{\gamma - 1}{\gamma R T_2}\frac{V_2^2}{2}\right) \tag{11.38}$$

which, when substituting $M = V/\sqrt{\gamma RT}$, may be put in the form

$$T_1\left(1 + \frac{\gamma - 1}{2} M_1^2\right) = T_2\left(1 + \frac{\gamma - 1}{2} M_2^2\right) \tag{11.39}$$

The continuity equation (Eq. 11.32), with the equation of state, $\rho = p/RT$, is written as

$$\frac{p_1 V_1}{RT_1} = \frac{p_2 V_2}{RT_2} \tag{11.40}$$

or, using $V = M\sqrt{\gamma RT}$,

$$\frac{p_1 M_1}{RT_1}\sqrt{\gamma RT_1} = \frac{p_2 M_2}{RT_2}\sqrt{\gamma RT_2} \tag{11.41}$$

Now, using Eq. 11.37 for the pressure ratio and Eq. 11.39 for the temperature ratio, Eq. 11.41 relates only the Mach numbers before and after the shock wave; the relationship is

$$\frac{M_1\sqrt{1 + \frac{\gamma - 1}{2} M_1^2}}{1 + \gamma M_1^2} = \frac{M_2\sqrt{1 + \frac{\gamma - 1}{2} M_2^2}}{1 + \gamma M_2^2} \tag{11.42}$$

This equation may be put in the following form, by solving for M_2^2,

$$M_2^2 = \frac{M_1^2 + \dfrac{2}{\gamma - 1}}{\dfrac{2\gamma}{\gamma - 1} M_1^2 - 1} \qquad (11.43)$$

This equation allows the momentum equation 11.37 to be expressed as

$$\frac{p_2}{p_1} = \frac{2\gamma}{\gamma + 1} M_1^2 - \frac{\gamma - 1}{\gamma + 1} \qquad (11.44)$$

and the energy equation 11.39 to become

$$\frac{T_2}{T_1} = \frac{\left(1 + \dfrac{\gamma - 1}{2} M_1^2\right)\left(\dfrac{2\gamma}{\gamma - 1} M_1^2 - 1\right)}{\dfrac{(\gamma + 1)^2}{2(\gamma - 1)} M_1^2} \qquad (11.45)$$

Let us use air ($\gamma = 1.4$) as the fluid and draw some conclusions concerning the shock wave. For air, Eq. 11.43 becomes

$$M_2^2 = \frac{M_1^2 + 5}{7M_1^2 - 1} \qquad (11.46)$$

Thus, we see that for $M_1 = 1$, $M_2 = 1$, which is a trivial solution corresponding to the case of no shock wave. If $M_1 > 1$ then $M_2 < 1$, and the sudden decrease in velocity proceeds from a supersonic state to a subsonic state. For $M_1 < 1$ and $M_2 > 1$, a sudden increase in velocity from subsonic to supersonic is suggested; see Fig. 11.4. This latter process may be eliminated as a possibility by considering the entropy change. It is found by using Eq. 11.12 to be

$$s_2 - s_1 = c_p \ln \frac{1 + \dfrac{\gamma - 1}{2} M_1^2}{1 + \dfrac{\gamma - 1}{2} M_2^2} - R \ln \frac{1 + \gamma M_1^2}{1 + \gamma M_2^2} \qquad (11.47)$$

where Eqs. 11.37 and 11.39 have been used. For air with $\gamma = 1.4$, this relationship is shown graphically in Fig. 11.4. For $M_1 < 1$, the entropy decreases across the possible shock wave, a violation of the second law of thermodynamics. The entropy must increase as the flow proceeds

Art. 11.3   THE NORMAL SHOCK WAVE   485

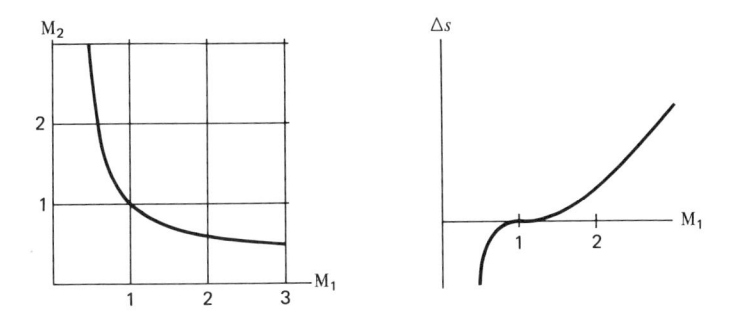

Fig. 11.4. M_2 and Δs vs. M_1 for a normal shock wave for air.

through the adiabatic shock. Hence, the flow through a shock wave always proceeds from a supersonic flow to a subsonic flow.

The dependence of the normal shock phenomenon on thermodynamic considerations is apparent in the equations above. A graphical way to demonstrate this dependence is to plot the process on a graph of the appropriate state variables. It will be convenient for this and for other compressible-flow situations to use the T-s diagram for the gas involved. Such a plot is shown in Fig. 11.5. The thermodynamic conditions upstream of the shock wave are identified as p_1, T_1; the intersection of these two curves identifies the initial point. In addition, the velocity V_1 must be specified to define the initial conditions. The information of the velocity magnitude can be represented on the T-s diagram if one considers the quantity termed the *adiabatic stagnation*

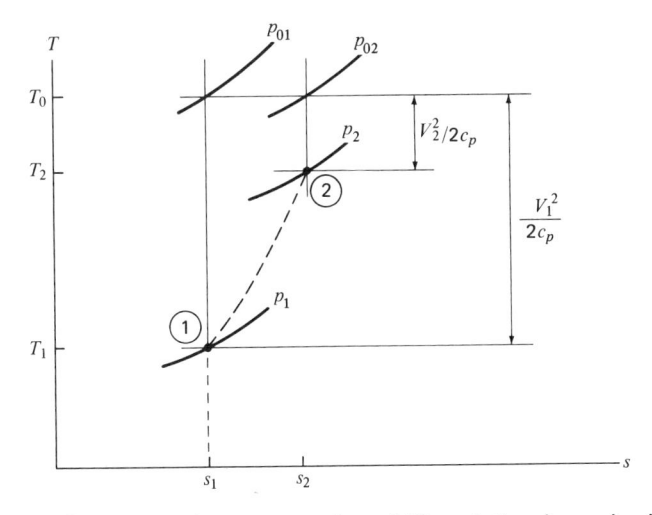

Fig. 11.5. Diagrammatic representation of T-s relation for a shock wave.

temperature T_0. This is defined as

$$T_0 = T + \frac{V^2}{2c_p} \tag{11.48}$$

and is identified as the temperature which would exist if the velocity were reduced to zero without heat transfer. The graphical interpretation of T_0 is shown on the figure.

The conditions downstream of the shock are given by Eqs. 11.43, 11.44, and 11.45. Only two of these relationships are required to specify the downstream conditions. For example, T_2 and p_2 are shown on the figure; the other properties may be considered fixed by the identification of the point at the intersection of the two curves. The magnitude of the velocity V_2 is indicated graphically by the vertical displacement between T_2 and T_0. Note that T_0 is also the adiabatic stagnation temperature for the conditions downstream of the shock.

The *isentropic stagnation pressure* is easily defined on this figure. It is the pressure which would be obtained if the gas were decelerated to zero velocity *isentropically*. Note that since the process represented by the shock wave involves an increase in the entropy, the stagnation pressure decreases across the shock wave. For example, the gas at 1 may be decelerated isentropically to obtain p_{01}; conversely, if it first flows through the shock and is then decelerated isentropically the stagnation pressure, p_{02} is less than p_{01}. This has considerable significance for the design of jet engines to operate at supersonic speeds. The combustion process is much more easily controlled at subsonic speeds, hence the gas must be decelerated. The magnitude of the stagnation pressure is the key factor in fixing the performance of the engine since it is this pressure which can be related to the magnitude of the velocity at the nozzle exit. Hence, the details of the deceleration from the supersonic to the subsonic state are quite important.

We have shown in this article that the changes which occur across a normal shock wave can be expressed in terms of the upstream Mach number M_1. The relations for the pressure ratio, the temperature ratio, and the downstream Mach number are presented in Eqs. 11.44, 11.45, and 11.43, respectively. By substituting in Eq. 11.40, the velocity ratio could be written. To aid in computations involving air a table has been prepared in which the various quantities are tabulated in terms of M_1. It is presented in Table D.2 of Appendix D. The ratio of stagnation pressures p_{02}/p_{01} is included in the table. The various ratios are also shown graphically in Fig. 11.6.

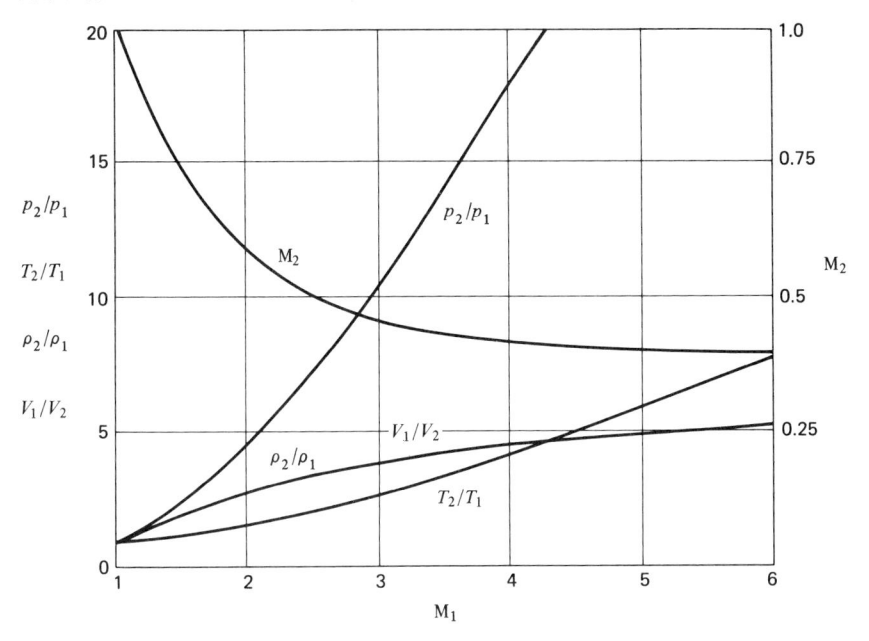

Fig. 11.6. Normal shock flow relations for air ($\gamma = 1.4$).

Example 11.2

Air at 40°F flows through a normal shock with an approach velocity of 1600 fps. Determine the pressure downstream of the shock wave if the pressure upstream of the shock is 10 psia. Use (a) the basic equations of continuity, momentum, and energy and (b) the Mach number relationships. $R_{air} = 1716$ ft-lb/slug-°R; $\gamma_{air} = 1.4$.

Solution. a. First, determine the density ρ_1 to be

$$\rho_1 = \frac{p_1}{RT_1} = \frac{10 \times 144}{1716 \times 500} = 0.00168 \text{ slug/ft}^3$$

Continuity requires that

$$\rho_1 V_1 = \rho_2 V_2$$

Using $V_1 = 1600$ fps,

$$0.00168 \times 1600 = \rho_2 V_2$$

Momentum requires that

$$p_1 - p_2 = \rho_1 V_1 (V_2 - V_1)$$

Using $p_1 = 10 \times 144$ psf,

$$1440 - p_2 = 1600 \times 0.00168 (V_2 - 1600)$$

The energy equation demands that

$$\frac{V_2^2 - V_1^2}{2} + \frac{\gamma}{\gamma - 1}\left(\frac{p_2}{\rho_2} - \frac{p_1}{\rho_1}\right) = 0$$

or

$$\frac{V_2^2 - 1600^2}{2} + \frac{1.4}{0.4}\left(\frac{p_2}{\rho_2} - 1716 \times 500\right) = 0$$

There are three unknowns, ρ_2, p_2, V_2, in these three independent equations. They are found as follows: substitute continuity and momentum in the energy equation and find that

$$\frac{V_2^2 - 1600^2}{2} + 3.5\left(\frac{1440 - 2.69(V_2 - 1600)}{2.69/V_2} - 858,000\right) = 0$$

This is a quadratic equation and can be solved to give

$$V_2 = 892 \text{ fps}$$

The momentum equation then yields

$$p_2 = 3340 \text{ psf} \quad \text{or} \quad 23.2 \text{ psi}$$

b. To use the Mach-number relationships, first determine M_1. It is

$$M_1 = \frac{V_1}{a_1} = \frac{1600}{\sqrt{1.4 \times 1716 \times 500}} = 1.46$$

Then, from Eq. 11.44, we simply have

$$p_2 = p_1\left[\frac{2\gamma}{\gamma + 1}M_1^2 - \frac{\gamma - 1}{\gamma + 1}\right] = 10\left(\frac{2.8}{2.4}1.46^2 - \frac{0.4}{2.4}\right) = 23.2 \text{ psi}$$

This is, of course, a much easier solution than that of part a; however, the fundamental equations are hidden in such a solution. It is important to note that continuity, momentum, and energy, with all the accompanying simplifications have been used.

An even easier procedure, which more completely obscures the fundamental relationships, is to use the normal shock table (Table D.2). At an initial Mach number of 1.46 the pressure ratio p_2/p_1 is given as 2.32; hence $p_2 = 10(2.32) = 23.2$ psi.*

*The student will note that the figures in text have been rounded off from those in Appendix D. The considerations guiding our procedure are given in the introduction to that appendix.

Art. 11.3 THE NORMAL SHOCK WAVE 489

Example 11.3

A shock wave is propagating in otherwise stagnant air at standard conditions at a speed of 1600 fps. Determine the pressure immediately after the shock wave passes. Also find the induced velocity in the stagnant air. Use the Mach number relations and check the results with Table D.2, the shock flow table.

Solution. We will treat the shock wave in a reference frame moving with the shock wave so that the steady-flow relationships developed in the preceding article are applicable. The incoming speed is $V_1 = 1600$ fps, $p_1 = 14.7$ psia, and $T_1 = 68°F$. The upstream Mach number is

$$M_1 = \frac{V_1}{a_1} = \frac{1600}{\sqrt{1.4 \times 1716 \times 528}} = 1.42$$

The downstream Mach number is found from Eq. 11.40 to be

$$M_2 = \sqrt{\frac{M_1^2 + 5}{7M_1^2 - 1}} = 0.73$$

This value is also found in the shock flow table (D.2) by interpolation. The pressure p_2, from Eq. 11.37, is

$$p_2 = p_1 \frac{1 + \gamma M_1^2}{1 + \gamma M_2^2}$$

$$= 14.7 \times \frac{1 + 1.4 \times 1.42^2}{1 + 1.4 \times .73^2} = 32.3 \text{ psi}$$

This ratio $p_1/p_2 = 2.20$ is verified in the shock table. The downstream temperature (see Eq. 11.39) is

$$T_2 = T_1 \frac{1 + \frac{\gamma - 1}{2} M_1^2}{1 + \frac{\gamma - 1}{2} M_2^2}$$

$$= 528 \times \frac{1 + .2 \times 1.42^2}{1 + .2 \times .73^2} = 670°R$$

giving a ratio of $T_2/T_1 = 1.265$, which may be checked in the shock flow table. The velocity V_2 is then

$$V_2 = M_2 a_2 = 0.73\sqrt{1.4 \times 1716 \times 670} = 925 \text{ fps}$$

To return to a stationary reference frame we simply superpose a velocity $-V_1$ and find the induced velocity to be in the same direction as the shock velocity. It is

$$V = V_1 - V_2 = 675 \text{ fps}$$

11.4 NOZZLE FLOW

The flow of a gas from a duct with a gradually varying cross-sectional area is of sufficient engineering interest to consider it in detail. A nozzle is of particular interest and will be used in this article to illustrate flow in a duct. The area must be gradually varying so that the component of velocity normal to the flow direction is small compared to the component of velocity in the flow direction. The assumption of uniform flow is then reasonable and the integral equations simplify to a set of equations which can be easily solved for a great number of flow situations. The results, however, are only valid for the average quantities at a particular cross section. To study the variation normal to the flow a two- or three-dimensional analysis would be necessary. Separation prediction and boundary-layer effects cannot be included in the one-dimensional flow study.

We will first consider the case of adiabatic, frictionless (isentropic) flow. Many of the important features of nozzle flow can be demonstrated quantitatively given these conditions; it is also a useful reference condition from which to consider the more general case of adiabatic frictional flow. The restriction to an isentropic process is essentially a statement that one thermodynamic property, the entropy, is known for all points in the flow. Consequently, only one additional property need be specified to fix the state of the flow. As an example, consider that room air is drawn into a large, evacuated reservoir. In the physical region where the isentropic process is approximated, the flow may be graphically represented on a $T-s$ diagram. The ambient pressure and temperature constitute the stagnation conditions; these curves intersect at a point on the $T-s$ diagram of Fig. 11.7 and the isentropic assumption constrains the state of the fluid to lie along the vertical line. If, for example, a second pressure is specified, then all of the properties are known at the second point. Recall that the velocity can be identified on this plot as shown.

The ratio of the temperature to the stagnation temperature may be written in terms of the Mach number, using the energy equation:

$$T_0 = T_1 + V_1^2/2c_p \qquad (11.49)$$

Rearranging,

$$\frac{T_0}{T_1} = 1 + \frac{V_1^2}{\gamma R T_1} \frac{\gamma R}{2c_p}$$

$$= 1 + \frac{\gamma - 1}{2} M_1^2 \qquad (11.50)$$

Art. 11.4 NOZZLE FLOW 491

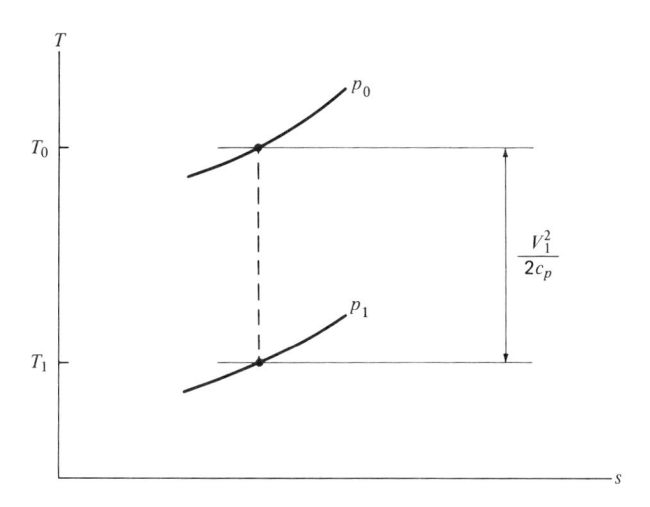

Fig. 11.7. The T-s diagram for an isentropic flow.

Note that this ratio is independent of the isentropic assumption. The isentropic ratios are

$$\frac{T_0}{T_1} = \left(\frac{p_0}{p_1}\right)^{\frac{\gamma-1}{\gamma}} = \left(\frac{\rho_0}{\rho_1}\right)^{\gamma-1} \tag{11.51}$$

Consequently,

$$\frac{p_0}{p_1} = \left(1 + \frac{\gamma-1}{2}\mathrm{M}_1^2\right)^{\frac{\gamma}{\gamma-1}} \tag{11.52}$$

$$\frac{\rho_0}{\rho_1} = \left(1 + \frac{\gamma-1}{2}\mathrm{M}_1^2\right)^{\frac{1}{\gamma-1}} \tag{11.53}$$

The values to obtain the condition of $M_1 = 1$ are of special interest. They are, for air with $\gamma = 1.4$,

$$\frac{p_1}{p_0} = 0.528 \qquad \frac{T_1}{T_0} = 0.833 \qquad \frac{\rho_1}{\rho_0} = 0.634 \tag{11.54}$$

The foregoing equations (11.51–11.53) provide the necessary pressure, temperature, and density ratios to achieve any given Mach number. However, the shape of the nozzle required to attain such a flow condition has not yet been established. Based upon our incompressible-flow experiences, it will be a strange and unexpected result that a converging-diverging passage is required to attain a supersonic flow.

492 **COMPRESSIBLE FLOWS** **Ch. 11**

A converging nozzle and a converging-diverging nozzle are shown in Fig. 11.8. The flow is assumed to initiate in a reservoir with conditions p_0, ρ_0, T_0, and exhaust into a receiver maintained at a pressure p_r. Often the receiver or the reservoir is simply the atmosphere. Assuming no significant frictional effects and no heat transfer, that is, isentropic flow (except, of course, across a shock wave), the governing equations are those for continuity,

$$\rho A V = \text{const} \qquad (11.55)$$

and for energy

$$\frac{V^2}{2} + \frac{\gamma}{\gamma - 1} \frac{p}{\rho} = \text{const} \qquad (11.56)$$

(a) Converging nozzle   (b) Converging-diverging nozzle

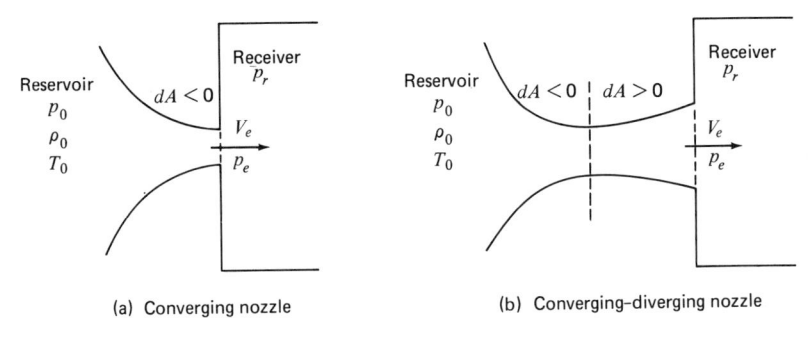

Fig. 11.8. Uniform nozzle flows.

The velocity V, pressure p, and density ρ are the average quantities at each cross section. Hence, they depend only on x, the axial location in the flow. If Eq. 11.55 is differentiated, there results

$$AV \frac{d\rho}{dx} + \rho V \frac{dA}{dx} + \rho A \frac{dV}{dx} = 0 \qquad (11.57)$$

This may be put in the form

$$\frac{d\rho}{\rho} + \frac{dA}{A} + \frac{dV}{V} = 0 \qquad (11.58)$$

Also, from Eq. 11.56, the energy equation may be written in the differential form

$$V \, dV + \frac{\gamma}{\gamma - 1} \left(\frac{dp}{\rho} - \frac{p \, d\rho}{\rho^2} \right) = 0 \qquad (11.59)$$

For an isentropic flow $p/\rho^\gamma = \text{constant}$; the differential energy equation may therefore be written as

$$V\,dV + \frac{\gamma p}{\rho}\frac{d\rho}{\rho} = 0 \tag{11.60}$$

Using Eq. 11.58 and $a = \sqrt{\gamma p/\rho}$, Eq. 11.60 becomes

$$V\,dV + a^2\left(-\frac{dA}{A} - \frac{dV}{V}\right) = 0 \tag{11.61}$$

or

$$\frac{dV}{V}(M^2 - 1) = \frac{dA}{A} \tag{11.62}$$

From this relationship we see that:

1. If $M < 1$ and $dA < 0$, then $dV > 0$, indicating an accelerating flow in a converging channel.
2. If $M > 1$ and $dA < 0$, then $dV < 0$, indicating a decelerating flow in a converging channel.
3. If $M < 1$ and $dA > 0$, then $dV < 0$: a decelerating flow in a diverging channel.
4. If $M > 1$ and $dA > 0$, then $dV > 0$: an accelerating flow in a diverging channel.
5. At a throat, $dA = 0$ and there exists either $dV = 0$, or $M = 1$.

The conditions $dV = 0$ and $M < 1$ at the throat describe the flow in a nozzle diffuser combination; a venturi meter is such a flow passage. If $M = 1$ and $dV > 0$ at the throat, a supersonic nozzle is obtained.

It is interesting to note that Eq. 11.62 was developed with only the continuity and energy equations. The reason, as emphasized above, is that the isentropic constraint provides a known thermodynamic property at any second point in the flow. The momentum equation could be used to determine the net force required to hold the nozzle in place.

The flow in the nozzle may now be considered in more detail. Consider a physical situation in which the receiver pressure p_r is decreased with respect to the reservoir pressure p_0 for the converging nozzle flow of Fig. 11.8a. The mass flux $\dot{m}$ will increase until the Mach number M_e at the exit, where $dA = 0$, is unity. As p_r is reduced below this particular value the mass flux will not increase (see Fig. 11.9), since M cannot exceed unity; in fact, none of the flow quantities in the nozzle will change; a *choked flow* condition results, with $p_r < p_e$. A change in conditions donwnstream of a throat where $M = 1$ cannot affect conditions upstream of the throat since disturbances travel at the speed of sound ($M = 1$). The upstream propagation speed is $(a - V)$

where V is the local velocity and a is the local speed of sound. Consequently the flow in the nozzle is not "aware" that the receiver pressure is less than that required for the M = 1 condition at the throat.

A second case of interest occurs if the reservoir pressure p_0 is increased for a fixed receiver pressure. A choked flow again occurs when $M_e = 1$; however, when p_0 is increased still further the mass flux will increase and p_e will again exceed p_r. See Fig. 11.9 for a graphical display of this mass flux variation.

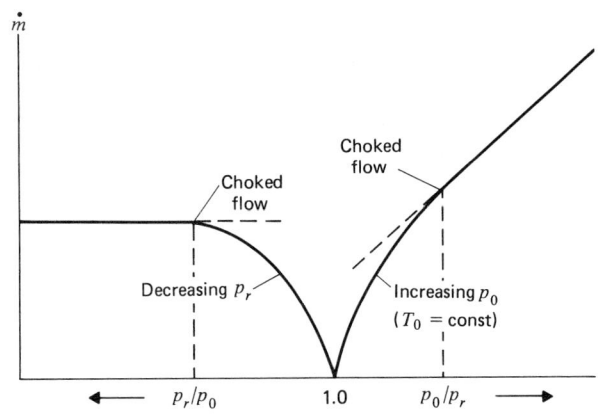

Fig. 11.9. Variation of $\dot{m}$ in a converging nozzle for an isentropic flow.

For the converging-diverging nozzle, called a *de Laval nozzle*, consider some special cases; first let p_r be slightly less than p_0. A subsonic isentropic flow would exist throughout the nozzle and the pressure would be as shown by curve a in Fig. 11.10. As p_r is further reduced, a critical pressure is reached when the Mach number at the throat reaches unity, as shown in c. If p_r is reduced still further, a supersonic flow will occur in the front part of the diverging section and a shock wave will occur at some position with the pressure curve d. The shock wave moves to the exit as p_r is reduced still further until the shock is located right at the exit; see curve e. When the pressure reaches point f the shock wave is "washed" out of the exit and isentropic flow exists throughout the channel. The flow quantities in the channel remain unchanged as p_r is reduced to some value indicated by g. If a shock wave occurs in the channel, the momentum equation must be used across the shock; isentropic flow would exist throughout the channel except across the shock. The flow can be completely solved with the appropriate equations.

Art. 11.4   NOZZLE FLOW   495

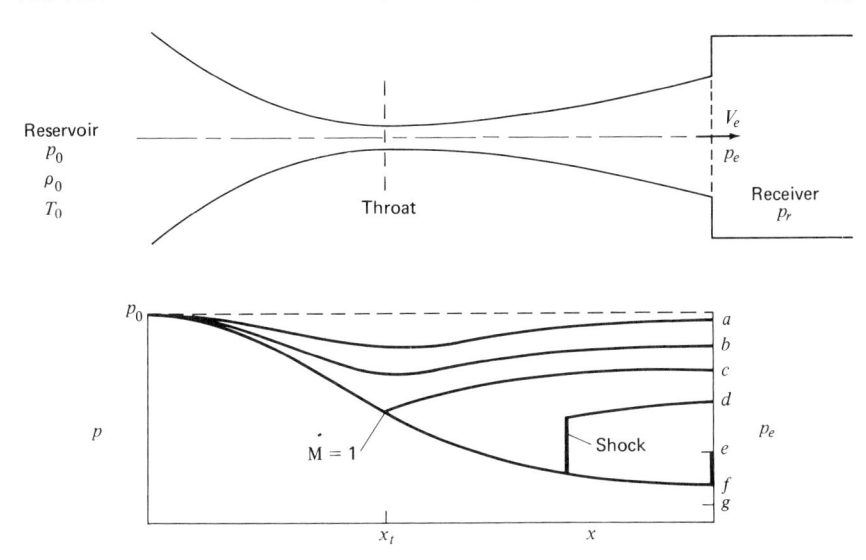

Fig. 11.10. The converging-diverging nozzle.

The flow may be represented on a T–s diagram, as shown in Fig. 11.11. The temperature and entropy of the various exit conditions are indicated. The entropy remains constant everywhere except for the increase across a shock wave.

We may apply the energy equation, continuity, and the isentropic relation between sections in the converging-diverging nozzle if a shock

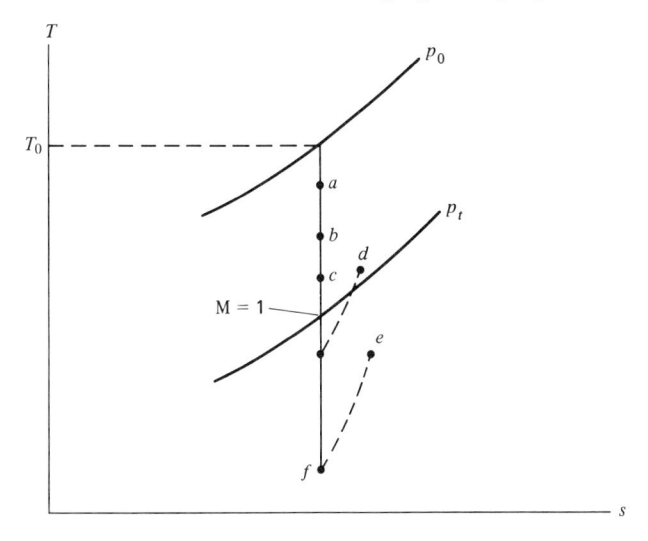

Fig. 11.11. A T–s diagram representation of an isentropic nozzle flow.

wave does not occur between the sections. The equations are

$$\frac{V_1^2}{2} + \frac{\gamma}{\gamma-1}\frac{p_1}{\rho_1} = \frac{V_2^2}{2} + \frac{\gamma}{\gamma-1}\frac{p_2}{\rho_2} \qquad (11.63)$$

$$\frac{p_1}{\rho_1^\gamma} = \frac{p_2}{\rho_2^\gamma} \qquad (11.64)$$

$$\rho_1 A_1 V_1 = \rho_2 A_2 V_2 \qquad (11.65)$$

The energy equation may be written as

$$\frac{T_1}{T_2} = \frac{M_2^2 + \dfrac{2}{\gamma-1}}{M_1^2 + \dfrac{2}{\gamma-1}} \qquad (11.66)$$

Using the isentropic relation (see Eq. 11.51), the pressure ratio is

$$\frac{p_1}{p_2} = \left[\frac{M_2^2 + \dfrac{2}{\gamma-1}}{M_1^2 + \dfrac{2}{\gamma-1}}\right]^{\frac{\gamma}{\gamma-1}} \qquad (11.67)$$

It follows that

$$\frac{\rho_1}{\rho_2} = \left[\frac{M_2^2 + \dfrac{2}{\gamma-1}}{M_1^2 + \dfrac{2}{\gamma-1}}\right]^{\frac{1}{\gamma-1}} \qquad (11.68)$$

A quantity that can be quite useful in calculations is the *critical area* A^*. It may be a fictitious area used only for computational purposes. It is that minimum area to which a channel can be reduced without affecting the mass flow rate; the Mach number at the critical area would be unity. Let us presume that the critical area is located at section ②; then, from Eqs. 11.67 and 11.68,

$$\frac{p_1}{p^*} = \left[\frac{\gamma+1}{(\gamma-1)M_1^2 + 2}\right]^{\frac{\gamma}{\gamma-1}} \qquad \frac{\rho_1}{\rho^*} = \left[\frac{\gamma+1}{(\gamma-1)M_1^2 + 2}\right]^{\frac{1}{\gamma-1}}$$

$$(11.69)$$

The continuity equation may be written as

$$\rho^* A^* V^* = \rho_1 A_1 V_1 \tag{11.70}$$

or, using $V = M\sqrt{\gamma p/\rho}$, we have

$$\frac{A_1}{A^*} = \frac{\rho^*}{\rho_1} \frac{\sqrt{\gamma p^*/\rho^*}}{\sqrt{\gamma p_1/\rho_1}} \frac{1}{M_1} = \frac{1}{M_1} \sqrt{\frac{p^* \rho^*}{p_1 \rho_1}} \tag{11.71}$$

With the use of Eqs. 11.69,

$$\frac{A_1}{A^*} = \frac{1}{M_1} \left[\frac{(\gamma - 1)M_1^2 + 2}{\gamma + 1} \right]^{\frac{\gamma + 1}{2(\gamma - 1)}} \tag{11.72}$$

The velocity ratio can be found by using Eq. 11.70. It is

$$\frac{V_1}{V^*} = \frac{\rho^*}{\rho_1} \frac{A^*}{A_1} = M_1 \left[\frac{\gamma + 1}{(\gamma - 1)M_1^2 + 2} \right]^{1/2} \tag{11.73}$$

The isentropic flow equations 11.50, 11.52, 11.53, 11.72, and 11.73 relate the various ratios to the Mach number M_1. To facilitate computations the ratios have been tabulated for air in Table D.1 in Appendix D. They are also presented in graphical form in Fig. 11.12.

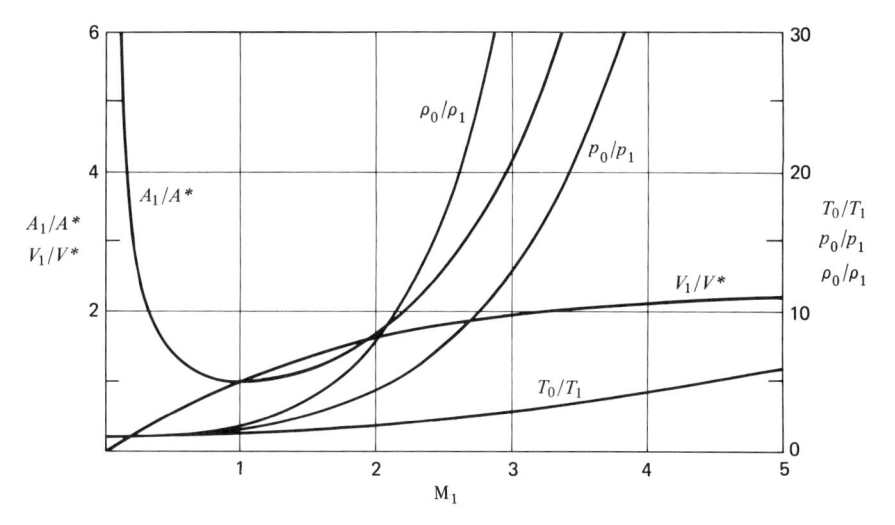

Fig. 11.12. Isentropic, one-dimensional flow relations for air ($\gamma = 1.4$).

Example 11.4

In previous chapters we have considered gas flows to be incompressible if the Mach number does not exceed 0.3. Determine the error involved in calculating the stagnation pressure (this would involve the maximum error) for an air flow with M = 0.3 by assuming the air to be incompressible.

Solution. For an incompressible flow, the stagnation pressure (Fig. E11.4) is related to the free-stream pressure by Bernoulli's equation,

$$\frac{V_1^2}{2} + \frac{p_1}{\rho} = \frac{p_0}{\rho}$$

The pressure difference is

$$p_0 - p_1 = \rho \frac{V_1^2}{2}$$

By using the isentropic flow equation (11.52), the pressure p_0 is

$$p_0 = p_1 \left(1 + \frac{\gamma - 1}{2} M_1^2\right)^{\frac{\gamma}{\gamma - 1}}$$

The quantity in parentheses may be expanded, using a binomial series, to give the stagnation pressure as

$$p_0 = p_1 \left(1 + \frac{\gamma}{2} M_1^2 + \frac{\gamma}{8} M_1^4 + \frac{(2 - \gamma)\gamma}{48} M_1^6 + \cdots\right)$$

Using $M_1^2 = V_1^2 \rho_1 / \gamma p_1$, this becomes

$$p_0 = p_1 + \rho_1 \frac{V_1^2}{2} + \rho_1 \frac{V_1^2}{2} \frac{M_1^2}{4} + \rho_1 \frac{V_1^2}{2} \frac{(2 - \gamma)\gamma}{24} M_1^4 + \cdots$$

or

$$p_0 - p_1 = \rho_1 \frac{V_1^2}{2} \left(1 + \frac{M_1^2}{4} + \frac{(2 - \gamma)\gamma}{24} M_1^4 + \cdots\right)$$

Setting $M_1 = 0.3$,

$$p_0 - p_1 = \rho_1 \frac{V_1^2}{2} (1 + 0.0228 + \cdots)$$

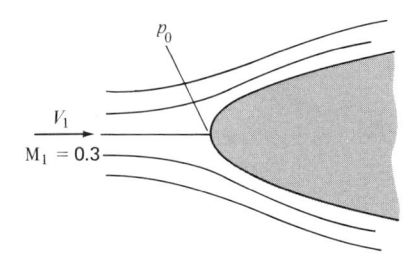

Fig. E11.4

Art. 11.4 NOZZLE FLOW 499

We see that the difference between stagnation and free-stream pressure differs from that of incompressible flow by approximately 2 percent. Thus the assumption that gas flows below $M = 0.3$ may be considered to be incompressible is justified.

Example 11.5

A converging nozzle with a 2-in.-dia. exit area exits to the atmosphere from a reservoir. Determine the mass flux when the condition of choked flow just occurs. The reservoir is maintained at a temperature of 70°F.

Solution. For a choked-flow condition the Mach number at the exit is $M_e = 1$. Then Eq. 11.50 may be used to find

$$T_e = \frac{T_0}{1 + \frac{\gamma - 1}{2} M_e^2} = \frac{530}{1 + 0.2} = 441°R$$

The exit density ρ_e is

$$\rho_e = \frac{p_e}{RT_e} = \frac{14.7 \times 144}{1716 \times 441} = 0.0028 \text{ slug/ft}^3$$

Using the energy equation (11.49) and $c_p = R\gamma/(\gamma - 1)$, the exit velocity is

$$V_e = \left[\frac{2\gamma R}{\gamma - 1} (T_0 - T_e) \right]^{1/2}$$

$$= \left[\frac{2 \times 1.4 \times 1716}{0.4} \times 89 \right]^{1/2} = 1030 \text{ fps}$$

This results in a mass flux of

$$\dot{m} = \rho_e A_e V_e = 0.0028 \times \frac{\pi}{144} \times 1030 = 0.0628 \text{ slugs/sec}$$

Extension 11.5.1. Assume that the nozzle empties into a low-pressure reservoir from a 70°F atmosphere. Now find the mass flux.
Ans. 0.0331 slugs/sec

Example 11.6

Air is flowing through a 4-in.-dia. pipe at 200 fps, a pressure of 20 psia, and a temperature of 70°F. The pipe is reduced in diameter such that choked flow just occurs. Determine this diameter. (a) Use the basic equations and (b) use the isentropic flow table D.1 in Appendix D.

Solution. a. Assuming an isentropic process and no losses, the energy equation may be written as

$$\frac{V_2^2}{2} - \frac{V_1^2}{2} + \frac{\gamma}{\gamma - 1} \left(\frac{p_2}{\rho_2} - \frac{p_1}{\rho_1} \right) = 0$$

500 COMPRESSIBLE FLOWS Ch. 11

The density in the 4-in. pipe is

$$\rho_1 = \frac{p_1}{RT_1} = \frac{20 \times 144}{1716 \times 530} = 0.00317 \text{ slug/ft}^3$$

The energy equation then gives

$$\frac{V_2^2}{2} + 3.5\frac{p_2}{\rho_2} = 20{,}000 + 3.5 \times \frac{20 \times 144}{0.00317} = 3.2 \times 10^6$$

For a choked flow the Mach number at the reduced section is unity; thus

$$M_2 = \frac{V_2}{a_2} = \frac{V_2}{\sqrt{\gamma p_2/\rho_2}} = 1$$

Substitute this in the energy equation to give

$$\frac{p_2}{\rho_2} = 7.62 \times 10^5$$

The isentropic relationship also relates p_2 and ρ_2 as

$$\frac{p_2}{\rho_2^{1.4}} = \frac{p_1}{\rho_1^{1.4}} = \frac{20 \times 144}{0.00317^{1.4}} = 0.900 \times 10^7$$

We thus find, by solving the equations above simultaneously,

$$p_2 = 1600 \text{ psf} \qquad \rho_2 = 0.0021 \text{ slug/ft}^3 \qquad V_2 = 1035 \text{ fps}$$

Using continuity,

$$\rho_1 V_1 A_1 = \rho_2 A_2 V_2$$

there results

$$A_2 = \frac{0.00317 \times 200 \times 4\pi}{0.00210 \times 1035 \times 144} = \frac{\pi d_2^2}{4 \times 144}$$

The diameter is

$$d_2 = 2.22 \text{ in.}$$

b. The Mach number M_1 in the pipe before the reduction in diameter is

$$M_1 = \frac{V_1}{a_1} = \frac{200}{\sqrt{1.4 \times 1716 \times 530}} = 0.177$$

From the isentropic flow table (D.1) we interpolate for A_1/A^* and find

$$\frac{A_1}{A^*} = 3.33$$

so that
$$A^* = \frac{\pi d^{*2}}{4} = \frac{16\pi}{4 \times 3.33}$$

We quickly find that
$$d^* = 2.19 \text{ in.}$$

The computational advantage of the isentropic flow table is obvious.

Example 11.7

Air flows from a 20-psia 70°F reservoir through a 2-in.-dia. throat and then through a diffuser. Calculate the exit pressure needed to orient the shock wave at a position where the diameter is 3 in. The exit diameter is 4 in. (a) Use the various equations and (b) use the gas tables in Appendix D.

Solution. If the flow increases from zero velocity in the reservoir to a supersonic velocity in the diffuser section it must pass through the throat with sonic velocity, $M_t = 1$. Thus from the reservoir to the throat we have (using Eqs. 11.52 and 11.50)

$$p_t = \frac{p_0}{\left[1 + \frac{\gamma - 1}{2} M_t^2\right]^{\frac{\gamma}{\gamma - 1}}} = \frac{20}{(1.2)^{3.5}} = 10.6 \text{ psia}$$

$$T_t = \frac{T_0}{1 + \frac{\gamma - 1}{2} M_t^2} = \frac{530}{1.2} = 442°R$$

The equation of state gives

$$\rho_t = \frac{p_t}{RT_t} = \frac{10.6 \times 144}{1716 \times 442} = 0.00203 \text{ slug/ft}^3$$

The velocity at the throat is

$$V_t = M_t a_t = 1 \times \sqrt{1.4 \times 1716 \times 442} = 1030 \text{ fps}$$

We can now go from the throat to a section ①just before the shock wave. Continuity, energy, and isentropy give

$$1030 \times 0.00203 \times 4\pi = \rho_1 V_1 \times 9\pi$$

$$\frac{V_1^2 - 1030^2}{2} + 3.5\left(\frac{p_1}{\rho_1} - \frac{10.6 \times 144}{0.00203}\right) = 0$$

$$\frac{p_1}{\rho_1^{1.4}} = \frac{10.6 \times 144}{0.00203^{1.4}}$$

From these three equations we find, by trial and error,

$$V_1 = 1820 \text{ fps} \qquad p_1 = 217 \text{ psfa} \qquad \rho_1 = 0.00051 \text{ slug/ft}^3$$

The Mach number and temperature are

$$M_1 = \frac{V_1}{\sqrt{\gamma p_1/\rho_1}} = \frac{1820}{\sqrt{1.4 \times 217/0.00051}} = 2.37$$

and

$$T_1 = \frac{p_1}{\rho_1 R} = \frac{217}{0.00051 \times 1716} = 248°R$$

Now, to go just across the shock wave to section ②, we use Eqs. 11.44 and 11.45 to give

$$p_2 = p_1 \left[\frac{2\gamma}{\gamma + 1} M_1^2 - \frac{\gamma - 1}{\gamma + 1} \right] = \left(\frac{2.8}{2.4} \times 2.37^2 - \frac{0.4}{2.4} \right) 217 = 1380 \text{ psfa}$$

$$T_2 = T_1 \frac{\left(1 + \frac{\gamma - 1}{2} M_1^2\right)\left(\frac{2\gamma}{\gamma - 1} M_1^2 - 1\right)}{\frac{(\gamma + 1)^2}{2(\gamma - 1)} M_1^2} = 248 \frac{2.12 \times 38.3}{40.4} = 498°R$$

The density is then

$$\rho_2 = \frac{p_2}{RT_2} = \frac{1380}{1716 \times 498} = 0.00161 \text{ slug/ft}^3$$

From continuity,

$$V_2 = \frac{\rho_1 V_1}{\rho_2} = \frac{0.00051 \times 1820}{0.00161} = 575 \text{ fps}$$

Now we go to the exit to determine the exit pressure, using

$$0.00161 \times 575 \times 9\pi = \rho_e V_e \times 16\pi$$

$$\frac{V_e^2 - 575^2}{2} + 3.5\left(\frac{p_e}{\rho_e} - \frac{1380}{0.00161}\right) = 0$$

$$\frac{p_e}{\rho_e^{1.4}} = \frac{1380}{0.00161^{1.4}}$$

The foregoing equations may then be solved to give

$$p_e = 1730 \text{ psfa} \qquad \text{or} \qquad 12 \text{ psia}$$

b. We will now use the tables. Since a supersonic flow occurs immediately after the throat, the Mach number must be unity there and the throat area is the critical area A^*. The area ratio from the throat to the shock wave is $A_1/A^* = 9/4$. We interpolate in table D.1 for the Mach number M_1 just before the shock, using $A_1/A^* = 2.25$, and find

$$M_1 = 2.33$$

Now the shock table, with $M_1 = 2.33$, may be interpolated to give

$$M_2 = 0.531 \qquad \frac{p_{02}}{p_{01}} = 0.570$$

where $p_{01} = 20$ psia is the upstream stagnation (reservoir) pressure. Then the downstream stagnation pressure is

$$p_{02} = 11.4 \text{ psia}$$

From the isentropic flow table at $M_2 = 0.531$,

$$\frac{A_2}{A^*} = 1.285$$

so that

$$\frac{A_e}{A^*} = \frac{A_2}{A^*} \times \frac{A_e}{A_2} = 1.285 \times \frac{16}{9} = 2.28$$

The Mach number and pressure ratio corresponding to this area ratio are

$$M_e = 0.265 \qquad \frac{p_e}{p_{0e}} = 0.952$$

For an isentropic flow $p_{02} = p_{0e}$ and there results

$$p_e = 0.952 \times 11.4 = 10.9 \text{ psia}$$

The answer generated here is slightly different from that of part a, owing to the inaccuracy in the trial-and-error solution in part a.

Example 11.8

A pitot probe and a static probe are used to measure the velocity in a supersonic airstream (Fig. E11.8). The pitot probe measures 40 psia, and the static probe, 10 psia. Determine the free-stream velocity. The temperature at the stagnation point of the probe is measured as $740°R$.

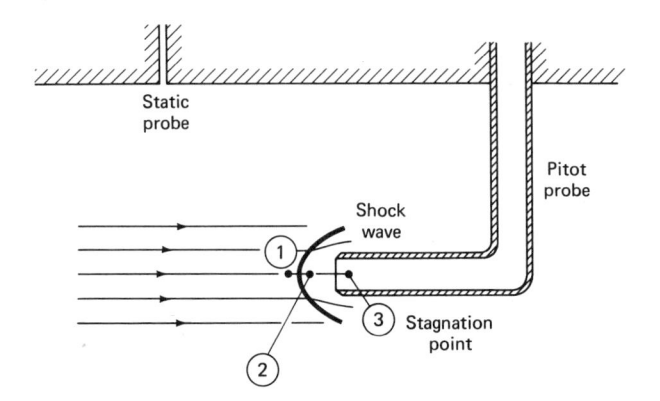

Fig. E11.8

Solution. When a pitot probe is inserted into a supersonic stream a shock wave will form around the front of the probe as shown. The flow approaching the stagnation point passes through a normal shock wave from points ① to ②. It then undergoes an isentropic process as it flows from ② to ③.

Let us use the energy equation (11.52) which is applicable for an isentropic process. Thus, between ② and ③

$$\frac{p_2}{p_3} = \left(1 + \frac{\gamma - 1}{2} M_2^2\right)^{\frac{\gamma}{1-\gamma}}$$

Across the shock wave we may use the momentum equation, in the form of Eq. 11.44, as

$$\frac{p_2}{p_1} = \frac{2\gamma}{\gamma + 1} M_1^2 - \frac{\gamma - 1}{\gamma + 1}$$

The Mach number M_2 is related to the Mach number M_1 by Eq. 11.43,

$$M_2^2 = \frac{M_1^2 + \frac{2}{\gamma - 1}}{\frac{2\gamma}{\gamma - 1} M_1^2 - 1}$$

Combining these three equations we find, after some algebraic manipulation, that

$$\frac{p_3}{p_1} = \frac{\left(\frac{\gamma + 1}{2} M_1^2\right)^{\frac{\gamma}{\gamma - 1}}}{\left(\frac{2\gamma M_1^2}{\gamma + 1} - \frac{\gamma - 1}{\gamma + 1}\right)^{\frac{1}{\gamma - 1}}}$$

This is known as the *Rayleigh pitot-tube formula*, good only for $M_1 > 1$.

Using $p_3 = 40$ and $p_1 = 10$, for the data presented in the example the formula above takes the form

$$\frac{40}{10} = \frac{(1.2M_1^2)^{3.5}}{(1.165M_1^2 - 0.167)^{2.5}}$$

This equation can be solved by trial and error (guess a value for M_1 and see if the r.h.s = l.h.s) to give

$$M_1 = 1.64$$

The Mach number after the shock is found from the shock table to be

$$M_2 = 0.657$$

The temperature behind the shock can be found from the isentropic table. It is

$$T_2 = T_3 \times 0.921 = 682°R$$

The temperature in front of the shock is then, from the shock table,

$$T_1 = \frac{T_2}{1.42} = 480°R$$

Finally, we can determine the velocity V_1 from

$$V_1 = M_1 a_1 = M_1 \sqrt{\gamma R T_1} = 1.64\sqrt{1.4 \times 1716 \times 480} = 1760 \text{ fps}$$

It should be pointed out that the procedure for measuring the static pressure before the shock wave may be carried out in a wind tunnel or shock tube as shown in Fig. E11.8; but to measure the static pressure with an instrument attached to a body moving at supersonic speeds is very difficult since the instrument would always involve a shock wave. It has been observed that the pressure approximately ten diameters downstream of a pitot probe nose is about equal to the static free-stream pressure.

11.5 STEADY, UNIFORM COMPRESSIBLE FLOW IN A CONSTANT-AREA CONDUIT

The steady flow of a compressible fluid through a conduit is a common problem when transporting gas or air through pipe lines. Often, for long piping systems, the gas line will approach a constant temperature so that heat generated in the pipe by friction is transferred to the surroundings; the resulting isothermal flow is of particular interest. It may be, however, that the pipe is insulated so that no heat transfer is allowed. In that case the flow is adiabatic and the temperature in the pipe changes due to the frictional effects. Problems involving a large heat transfer rate and insignificant losses are also of some interest, especially in the combuster cam of a jet engine.

The fundamental equations useful in treating the compressible-flow problem in the pipe are the integral continuity, energy, and momentum equations. We will again consider the velocity to be uniform over a cross-sectional area. A process equation $\rho = \rho(p)$ is sometimes specified; the equation of state, $p = \rho RT$, is assumed valid for the pressures and temperatures of interest.

Entrance effects will be ignored and the velocity profile will be assumed uniform at each cross-sectional area. The momentum equation, applied to a control volume of length L, is

$$(p_1 - p_2)A - \pi D \int_0^L \tau_0 \, dx = \rho_1 A_1 V_1 (V_2 - V_1) \tag{11.74}$$

where τ_0 is the viscous stress acting on the wall. It is related to the friction factor f by

$$\tau_0 = \tfrac{1}{8} f \rho V^2 \tag{11.75}$$

where the friction factor from the Moody diagram may be used. The friction factor is essentially independent of the Mach number and for pipe flows involving gases the Reynolds number is typically large enough so that for a particular e/D the friction factor is constant. But, it is noted that the shearing stress τ_0 depends on V^2 and that, for a compressible flow, $V = V(x)$. Thus, the integration in Eq. 11.74 cannot be performed since $V(x)$ is not known. This differs from an incompressible flow in a pipe for which the average velocity is constant.

The momentum equation written in differential form is needed for the subsequent analysis. We obtain such a form by applying the equation to an elemental control volume of length dx, shown in Fig. 11.13. The quantities at position x are denoted by ρ, V, p and at $x + dx$ by $\rho + d\rho, V + dV, p + dp$. Hence, the differential momentum equation becomes

$$-\tau_0 \pi D \, dx - dp \frac{\pi D^2}{4} = \rho \frac{\pi D^2}{4} V[(V + dV) - V] \tag{11.76}$$

or

$$-\frac{4\tau_0}{D} dx - dp = \rho V \, dV \tag{11.77}$$

where D is the diameter* of the pipe. In this equation $\tau_0, \rho, p,$ and V

---

*For a noncircular cross-section the hydraulic diameter $d_H = 4A/P$ (where P is the perimeter) could be used.

Art. 11.5   CONSTANT-AREA CONDUITS   507

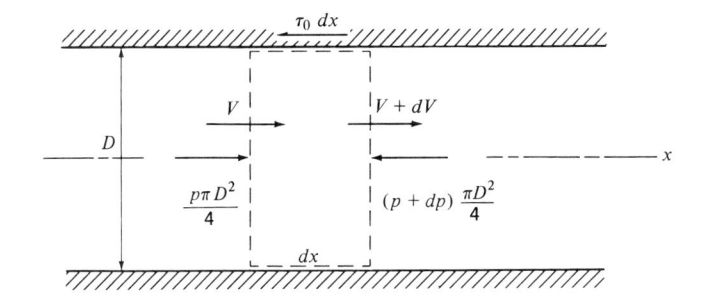

Fig. 11.13. Elemental control volume for momentum equation.

are all functions of x. The continuity equation and energy equation are necessary to obtain a solution. Let us first consider adiabatic flow.

1. Adiabatic Flow with Friction

For adiabatic flow ($\dot{Q} = 0$) and no shaft work, the energy equation is

$$\frac{V_1^2}{2} + \frac{p_1}{\rho_1} + \tilde{u}_1 = \frac{V_2^2}{2} + \frac{p_2}{\rho_2} + \tilde{u}_2 \tag{11.78}$$

or, for any position x along the pipe,

$$\frac{V^2}{2} + \frac{p}{\rho} + \tilde{u} = \text{const} \tag{11.79}$$

This can be written in an equivalent form as, using $h = \frac{p}{\rho} + \tilde{u} = C_p T$ and $C_p = \gamma R/(\gamma - 1)$,

$$\frac{V^2}{2} + \frac{\gamma R}{\gamma - 1} T = \text{const} \tag{11.80}$$

Introducing the Mach number with $V^2 = M^2 \gamma R T$ the energy equation can be written as

$$[(\gamma - 1)M^2 + 2]T = \text{const} \tag{11.81}$$

Differentiating and rearranging gives

$$\frac{dT}{T} = -\frac{2M}{\frac{2}{\gamma - 1} + M^2} dM \tag{11.82}$$

The continuity equation is

$$\rho V = \text{const} \tag{11.83}$$

or, in terms of the Mach number,

$$\rho^2 M^2 T = \text{const} \tag{11.84}$$

Differentiate this equation and divide by $\rho^2 M^2 T$ to obtain

$$2\frac{d\rho}{\rho} + 2\frac{dM}{M} + \frac{dT}{T} = 0 \tag{11.85}$$

Using the differential energy equation (11.82), the above differential continuity equation can be put in the form

$$\frac{d\rho}{\rho} = \left[-\frac{1}{M} + \frac{M}{\frac{2}{\gamma - 1} + M^2} \right] dM \tag{11.86}$$

Finally, the equation of state, $\dfrac{p}{\rho T} = R$, may be differentiated to yield

$$\frac{dp}{p} - \frac{d\rho}{\rho} - \frac{dT}{T} = 0 \tag{11.87}$$

Substituting Eqs. 11.82 and 11.86 in 11.87, we obtain

$$\frac{dp}{p} = -\left[\frac{1}{M} + \frac{M}{\frac{2}{\gamma - 1} + M^2} \right] dM \tag{11.88}$$

The continuity equation, in differential form, allows us to write

$$\frac{dV}{V} = -\frac{d\rho}{\rho} = \left[\frac{1}{M} - \frac{M}{\frac{2}{\gamma - 1} + M^2} \right] dM \tag{11.89}$$

The relationships above may be integrated by referring to the critical condition $M^* = 1$ with associated values V^*, p^*, T^*, and ρ^*. Thus, integrating from some general location to the critical location, we have

$$\frac{T}{T^*} = \frac{\gamma + 1}{2 + (\gamma - 1)M^2} \tag{11.90}$$

$$\frac{p}{p^*} = \frac{1}{M}\left[\frac{\gamma + 1}{2 + (\gamma - 1)M^2} \right]^{1/2} \tag{11.91}$$

$$\frac{\rho}{\rho^*} = \frac{V^*}{V} = \frac{1}{M}\left[\frac{2 + (\gamma - 1)M^2}{\gamma + 1} \right]^{1/2} \tag{11.92}$$

Returning to the momentum equation (11.77), and using Eq. 11.75, we may write, after some manipulation,

$$\frac{f}{D} dx = \frac{1 - M^2}{\gamma M^4 \left(1 + \frac{\gamma - 1}{2} M^2\right)} dM^2 \quad (11.93)$$

Let the critical location where $M = 1$ be located by x_{crit}. The foregoing may integrated, by partial fractions, to yield,

$$\frac{\bar{f} L_{\text{crit}}}{D} = \frac{1}{\gamma} \frac{1 - M^2}{M^2} + \frac{\gamma + 1}{2\gamma} \ln \frac{(\gamma + 1)M^2}{2 + (\gamma - 1)M^2} \quad (11.94)$$

where $L_{\text{crit}} = x_{\text{crit}} - x$.

The ratios above are all functions of the Mach number M at a particular location x. They are tabulated for air in Table D.3. Sample calculation techniques are illustrated in the examples.

Before proceeding with sample calculations, let us make two important observations. First, there is a critical length L_{crit}, at which the Mach number reaches unity. For lengths greater than L_{crit}, a choked flow results, with associated reductions in the mass flow rate. If the flow exists in a fixed-length pipe connecting a reservoir with a receiver, and if the flow is caused by a decreasing receiver pressure, the mass flow rate will continue to increase until the receiver pressure $p_r = p^*$; choked flow then occurs and further reduction in p_r has no influence on the flow rate since $M = 1$ at the receiver end of the pipe.

The second observation concerns the entropy. Using the relation

$$s = \frac{\gamma R}{\gamma - 1} \ln T - R \ln p + \text{const} \quad (11.95)$$

we can show that

$$ds = \frac{2RM(1 - M^2)}{M^2[2 + (\gamma - 1)M^2]} dM \quad (11.96)$$

For a frictional process the entropy change $ds > 0$. Thus, we observe:

1. For $M > 1$, dM is negative, so that the Mach number decreases.
2. For $M < 1$, dM is positive, so that the Mach number increases.
3. For $M = 1$, ds is zero.

This may be graphically demonstrated by the T–s diagram of Fig. 11.14. This diagram may be constructed by setting an arbitrary s_1 for a set of conditions T_1, V_1, p_1; choose a T and, by using the energy and continuity equations and the equation of state, calculate s. For subsonic flows, a somewhat surprising result is observed: the temperature decreases in an adiabatic, frictional flow. Also, we observe that supersonic flows are not possible from the subsonic state since this would indicate a decrease in entropy below the M = 1 state, a violation of the second law of thermodynamics. The line formed on the T-s diagram is called a *Fanno line*.

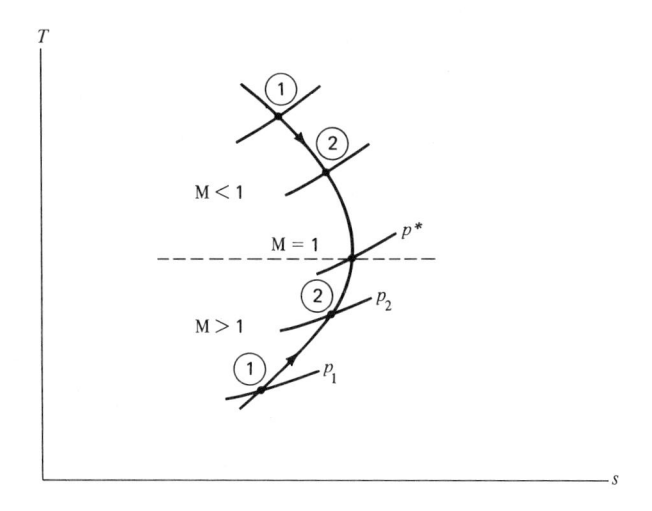

Fig. 11.14. The T–s diagram for an adiabatic pipe flow (Fanno line).

A second aspect of the notion of choked flow may be identified. Consider that the receiver pressure is lower than the p^* value shown in the figure. The choked flow involves a particular value of the flow rate and a particular critical length to attain the choked condition. If an additional length of pipe is added to the existing conduit, then a second Fanno line will be formed, to the right of the present one; p_1 will be larger, indicating that V_1 is less than the former value. Consequently, the flow rate is still choked but at the cost of a reduced flow rate.

The discussion above is sufficient to introduce the notions of Fanno-line flow; other cases involving supersonic flow in the duct, the corresponding critical lengths, and the presence of shock waves are extensions which will not be considered. The reader is referred to more advanced texts for these considerations.*

---

*See for example, A. H. Shapiro, *The Dynamics and Thermodynamics of Compressible Fluid Flow*, Ronald Press, New York, 1954.

2. Isothermal Flow with Friction

For isothermal flow with friction, $T = $ constant and $dT = 0$. The continuity equation $\rho V =$ constant, using $V = \mathrm{M}a$, becomes

$$\rho \mathrm{M} = \mathrm{const} \tag{11.97}$$

where the speed of sound a has been absorbed into the constant. In differential form,

$$-\frac{dV}{V} = \frac{d\rho}{\rho} = -\frac{d\mathrm{M}}{\mathrm{M}} \tag{11.98}$$

The equation of state, $p/\rho = RT = $ constant, may be differentiated to yield

$$\frac{dp}{p} = \frac{d\rho}{\rho} = -\frac{d\mathrm{M}}{\mathrm{M}} \tag{11.99}$$

The differential momentum equation (11.77) is written in terms of the friction factor as

$$\frac{f\,dx}{D} = -\frac{2\,dp}{\rho V^2} - \frac{2\,dV}{V} \tag{11.100}$$

This may now be put in the form

$$\frac{f\,dx}{2D} = -\frac{\gamma \mathrm{M}^2 - 1}{\gamma \mathrm{M}^2}\frac{d\mathrm{M}}{\mathrm{M}} \tag{11.101}$$

This equation shows that if $\mathrm{M} > 1/\sqrt{\gamma}$, $d\mathrm{M} < 0$; that is, the Mach number and thus the velocity are decreasing with x. For $\mathrm{M} < 1/\sqrt{\gamma}$, $d\mathrm{M} > 0$ indicating an increasing Mach number and velocity with x. We observe that the critical Mach number for isothermal flow is $\mathrm{M}_{\mathrm{crit}} = 1/\sqrt{\gamma}$. By integrating Eq. 11.101 we can find L_{crit}. It is found from

$$\frac{\bar{f}L_{\mathrm{crit}}}{D} = \ln(\gamma \mathrm{M}_1^2) + \frac{1 - \gamma \mathrm{M}_1^2}{\gamma \mathrm{M}_1^2} \tag{11.102}$$

This would be the condition of choked flow.

The preceding equations can all be solved directly without laborious computations. Hence, no tables are presented for isothermal flows. We simply integrate the equations and solve them analytically for any particular flow situations. Integrating from section ① to ②, the various ratios become

$$\frac{\rho_1}{\rho_2} = \frac{V_2}{V_1} = \frac{p_1}{p_2} = \frac{\mathrm{M}_2}{\mathrm{M}_1} \tag{11.103}$$

and
$$\frac{\bar{f}L}{D} = \frac{1}{\gamma M_1^2} - \frac{1}{\gamma M_2^2} + \ln \frac{M_1^2}{M_2^2} \tag{11.104}$$

For any set of conditions in an isothermal flow, the equations above provide the solution.

3. Frictionless Flow with Heat Transfer

The isothermal and the adiabatic frictional flows just considered in Sections 2 and 3 are rather easily realized in an appropriate experimental situation. The topic of this section is frictionless flow with heat transfer, which is not readily achieved; that is, it is not feasible to introduce the energy without effecting other changes such as a change of chemical composition, mass flux (e.g., evaporation) or influencing the wall friction. However, the relative ease with which this type of flow can be computed and the instructive nature of the results makes it of interest. For example, if for flows of engineering interest (such as a jet engine combustor) the heat transfer effects dominate over the frictional and composition change effects, then approximate answers are available from the following relationships. For a frictionless flow the differential momentum equation (11.77) takes the form

$$dp = -\rho V dV \tag{11.105}$$

Using the continuity equation, the equation of state, and the relation between Mach number and velocity, all in differential form, we find that

$$\frac{dp}{p} = -\frac{2\gamma M}{1 + \gamma M^2} dM \tag{11.106}$$

$$\frac{dV}{V} = -\frac{d\rho}{\rho} = \frac{2}{1 + \gamma M^2} \frac{dM}{M} \tag{11.107}$$

$$\frac{dT}{T} = \frac{2(1 - \gamma M^2)}{1 + \gamma M^2} \frac{dM}{M} \tag{11.108}$$

The differential entropy change

$$ds = \frac{\gamma R}{\gamma - 1} \frac{dT}{T} - R \frac{dp}{p} \tag{11.109}$$

can be written as

$$ds = \frac{\gamma R(1 - M^2)2M}{(\gamma - 1)M^2(1 + \gamma M^2)} dM \tag{11.110}$$

Art. 11.5   CONSTANT-AREA CONDUITS   513

For the condition of adding heat, $ds > 0$, and we make the following observations:

1. For $M < 1$, dM is positive, so that the Mach number is increasing.
2. For $M > 1$, dM is negative, so that the Mach number is decreasing.
3. For $M = 1$, ds is zero.

If heat were transferred from the fluid then $ds < 0$ and the Mach number would decrease for a subsonic flow and increase in a supersonic flow.

We again observe that choking may result whenever the exit Mach number equals unity for the condition of heat addition. Thus, we integrate between conditions at a particular point and the critical condition of $M^* = 1$. The various ratios become

$$\frac{p}{p^*} = -\frac{1+\gamma}{1+\gamma M^2} \tag{11.111}$$

$$\frac{T}{T^*} = \left[\frac{1+\gamma}{1+\gamma M^2}\right]^2 M^2 \tag{11.112}$$

$$\frac{V}{V^*} = \frac{\rho^*}{\rho} = \frac{(1+\gamma)M^2}{1+\gamma M^2} \tag{11.113}$$

The ratios are tabulated in Table D–4.

Example 11.9

Air enters a 2-in.-dia. cast iron pipe at $p_1 = 50$ psia, $T_1 = 200°F$, and $V_1 = 200$ fps. For a 100-ft-long insulated pipe exiting to a receiver, find the conditions at the exit.

Solution. The Reynolds number at the inlet is

$$\text{Re} = \frac{200 \times 2/12}{1.89 \times 10^{-4}} = 1.76 \times 10^5$$

For the cast iron pipe $e/D = 0.0051$. For this Reynolds number and e/D, the friction factor is nearly constant at $f = 0.03$. Thus,

$$\frac{\bar{f}L}{D} = \frac{0.03 \times 100}{2/12} = 18$$

The inlet Mach number is

$$M_1 = \frac{V_1}{a_1} = \frac{200}{\sqrt{1.4 \times 1716 \times 660}} = 0.16$$

From the adiabatic flow table (D.3) we find

$$\left(\frac{\bar{f}L_{crit}}{D}\right)_1 = 24.2$$

This would indicate that the pipe is shorter than the critical length necessary to give choked flow. To find M_2, consider Fig. E11.9, and observe that

$$\left(\frac{\bar{f}L_{crit}}{D}\right)_2 = \left(\frac{\bar{f}L_{crit}}{D}\right)_1 - \frac{\bar{f}L}{D}$$

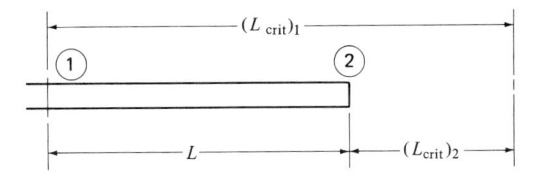

Fig. E11.9

assuming that $\bar{f}$ is constant over the length of the pipe. This gives

$$\left(\frac{\bar{f}L_{crit}}{D}\right)_2 = 6.2$$

From the adiabatic flow table the corresponding value of M_2 is

$$M_2 = 0.283$$

We find the pressure at the exit by forming the product

$$P_2 = \frac{p_2}{p^*} \times \frac{p^*}{p_1} \times p_1$$

$$= 3.84 \times \frac{1}{6.83} \times 50 = 28.1 \text{ psia}$$

This would be the receiver pressure. The temperature is

$$T_2 = \frac{T_2}{T^*} \times \frac{T^*}{T_1} \times T_1$$

$$= 1.18 \times \frac{1}{1.194} \times 660 = 652°R$$

The temperature at the exit is lower than T_1, a rather surprising result considering the frictional effects in the flow. The exit velocity is

$$V_2 = M_2 a_2 = 0.244\sqrt{1.4 \times 1716 \times 652} = 306 \text{ fps}$$

We note that the velocity increased significantly along the pipe.

Example 11.10

Air flows from a reservoir maintained at 70°F through a 100 ft long insulated 2-in.-dia. pipe to the atmosphere. Determine the mass flow rate from the pipe if $\bar{f} = 0.00575$ and the pressure in the reservoir is (a) 75 psia and (b) 37.5 psia.

Solution. **a.** Let us assume choked flow; that is, the pipe is of length L_{crit}. Then, the Mach number at the exit is unity and the exit pressure is $p_e > p_{atm}$. From the given quantities

$$\frac{\bar{f}L}{D} = \frac{0.00575 \times 100}{2/12} = 3.45$$

From the adiabatic flow table (D.3) we find that the Mach number at the inlet to the pipe is

$$M_1 = 0.35$$

For this Mach number $p_0/p_0^* = 1.78$ so that

$$p_0^* = \frac{75}{1.78} = 42.1 \text{ psia}$$

Then from the isentropic flow table, which relates the local pressure to the local isentropic stagnation pressure, at the exit where $M^* = 1$,

$$p^* = 0.528 \times 42.1 = 22.2 \text{ psia}$$

This pressure is larger than the receiver pressure p_{atm}; hence, the flow is choked as assumed. Now, assuming an isentropic flow from the reservoir to the inlet, the isentropic flow table gives us, at $M_1 = 0.35$,

$$p_1 = 0.919 p_0 = 0.919 \times 75 = 68.9 \text{ psia}$$

$$T_1 = 0.976 T_0 = 0.976 \times 530 = 517°R$$

The density and velocity are calculated to be

$$\rho_1 = \frac{p_1}{RT_1} = \frac{68.9 \times 144}{1716 \times 517} = 0.0112 \text{ slug/ft}^3$$

$$V_1 = M_1 a_1 = 0.35 \times \sqrt{1.4 \times 1716 \times 517} = 1243 \text{ fps}$$

The mass flux is finally,

$$\dot{m} = \rho_1 A_1 V_1 = 0.0112 \times \frac{\pi}{144} \times 1243 = 0.304 \text{ slug/sec}$$

b. Assuming choked flow again we find $p_0^* = 21.1$ psia and $p^* = 11.1$ psia. This pressure is lower than the atmospheric pressure of 14.7 psia; hence, choked flow does not occur. This problem is solved by trial and error. We guess a value for M_1 and calculate p_2. This is repeated until $p_2 \cong 14.7$.

Guess $M_1 = 0.30$. From the adiabatic flow table, $(\bar{f}L_{\text{crit}}/D)_1 = 5.30$. From the given information $\bar{f}L/D = 3.45$. The difference leads to

$$\left(\frac{\bar{f}L_{\text{crit}}}{D}\right)_2 = \left(\frac{\bar{f}L_{\text{crit}}}{D}\right)_1 - \frac{\bar{f}L}{D} = 5.30 - 3.45 = 1.85$$

The adiabatic flow table is interpolated to give

$$M_2 = 0.429$$

The pressure p_2 is found to be

$$p_2 = p_0 \left(\frac{p_1}{p_0}\right)_{M_1} \left(\frac{p^*}{p_1}\right)_{M_1} \left(\frac{p_2}{p^*}\right)_{M_2}$$

$$= 37.5 \times 0.939 \times \frac{1}{3.62} \times 2.51 = 24.4 \text{ psia}$$

This pressure is much too high. Guess a larger value for M_1. Guess $M_1 = 0.34$; then $(\bar{f}L_{\text{crit}}/D)_2 = 0.369$, $M_2 = 0.660$ giving

$$p_2 = 37.5 \times 0.923 \times \frac{1}{3.197} \times 1.59 = 17.3 \text{ psia}$$

Guess $M_1 = 0.347$; then $(\bar{f}L_{\text{crit}}/D)_2 = 0.09$, $M_2 = 0.782$, giving

$$p_2 = 37.5 \times 0.920 \times \frac{1}{3.123} \times 1.32 = 14.6 \text{ psia}$$

This is an acceptable value for M_1. With this value we find that $p_1 = 33.8$ psia, $T_1 = 518°$R, $\rho_1 = 0.00549$ slug/ft^3 and $V_1 = 387$ fps. The mass flux rate is then

$$\dot{m} = 0.00549 \times \frac{\pi}{144} \times 387 = 0.0463 \text{ slug/sec}$$

Example 11.11

Air enters a 2-in.-dia., 100-ft-long cast iron pipe at a pressure of 75 psia. It exits to a receiver at 15 psia and remains essentially isothermal at 50°F. Determine the necessary heat addition over the length of the pipe.

Solution. The integral energy equation, with $T_1 = T_2$ and zero shaft work, is simply

$$\dot{Q} = \left(\frac{V_2^2}{2} - \frac{V_1^2}{2}\right)\dot{m}$$

To find the heat-transfer rate we must find V_1, V_2, and ρ_2.

Using the Moody diagram with $e/D = 0.0051$ as a guide, assume $\bar{f} = 0.031$. Then, using Eq. 11.104, we have, with $M_2/M_1 = p_1/p_2 = 5$,

$$\frac{0.031 \times 1000}{2/12} = \frac{1}{\gamma M_1^2} - \frac{1}{25\gamma M_1^2} - \ln 25$$

Art. 11.5 THE NORMAL SHOCK WAVE 517

giving $M_1 = 0.0619$. Then
$$M_2 = 5M_1 = 0.310$$
$$V_2 = M_2 a_2 = 0.310\sqrt{1.4 \times 1716 \times 510} = 344 \text{ fps}$$
$$V_1 = \frac{V_2}{5} = 68.8 \text{ fps}$$
$$\rho_2 = \frac{p_2}{RT_2} = \frac{15 \times 144}{1716 \times 510} = 0.00246 \text{ slugs/ft}^3$$

Thus,
$$\dot{m} = 0.00246 \times \frac{\pi}{144} \times 344 = 0.0185 \text{ slugs/sec}$$

Now check $\bar{f}$ with the Moody diagram. Calculate the Reynolds numbers:
$$\text{Re}_1 = \frac{V_1 D \rho_1}{\mu_1} = \frac{68.8 \times 2/12 \times 0.0123}{3.67 \times 10^{-7}} = 2.98 \times 10^5$$
$$\text{Re}_2 = \frac{V_2 D \rho_2}{\mu_2} = \frac{344 \times 2/12 \times 0.00246}{3.67 \times 10^{-7}} = 14.9 \times 10^5$$

From the Moody diagram, with $e/D = 0.0051$, $\bar{f} \cong 0.031$. Hence this is an acceptable value for $\bar{f}$. The heat transfer is then
$$\dot{Q} = \left(\frac{344^2}{2} - \frac{68.8^2}{2}\right) \times 0.0185 = 960 \frac{\text{ft-lb}}{\text{sec}} \quad \text{or} \quad 1.23 \text{ Btu/sec}$$

Example 11.12

Air at 70°F and 15 psia enters a 6-in.-dia. combustion chamber at 170 fps. We wish to add enough heat in the combustion process so that choked flow just occurs at the exit. Estimate this heat-transfer rate.

Solution. The energy equation relates the heat transfer rate in ft-lb/sec to the flow variables. It is
$$\dot{Q} = \dot{m}\left(\frac{V_2^2}{2} + \frac{\gamma R}{\gamma - 1} T_2 - \frac{V_1^2}{2} - \frac{\gamma R}{\gamma - 1} T_1\right)$$

At the exit the Mach number is unity so that $T_2 = T^*$ and $V_2 = V^*$. The Mach number at the inlet is
$$M_1 = \frac{V_1}{a_1} = \frac{170}{\sqrt{1.4 \times 1716 \times 530}} = 0.15$$

From the frictionless gas flow table D.4,
$$T_2 = T^* = \frac{T_1}{0.122} = \frac{530}{0.122} = 4350°R$$
$$V_2 = V^* = \frac{V_1}{0.0524} = \frac{170}{0.0524} = 3250 \text{ fps}$$

Now, before we calculate $\dot{Q}$ we must know $\dot{m}$. It is given by

$$\dot{m} = \rho_1 A_1 V_1 = \frac{15 \times 144}{1716 \times 530} \times \frac{\pi 9}{144} \times 170 = 0.0795 \text{ slug/sec}$$

Finally,

$$\dot{Q} = 0.0795\left[\frac{3250^2}{2} + \frac{1.4 \times 1716}{0.4}(4350 - 530) - \frac{170^2}{2}\right]$$

$$= 2{,}240{,}000 \text{ ft-lb/sec}$$

or

$$\dot{Q} = 2880 \text{ Btu/sec}$$

11.6 THE OBLIQUE SHOCK WAVE

In the previous articles in this chapter one-dimensional flow situations have been considered in which the velocity profile was uniform over the cross-sectional area. The oblique shock wave, which will be studied in this article, involves two velocity components and thus is a departure from the one-dimensional flow. The departure is not very significant though since we will assume the flow before and after the oblique shock to be uniform and one-dimensional with the oblique shock providing an abrupt change in both the direction and magnitude of the velocity vector. Oblique shock waves form on the leading edge of a sharp object such as a supersonic airfoil and in a corner as shown in Fig. 11.15. Oblique shock waves also form on axisymmetric bodies such as a nose cone or bullet, as shown in Fig. 11.16. However, we will only consider two-dimensional flows herein. The function of the oblique

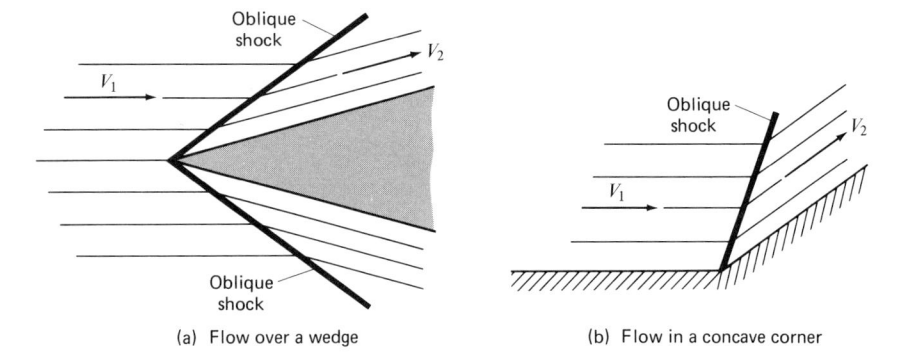

(a) Flow over a wedge (b) Flow in a concave corner

Fig. 11.15. Oblique shock waves in a supersonic flow.

Fig. 11.16. Oblique shock waves.

shock is to turn the flow so that it flows parallel to the wall. The angle that the velocity vector V_2 makes with respect to V_1 introduces a new variable into the analysis. With the additional component-momentum equation tangential to the shock the problem may remain solvable.

A control volume enclosing a portion of the oblique shock wave is shown in Fig. 11.17. It moves with a moving shock wave so that a steady flow is observed. The velocity vector for the upstream flow is assumed to be in the x-direction only. The oblique shock wave is oriented at an angle β and turns the flow through the *deflection angle* θ,

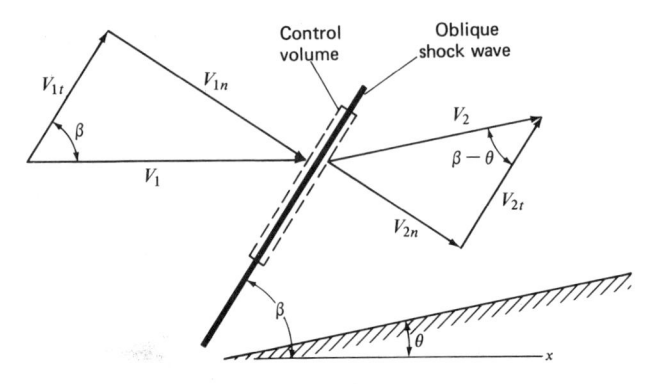

Fig. 11.17. Control volume enclosing a portion of an oblique shock wave.

often referred to as the *wedge angle*. The normal and tangential components are as shown. The continuity equation is, with $A_1 = A_2$,

$$\rho_1 V_{1n} = \rho_2 V_{2n} \qquad (11.114)$$

The tangential component momentum equation is, in general form,

$$\Sigma F_t = \int_{c.s.} V_t \rho \mathbf{V} \cdot \hat{n} \, dA \qquad (11.115)$$

Applied to the control volume enclosing the oblique shock wave, there results

$$0 = V_{2t} \rho_2 V_{2n} A_2 - V_{1t} \rho_1 V_{1n} A_1$$

showing that

$$V_{2t} = V_{1t} \qquad (11.116)$$

The normal component momentum equation is

$$p_1 - p_2 = \rho_2 V_{2n}^2 - \rho_1 V_{1n}^2 \qquad (11.117)$$

Using $V_1^2 = V_{1n}^2 + V_{1t}^2$, the energy equation may be written as

$$\frac{V_{1n}^2}{2} + \frac{\gamma R}{\gamma - 1} T_1 = \frac{V_{2n}^2}{2} + \frac{\gamma R}{\gamma - 1} T_2 \qquad (11.118)$$

where the tangential components have subtracted from both sides. We note that the tangential components do not enter the continuity, energy, and normal component momentum equations. Hence, the normal shock wave equations, developed in Art. 11.3, may be used, with the normal components of velocity V_{1n} and V_{2n} substituted for V_1 and V_2, respectively. The shock table in Appendix D is useful for computations with M_{1n} replacing M_1 and M_{2n} replacing M_2.

Art. 11.6   THE OBLIQUE SHOCK WAVE   521

It is often convenient to relate the oblique shock angle β to the deflection angle θ. This is done by forming the density ratio and using Eqs. 11.114, and 11.117, in addition to the geometry of Fig. 11.17, to obtain

$$\frac{\rho_2}{\rho_1} = \frac{V_{1n}}{V_{2n}} = \frac{V_{1t} \tan \beta}{V_{2t} \tan (\beta - \theta)} = \frac{\tan \beta}{\tan (\beta - \theta)} \quad (11.119)$$

From the normal shock wave equations (11.44 and 11.45), we can find that

$$\frac{\rho_2}{\rho_1} = \frac{(1 + \gamma)M_{1n}}{(\gamma - 1)M_{1n}^2 + 2} \quad (11.120)$$

Then, using Eq. 11.119 and $M_{1n} = M_1 \sin \beta$, we can arrive at

$$\tan (\beta - \theta) = \frac{2 \tan \beta}{\gamma + 1} \left[\frac{\gamma - 1}{2} + \frac{1}{M_1^2 \sin^2 \beta} \right] \quad (11.121)$$

For a particular M_1, Eq. 11.121 relates θ to β. This relationship is plotted in Fig. 11.18. Several interesting observations are made in reference to this figure:

1. For a particular upstream Mach number M_1, and a given angle θ, there are two possible angles β, the larger β, corresponding to a "strong" shock, and the smaller β, corresponding to a "weak" shock.

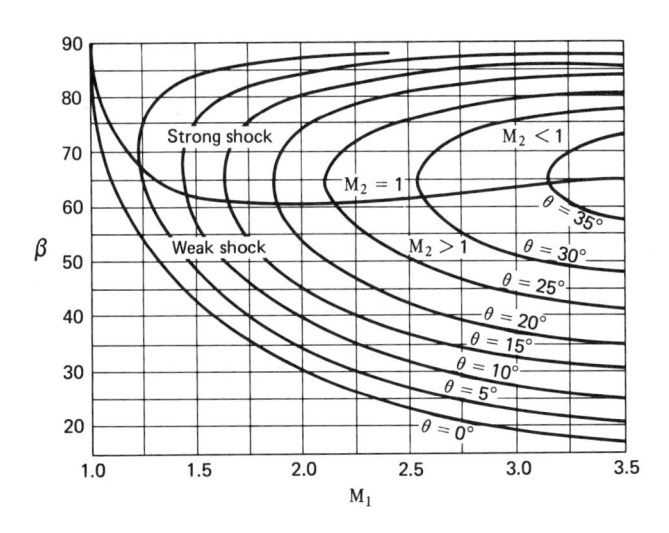

Fig. 11.18. Oblique shock wave relationships for air ($\gamma = 1.4$). (Adapted from S. W. Yuan, *Foundations of Fluid Mechanics*, Prentice-Hall, New York, 1967.)

2. For a particular deflection angle θ there is a minimum Mach number M_1 for which there is only one shock angle β.
3. For a particular deflection angle θ if M_1 is less than the minimum referred to in point 2 above then no oblique shock wave exists. The shock wave would become detached from the wedge, as shown in Fig. 11.19. A detached shock would also be present for a particular M_1 if θ were too large.

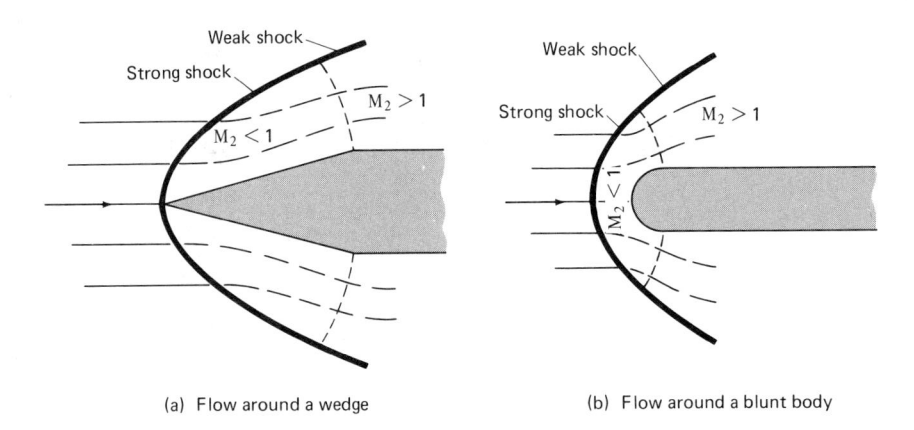

Fig. 11.19. A detached shock wave.

(a) Flow around a wedge (b) Flow around a blunt body

Whether an oblique shock wave is weak or strong depends on the pressure condition downstream of the oblique shock. If the pressure rise is large, such as that encountered near a stagnation region, a strong shock results. For a small pressure rise, encountered in flow around sharp edged supersonic airfoils, a weak shock occurs. A strong or weak oblique shock wave may occur in a concave corner depending on the magnitude of the back pressure.

For the detached shock of Fig. 11.19, the flow along the dividing streamline would undergo a normal shock wave so that $M_2 < 1$. Away from the body, the velocity V_2 would not change direction substantially so that the shock wave must become weak far from the body. If V_1 is supersonic this means that V_2 must be supersonic away from the body so there must be an acceleration of the subsonic flow near the nose to a supersonic flow as shown. Flow around a blunt body is also shown. It would not be very different from the detached wedge flow. However, for the blunt body a detached shock would always exist.

Art. 11.7　　　　　　　　ISENTROPIC TURNING　　　　　　　　523

Example 11.13

An air flow strikes a wedge as shown in Fig. E11.13. A weak shock is formed which reflects from a wall. Estimate the value of M_3 and β_3.

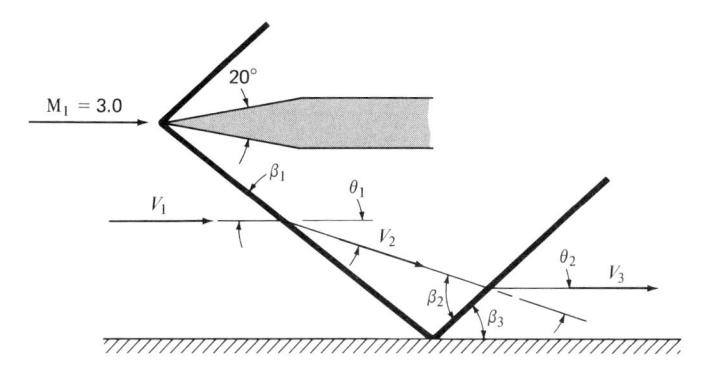

Fig. E11.13

Solution. From Fig. 11.18 for $\theta = 10°$ and $M_1 = 3.0$ we find that $\beta_1 = 27.5°$. Hence,

$$M_{1n} = M_1 \sin 27.5° = 1.39$$

From the shock table,

$$M_{2n} = 0.744 = M_2 \sin (27.5° - 10°)$$

This gives

$$M_2 = 2.48$$

The reflected shock wave must cause the air to flow parallel to the wall so that $\theta_2 = 10°$. Using this angle and $M_2 = 2.48$, we find β_2 from Fig. 11.18 to be $\beta_2 = 33$ degrees. This results in

$$M_{2n} = 2.48 \sin 33° = 1.35$$

From the shock table

$$M_{3n} = 0.762 = M_3 \sin 23°$$

giving

$$M_3 = 1.95$$

The angle β_3 is then

$$\beta_3 = \beta_2 - 10° = 23°$$

11.7　ISENTROPIC TURNING BY COMPRESSION AND EXPANSION WAVES

Often there is no abrupt change in the geometry of a corner so that a sudden oblique shock wave is not produced in the flow. Consider a

524   COMPRESSIBLE FLOWS   Ch. 11

smooth curve connecting the walls, as shown in Fig. 11.20. Near the wall an infinite number of Mach waves would turn the flow the prescribed angle θ. The Mach waves would converge away from the wall as shown and form an oblique shock wave which would turn the flow through the angle θ. This oblique shock wave has been studied in the previous article.

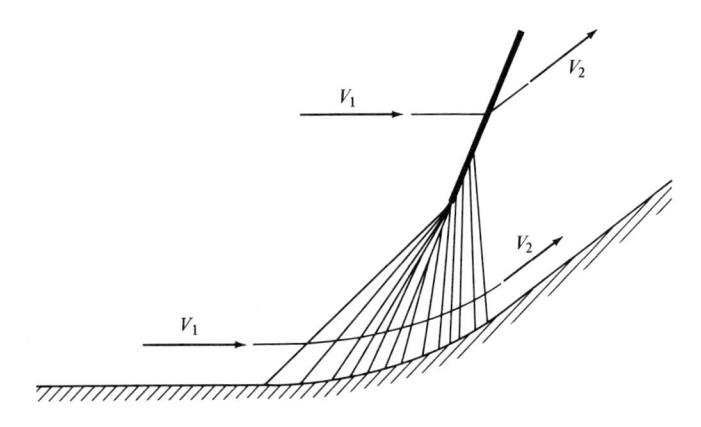

Fig. 11.20. Flow in a corner.

(a) A finite wave

(b) An infinite number of Mach waves

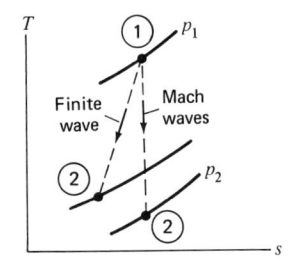

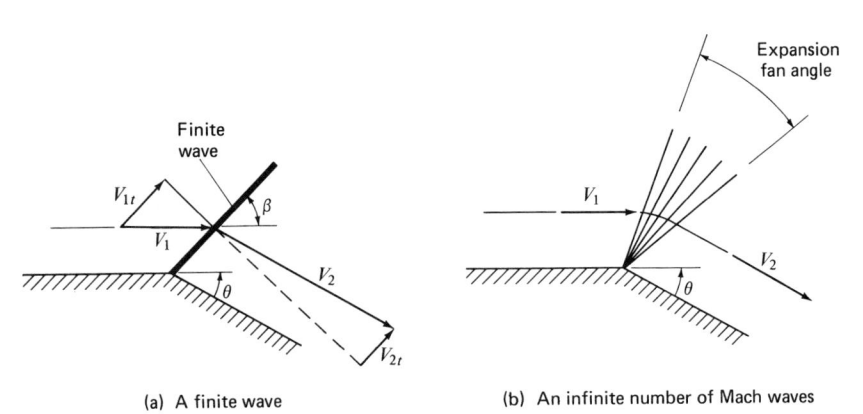

Fig. 11.21. Flow over a convex corner.

ISENTROPIC TURNING

For a supersonic flow over a convex corner as shown in Fig. 11.21a, let us assume that a single wave turns the flow through the angle θ. If we require the tangential velocity to be conserved it is obvious that $V_2 > V_1$. The normal component of such a wave would necessarily change from subsonic to supersonic, as discussed in Art. 11.3, on the normal shock wave. This would require a decrease in entropy, and thus such a wave is ruled out as a possible turning mechanism.

Consider another possible turning mechanism, an infinite number of Mach waves, as shown in Fig. 11.21b. This method of turning would not violate the second law of thermodynamics since the Mach waves are isentropic waves. Let us determine the effect of a single Mach wave on the flow. For a control volume enclosing a Mach wave (see Fig. 11.22), continuity is, for infinitesimal changes,

$$\rho V_n = (\rho + d\rho)(V_n + dV_n) \tag{11.122}$$

The tangential momentum equation is

$$0 = (\rho + d\rho)(V_n + dV_n)(V_t + dV_t) - \rho V_n V_t \tag{11.123}$$

A simultaneous solution of the above two equations shows that

$$dV_t = 0 \tag{11.124}$$

The tangential velocity component remains unchanged. From the triangles in Fig. 11.22 we observe that

$$V_t = V \cos \mu = (V + dV) \cos(\mu + d\theta) \tag{11.125}$$

The deflection angle $d\theta$ is small so that this equation, using $\cos(\mu + d\theta) \simeq \cos \mu - d\theta \sin \mu$, $\sin \mu = 1/M$, and $\cos \mu = \sqrt{M^2 - 1}/M$,

$$d\theta = \frac{\cos \mu}{\sin \mu} \frac{dV}{V} = \sqrt{M^2 - 1} \frac{dV}{V} \tag{11.126}$$

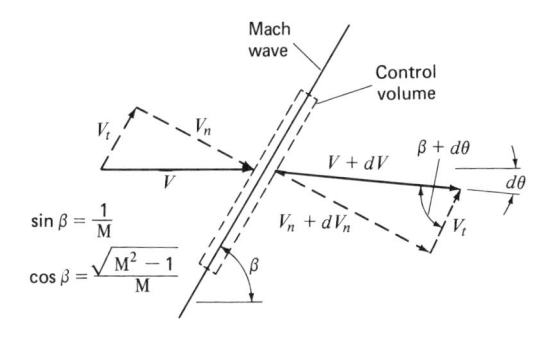

Fig. 11.22. The Mach wave.

To express V in terms of M, we differentiate $V = M\sqrt{\gamma RT}$ to obtain

$$\frac{dV}{V} = \frac{dM}{M} + \frac{1}{2}\frac{dT}{T} \tag{11.127}$$

The energy equation $\dfrac{V^2}{2} + \dfrac{\gamma R}{\gamma - 1} T = \text{constant}$ may be differentiated to give

$$\frac{dV}{V} + \frac{1}{(\gamma - 1)M^2} \frac{dT}{T} = 0 \tag{11.128}$$

Combining the two foregoing equations, we find

$$\frac{dV}{V} = \frac{1}{1 + \frac{\gamma - 1}{2}M^2} \frac{dM}{M} \tag{11.129}$$

Substituting in Eq. 11.126 results in this relationship between $d\theta$ and M:

$$d\theta = \frac{\sqrt{M^2 - 1}}{1 + \frac{\gamma - 1}{2}M^2} \frac{dM}{M} \tag{11.130}$$

This equation may be integrated, using $\theta = 0$ at $M = 1$, to give

$$\theta = \sqrt{\frac{\gamma + 1}{\gamma - 1}} \tan^{-1} \sqrt{\frac{\gamma - 1}{\gamma + 1}(M^2 - 1)} - \tan^{-1} \sqrt{M^2 - 1}$$

$$\tag{11.131}$$

The above function $\theta(M)$ is generally referred to as the *Prandtl-Meyer function*. Tabulated values are presented in Table D.5. The changes in pressure, temperature, and other thermodynamic properties can be found from the isentropic-flow relations.

It is observed that as a compressible fluid flows around a convex corner the velocity differential dV is positive so that the velocity increases. The velocity vector turns away from the Mach wave. Conversely, as the fluid flows into a smooth concave wall change the velocity decreases and the velocity vector turns toward the Mach wave.

Example 11.14

Air at a Mach number of 2.0 and a pressure of 20 psia flows around (a) a convex angle of 10° and (b) a smooth concave wall change of 10°. Determine M_2 and p_2 for each angle.

Art. 11.7   ISENTROPIC TURNING

Solution. a. The air undergoes an expansion as it turns the convex corner illustrated in Fig. E11.14a. Since Table D.5 uses $M = 1$ as a reference, we show the flow with $M_1 = 2$ as having originated from a flow with $M = 1$. From the table

$$\theta_1 = 26.4°$$

Adding an additional 10 degrees to the deflection angle, $\theta_2 = 36.4°$. From Table D.5,

$$M_2 = 2.38$$

The pressure may be found from the isentropic gas table to be

$$p_2 = p_1 \frac{p_0}{p_1} \frac{p_2}{p_0}$$

$$= 20 \times \frac{1}{0.128} \times 0.07 = 10.9 \text{ psia}$$

b. For the compression we again find $\theta_1 = 26.4°$. Now, we subtract the 10 degrees compression turn and $\theta_2 = 16.4°$. From Table D.5 we find that

$$M_2 = 1.65$$

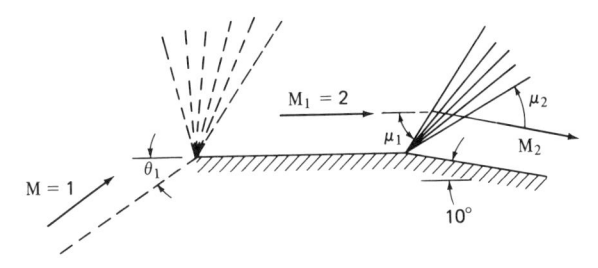

(a) Expansion angle

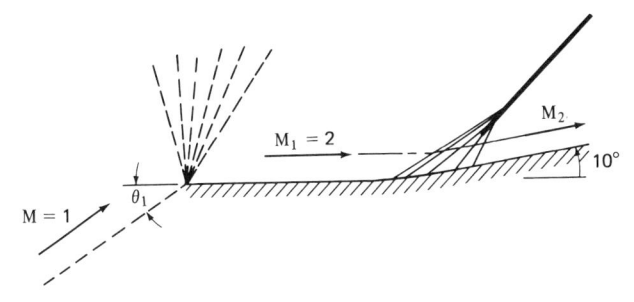

(b) Compression angle

Fig. E11.14

From the isentropic flow table the pressure after the compression fan is

$$p_2 = p_1 \frac{p_0}{p_1} \frac{p_2}{p_0}$$

$$= 20 \times \frac{1}{0.128} \times 0.218 = 34 \text{ psia}$$

11.8 CLOSURE

We have attempted in this chapter to introduce the student to some of the phenomena encountered in compressible-fluid flow. The flows have been one-dimensional except for the simple extension to oblique shocks and isentropic turning fans. More complex one-dimensional flow situations may be encountered; an example would be heat added to a converging frictionless flow. Texts on gas dynamics include treatment of such topics.

An obvious extension of compressible flow theory would be to consider the compressible-flow Navier-Stokes equations. Since velocities in compressible flows are usually quite large (M > 0.3) the viscous terms are often negligible outside the boundary layers so that, in flow around bodies, a potential flow results. It is in this subject of fluid mechanics for supersonic flows that the hyperbolic partial differential equation is encountered with the associated solution method termed the method of characteristics. A boundary layer in a supersonic flow represents the interesting mixture of high-speed compressible effects added to the other boundary-layer phenomena described in Chapter 9. Note that the flow inside the boundary layer will be supersonic, subsonic, and, near the surface, incompressible as the zero wall velocity is approached.

The complexity of two- and three-dimensional flows with viscous effects and and heat transfer makes these topics suitable for advanced, graduate-level courses. The study of the incompressible Navier-Stokes equations in Chapters 3 through 10, along with the introduction to compressible-flow phenomena presented in this chapter, should provide a firm foundation for such advanced study in compressible, viscous fluids, for those readers who wish to pursue such study.

PROBLEMS

11.1 Express c_p in terms of R and γ.

11.2 A pitot-static probe is used to measure the speed of an airplane flying at a subsonic speed. The difference between the stagnation pressure and the static pressure is 1.2 psi. If the airplane is at an altitude of 35,000 ft, determine its speed. Assume an isentropic process from free-stream to the stagnation point.

11.3 A surface test vehicle uses a pitot-static probe to determine its speed. A reading of 1.8 psi is measured by the instrument. Calculate the vehicle speed, assuming (a) incompressible flow and (b) compressible, isentropic flow. Compare the results.

11.4 Calculate the Mach number for an aircraft traveling at 600 mph in the atmosphere at sea level, 30,000 ft, 60,000 ft, and 100,000 ft.

11.5 Two rocks are hammered together under water on one side of a lake. An observer on the opposite bank with his head under water picks up the propagating wave 0.8 sec later. Determine the distance across the lake.

11.6 A wave propagating isentropically through air at sea level at standard conditions, has an associated pressure rise of 0.006 psi. (a) Estimate the velocity induced in the air behind the wave. (b) Estimate the temperature rise.

11.7 A needle-nosed projectile is flying at an elevation of 3,000 ft with a speed of 4000 mph. The wave generated at the needle nose approximates a Mach wave. The projectile passes overhead of an observer on the ground. How long after it passes overhead will the observer hear it? How far away will the projectile be?

11.8 The pressure, temperature, and velocity before a normal stationary shock are 10 psia, 40°F, and 2000 fps, respectively. Determine the flow conditions of the air after the shock, as well as the Mach number before and after the shock. (a) Use the basic equations. (b) Use the Mach relations. (c) Use the shock table in Appendix D.

11.9 A shock wave is created in a shock tube by bursting a membrane that separates a high-pressure region from a low-pressure region. Determine the velocity induced in the shock tube immediately behind the shock wave if the shock wave is traveling at M = 3.0. Air is in the tube.

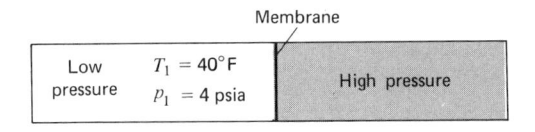

Prob. 11.9

11.10 Derive the Rankine-Hugoniot relationship

$$\frac{\rho_2}{\rho_1} = \frac{\frac{\gamma+1}{\gamma-1}\frac{p_2}{p_1}+1}{\frac{p_2}{p_1}+\frac{\gamma+1}{\gamma-1}}$$

which relates the density ratio to the pressure ratio across a shock wave. For a very strong shock wave, $p_2/p_1 \gg 1$, find the limiting density ratio.

11.11 A bomb explodes just above the surface of the earth, producing a shock wave which travels with a Mach number of 2.5. Determine the pressure rise associated with the shock wave and the wind velocity immediately behind the shock.

11.12 Air at 12 psia and 30°F passes through a shock wave. The Mach number after the shock is 0.5. Determine V_1, V_2, p_2, T_2, and ρ_2. Also determine the losses associated with the shock wave.

11.13 A venturi tube is used to measure an airflow rate. It reduces a 4-in.-dia. pipe to 2 in. and then back to the 4-in.-dia. The pressures measured at the two sections are 40 psia and 35 psia, respectively. Also, $T_1 = 40°F$. Calculate the flow rate in cubic feet per minute.

11.14 Air flows from a reservoir ($T_0 = 100°F$) out a converging nozzle with a 1-in.-dia. exit to the atmosphere. Calculate the reservoir pressure necessary to just achieve a Mach number of one at the exit. Also calculate the mass flow rate in slugs per second. (a) Use the basic equations. (b) Use the isentropic flow table.

11.15 Air flows from a 100°F reservoir out a converging nozzle and exits to the atmosphere. Determine the reservoir pressure so that choked flow just occurs. Calculate the mass flux for a 2-in.-dia. nozzle. Now double this reservoir pressure, holding the temperature constant, and calculate the new mass flux. Also determine the new exit pressure.

11.16 Air flows through a converging-diverging nozzle from the atmosphere at 70°F into a receiver with pressure maintained at 13 psia. Calculate the maximum velocity in the nozzle if the throat diameter is 1.8 in. and the exit diameter at the receiver is 2-in. Also determine the mass flux.

11.17 Air at 100°F is flowing in a 4-in. dia. pipe at a speed of 400 fps. A venturi tube is used to monitor the flow rate. What should be the minimum diameter of the tube so that choked flow would not occur?

11.18 Air flows from a reservoir maintained at 20 psia and 70°F through a converging-diverging nozzle and exits to the atmosphere at a nozzle diameter of 6 in. The flow is everywhere subsonic. The throat is now reduced in area. Determine the least diameter of the throat at which no further reduction in throat density would occur.

11.19 Air flows from a reservoir held at 70°F to the atmosphere through a converging-diverging nozzle with a 2-in.-dia. throat and a 4-in.-dia. exit. Determine the reservoir pressure which will just give M = 1 at the throat. Also, find the mass flux. Now decrease the throat diameter to 1.5 in. and calculate the new mass flux, holding the reservoir pressure and temperature constant. Sketch the two flows on a p-vs.-x plot, as in Fig. 11.10.

11.20 A converging-diverging nozzle with throat diameter of 2 in. and exit diameter of 4 in. is attached to a receiver. (a) What receiver pressure is necessary to locate a normal shock at the nozzle exit? The atmosphere temperature is 70°F. Also determine the velocity and pressure (b) at the throat, (c) before the shock wave, and (d) after the shock wave.

11.21 A receiver is maintained at a pressure of 7 psia. Air is drawn into the receiver from the 70°F atmosphere through a converging-diverging nozzle with throat area 1 × 2 in. and exit area 1 × 8 in. Determine the location of the shock wave. (This is a trial-and-error solution. Guess a location of the shock.)

11.22 A converging-diverging nozzle is bolted into a reservoir. Air flows through the 2-in.-dia. throat and out the 4-in.-dia. exit. For a reservoir temperature of 100°F, determine the force necessary to hold the nozzle in place if the flow is isentropic throughout but supersonic in the diverging section.

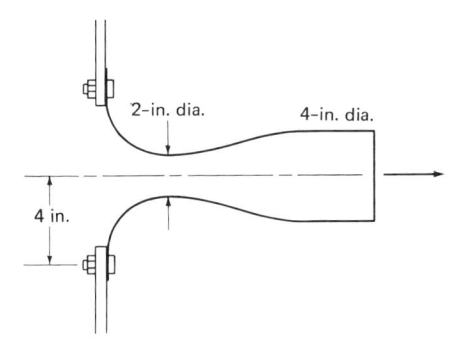

Prob. 11.22

11.23 Determine the two possible receiver pressures that will result in isentropic flow of air from a converging-diverging nozzle if the throat Mach number is unity. The reservoir pressure and temperature are 75 psia and 100°F, respectively. The throat diameter is 1 in. and the exit diameter is 3 in. Determine the mass flow rate for both pressures.

11.24 A converging-diverging nozzle, with a 1.5-in. throat diameter and 3-in. exit diameter, is attached to a small rocket. The combustion chamber is designed so that the combustion chamber pressure causes an isentropic flow throughout the nozzle. Calculate the thrust at liftoff ($p_e = 14.7$ psia). Assume a combustion chamber temperature of 6000°R and consider the combustion gases to have the same properties as air.

11.25 A blunt object travels at a speed of 3000 fps near the surface of the earth where the temperature is 40°F. The flow approaching the stagnation point passes through a shock wave and then decelerates isentropically to the stagnation point. Calculate the stagnation pressure and temperature.

11.26 Calculate the manometer reading if the velocity in the pipe shown (*next page*) is 1500 fps. Mercury has a density of 13.6 ρ_{H_2O}.

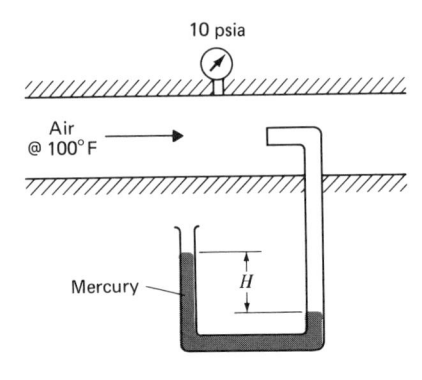

Prob. 11.26

11.27 An aircraft at an altitude of 35,000 ft is flying at a speed of 2400 fps. Determine the reading in psia the pilot would read from a pitot-probe instrument.

11.28 A 6-in.-dia. insulated pipe is connected to the end of a diverging nozzle. Air leaves the nozzle at a Mach number of 2.3. Assume $\bar{f} = 0.02$. (a) How long should the pipe be if the Mach number at the exit is just unity? (b) For a pipe one half the length of part a, determine the exit Mach number.

11.29 A reservoir of air, maintained at 75 psia and 120°F, flows to the atmosphere through a short 1-in.-dia. pipe. Estimate the mass flux in slugs per second. Now, add on a 20-ft-long, 1-in.-dia. insulated pipe and estimate the flux. Assume $\bar{f} = 0.03$.

11.30 Determine the Mach number above which the friction factor is constant for air flowing in a 2-in.-dia. pipe at 60°F and 14.7 psia if the pipe is (a) wrought iron and (b) cast iron.

11.31 Determine the mass flux exiting from a 10 ft-long, $\frac{1}{2}$-in.-dia. steel pipe transporting air from a reservoir maintained at 100 psia and 90°F. Assume (a) the pipe is insulated and (b) the flow in the pipe is isothermal.

11.32 Air is flowing from a high-pressure reservoir through a 1-in.-dia. cast iron pipe. At section 1 the pressure is 70 psia and the temperature is 100°F. The pipe supplies air at 20 psia to a device 100 ft downstream. Determine the heat transfer necessary over the 100-ft length for an isothermal flow.

11.33 The total temperature in an airflow at the inlet to a 6-in.-dia. pipe is 550°R. The inlet Mach number is 0.25 and the inlet pressure is 30 psia. Heat is added at the rate of 300 Btu/sec. Assume a frictionless flow and calculate $\dot{m}$, V_2, p_2, T_2, and M_2 downstream of the heat-addition region.

11.34 Air enters a 10-in.-dia. combustion chamber at a velocity of 100 fps, a pressure of 10 psia, and a temperature of 30°F. After 1000 Btu/sec of heat are added during the combustion process the pipe is reduced to a throat, where the Mach number is unity. Assuming a frictionless flow, determine the diameter of the throat.

PROBLEMS

11.35 Air enters a jet engine as shown. Just in front of the combustion chamber the flow area is 4 ft^2, the velocity is 300 fps, the pressure is 10 psia, and the temperature is 400°R. After the combustion process the flow passes a throat area, where the Mach number is unity, and is exhausted to the atmosphere isentropically at a supersonic speed. Determine the exit velocity and estimate the thrust produced by the engine.

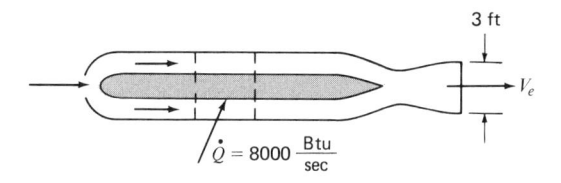

Prob. 11.35

11.36 A uniform airflow approaches a concave corner with $\theta = 10°$ and is turned by means of an oblique shock wave. The velocity, temperature, and pressure are 2000 fps, 100°F, and 10 psia, respectively. (a) Find the Mach number, pressure, and stagnation pressure downstream of the wave for (1) a weak shock and (2) a strong shock. (b) Sketch the shock wave for a concave corner angle of 25°.

11.37 A supersonic inlet can be designed to have a normal shock wave oriented at the inlet or a wedge can be used to provide a weak oblique shock before the flow passes into the inlet. (a) Compare the pressures just after the normal shocks if $M_1 = 3.0$ and $p_1 = 5$ psia. (b) Compare the losses from the free stream to sections after the normal shocks if $T_1 = 420°R$.

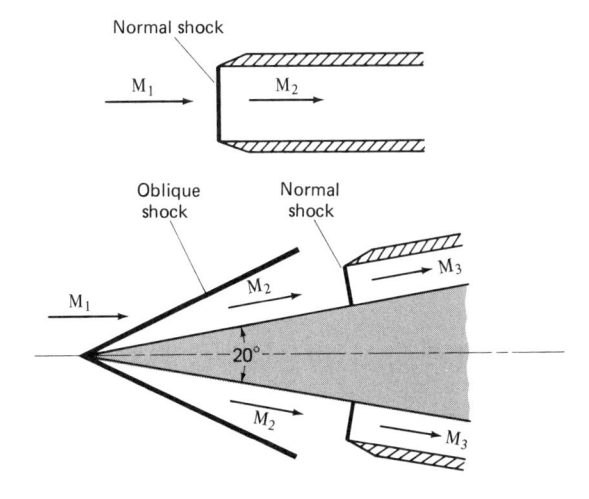

Prob. 11.37

11.38 Two oblique shocks intersect as shown. Determine the angle β of the reflected shocks if the airflow leaves parallel to its original direction. Also determine M_3. Sketch a streamline.

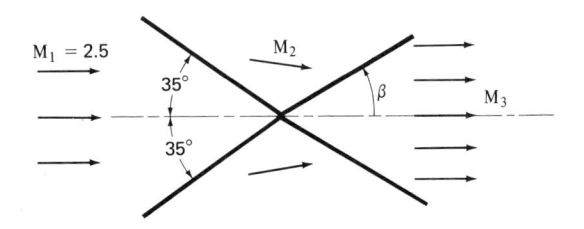

Prob. 11.38

11.39 An oblique shock at a 35° angle is reflected from a plane wall. The incoming Mach number M_1 is 3.5 and the temperature T_1 is 460°R. For the airflow, find the velocity V_3 after the reflected wave.

11.40 A supersonic airflow traveling at $M_1 = 3.0$ turns a convex corner (smooth or sharp) of 20°. The pressure and temperature are 5 psia and 400°R, respectively. Determine p_2, T_2, and V_2 after the expansion. Also calculate the included angle of the expansion fan for a sharp corner.

11.41 The flat plate shown is used as an airfoil at a 5° angle of attack. Oblique shocks and expansion fans allow the air to remain attached to the airfoil with the flow after the airfoil parallel to the original direction. (a) Calculate the pressure on the upper and lower sides of the plate. (b) Find the downstream Mach numbers M_{2u} and M_{2l}. (c) Find the lift coefficient $C_L = \text{Lift}/\frac{1}{2}\rho_1 V_1^2 A$, where A is the area of one side of the flat plate. (d) Find the drag coefficient $C_D = \text{Drag}/\frac{1}{2}\rho_1 V_1^2 A$. Note that $\rho_1 V_1^2 = \gamma M_1^2 p_1$.

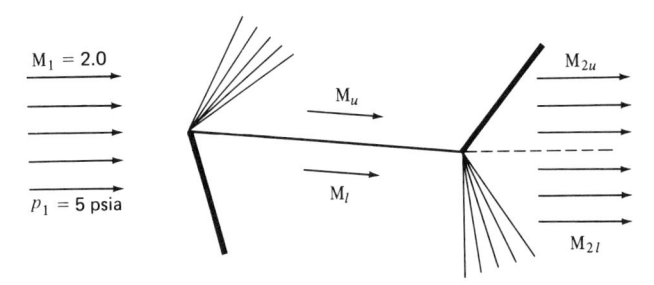

Prob. 11.41

11.42 The supersonic airfoil shown is flying at zero angle of attack. Find the drag coefficient, $C_D = \text{Drag}/\frac{1}{2}\rho_1 V_1^2 A$. Note that $\rho_1 V_1^2 = \gamma M_1^2 p_1$.

SELECTED REFERENCES

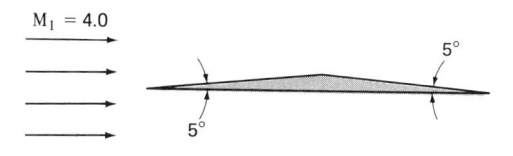

Prob. 11.42

11.43 The supersonic airfoil shown is flying at an angle of attack of 5°. (a) Calculate the lift and drag coefficients. (b) Compare the coefficients of part a with those of a flat plate at a 5° angle of attack.

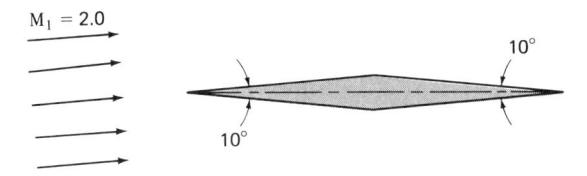

Prob. 11.43

11.44 Sketch the flow situation you would expect from the exit of an underexpanded nozzle ($p_{exit} > p_{receiver}$) and an overexpanded nozzle ($p_{exit} < p_{receiver}$).

11.45 The Mach number at the exit from the supersonic nozzle shown is $M_e = 3.0$. The exit pressure is less than the receiver pressure so oblique shocks form as shown. Determine the receiver pressure if $p_e = 2$ psia. Also, estimate M_2 and p_2.

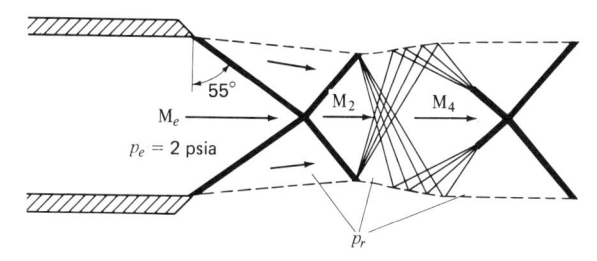

Prob. 11.45

SELECTED REFERENCES

The motion picture *Channel Flow of a Compressible Fluid* (No. 21616, D. Coles, film principal) is recommended as a supplement to this chapter.

For a discussion of compressible-flow phenomena at approximately the same level as that of the present text, see A. G. Hansen, *Fluid Mechanics*, John

Wiley & Sons, 1967, New York, and Y. S. Yuan, *Foundations of Fluid Mechanics*, Prentice-Hall, Englewood Cliffs, N.J., 1967. A classical work providing more advanced coverage is A. H. Shapiro, *The Dynamics and Thermodynamics of Compressible Fluid Flow* (2 vols.), Ronald Press, 1953. More recent advanced texts are J. A. Owczarek, *Gas Dynamics*, International Textbook Co., Scranton, Pa., 1964, and E. A. J. John, *Gas Dynamics*, Allyn & Bacon, Boston, 1969.

APPENDIX

A

Fluid Properties

TABLE A.1
Properties of Water

Temperature (°F)	Density (slugs/ft^3)	Viscosity (lb-sec/ft^2)	Surface Tension (lb/ft)	Vapor Pressure (lb/in.^{2})	Bulk Modulus (lb/in.^{2})
32	1.94	3.75×10^{-5}	0.518×10^{-2}	0.089	293,000
40	1.94	3.23	0.514	0.122	294,000
50	1.94	2.74	0.509	0.178	305,000
60	1.94	2.36	0.504	0.256	311,000
70	1.94	2.05	0.500	0.340	320,000
80	1.93	1.80	0.492	0.507	322,000
90	1.93	1.60	0.486	0.698	323,000
100	1.93	1.42	0.480	0.949	327,000
120	1.92	1.17	0.465	1.69	333,000
140	1.91	0.98	0.454	2.89	330,000
160	1.90	0.84	0.441	4.74	326,000
180	1.88	0.73	0.426	7.51	318,000
200	1.87	0.64	0.412	11.53	308,000
212	1.86	0.59×10^{-5}	0.404×10^{-2}	14.7	300,000

TABLE A.2
Properties of Air

Temperature (°F)	Density (slugs/ft^3)	Viscosity (lb-sec/ft^2)	Kinematic Viscosity (ft^2/sec)
0	0.00268	3.28×10^{-7}	12.6×10^{-5}
20	0.00257	3.50	13.6
40	0.00247	3.62	14.6
60	0.00237	3.74	15.8
68	0.00233	3.81	16.0
80	0.00228	3.85	16.9
100	0.00220	3.96	18.0
120	0.00215	4.07×10^{-7}	18.9×10^{-5}

$$R = 1716 \frac{\text{ft}^2}{\text{sec}^2\text{-}°R} \; ; \; p = 14.7 \text{ psia}$$

TABLE A.3
Properties of the Atmosphere

Altitude (ft)	Temperature (°F)	Pressure (lb/ft^2)	Density (slugs/ft^3)	Kinematic Viscosity (ft^2/sec)	Velocity of Sound (ft/sec)
0	59.0	2116	0.00237	1.56×10^{-4}	1117
1,000	55.4	2041	0.00231	1.60	1113
2,000	51.9	1968	0.00224	1.64	1109
5,000	41.2	1760	0.00205	1.77	1098
10,000	23.4	1455	0.00176	2.00	1078
15,000	5.54	1194	0.00150	2.28	1058
20,000	-12.3	973	0.00127	2.61	1037
25,000	-30.1	785	0.00107	3.00	1016
30,000	-48.0	628	0.000890	3.47	995
35,000	-65.8	498	0.000737	4.04	973
36,000	-67.6	475	0.000709	4.18	971
40,000	-67.6	392	0.000586	5.06	971
50,000	-67.6	242	0.000362	8.18	971
100,000	-67.6	22.4	3.31×10^{-5}	89.5×10^{-4}	971
110,000	-47.4	13.9	1.97×10^{-5}	1.57×10^{-6}	996
150,000	113.5	3.00	3.05×10^{-6}	13.2	1174
200,000	160.0	0.665	6.20×10^{-7}	68.4	1220
260,000	-28	0.0742	1.0×10^{-7}	321×10^{-6}	1019

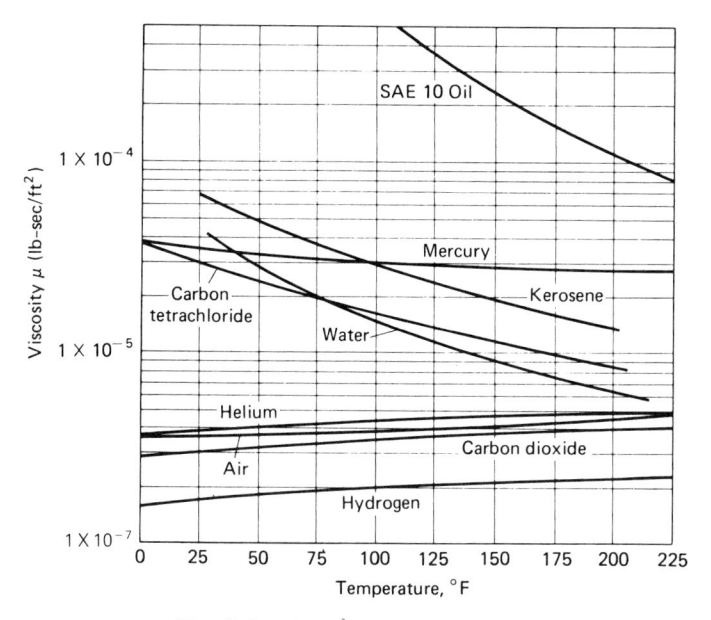

Fig. A.1. Viscosity of several fluids.

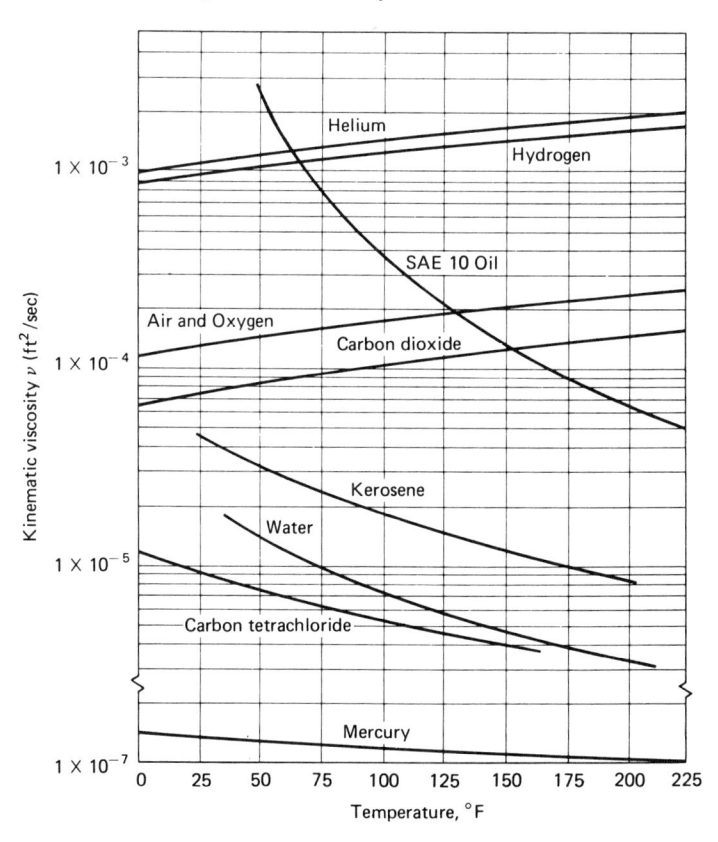

Fig. A.2. Kinematic viscosity of several fluids.

B

Dimensions, Units, Conversions

TABLE B.1
Dimensions and Units of Quantities

Quantity	Symbol	Dimensions	Unit
Length	L	L	ft
Force	F	F	lb
Time	t	t	sec
Mass	M	$M = \dfrac{Ft^2}{L}$	$\text{slug} = \dfrac{\text{lb-sec}^2}{\text{ft}}$
Velocity	V	L/t	ft/sec
Acceleration	a	L/t^2	ft/sec^2
Pressure	p	F/L^2	lb/ft^2
Density	ρ	$\dfrac{M}{L^3} = \dfrac{Ft^2}{L^4}$	$\dfrac{\text{slug}}{\text{ft}^3} = \dfrac{\text{lb-sec}^2}{\text{ft}^4}$
Viscosity	μ	Ft/L^2	lb-sec/ft^2
Kinematic viscosity	ν	L^2/t	ft^2/sec
Bulk modulus	B	F/L^2	lb/ft^2
Surface tension	σ	F/L	lb/ft
Vapor pressure	p_v	F/L^2	lb/ft^2
Temperature	T	T	°F or °R
Gas constant	R	$\dfrac{LF}{MT} = \dfrac{L^2}{t^2 T}$	$\dfrac{\text{ft-lb}}{\text{slug-°R}} = \dfrac{\text{ft}^2}{\text{sec}^2\text{-°R}}$
Specific heat	c	$\dfrac{LF}{MT} = \dfrac{L^2}{t^2 T}$	$\dfrac{\text{ft-lb}}{\text{slug-°R}} = \dfrac{\text{ft}^2}{\text{sec}^2\text{-°R}}$

TABLE B.2
Conversions

Length	Force	Mass
1 cm = 0.3937 in.	1 lb = 0.4536 kg	1 oz = 28.35 g
1 m = 3.281 ft	1 lb = 0.4448 × 10^6 dyn	1 lb = 0.4536 kg
1 km = 0.6214 mi	1 lb = 32.17 pdl	1 slug = 32.17 lb
1 in. = 2.54 cm	1 kg = 2.205 lb	1 slug = 14.59 kg
1 ft = 0.3048 m	1 pdl = 0.03108 lb	1 kg = 2.205 lb
1 m = 1.609 km	1 dyn = 2.248 × 10^{-6} lb	
1 m = 5280 ft		
1 m = 1760 yd		

Work and Power	Pressure
1 Btu = 778.2 ft-lb	1 psi = 2.036 in. Hg
1 hp = 550 ft-lb/sec	1 psi = 27.7 in. H_2O
1 hp = 0.7067 Btu/sec	14.7 psi = 29.92 in. Hg
1 hp = 0.7455 kW	14.7 psi = 33.93 ft H_2O
1 hp = 76.04 kg-m/sec	14.7 psi = 1.0332 kg/cm^2
1 W = 1 J/sec	14.7 psi = 1.0133 Megabars
1 W = 1.0 × 10^7 dyn-cm/sec	1 kg/cm^2 = 14.22 psi
1 J = 10^7 ergs	1 in. Hg = 0.4912 psi
1 calorie = 3.088 ft-lb	1 ft H_2O = 0.4331 psi
1 calorie = 0.003968 Btu	

Velocity	Viscosity	Volume
1 mph = 1.467 fps	1 poise = 0.00209 $\frac{\text{lb-sec}}{\text{ft}^2}$	1 ft^3 = 28.32 liters
1 mph = 0.8684 knot	1 $\frac{\text{lb-sec}}{\text{ft}^2}$ = 478 poises	1 ft^3 = 7.481 gal (U.S.)
1 fps = 0.3048 m/sec	1 poise = 1 g/cm-sec	1 gal (U.S.) = 231 $in.^3$
		1 gal (Brit.) = 1.2 gal (U.S.)

TABLE B.3
United States Engineering Units, SI Units, and Their Conversion Factors

Quantity	Engineering Units (U.S. System)	International System (SI)*	Conversion Factor
Length	inch	millimeter	1 in. = 25.4 mm
	foot	meter	1 ft = 0.3048 m
	mile	kilometer	1 mi = 1.609 km
Area	square inch	square centimeter	1 in.^{2} = 6.452 cm^2
	square foot	square meter	1 ft^2 = 0.09290 m^2
Volume	cubic inch	cubic centimeter	1 in.^{3} = 16.39 cm^3
	cubic foot	cubic meter	1 ft^3 = 0.02832 m^3
	gallon		1 gal = 0.003786 m^3
Mass	pound-mass	kilogram	1 lb$_m$ = 0.4536 kg
	slug		1 slug = 14.59 kg
Density	pound/cubic foot	kilogram/cubic meter	1 lb$_m$/ft^3 = 16.02 kg/m^3
Force	pound-force	newton	1 lb = 4.448 N
Work or Torque	foot-pound	newton-meter	1 ft-lb = 1.356 N·m
Pressure	pound/square inch	newton/square meter	1 psi = 6895 N/m^2
	pound/square foot		1 psf = 47.88 N/m^2
Temperature	degree Fahrenheit	degree Celsius	1°F = $\frac{9}{5}$ °C + 32
	degree Rankine	degree Kelvin	1°R = $\frac{9}{5}$ °K
Energy	British thermal unit	joule	1 Btu = 1055 J
	calorie		1 cal = 4.186 J
	foot-pound		1 ft-lb = 1.356 J
Power	horsepower	watt	1 hp = 745.7 W
	foot-pound/second		1 ft-lb/sec = 1.356 W
Velocity	foot/second	meter/second	1 fps = 0.3048 m/sec
Acceleration	foot/second squared	meter/second squared	1 ft/sec^2 = 0.3048 m/sec^2
Frequency	cycle/second	hertz	1 cps = 1.000 Hz
Viscosity	pound-second/square foot	newton-second/square meter	1 lb-sec/ft^2 = 47.88 N·s/m^2
Kinematic viscosity	square foot/second	square meter/second	1 ft^2/sec = 0.09290 m^2/sec
Surface tension	pound/foot	newton/meter	1 lb/ft = 14.59 N/m
Thermal conductivity	Btu/hour-foot-°F	watt/meter-°K	1 Btu/hr-ft-°F = 1.731 W/m-°K

C

Vector Relationships

TABLE C.1
Vector Functions

Cartesian Coordinates

$$\nabla = \frac{\partial}{\partial x}\hat{i} + \frac{\partial}{\partial y}\hat{j} + \frac{\partial}{\partial z}\hat{k}$$

$$\frac{D}{Dt} = \frac{\partial}{\partial t} + u\frac{\partial}{\partial x} + v\frac{\partial}{\partial y} + w\frac{\partial}{\partial z}$$

$$\nabla \cdot \mathbf{V} = \frac{\partial u}{\partial x} + \frac{\partial v}{\partial y} + \frac{\partial w}{\partial z}$$

$$\nabla^2 = \frac{\partial^2}{\partial x^2} + \frac{\partial^2}{\partial y^2} + \frac{\partial^2}{\partial z^2}$$

Cylindrical Coordinates

$$\nabla = \frac{\partial}{\partial r}\hat{e}_r + \frac{1}{r}\frac{\partial}{\partial \theta}\hat{e}_\theta + \frac{\partial}{\partial z}\hat{e}_z$$

$$\frac{D}{Dt} = \frac{\partial}{\partial t} + v_r\frac{\partial}{\partial r} + \frac{v_\theta}{r}\frac{\partial}{\partial \theta} + v_z\frac{\partial}{\partial z}$$

$$\nabla \cdot \mathbf{V} = \frac{1}{r}\frac{\partial}{\partial r}(rv_r) + \frac{1}{r}\frac{\partial v_\theta}{\partial \theta} + \frac{\partial v_z}{\partial z}$$

$$\nabla^2 = \frac{\partial^2}{\partial r^2} + \frac{1}{r}\frac{\partial}{\partial r} + \frac{1}{r^2}\frac{\partial^2}{\partial \theta^2} + \frac{\partial^2}{\partial z^2}$$

Spherical Coordinates

$$\nabla = \frac{\partial}{\partial r}\hat{e}_r + \frac{1}{r}\frac{\partial}{\partial \theta}\hat{e}_\theta + \frac{1}{r \sin \theta}\frac{\partial}{\partial \phi}\hat{e}_\phi$$

$$\frac{D}{Dt} = \frac{\partial}{\partial t} + v_r\frac{\partial}{\partial r} + \frac{v_\theta}{r}\frac{\partial}{\partial \theta} + \frac{v_\phi}{r \sin \theta}\frac{\partial}{\partial \phi}$$

$$\nabla \cdot \mathbf{V} = \frac{1}{r^2}\frac{\partial}{\partial r}(r^2 v_r) + \frac{1}{r \sin \theta}\frac{\partial}{\partial \theta}(v_\theta \sin \theta) + \frac{1}{r \sin \theta}\frac{\partial v_\phi}{\partial \phi}$$

$$\nabla^2 = \frac{1}{r^2}\frac{\partial}{\partial r}\left(r^2\frac{\partial}{\partial r}\right) + \frac{1}{r^2 \sin \theta}\frac{\partial}{\partial \theta}\left(\sin \theta \frac{\partial}{\partial \theta}\right) + \frac{1}{r^2 \sin^2 \theta}\frac{\partial^2}{\partial \phi^2}$$

TABLE C.2
Vector Identities

$$\nabla \times \nabla \phi = 0$$

$$\nabla \cdot \nabla \times \mathbf{A} = 0$$

$$\nabla \times (\phi \mathbf{A}) = \nabla \phi \times \mathbf{A} + \phi \nabla \times \mathbf{A}$$

$$\nabla \times (\mathbf{A} \times \mathbf{B}) = \mathbf{A}(\nabla \cdot \mathbf{B}) - \mathbf{B}(\nabla \cdot \mathbf{A}) + (\mathbf{B} \cdot \nabla)\mathbf{A} - (\mathbf{A} \cdot \nabla)\mathbf{B}$$

$$\nabla \times (\nabla \times \mathbf{A}) = \nabla(\nabla \cdot \mathbf{A}) - (\nabla \cdot \nabla)\mathbf{A}$$

$$\nabla(\mathbf{A} \cdot \mathbf{B}) = (\mathbf{A} \cdot \nabla)\mathbf{B} + (\mathbf{B} \cdot \nabla)\mathbf{A} + \mathbf{A} \times (\nabla \times \mathbf{B}) + \mathbf{B} \times (\nabla \times \mathbf{A})$$

$$\mathbf{A} \times (\nabla \times \mathbf{A}) = \tfrac{1}{2}\nabla(\mathbf{A} \cdot \mathbf{A}) - (\mathbf{A} \cdot \nabla)\mathbf{A}$$

$$\nabla \cdot (\mathbf{A} \times \mathbf{B}) = (\nabla \times \mathbf{A}) \cdot \mathbf{B} - \mathbf{A} \cdot (\nabla \times \mathbf{B})$$

D

Compressible-Flow Tables for Air

Values of thermodynamic and flow variables for the uniform-, compressible-flow field, under five different sets of conditions, are tabulated in this appendix.* The values have been computed exclusively for air, with the restriction that $\gamma = 1.40$. They are expressed with a precision (up to six places) that is not to be mistaken for their accuracy in describing the actual behavior of an airstream; the accuracy in determining c_p and c_v does not support values to more than three significant digits. The equations used to calculate these values are fully described in Chapter 11.

*Tables D.1, D.2, and D.5 are adapted from NACA Report 1135, "Equations, Tables, and Charts for Compressible Flow," Ames Research Staff (1953), as reproduced in J. E. A. John, *Gas Dynamics* (Boston, Mass.: Allyn & Bacon, 1969). Tables D.2 and D.3 are adapted from J. H. Keenan and J. Kaye, *Gas Tables* (New York: John Wiley & Sons), as reproduced in J. E. A. John, *Gas Dynamics* (Boston, Mass.: Allyn & Bacon, 1969).

TABLE D.1
Isentropic Flow

M	p/p_0	T/T_0	A/A^*	M	p/p_0	T/T_0	A/A^*
0	1.0000	1.0000	∞	.60	.7840	.9328	1.1882
.01	.9999	1.0000	57.8738	.61	.7778	.9307	1.1767
.02	.9997	.9999	28.9421	.62	.7716	.9286	1.1657
.03	.9994	.9998	19.3005	.63	.7654	.9265	1.1552
.04	.9989	.9997	14.4815	.64	.7591	.9243	1.1452
.05	.9983	.9995	11.5914	.65	.7528	.9221	1.1356
.06	.9975	.9993	9.6659	.66	.7465	.9199	1.1265
.07	.9966	.9990	8.2915	.67	.7401	.9176	1.1179
.08	.9955	.9987	7.2616	.68	.7338	.9153	1.1097
.09	.9944	.9984	6.4613	.69	.7274	.9131	1.1018
.10	.9930	.9980	5.8218	.70	.7209	.9107	1.0944
.11	.9916	.9976	5.2992	.71	.7145	.9084	1.0873
.12	.9900	.9971	4.8643	.72	.7080	.9061	1.0806
.13	.9883	.9966	4.4969	.73	.7016	.9037	1.0742
.14	.9864	.9961	4.1824	.74	.6951	.9013	1.0681
.15	.9844	.9955	3.9103	.75	.6886	.8989	1.0624
.16	.9823	.9949	3.6727	.76	.6821	.8964	1.0570
.17	.9800	.9943	3.4635	.77	.6756	.8940	1.0519
.18	.9776	.9936	3.2779	.78	.6691	.8915	1.0471
.19	.9751	.9928	3.1123	.79	.6625	.8890	1.0425
.20	.9725	.9921	2.9635	.80	.6560	.8865	1.0382
.21	.9697	.9913	2.8293	.81	.6495	.8840	1.0342
.22	.9668	.9904	2.7076	.82	.6430	.8815	1.0305
.23	.9638	.9895	2.5968	.83	.6365	.8789	1.0270
.24	.9607	.9886	2.4956	.84	.6300	.8763	1.0237
.25	.9575	.9877	2.4027	.85	.6235	.8737	1.0207
.26	.9541	.9867	2.3173	.86	.6170	.8711	1.0179
.27	.9506	.9856	2.2385	.87	.6106	.8685	1.0153
.28	.9470	.9846	2.1656	.88	.6041	.8659	1.0129
.29	.9433	.9835	2.0979	.89	.5977	.8632	1.0108
.30	.9395	.9823	2.0351	.90	.5913	.8606	1.0089
.31	.9355	.9811	1.9765	.91	.5849	.8579	1.0071
.32	.9315	.9799	1.9219	.92	.5785	.8552	1.0056
.33	.9274	.9787	1.8707	.93	.5721	.8525	1.0043
.34	.9231	.9774	1.8229	.94	.5658	.8498	1.0031
.35	.9188	.9761	1.7780	.95	.5595	.8471	1.0022
.36	.9143	.9747	1.7358	.96	.5532	.8444	1.0014
.37	.9098	.9733	1.6961	.97	.5469	.8416	1.0008
.38	.9052	.9719	1.6587	.98	.5407	.8389	1.0003
.39	.9004	.9705	1.6234	.99	.5345	.8361	1.0001
.40	.8956	.9690	1.5901	1.00	.5283	.8333	1.000
.41	.8907	.9675	1.5587	1.01	.5221	.8306	1.000
.42	.8857	.9659	1.5289	1.02	.5160	.8278	1.000
.43	.8807	.9643	1.5007	1.03	.5099	.8250	1.001
.44	.8755	.9627	1.4740	1.04	.5039	.8222	1.001
.45	.8703	.9611	1.4487	1.05	.4979	.8193	1.002
.46	.8650	.9594	1.4246	1.06	.4919	.8165	1.003
.47	.8596	.9577	1.4018	1.07	.4860	.8137	1.004
.48	.8541	.9560	1.3801	1.08	.4800	.8108	1.005
.49	.8486	.9542	1.3595	1.09	.4742	.8080	1.006
.50	.8430	.9524	1.3398	1.10	.4684	.8052	1.008
.51	.8374	.9506	1.3212	1.11	.4626	.8023	1.010
.52	.8317	.9487	1.3034	1.12	.4568	.7994	1.011
.53	.8259	.9468	1.2865	1.13	.4511	.7966	1.013
.54	.8201	.9449	1.2703	1.14	.4455	.7937	1.015
.55	.8142	.9430	1.2550	1.15	.4398	.7908	1.017
.56	.8082	.9410	1.2403	1.16	.4343	.7879	1.020
.57	.8022	.9390	1.2263	1.17	.4287	.7851	1.022
.58	.7962	.9370	1.2130	1.18	.4232	.7822	1.025
.59	.7901	.9349	1.2003	1.19	.4178	.7793	1.026

TABLE D.1 (Continued)

M	p/p_0	T/T_0	A/A^*	M	p/p_0	T/T_0	A/A^*
1.20	.4124	.7764	1.030	1.85	.1612	.5936	1.495
1.21	.4070	.7735	1.033	1.86	.1587	.5910	1.507
1.22	.4017	.7706	1.037	1.87	.1563	.5884	1.519
1.23	.3964	.7677	1.040	1.88	.1539	.5859	1.531
1.24	.3912	.7648	1.043	1.89	.1516	.5833	1.543
1.25	.3861	.7619	1.047	1.90	.1492	.5807	1.555
1.26	.3809	.7590	1.050	1.91	.1470	.5782	1.568
1.27	.3759	.7561	1.054	1.92	.1447	.5756	1.580
1.28	.3708	.7532	1.058	1.93	.1425	.5731	1.593
1.29	.3658	.7503	1.062	1.94	.1403	.5705	1.606
1.30	.3609	.7474	1.066	1.95	.1381	.5680	1.619
1.31	.3560	.7445	1.071	1.96	.1360	.5655	1.633
1.32	.3512	.7416	1.075	1.97	.1339	.5630	1.646
1.33	.3464	.7387	1.080	1.98	.1318	.5605	1.660
1.34	.3417	.7358	1.084	1.99	.1298	.5580	1.674
1.35	.3370	.7329	1.089	2.00	.1278	.5556	1.688
1.36	.3323	.7300	1.094	2.01	.1258	.5531	1.702
1.37	.3277	.7271	1.099	2.02	.1239	.5506	1.716
1.38	.3232	.7242	1.104	2.03	.1220	.5482	1.730
1.39	.3187	.7213	1.109	2.04	.1201	.5458	1.745
1.40	.3142	.7184	1.115	2.05	.1182	.5433	1.760
1.41	.3098	.7155	1.120	2.06	.1164	.5409	1.775
1.42	.3055	.7126	1.126	2.07	.1146	.5385	1.790
1.43	.3012	.7097	1.132	2.08	.1128	.5361	1.806
1.44	.2969	.7069	1.138	2.09	.1111	.5337	1.821
1.45	.2927	.7040	1.144	2.10	.1094	.5313	1.837
1.46	.2886	.7011	1.150	2.11	.1077	.5290	1.853
1.47	.2845	.6982	1.156	2.12	.1060	.5266	1.869
1.48	.2804	.6954	1.163	2.13	.1043	.5243	1.885
1.49	.2764	.6925	1.169	2.14	.1027	.5219	1.902
1.50	.2724	.6897	1.176	2.15	.1011	.5196	1.919
1.51	.2685	.6868	1.183	2.16	$.9956 \times 10^{-1}$	.5173	1.935
1.52	.2646	.6840	1.190	2.17	$.9802 \times 10^{-1}$	.5150	1.953
1.53	.2608	.6811	1.197	2.18	$.9649 \times 10^{-1}$	.5127	1.970
1.54	.2570	.6783	1.204	2.19	$.9500 \times 10^{-1}$	.5104	1.987
1.55	.2533	.6754	1.212	2.20	$.9352 \times 10^{-1}$	.5081	2.005
1.56	.2496	.6726	1.219	2.21	$.9207 \times 10^{-1}$	.5059	2.023
1.57	.2459	.6698	1.227	2.22	$.9064 \times 10^{-1}$	.5036	2.041
1.58	.2423	.6670	1.234	2.23	$.8923 \times 10^{-1}$	.5014	2.059
1.59	.2388	.6642	1.242	2.24	$.8785 \times 10^{-1}$	.4991	2.078
1.60	.2353	.6614	1.250	2.25	$.8648 \times 10^{-1}$	.4969	2.096
1.61	.2318	.6586	1.258	2.26	$.8514 \times 10^{-1}$	.4947	2.115
1.62	.2284	.6558	1.267	2.27	$.8382 \times 10^{-1}$	.4925	2.134
1.63	.2250	.6530	1.275	2.28	$.8251 \times 10^{-1}$	.4903	2.154
1.64	.2217	.6502	1.284	2.29	$.8123 \times 10^{-1}$	.4881	2.173
1.65	.2184	.6475	1.292	2.30	$.7997 \times 10^{-1}$	.4859	2.193
1.66	.2151	.6447	1.301	2.31	$.7873 \times 10^{-1}$	.4837	2.213
1.67	.2119	.6419	1.310	2.32	$.7751 \times 10^{-1}$	.4816	2.233
1.68	.2088	.6392	1.319	2.33	$.7631 \times 10^{-1}$	.4795	2.254
1.69	.2057	.6364	1.328	2.34	$.7512 \times 10^{-1}$	.4773	2.274
1.70	.2026	.6337	1.338	2.35	$.7396 \times 10^{-1}$	.4752	2.295
1.71	.1996	.6310	1.347	2.36	$.7281 \times 10^{-1}$	.4731	2.316
1.72	.1966	.6283	1.357	2.37	$.7168 \times 10^{-1}$	.4709	2.338
1.73	.1936	.6256	1.367	2.38	$.7057 \times 10^{-1}$	.4688	2.359
1.74	.1907	.6229	1.376	2.39	$.6948 \times 10^{-1}$	.4668	2.381
1.75	.1878	.6202	1.386	2.40	$.6840 \times 10^{-1}$	.4647	2.403
1.76	.1850	.6175	1.397	2.41	$.6734 \times 10^{-1}$	.4626	2.425
1.77	.1822	.6148	1.407	2.42	$.6630 \times 10^{-1}$	.4606	2.448
1.78	.1794	.6121	1.418	2.43	$.6527 \times 10^{-1}$	.4585	2.471
1.79	.1767	.6095	1.428	2.44	$.6426 \times 10^{-1}$	.4565	2.494
1.80	.1740	.6068	1.439	2.45	$.6327 \times 10^{-1}$	.4544	2.517
1.81	.1714	.6041	1.450	2.46	$.6229 \times 10^{-1}$	.4524	2.540
1.82	.1688	.6015	1.461	2.47	$.6133 \times 10^{-1}$	.4504	2.564
1.83	.1662	.5989	1.472	2.48	$.6038 \times 10^{-1}$	.4484	2.588
1.84	.1637	.5963	1.484	2.49	$.5945 \times 10^{-1}$	.4464	2.612

TABLE D.1 (Continued)

M	p/p_0	T/T_0	A/A^*	M	p/p_0	T/T_0	A/A^*
2.50	.5853 -1	.4444	2.637	3.15	.2177 -1	.3351	4.884
2.51	.5762 -1	.4425	2.661	3.16	.2146 -1	.3337	4.930
2.52	.5674 -1	.4405	2.686	3.17	.2114 -1	.3323	4.977
2.53	.5586 -1	.4386	2.712	3.18	.2083 -1	.3309	5.025
2.54	.5500 -1	.4366	2.737	3.19	.2053 -1	.3295	5.073
2.55	.5415 -1	.4347	2.763	3.20	.2023 -1	.3281	5.121
2.56	.5332 -1	.4328	2.789	3.21	.1993 -1	.3267	5.170
2.57	.5250 -1	.4309	2.815	3.22	.1964 -1	.3253	5.219
2.58	.5169 -1	.4289	2.842	3.23	.1936 -1	.3240	5.268
2.59	.5090 -1	.4271	2.869	3.24	.1908 -1	.3226	5.319
2.60	.5012 -1	.4252	2.896	3.25	.1880 -1	.3213	5.369
2.61	.4935 -1	.4233	2.923	3.26	.1853 -1	.3199	5.420
2.62	.4859 -1	.4214	2.951	3.27	.1826 -1	.3186	5.472
2.63	.4784 -1	.4196	2.979	3.28	.1799 -1	.3173	5.523
2.64	.4711 -1	.4177	3.007	3.29	.1773 -1	.3160	5.576
2.65	.4639 -1	.4159	3.036	3.30	.1748 -1	.3147	5.629
2.66	.4568 -1	.4141	3.065	3.31	.1722 -1	.3134	5.682
2.67	.4498 -1	.4122	3.094	3.32	.1698 -1	.3121	5.736
2.68	.4429 -1	.4104	3.123	3.33	.1673 -1	.3108	5.790
2.69	.4362 -1	.4086	3.153	3.34	.1649 -1	.3095	5.845
2.70	.4295 -1	.4068	3.183	3.35	.1625 -1	.3082	5.900
2.71	.4229 -1	.4051	3.213	3.36	.1602 -1	.3069	5.956
2.72	.4165 -1	.4033	3.244	3.37	.1579 -1	.3057	6.012
2.73	.4102 -1	.4015	3.275	3.38	.1557 -1	.3044	6.069
2.74	.4039 -1	.3998	3.306	3.39	.1534 -1	.3032	6.126
2.75	.3978 -1	.3980	3.338	3.40	.1512 -1	.3019	6.184
2.76	.3917 -1	.3963	3.370	3.41	.1491 -1	.3007	6.242
2.77	.3858 -1	.3945	3.402	3.42	.1470 -1	.2995	6.301
2.78	.3799 -1	.3928	3.434	3.43	.1449 -1	.2982	6.360
2.79	.3742 -1	.3911	3.467	3.44	.1428 -1	.2970	6.420
2.80	.3685 -1	.3894	3.500	3.45	.1408 -1	.2958	6.480
2.81	.3629 -1	.3877	3.534	3.46	.1388 -1	.2946	6.541
2.82	.3574 -1	.3860	3.567	3.47	.1368 -1	.2934	6.602
2.83	.3520 -1	.3844	3.601	3.48	.1349 -1	.2922	6.664
2.84	.3467 -1	.3827	3.636	3.49	.1330 -1	.2910	6.727
2.85	.3415 -1	.3810	3.671	3.50	.1311 -1	.2899	6.790
2.86	.3363 -1	.3794	3.706	3.51	.1293 -1	.2887	6.853
2.87	.3312 -1	.3777	3.741	3.52	.1274 -1	.2875	6.917
2.88	.3263 -1	.3761	3.777	3.53	.1256 -1	.2864	6.982
2.89	.3213 -1	.3745	3.813	3.54	.1239 -1	.2852	7.047
2.90	.3165 -1	.3729	3.850	3.55	.1221 -1	.2841	7.113
2.91	.3118 -1	.3712	3.887	3.56	.1204 -1	.2829	7.179
2.92	.3071 -1	.3696	3.924	3.57	.1188 -1	.2818	7.246
2.93	.3025 -1	.3681	3.961	3.58	.1171 -1	.2806	7.313
2.94	.2980 -1	.3665	3.999	3.59	.1155 -1	.2795	7.382
2.95	.2935 -1	.3649	4.038	3.60	.1138 -1	.2784	7.450
2.96	.2891 -1	.3633	4.076	3.61	.1123 -1	.2773	7.519
2.97	.2848 -1	.3618	4.115	3.62	.1107 -1	.2762	7.589
2.98	.2805 -1	.3602	4.155	3.63	.1092 -1	.2751	7.659
2.99	.2764 -1	.3587	4.194	3.64	.1076 -1	.2740	7.730
3.00	.2722 -1	.3571	4.235	3.65	.1062 -1	.2729	7.802
3.01	.2682 -1	.3556	4.275	3.66	.1047 -1	.2718	7.874
3.02	.2642 -1	.3541	4.316	3.67	.1032 -1	.2707	7.947
3.03	.2603 -1	.3526	4.357	3.68	.1018 -1	.2697	8.020
3.04	.2564 -1	.3511	4.399	3.69	.1004 -1	.2686	8.094
3.05	.2526 1	.3496	4.441	3.70	.9903 -2	.2675	8.169
3.06	.2489 -1	.3481	4.483	3.71	.9767 -2	.2665	8.244
3.07	.2452 -1	.3466	4.526	3.72	.9633 -2	.2654	8.320
3.08	.2416 -1	.3452	4.570	3.73	.9500 -2	.2644	8.397
3.09	.2380 -1	.3437	4.613	3.74	.9370 -2	.2633	8.474
3.10	.2345 -1	.3422	4.657	3.75	.9242 -2	.2623	8.552
3.11	.2310 -1	.3408	4.702	3.76	.9116 -2	.2613	8.630
3.12	.2276 -1	.3393	4.747	3.77	.8991 -2	.2602	8.709
3.13	.2243 -1	.3379	4.792	3.78	.8869 -2	.2592	8.789
3.14	.2210 -1	.3365	4.838	3.79	.8748 -2	.2582	8.870

TABLE D.1 (Continued)

M	p/p_0	T/T_0	A/A^*	M	p/p_0	T/T_0	A/A^*
3.80	.8629 −2	.2572	8.951	4.45	.3678 −2	.2016	15.87
3.81	.8512 −2	.2562	9.032	4.46	.3613 −2	.2009	16.01
3.82	.8396 −2	.2552	9.115	4.47	.3587 −2	.2002	16.15
3.83	.8283 −2	.2542	9.198	4.48	.3543 −2	.1994	16.28
3.84	.8171 −2	.2532	9.282	4.49	.3499 −2	.1987	16.42
3.85	.8060 −2	.2522	9.366	4.50	.3455 −2	.1980	16.56
3.86	.7951 −2	.2513	9.451	4.51	.3412 −2	.1973	16.70
3.87	.7844 −2	.2503	9.537	4.52	.3370 −2	.1966	16.84
3.88	.7739 −2	.2493	9.624	4.53	.3329 −2	.1959	16.99
3.89	.7635 −2	.2484	9.711	4.54	.3288 −2	.1952	17.13
3.90	.7532 −2	.2474	9.799	4.55	.3247 −2	.1945	17.28
3.91	.7431 −2	.2464	9.888	4.56	.3207 −2	.1938	17.42
3.92	.7332 −2	.2455	9.977	4.57	.3168 −2	.1932	17.57
3.93	.7233 −2	.2446	10.07	4.58	.3129 −2	.1925	17.72
3.94	.7137 −2	.2436	10.16	4.59	.3090 −2	.1918	17.87
3.95	.7042 −2	.2427	10.25	4.60	.3053 −2	.1911	18.02
3.96	.6948 −2	.2418	10.34	4.61	.3015 −2	.1905	18.17
3.97	.6855 −2	.2408	10.44	4.62	.2978 −2	.1898	18.32
3.98	.6764 −2	.2399	10.53	4.63	.2942 −2	.1891	18.48
3.99	.6675 −2	.2390	10.62	4.64	.2906 −2	.1885	18.63
4.00	.6586 −2	.2381	10.72	4.65	.2871 −2	.1878	18.79
4.01	.6499 −2	.2372	10.81	4.66	.2836 −2	.1872	18.94
4.02	.6413 −2	.2363	10.91	4.67	.2802 −2	.1865	19.10
4.03	.6328 −2	.2354	11.01	4.68	.2768 −2	.1859	19.26
4.04	.6245 −2	.2345	11.11	4.69	.2734 −2	.1852	19.42
4.05	.6163 −2	.2336	11.21	4.70	.2701 −2	.1846	19.58
4.06	.6082 −2	.2327	11.31	4.71	.2669 −2	.1839	19.75
4.07	.6002 −2	.2319	11.41	4.72	.2637 −2	.1833	19.91
4.08	.5923 −2	.2310	11.51	4.73	.2605 −2	.1827	20.07
4.09	.5845 −2	.2301	11.61	4.74	.2573 −2	.1820	20.24
4.10	.5769 −2	.2293	11.71	4.75	.2543 −2	.1814	20.41
4.11	.5694 −2	.2284	11.82	4.76	.2512 −2	.1808	20.58
4.12	.5619 −2	.2275	11.92	4.77	.2482 −2	.1802	20.75
4.13	.5546 −2	.2267	12.03	4.78	.2452 −2	.1795	20.92
4.14	.5474 −2	.2258	12.14	4.79	.2423 −2	.1789	21.09
4.15	.5403 −2	.2250	12.24	4.80	.2394 −2	.1783	21.26
4.16	.5333 −2	.2242	12.35	4.81	.2366 −2	.1777	21.44
4.17	.5264 −2	.2233	12.46	4.82	.2338 −2	.1771	21.61
4.18	.5195 −2	.2225	12.57	4.83	.2310 −2	.1765	21.79
4.19	.5128 −2	.2217	12.68	4.84	.2283 −2	.1759	21.97
4.20	.5062 −2	.2208	12.79	4.85	.2255 −2	.1753	22.15
4.21	.4997 −2	.2200	12.90	4.86	.2229 −2	.1747	22.33
4.22	.4932 −2	.2192	13.02	4.87	.2202 −2	.1741	22.51
4.23	.4869 −2	.2184	13.13	4.88	.2177 −2	.1735	22.70
4.24	.4806 −2	.2176	13.25	4.89	.2151 −2	.1729	22.88
4.25	.4745 −2	.2168	13.36	4.90	.2126 −2	.1724	23.07
4.26	.4684 −2	.2160	13.48	4.91	.2101 −2	.1718	23.25
4.27	.4624 −2	.2152	13.60	4.92	.2076 −2	.1712	23.44
4.28	.4565 −2	.2144	13.72	4.93	.2052 −2	.1706	23.63
4.29	.4507 −2	.2136	13.83	4.94	.2028 −2	.1700	23.82
4.30	.4449 −2	.2129	13.95	4.95	.2004 −2	.1695	24.02
4.31	.4393 −2	.2121	14.08	4.96	.1981 −2	.1689	24.21
4.32	.4337 −2	.2113	14.20	4.97	.1957 −2	.1683	24.41
4.33	.4282 −2	.2105	14.32	4.98	.1935 −2	.1678	24.60
4.34	.4228 −2	.2098	14.45	4.99	.1912 −2	.1672	24.80
4.35	.4174 −2	.2090	14.57	5.00	.1890 −2	.1667	25.00
4.36	.4121 −2	.2083	14.70	6.00	.0633 −2	.1219	53.19
4.37	.4069 −2	.2075	14.82	7.00	.0242 −2	.0926	104.14
4.38	.4018 −2	.2067	14.95	8.00	.0102 −2	.0725	190.11
4.39	.3968 −2	.2060	15.08	9.00	.0474 −3	.0582	327.19
4.40	.3918 −2	.2053	15.21	10.00	.0236 −3	.0476	535.94
4.41	.3868 −2	.2045	15.34	∞	0	0	∞
4.42	.3820 −2	.2038	15.47				
4.43	.3772 −2	.2030	15.61				
4.44	.3725 −2	.2023	15.74				

TABLE D.2

Normal-Shock Flow

M_1	M_2	p_2/p_1	T_2/T_1	p_{02}/p_{01}
1.00	1.000	1.000	1.000	1.000
1.01	.9901	1.023	1.007	1.000
1.02	.9805	1.047	1.013	1.000
1.03	.9712	1.071	1.020	1.000
1.04	.9620	1.095	1.026	.9999
1.05	.9531	1.120	1.033	.9999
1.06	.9444	1.144	1.039	.9997
1.07	.9360	1.169	1.046	.9996
1.08	.9277	1.194	1.052	.9994
1.09	.9196	1.219	1.059	.9992
1.10	.9118	1.245	1.065	.9989
1.11	.9041	1.271	1.071	.9986
1.12	.8966	1.297	1.078	.9982
1.13	.8892	1.323	1.084	.9978
1.14	.8820	1.350	1.090	.9973
1.15	.8750	1.376	1.097	.9967
1.16	.8682	1.403	1.103	.9961
1.17	.8615	1.430	1.109	.9953
1.18	.8549	1.458	1.115	.9946
1.19	.8485	1.485	1.122	.9937
1.20	.8422	1.513	1.128	.9928
1.21	.8360	1.541	1.134	.9918
1.22	.8300	1.570	1.141	.9907
1.23	.8241	1.598	1.147	.9896
1.24	.8183	1.627	1.153	.9884
1.25	.8126	1.656	1.159	.9871
1.26	.8071	1.686	1.166	.9857
1.27	.8016	1.715	1.172	.9842
1.28	.7963	1.745	1.178	.9827
1.29	.7911	1.775	1.185	.9811
1.30	.7860	1.805	1.191	.9794
1.31	.7809	1.835	1.197	.9776
1.32	.7760	1.866	1.204	.9758
1.33	.7712	1.897	1.210	.9738
1.34	.7664	1.928	1.216	.9718
1.35	.7618	1.960	1.223	.9697
1.36	.7572	1.991	1.229	.9676
1.37	.7527	2.023	1.235	.9653
1.38	.7483	2.055	1.242	.9630
1.39	.7440	2.087	1.248	.9607
1.40	.7397	2.120	1.255	.9582
1.41	.7355	2.153	1.261	.9557
1.42	.7314	2.186	1.268	.9531
1.43	.7274	2.219	1.274	.9504
1.44	.7235	2.253	1.281	.9476
1.45	.7196	2.286	1.287	.9448
1.46	.7157	2.320	1.294	.9420
1.47	.7120	2.354	1.300	.9390
1.48	.7083	2.389	1.307	.9360
1.49	.7047	2.423	1.314	.9329
1.50	.7011	2.458	1.320	.9298
1.51	.6976	2.493	1.327	.9266
1.52	.6941	2.529	1.334	.9233
1.53	.6907	2.564	1.340	.9200
1.54	.6874	2.600	1.347	.9166
1.55	.6841	2.636	1.354	.9132
1.56	.6809	2.673	1.361	.9097
1.57	.6777	2.709	1.367	.9061
1.58	.6746	2.746	1.374	.9026
1.59	.6715	2.783	1.381	.8989

TABLE D.2 (Continued)

M_1	M_2	p_2/p_1	T_2/T_1	p_{02}/p_{01}
1.60	.6684	2.820	1.388	.8952
1.61	.6655	2.857	1.395	.8915
1.62	.6625	2.895	1.402	.8877
1.63	.6596	2.933	1.409	.8838
1.64	.6568	2.971	1.416	.8799
1.65	.6540	3.010	1.423	.8760
1.66	.6512	3.048	1.430	.8720
1.67	.6485	3.087	1.437	.8680
1.68	.6458	3.126	1.444	.8640
1.69	.6431	3.165	1.451	.8598
1.70	.6405	3.205	1.458	.8557
1.71	.6380	3.245	1.466	.8516
1.72	.6355	3.285	1.473	.8474
1.73	.6330	3.325	1.480	.8431
1.74	.6305	3.366	1.487	.8389
1.75	.6281	3.406	1.495	.8346
1.76	.6257	3.447	1.502	.8302
1.77	.6234	3.488	1.509	.8259
1.78	.6210	3.530	1.517	.8215
1.79	.6188	3.571	1.524	.8171
1.80	.6165	3.613	1.532	.8127
1.81	.6143	3.655	1.539	.8082
1.82	.6121	3.698	1.547	.8038
1.83	.6099	3.740	1.554	.7993
1.84	.6078	3.783	1.562	.7948
1.85	.6057	3.826	1.569	.7902
1.86	.6036	3.870	1.577	.7857
1.87	.6016	3.913	1.585	.7811
1.88	.5996	3.957	1.592	.7765
1.89	.5976	4.001	1.600	.7720
1.90	.5956	4.045	1.608	.7674
1.91	.5937	4.089	1.616	.7627
1.92	.5918	4.134	1.624	.7581
1.93	.5899	4.179	1.631	.7535
1.94	.5880	4.224	1.639	.7488
1.95	.5862	4.270	1.647	.7442
1.96	.5844	4.315	1.655	.7395
1.97	.5826	4.361	1.663	.7349
1.98	.5808	4.407	1.671	.7302
1.99	.5791	4.453	1.679	.7255
2.00	.5774	4.500	1.688	.7209
2.01	.5757	4.547	1.696	.7162
2.02	.5740	4.594	1.704	.7115
2.03	.5723	4.641	1.712	.7069
2.04	.5707	4.689	1.720	.7022
2.05	.5691	4.736	1.729	.6975
2.06	.5675	4.784	1.737	.6928
2.07	.5659	4.832	1.745	.6882
2.08	.5643	4.881	1.754	.6835
2.09	.5628	4.929	1.762	.6789
2.10	.5613	4.978	1.770	.6742
2.11	.5598	5.027	1.779	.6696
2.12	.5583	5.077	1.787	.6649
2.13	.5568	5.126	1.796	.6603
2.14	.5554	5.176	1.805	.6557
2.15	.5540	5.226	1.813	.6511
2.16	.5525	5.277	1.822	.6464
2.17	.5511	5.327	1.831	.6419
2.18	.5498	5.378	1.839	.6373
2.19	.5484	5.429	1.848	.6327
2.20	.5471	5.480	1.857	.6281
2.21	.5457	5.531	1.866	.6236
2.22	.5444	5.583	1.875	.6191
2.23	.5431	5.636	1.883	.6145
2.24	.5418	5.687	1.892	.6100

TABLE D.2 (Continued)

M_1	M_2	p_2/p_1	T_2/T_1	p_{02}/p_{01}
2.25	.5406	5.740	1.901	.6055
2.26	.5393	5.792	1.910	.6011
2.27	.5381	5.845	1.919	.5966
2.28	.5368	5.898	1.929	.5921
2.29	.5356	5.951	1.938	.5877
2.30	.5344	6.005	1.947	.5833
2.31	.5332	6.059	1.956	.5789
2.32	.5321	6.113	1.965	.5745
2.33	.5309	6.167	1.974	.5702
2.34	.5297	6.222	1.984	.5658
2.35	.5286	6.276	1.993	.5615
2.36	.5275	6.331	2.002	.5572
2.37	.5264	6.386	2.012	.5529
2.38	.5253	6.442	2.021	.5486
2.39	.5242	6.497	2.031	.5444
2.40	.5231	6.553	2.040	.5401
2.41	.5221	6.609	2.050	.5359
2.42	.5210	6.666	2.059	.5317
2.43	.5200	6.722	2.069	.5276
2.44	.5189	6.779	2.079	.5234
2.45	.5179	6.836	2.088	.5193
2.46	.5169	6.894	2.098	.5152
2.47	.5159	6.951	2.108	.5111
2.48	.5149	7.009	2.118	.5071
2.49	.5140	7.067	2.128	.5030
2.50	.5130	7.125	2.138	.4990
2.51	.5120	7.183	2.147	.4950
2.52	.5111	7.242	2.157	.4911
2.53	.5102	7.301	2.167	.4871
2.54	.5092	7.360	2.177	.4832
2.55	.5083	7.420	2.187	.4793
2.56	.5074	7.479	2.198	.4754
2.57	.5065	7.539	2.208	.4715
2.58	.5056	7.599	2.218	.4677
2.59	.5047	7.659	2.228	.4639
2.60	.5039	7.720	2.238	.4601
2.61	.5030	7.781	2.249	.4564
2.62	.5022	7.842	2.259	.4526
2.63	.5013	7.903	2.269	.4489
2.64	.5005	7.965	2.280	.4452
2.65	.4996	8.026	2.290	.4416
2.66	.4988	8.088	2.301	.4379
2.67	.4980	8.150	2.311	.4343
2.68	.4972	8.213	2.322	.4307
2.69	.4964	8.275	2.332	.4271
2.70	.4956	8.338	2.343	.4236
2.71	.4949	8.401	2.354	.4201
2.72	.4941	8.465	2.364	.4166
2.73	.4933	8.528	2.375	.4131
2.74	.4926	8.592	2.386	.4097
2.75	.4918	8.656	2.397	.4062
2.76	.4911	8.721	2.407	.4028
2.77	.4903	8.785	2.418	.3994
2.78	.4896	8.850	2.429	.3961
2.79	.4889	8.915	2.440	.3928
2.80	.4882	8.980	2.451	.3895
2.81	.4875	9.045	2.462	.3862
2.82	.4868	9.111	2.473	.3829
2.83	.4861	9.177	2.484	.3797
2.84	.4854	9.243	2.496	.3765
2.85	.4847	9.310	2.507	.3733
2.86	.4840	9.376	2.518	.3701
2.87	.4833	9.443	2.529	.3670
2.88	.4827	9.510	2.540	.3639
2.89	.4820	9.577	2.552	.3608

TABLE D.2 (Continued)

M_1	M_2	p_2/p_1	T_2/T_1	p_{02}/p_{01}
2.90	.4814	9.645	2.563	.3577
2.91	.4807	9.713	2.575	.3547
2.92	.4801	9.781	2.586	.3517
2.93	.4795	9.849	2.598	.3487
2.94	.4788	9.918	2.609	.3457
2.95	.4782	9.986	2.621	.3428
2.96	.4776	10.06	2.632	.3398
2.97	.4770	10.12	2.644	.3369
2.98	.4764	10.19	2.656	.3340
2.99	.4758	10.26	2.667	.3312
3.00	.4752	10.33	2.679	.3283
3.01	.4746	10.40	2.691	.3255
3.02	.4740	10.47	2.703	.3227
3.03	.4734	10.54	2.714	.3200
3.04	.4729	10.62	2.726	.3172
3.05	.4723	10.69	2.738	.3145
3.06	.4717	10.76	2.750	.3118
3.07	.4712	10.83	2.762	.3091
3.08	.4706	10.90	2.774	.3065
3.09	.4701	10.97	2.786	.3038
3.10	.4695	11.05	2.799	.3012
3.11	.4690	11.12	2.811	.2986
3.12	.4685	11.19	2.823	.2960
3.13	.4679	11.26	2.835	.2935
3.14	.4674	11.34	2.848	.2910
3.15	.4669	11.41	2.860	.2885
3.16	.4664	11.48	2.872	.2860
3.17	.4659	11.56	2.885	.2835
3.18	.4654	11.63	2.897	.2811
3.19	.4648	11.71	2.909	.2786
3.20	.4643	11.78	2.922	.2762
3.21	.4639	11.85	2.935	.2738
3.22	.4634	11.93	2.947	.2715
3.23	.4629	12.01	2.960	.2691
3.24	.4624	12.08	2.972	.2668
3.25	.4619	12.16	2.985	.2645
3.26	.4614	12.23	2.998	.2622
3.27	.4610	12.31	3.011	.2600
3.28	.4605	12.38	3.023	.2577
3.29	.4600	12.46	3.036	.2555
3.30	.4596	12.54	3.049	.2533
3.31	.4591	12.62	3.062	.2511
3.32	.4587	12.69	3.075	.2489
3.33	.4582	12.77	3.088	.2468
3.34	.4578	12.85	3.101	.2446
3.35	.4573	12.93	3.114	.2425
3.36	.4569	13.00	3.127	.2404
3.37	.4565	13.08	3.141	.2383
3.38	.4560	13.16	3.154	.2363
3.39	.4556	13.24	3.167	.2342
3.40	.4552	13.32	3.180	.2322
3.41	.4548	13.40	3.194	.2302
3.42	.4544	13.48	3.207	.2282
3.43	.4540	13.56	3.220	.2263
3.44	.4535	13.64	3.234	.2243
3.45	.4531	13.72	3.247	.2224
3.46	.4527	13.80	3.261	.2205
3.47	.4523	13.88	3.274	.2186
3.48	.4519	13.96	3.288	.2167
3.49	.4515	14.04	3.301	.2148
3.50	.4512	14.13	3.315	.2129
3.51	.4508	14.21	3.329	.2111
3.52	.4504	14.29	3.343	.2093
3.53	.4500	14.37	3.356	.2075
3.54	.4496	14.45	3.370	.2057

TABLE D.2 (Continued)

M_1	M_2	p_2/p_1	T_2/T_1	p_{02}/p_{01}
3.55	.4492	14.54	3.384	.2039
3.56	.4489	14.62	3.398	.2022
3.57	.4485	14.70	3.412	.2004
3.58	.4481	14.79	3.426	.1987
3.59	.4478	14.87	3.440	.1970
3.60	.4474	14.95	3.454	.1953
3.61	.4471	15.04	3.468	.1936
3.62	.4467	15.12	3.482	.1920
3.63	.4463	15.21	3.496	.1903
3.64	.4460	15.29	3.510	.1887
3.65	.4456	15.38	3.525	.1871
3.66	.4453	15.46	3.539	.1855
3.67	.4450	15.55	3.553	.1839
3.68	.4446	15.63	3.568	.1823
3.69	.4443	15.72	3.582	.1807
3.70	.4439	15.81	3.596	.1792
3.71	.4436	15.89	3.611	.1777
3.72	.4433	15.98	3.625	.1761
3.73	.4430	16.07	3.640	.1746
3.74	.4426	16.15	3.654	.1731
3.75	.4423	16.24	3.669	.1717
3.76	.4420	16.33	3.684	.1702
3.77	.4417	16.42	3.698	.1687
3.78	.4414	16.50	3.713	.1673
3.79	.4410	16.59	3.728	.1659
3.80	.4407	16.68	3.743	.1645
3.81	.4404	16.77	3.758	.1631
3.82	.4401	16.86	3.772	.1617
3.83	.4398	16.95	3.787	.1603
3.84	.4395	17.04	3.802	.1589
3.85	.4392	17.13	3.817	.1576
3.86	.4389	17.22	3.832	.1563
3.87	.4386	17.31	3.847	.1549
3.88	.4383	17.40	3.863	.1536
3.89	.4380	17.49	3.878	.1523
3.90	.4377	17.58	3.893	.1510
3.91	.4375	17.67	3.908	.1497
3.92	.4372	17.76	3.923	.1485
3.93	.4369	17.85	3.939	.1472
3.94	.4366	17.94	3.954	.1460
3.95	.4363	18.04	3.969	.1448
3.96	.4360	18.13	3.985	.1435
3.97	.4358	18.22	4.000	.1423
3.98	.4355	18.31	4.016	.1411
3.99	.4352	18.41	4.031	.1399
4.00	.4350	18.50	4.047	.1388
4.01	.4347	18.59	4.062	.1376
4.02	.4344	18.69	4.078	.1364
4.03	.4342	18.78	4.094	.1353
4.04	.4339	18.88	4.110	.1342
4.05	.4336	18.97	4.125	.1330
4.06	.4334	19.06	4.141	.1319
4.07	.4331	19.16	4.157	.1308
4.08	.4329	19.25	4.173	.1297
4.09	.4326	19.35	4.189	.1286
4.10	.4324	19.45	4.205	.1276
4.11	.4321	19.54	4.221	.1265
4.12	.4319	19.64	4.237	.1254
4.13	.4316	19.73	4.253	.1244
4.14	.4314	19.83	4.269	.1234
4.15	.4311	19.93	4.285	.1223
4.16	.4309	20.02	4.301	.1213
4.17	.4306	20.12	4.318	.1203
4.18	.4304	20.22	4.334	.1193
4.19	.4302	20.32	4.350	.1183

TABLE D.2 (Continued)

M_1	M_2	p_2/p_1	T_2/T_1	p_{02}/p_{01}
4.20	.4299	20.41	4.367	.1173
4.21	.4297	20.51	4.383	.1164
4.22	.4295	20.61	4.399	.1154
4.23	.4292	20.71	4.416	.1144
4.24	.4290	20.81	4.432	.1135
4.25	.4288	20.91	4.449	.1126
4.26	.4286	21.01	4.466	.1116
4.27	.4283	21.11	4.482	.1107
4.28	.4281	21.20	4.499	.1098
4.29	.4279	21.30	4.516	.1089
4.30	.4277	21.41	4.532	.1080
4.31	.4275	21.51	4.549	.1071
4.32	.4272	21.61	4.566	.1062
4.33	.4270	21.71	4.583	.1054
4.34	.4268	21.81	4.600	.1045
4.35	.4266	21.91	4.617	.1036
4.36	.4264	22.01	4.633	.1028
4.37	.4262	22.11	4.651	.1020
4.38	.4260	22.22	4.668	.1011
4.39	.4258	22.32	4.685	.1003
4.40	.4255	22.42	4.702	.9948 $^{-1}$
4.41	.4253	22.52	4.719	.9867 $^{-1}$
4.42	.4251	22.63	4.736	.9787 $^{-1}$
4.43	.4249	22.73	4.753	.9707 $^{-1}$
4.44	.4247	22.83	4.771	.9628 $^{-1}$
4.45	.4245	22.94	4.788	.9550 $^{-1}$
4.46	.4243	23.04	4.805	.9473 $^{-1}$
4.47	.4241	23.14	4.823	.9396 $^{-1}$
4.48	.4239	23.25	4.840	.9320 $^{-1}$
4.49	.4237	23.35	4.858	.9244 $^{-1}$
4.50	.4236	23.46	4.875	.9170 $^{-1}$
4.51	.4234	23.56	4.893	.9096 $^{-1}$
4.52	.4232	23.67	4.910	.9022 $^{-1}$
4.53	.4230	23.77	4.928	.8950 $^{-1}$
4.54	.4228	23.88	4.946	.8878 $^{-1}$
4.55	.4226	23.99	4.963	.8806 $^{-1}$
4.56	.4224	24.09	4.981	.8735 $^{-1}$
4.57	.4222	24.20	4.999	.8665 $^{-1}$
4.58	.4220	24.31	5.017	.8596 $^{-1}$
4.59	.4219	24.41	5.034	.8527 $^{-1}$
4.60	.4217	24.52	5.052	.8459 $^{-1}$
4.61	.4215	24.63	5.070	.8391 $^{-1}$
4.62	.4213	24.74	5.088	.8324 $^{-1}$
4.63	.4211	24.84	5.106	.8257 $^{-1}$
4.64	.4210	24.95	5.124	.8192 $^{-1}$
4.65	.4208	25.06	5.143	.8126 $^{-1}$
4.66	.4206	25.17	5.160	.8062 $^{-1}$
4.67	.4204	25.28	5.179	.7998 $^{-1}$
4.68	.4203	25.39	5.197	.7934 $^{-1}$
4.69	.4201	25.50	5.215	.7871 $^{-1}$
4.70	.4199	25.61	5.233	.7809 $^{-1}$
4.71	.4197	25.71	5.252	.7747 $^{-1}$
4.72	.4196	25.82	5.270	.7685 $^{-1}$
4.73	.4194	25.94	5.289	.7625 $^{-1}$
4.74	.4192	26.05	5.307	.7564 $^{-1}$
4.75	.4191	26.16	5.325	.7505 $^{-1}$
4.76	.4189	26.27	5.344	.7445 $^{-1}$
4.77	.4187	26.38	5.363	.7387 $^{-1}$
4.78	.4186	26.49	5.381	.7329 $^{-1}$
4.79	.4184	26.60	5.400	.7271 $^{-1}$
4.80	.4183	26.71	5.418	.7214 $^{-1}$
4.81	.4181	26.83	5.437	.7157 $^{-1}$
4.82	.4179	26.94	5.456	.7101 $^{-1}$
4.83	.4178	27.05	5.475	.7046 $^{-1}$
4.84	.4176	27.16	5.494	.6991 $^{-1}$

TABLE D.2 (Continued)

M_1	M_2	p_2/p_1	T_2/T_1	p_{02}/p_{01}
4.85	.4175	27.28	5.512	.6936 -1
4.86	.4173	27.39	5.531	.6882 -1
4.87	.4172	27.50	5.550	.6828 -1
4.88	.4170	27.62	5.569	.6775 -1
4.89	.4169	27.73	5.588	.6722 -1
4.90	.4167	27.85	5.607	.6670 -1
4.91	.4165	27.96	5.626	.6618 -1
4.92	.4164	28.07	5.646	.6567 -1
4.93	.4163	28.19	5.665	.6516 -1
4.94	.4161	28.30	5.684	.6465 -1
4.95	.4160	28.42	5.703	.6415 -1
4.96	.4158	28.54	5.723	.6366 -1
4.97	.4157	28.65	5.742	.6317 -1
4.98	.4155	28.77	5.761	.6268 -1
4.99	.4154	28.88	5.781	.6220 -1
5.00	.4152	29.00	5.800	.6172 -1
6.00	.4042	41.83	7.941	.2965 -1
7.00	.3974	57.00	10.469	.1535 -1
8.00	.3929	74.50	13.387	.0849 -1
9.00	.3898	94.33	16.693	.0496 -1
10.00	.3875	116.50	20.388	.0304 -1
∞	.3780	∞	∞	0

TABLE D.3
Adiabatic, Frictional Flow

M	T/T^*	p/p^*	p_0/p_0^*	V/V^*	$\bar{f}l_{\text{crit}}/D$
0	1.2000	∞	∞	0	∞
0.01	1.2000	109.544	57.874	.01095	7134.40
.02	1.1999	54.770	28.942	.02191	1778.45
.03	1.1998	36.511	19.300	.03286	787.08
.04	1.1996	27.382	14.482	.04381	440.35
.05	1.1994	21.903	11.5914	.05476	280.02
.06	1.1991	18.251	9.6659	.06570	193.03
.07	1.1988	15.642	8.2915	.07664	140.66
.08	1.1985	13.684	7.2616	.08758	106.72
.09	1.1981	12.162	6.4614	.09851	83.496
.10	1.1976	10.9435	5.8218	.10943	66.922
.11	1.1971	9.9465	5.2992	.12035	54.688
.12	1.1966	9.1156	4.8643	.13126	45.408
.13	1.1960	8.4123	4.4968	.14216	38.207
.14	1.1953	7.8093	4.1824	.15306	32.511
.15	1.1946	7.2866	3.9103	.16395	27.932
.16	1.1939	6.8291	3.6727	.17482	24.198
.17	1.1931	6.4252	3.4635	.18568	21.115
.18	1.1923	6.0662	3.2779	.19654	18.543
.19	1.1914	5.7448	3.1123	.20739	16.375
.20	1.1905	5.4555	2.9635	.21822	14.533
.21	1.1895	5.1936	2.8293	.22904	12.956
.22	1.1885	4.9554	2.7076	.23984	11.596
.23	1.1874	4.7378	2.5968	.25063	10.416
.24	1.1863	4.5383	2.4956	.26141	9.3865

TABLE D.3 (Continued)

M	T/T^*	p/p^*	p_0/p_0^*	V/V^*	$\bar{f}l_{\text{crit}}/D$
.25	1.1852	4.3546	2.4027	.27217	8.4834
.26	1.1840	4.1850	2.3173	.28291	7.6876
.27	1.1828	4.0280	2.2385	.29364	6.9832
.28	1.1815	3.8820	2.1656	.30435	6.3572
.29	1.1802	3.7460	2.0979	.31504	5.7989
.30	1.1788	3.6190	2.0351	.32572	5.2992
.31	1.1774	3.5002	1.9765	.33637	4.8507
.32	1.1759	3.3888	1.9219	.34700	4.4468
.33	1.1744	3.2840	1.8708	.35762	4.0821
.34	1.1729	3.1853	1.8229	.36822	3.7520
.35	1.1713	3.0922	1.7780	.37880	3.4525
.36	1.1697	3.0042	1.7358	.38935	3.1801
.37	1.1680	2.9209	1.6961	.39988	2.9320
.38	1.1663	2.8420	1.6587	.41039	2.7055
.39	1.1646	2.7671	1.6234	.42087	2.4983
.40	1.1628	2.6958	1.5901	.43133	2.3085
.41	1.1610	2.6280	1.5587	.44177	2.1344
.42	1.1591	2.5634	1.5289	.45218	1.9744
.43	1.1572	2.5017	1.5007	.46257	1.8272
.44	1.1553	2.4428	1.4739	.47293	1.6915
.45	1.1533	2.3865	1.4486	.48326	1.5664
.46	1.1513	2.3326	1.4246	.49357	1.4509
.47	1.1492	2.2809	1.4018	.50385	1.3442
.48	1.1471	2.2314	1.3801	.51410	1.2453
.49	1.1450	2.1838	1.3595	.52433	1.1539
.50	1.1429	2.1381	1.3399	.53453	1.06908
.51	1.1407	2.0942	1.3212	.54469	.99042
.52	1.1384	2.0519	1.3034	.55482	.91741
.53	1.1362	2.0112	1.2864	.56493	.84963
.54	1.1339	1.9719	1.2702	.57501	.78662
.55	1.1315	1.9341	1.2549	.58506	.72805
.56	1.1292	1.8976	1.2403	.59507	.67357
.57	1.1268	1.8623	1.2263	.60505	.62286
.58	1.1244	1.8282	1.2130	.61500	.57568
.59	1.1219	1.7952	1.2003	.62492	.53174
.60	1.1194	1.7634	1.1882	.63481	.49081
.61	1.1169	1.7325	1.1766	.64467	.45270
.62	1.1144	1.7026	1.1656	.65449	.41720
.63	1.1118	1.6737	1.1551	.66427	.38411
.64	1.1091	1.6456	1.1451	.67402	.35330
.65	1.10650	1.6183	1.1356	.68374	.32460
.66	1.10383	1.5919	1.1265	.69342	.29785
.67	1.10114	1.5662	1.1179	.70306	.27295
.68	1.09842	1.5413	1.1097	.71267	.24978
.69	1.09567	1.5170	1.1018	.72225	.22821
.70	1.09290	1.4934	1.09436	.73179	.20814
.71	1.09010	1.4705	1.08729	.74129	.18949
.72	1.08727	1.4482	1.08057	.75076	.17215
.73	1.08442	1.4265	1.07419	.76019	.15606
.74	1.08155	1.4054	1.06815	.76958	.14113

TABLE D.3 (Continued)

M	T/T^*	p/p^*	p_0/p_0^*	V/V^*	$\bar{fl}_{\text{crit}}/D$
.75	1.07865	1.3848	1.06242	.77893	.12728
.76	1.07573	1.3647	1.05700	.78825	.11446
.77	1.07279	1.3451	1.05188	.79753	.10262
.78	1.06982	1.3260	1.04705	.80677	.09167
.79	1.06684	1.3074	1.04250	.81598	.08159
.80	1.06383	1.2892	1.03823	.82514	.07229
.81	1.06080	1.2715	1.03422	.83426	.06375
.82	1.05775	1.2542	1.03047	.84334	.05593
.83	1.05468	1.2373	1.02696	.85239	.04878
.84	1.05160	1.2208	1.02370	.86140	.04226
.85	1.04849	1.2047	1.02067	.87037	.03632
.86	1.04537	1.1889	1.01787	.87929	.03097
.87	1.04223	1.1735	1.01529	.88818	.02613
.88	1.03907	1.1584	1.01294	.89703	.02180
.89	1.03589	1.1436	1.01080	.90583	.01793
.90	1.03270	1.12913	1.00887	.91459	.014513
.91	1.02950	1.11500	1.00714	.92332	.011519
.92	1.02627	1.10114	1.00560	.93201	.008916
.93	1.02304	1.08758	1.00426	.94065	.006694
.94	1.01978	1.07430	1.00311	.94925	.004815
.95	1.01652	1.06129	1.00215	.95782	.003280
.96	1.01324	1.04854	1.00137	.96634	.002056
.97	1.00995	1.03605	1.00076	.97481	.001135
.98	1.00664	1.02379	1.00033	.98324	.000493
.99	1.00333	1.01178	1.00008	.99164	.000120
1.00	1.00000	1.00000	1.00000	1.00000	0
1.01	.99666	.98844	1.00008	1.00831	.000114
1.02	.99331	.97711	1.00033	1.01658	.000458
1.03	.98995	.96598	1.00073	1.02481	.001013
1.04	.98658	.95506	1.00130	1.03300	.001771
1.05	.98320	.94435	1.00203	1.04115	.002712
1.06	.97982	.93383	1.00291	1.04925	.003837
1.07	.97642	.92350	1.00394	1.05731	.005129
1.08	.97302	.91335	1.00512	1.06533	.006582
1.09	.96960	.90338	1.00645	1.07331	.008185
1.10	.96618	.89359	1.00793	1.08124	.009933
1.11	.96276	.88397	1.00955	1.08913	.011813
1.12	.95933	.87451	1.01131	1.09698	.013824
1.13	.95589	.86522	1.01322	1.10479	.015949
1.14	.95244	.85608	1.01527	1.11256	.018187
1.15	.94899	.84710	1.01746	1.1203	.02053
1.16	.94554	.83827	1.01978	1.1280	.02298
1.17	.94208	.82958	1.02224	1.1356	.02552
1.18	.93862	.82104	1.02484	1.1432	.02814
1.19	.93515	.81263	1.02757	1.1508	.03085

TABLE D.3 (Continued)

M	T/T^*	p/p^*	p_0/p_0^*	V/V^*	$\bar{fl}_{\text{crit}}/D$
1.20	.93168	.80436	1.03044	1.1583	.03364
1.21	.92820	.79623	1.03344	1.1658	.03650
1.22	.92473	.78822	1.03657	1.1732	.03942
1.23	.92125	.78034	1.03983	1.1806	.04241
1.24	.91777	.77258	1.04323	1.1879	.04547
1.25	.91429	.76495	1.04676	1.1952	.04858
1.26	.91080	.75743	1.05041	1.2025	.05174
1.27	.90732	.75003	1.05419	1.2097	.05494
1.28	.90383	.74274	1.05809	1.2169	.05820
1.29	.90035	.73556	1.06213	1.2240	.06150
1.30	.89686	.72848	1.06630	1.2311	.06483
1.31	.89338	.72152	1.07060	1.2382	.06820
1.32	.88989	.71465	1.07502	1.2452	.07161
1.33	.88641	.70789	1.07957	1.2522	.07504
1.34	.88292	.70123	1.08424	1.2591	.07850
1.35	.87944	.69466	1.08904	1.2660	.08199
1.36	.87596	.68818	1.09397	1.2729	.08550
1.37	.87249	.68180	1.09902	1.2797	.08904
1.38	.86901	.67551	1.10419	1.2864	.09259
1.39	.86554	.66931	1.10948	1.2932	.09616
1.40	.86207	.66320	1.1149	1.2999	.09974
1.41	.85860	.65717	1.1205	1.3065	.10333
1.42	.85514	.65122	1.1262	1.3131	.10694
1.43	.85168	.64536	1.1320	1.3197	.11056
1.44	.84822	.63958	1.1379	1.3262	.11419
1.45	.84477	63387	1.1440	1.3327	.11782
1.46	.84133	.62824	1.1502	1.3392	.12146
1.47	.83788	.62269	1.1565	1.3456	.12510
1.48	.83445	.61722	1.1629	1.3520	.12875
1.49	.83101	.61181	1.1695	1.3583	.13240
1.50	.82759	.60648	1.1762	1.3646	.13605
1.51	.82416	.60122	1.1830	1.3708	.13970
1.52	.82075	.59602	1.1899	1.3770	.14335
1.53	.81734	.59089	1.1970	1.3832	.14699
1.54	.81394	.58583	1.2043	1.3894	.15063
1.55	.81054	.58084	1.2116	1.3955	.15427
1.56	.80715	.57591	1.2190	1.4015	.15790
1.57	.80376	.57104	1.2266	1.4075	.16152
1.58	.80038	.56623	1.2343	1.4135	.16514
1.59	.79701	.56148	1.2422	1.4195	.16876
1.60	.79365	.55679	1.2502	1.4254	.17236
1.61	.79030	.55216	1.2583	1.4313	.17595
1.62	.78695	.54759	1.2666	1.4371	.17953
1.63	.78361	.54308	1.2750	1.4429	.18311
1.64	.78028	.53862	1.2835	1.4487	.18667

TABLE D.3 (Continued)

M	T/T^*	p/p^*	p_0/p_0^*	V/V^*	$\bar{fl}_{\text{crit}}/D$
1.65	.77695	.53421	1.2922	1.4544	.19022
1.66	.77363	.52986	1.3010	1.4601	.19376
1.67	.77033	.52556	1.3099	1.4657	.19729
1.68	.76703	.52131	1.3190	1.4713	.20081
1.69	.76374	.51711	1.3282	1.4769	.20431
1.70	.76046	.51297	1.3376	1.4825	.20780
1.71	.75718	.50887	1.3471	1.4880	.21128
1.72	.75392	.50482	1.3567	1.4935	.21474
1.73	.75067	.50082	1.3665	1.4989	.21819
1.74	.74742	.49686	1.3764	1.5043	.22162
1.75	.74419	.49295	1.3865	1.5097	.22504
1.76	.74096	.48909	1.3967	1.5150	.22844
1.77	.73774	.48527	1.4070	1.5203	.23183
1.78	.73453	.48149	1.4175	1.5256	.23520
1.79	.73134	.47776	1.4282	1.5308	.23855
1.80	.72816	.47407	1.4390	1.5360	.24189
1.81	.72498	.47042	1.4499	1.5412	.24521
1.82	.72181	.46681	1.4610	1.5463	.24851
1.83	.71865	.46324	1.4723	1.5514	.25180
1.84	.71551	.45972	1.4837	1.5564	.25507
1.85	.71238	.45623	1.4952	1.5614	.25832
1.86	.70925	.45278	1.5069	1.5664	.26156
1.87	.70614	.44937	1.5188	1.5714	.26478
1.88	.70304	.44600	1.5308	1.5763	.26798
1.89	.69995	.44266	1.5429	1.5812	.27116
1.90	.69686	.43936	1.5552	1.5861	.27433
1.91	.69379	.43610	1.5677	1.5909	.27748
1.92	.69074	.43287	1.5804	1.5957	.28061
1.93	.68769	.42967	1.5932	1.6005	.28372
1.94	.68465	.42651	1.6062	1.6052	.28681
1.95	.68162	.42339	1.6193	1.6099	.28989
1.96	.67861	.42030	1.6326	1.6146	.29295
1.97	.67561	.41724	1.6461	1.6193	.29599
1.98	.67262	.41421	1.6597	1.6239	.29901
1.99	.66964	.41121	1.6735	1.6284	.30201
2.00	.66667	.40825	1.6875	1.6330	.30499
2.01	.66371	.40532	1.7017	1.6375	.30796
2.02	.66076	.40241	1.7160	1.6420	.31091
2.03	.65783	.39954	1.7305	1.6465	.31384
2.04	.65491	.39670	1.7452	1.6509	.31675
2.05	.65200	.39389	1.7600	1.6553	.31965
2.06	.64910	.39110	1.7750	1.6597	.32253
2.07	.64621	.38834	1.7902	1.6640	.32538
2.08	.64333	.38562	1.8056	1.6683	.32822
2.09	.64047	.38292	1.8212	1.6726	.33104

TABLE D.3 (Continued)

M	T/T^*	p/p^*	p_0/p_0^*	V/V^*	$\bar{f}l_{\text{crit}}/D$
2.10	.63762	.38024	1.8369	1.6769	.33385
2.11	.63478	.37760	1.8528	1.6811	.33664
2.12	.63195	.37498	1.8690	1.6853	.33940
2.13	.62914	.37239	1.8853	1.6895	.34215
2.14	.62633	.36982	1.9018	1.6936	.34488
2.15	.62354	.36728	1.9185	1.6977	.34760
2.16	.62076	.36476	1.9354	1.7018	.35030
2.17	.61799	.36227	1.9525	1.7059	.35298
2.18	.61523	.35980	1.9698	1.7099	.35564
2.19	.61249	.35736	1.9873	1.7139	.35828
2.20	.60976	.35494	2.0050	1.7179	.36091
2.21	.60704	.35254	2.0228	1.7219	.36352
2.22	.60433	.35017	2.0409	1.7258	.36611
2.23	.60163	.34782	2.0592	1.7297	.36868
2.24	.59895	.34550	2.0777	1.7336	.37124
2.25	.59627	.34319	2.0964	1.7374	.37378
2.26	.59361	.34091	2.1154	1.7412	.37630
2.27	.59096	.33865	2.1345	1.7450	.37881
2.28	.58833	.33641	2.1538	1.7488	.38130
2.29	.58570	.33420	2.1733	1.7526	.38377
2.30	.58309	.33200	2.1931	1.7563	.38623
2.31	.58049	.32983	2.2131	1.7600	.38867
2.32	.57790	.32767	2.2333	1.7637	.39109
2.33	.57532	.32554	2.2537	1.7673	.39350
2.34	.57276	.32342	2.2744	1.7709	.39589
2.35	.57021	.32133	2.2953	1.7745	.39826
2.36	.56767	.31925	2.3164	1.7781	.40062
2.37	.56514	.31720	2.3377	1.7817	.40296
2.38	.56262	.31516	2.3593	1.7852	.40528
2.39	.56011	.31314	2.3811	1.7887	.40760
2.40	.55762	.31114	2.4031	1.7922	.40989
2.41	.55514	.30916	2.4254	1.7956	.41216
2.42	.55267	.30720	2.4479	1.7991	.41442
2.43	.55021	.30525	2.4706	1.8025	.41667
2.44	.54776	.30332	2.4936	1.8059	.41891
2.45	.54533	.30141	2.5168	1.8092	.42113
2.46	.54291	.29952	2.5403	1.8126	.42333
2.47	.54050	.29765	2.5640	1.8159	.42551
2.48	.53810	.29579	2.5880	1.8192	.42768
2.49	.53571	.29395	2.6122	1.8225	.42983
2.50	.53333	.29212	2.6367	1.8257	.43197
2.51	.53097	.29031	2.6615	1.8290	.43410
2.52	.52862	.28852	2.6865	1.8322	.43621
2.53	.52627	.28674	2.7117	1.8354	.43831
2.54	.52394	.28498	2.7372	1.8386	.44040

TABLE D.3 (Continued)

M	T/T^*	p/p^*	p_0/p_0^*	V/V^*	$\bar{f}l_{\text{crit}}/D$
2.55	.52163	.28323	2.7630	1.8417	.44247
2.56	.51932	.28150	2.7891	1.8448	.44452
2.57	.51702	.27978	2.8154	1.8479	.44655
2.58	.51474	.27808	2.8420	1.8510	.44857
2.59	.51247	.27640	2.8689	1.8541	.45059
2.60	.51020	.27473	2.8960	1.8571	.45259
2.61	.50795	.27307	2.9234	1.8602	.45457
2.62	.50571	.27143	2.9511	1.8632	.45654
2.63	.50349	.26980	2.9791	1.8662	.45850
2.64	.50127	.26818	3.0074	1.8691	.46044
2.65	.49906	.26658	3.0359	1.8721	.46237
2.66	.49687	.26499	3.0647	1.8750	.46429
2.67	.49469	.26342	3.0938	1.8779	.46619
2.68	.49251	.26186	3.1234	1.8808	.46807
2.69	.49035	.26032	3.1530	1.8837	.46996
2.70	.48820	.25878	3.1830	1.8865	.47182
2.71	.48606	.25726	3.2133	1.8894	.47367
2.72	.48393	.25575	3.2440	1.8922	.47551
2.73	.48182	.25426	3.2749	1.8950	.47734
2.74	.47971	.25278	3.3061	1.8978	.47915
2.75	.47761	.25131	3.3376	1.9005	.48095
2.76	.47553	.24985	3.3695	1.9032	.48274
2.77	.47346	.24840	3.4017	1.9060	.48452
2.78	.47139	.24697	3.4342	1.9087	.48628
2.79	.46933	.24555	3.4670	1.9114	.48803
2.80	.46729	.24414	3.5001	1.9140	.48976
2.81	.46526	.24274	3.5336	1.9167	.49148
2.82	.46324	.24135	3.5674	1.9193	.49321
2.83	.46122	.23997	3.6015	1.9220	.49491
2.84	.45922	.23861	3.6359	1.9246	.49660
2.85	.45723	.23726	3.6707	1.9271	.49828
2.86	.45525	.23592	3.7058	1.9297	.49995
2.87	.45328	.23458	3.7413	1.9322	.50161
2.88	.45132	.23326	3.7771	1.9348	.50326
2.89	.44937	.23196	3.8133	1.9373	.50489
2.90	.44743	.23066	3.8498	1.9398	.50651
2.91	.44550	.22937	3.8866	1.9423	.50812
2.92	.44358	.22809	3.9238	1.9448	.50973
2.93	.44167	.22682	3.9614	1.9472	.51133
2.94	.43977	.22556	3.9993	1.9497	.51291
2.95	.43788	.22431	4.0376	1.9521	.51447
2.96	.43600	.22307	4.0763	1.9545	.51603
2.97	.43413	.22185	4.1153	1.9569	.51758
2.98	.43226	.22063	4.1547	1.9592	.51912
2.99	.43041	.21942	4.1944	1.9616	.52064

TABLE D.3 (Continued)

M	T/T^*	p/p^*	p_0/p_0^*	V/V^*	$\bar{f}l_{\mathrm{crit}}/D$
3.0	.42857	.21822	4.2346	1.9640	.52216
3.5	.34783	.16850	6.7896	2.0642	.58643
4.0	.28571	.13363	10.719	2.1381	.63306
4.5	.23762	.10833	16.562	2.1936	.66764
5.0	.20000	.08944	25.000	2.2361	.69381
6.0	.14634	.06376	53.180	2.2953	.72987
7.0	.11111	.04762	104.14	2.3333	.75281
8.0	.08696	.03686	190.11	2.3591	.76820
9.0	.06977	.02935	327.19	2.3772	.77898
10.0	.05714	.02390	535.94	2.3905	.78683
∞	0	0	∞	2.4495	.82153

TABLE D.4
Frictionless Flow with Heat Transfer

M	T_0/T_0^*	T/T^*	p/p^*	p_0/p_0^*	V/V^*
0	0	0	2.4000	1.2679	0
0.01	.000480	.000576	2.3997	1.2678	.000240
.02	.00192	.00230	2.3987	1.2675	.000959
.03	.00431	.00516	2.3970	1.2671	.00216
.04	.00765	.00917	2.3946	1.2665	.00383
.05	.01192	.01430	2.3916	1.2657	.00598
.06	.01712	.02053	2.3880	1.2647	.00860
.07	.02322	.02784	2.3837	1.2636	.01168
.08	.03021	.03621	2.3787	1.2623	.01522
.09	.03807	.04562	2.3731	1.2608	.01922
.10	.04678	.05602	2.3669	1.2591	.02367
.11	.05630	.06739	2.3600	1.2573	.02856
.12	.06661	.07970	2.3526	1.2554	.03388
.13	.07768	.09290	2.3445	1.2533	.03962
.14	.08947	.10695	2.3359	1.2510	.04578
.15	.10196	.12181	2.3267	1.2486	.05235
.16	.11511	.13743	2.3170	1.2461	.05931
.17	.12888	.15377	2.3067	1.2434	.06666
.18	.14324	.17078	2.2959	1.2406	.07438
.19	.15814	.18841	2.2845	1.2377	.08247
.20	.17355	.20661	2.2727	1.2346	.09091
.21	.18943	.22533	2.2604	1.2314	.09969
.22	.20574	.24452	2.2477	1.2281	.10879
.23	.22244	.26413	2.2345	1.2248	.11820
.24	.23948	.28411	2.2209	1.2213	.12792

TABLE D.4 (Continued)

M	T_0/T_0^*	T/T^*	p/p^*	p_0/p_0^*	V/V^*
.25	.25684	.30440	2.2069	1.2177	.13793
.26	.27446	.32496	2.1925	1.2140	.14821
.27	.29231	.34573	2.1777	1.2102	.15876
.28	.31035	.36667	2.1626	1.2064	.16955
.29	.32855	.38773	2.1472	1.2025	.18058
.30	.34686	.40887	2.1314	1.1985	.19183
.31	.36525	.43004	2.1154	1.1945	.20329
.32	.38369	.45119	2.0991	1.1904	.21494
.33	.40214	.47228	2.0825	1.1863	.22678
.34	.42057	.49327	2.0657	1.1821	.23879
.35	.43894	.51413	2.0487	1.1779	.25096
.36	.45723	.53482	2.0314	1.1737	.26327
.37	.47541	.55530	2.0140	1.1695	.27572
.38	.49346	.57553	1.9964	1.1652	.28828
.39	.51134	.59549	1.9787	1.1609	.30095
.40	.52903	.61515	1.9608	1.1566	.31372
.41	.54651	.63448	1.9428	1.1523	.32658
.42	.56376	.65345	1.9247	1.1480	.33951
.43	.58075	.67205	1.9065	1.1437	.35251
.44	.59748	.69025	1.8882	1.1394	.36556
.45	.61393	.70803	1.8699	1.1351	.37865
.46	.63007	.72538	1.8515	1.1308	.39178
.47	.64589	.74228	1.8331	1.1266	.40493
.48	.66139	.75871	1.8147	1.1224	.41810
.49	.67655	.77466	1.7962	1.1182	.43127
.50	.69136	.79012	1.7778	1.1140	.44445
.51	.70581	.80509	1.7594	1.1099	.45761
.52	.71990	.81955	1.7410	1.1059	.47075
.53	.73361	.83351	1.7226	1.1019	.48387
.54	.74695	.84695	1.7043	1.0979	.49696
.55	.75991	.85987	1.6860	1.09397	.51001
.56	.77248	.87227	1.6678	1.09010	.52302
.57	.78467	.88415	1.6496	1.08630	.53597
.58	.79647	.89552	1.6316	1.08255	.54887
.59	.80789	.90637	1.6136	1.07887	.56170
.60	.81892	.91670	1.5957	1.07525	.57447
.61	.82956	.92653	1.5780	1.07170	.58716
.62	.83982	.93585	1.5603	1.06821	.59978
.63	.84970	.94466	1.5427	1.06480	.61232
.64	.85920	.95298	1.5253	1.06146	.62477
.65	.86833	.96081	1.5080	1.05820	.63713
.66	.87709	.96816	1.4908	1.05502	.64941
.67	.88548	.97503	1.4738	1.05192	.66159
.68	.89350	.98144	1.4569	1.04890	.67367
.69	.90117	.98739	1.4401	1.04596	.68564

TABLE D.4 (Continued)

M	T_0/T_0^*	T/T^*	p/p^*	p_0/p_0^*	V/V^*
.70	.90850	.99289	1.4235	1.04310	.69751
.71	.91548	.99796	1.4070	1.04033	.70927
.72	.92212	1.00260	1.3907	1.03764	.72093
.73	.92843	1.00682	1.3745	1.03504	.73248
.74	.93442	1.01062	1.3585	1.03253	.74392
.75	.94009	1.01403	1.3427	1.03010	.75525
.76	.94546	1.01706	1.3270	1.02776	.76646
.77	.95052	1.01971	1.3115	1.02552	.77755
.78	.95528	1.02198	1.2961	1.02337	.78852
.79	.95975	1.02390	1.2809	1.02131	.79938
.80	.96394	1.02548	1.2658	1.01934	.81012
.81	.96786	1.02672	1.2509	1.01746	.82075
.82	.97152	1.02763	1.2362	1.01569	.83126
.83	.97492	1.02823	1.2217	1.01399	.84164
.84	.97807	1.02853	1.2073	1.01240	.85190
.85	.98097	1.02854	1.1931	1.01091	.86204
.86	.98363	1.02826	1.1791	1.00951	.87206
.87	.98607	1.02771	1.1652	1.00819	.88196
.88	.98828	1.02690	1.1515	1.00698	.89175
.89	.99028	1.02583	1.1380	1.00587	.90142
.90	.99207	1.02451	1.1246	1.04485	.91097
.91	.99366	1.02297	1.1114	1.00393	.92039
.92	.99506	1.02120	1.09842	1.00310	.92970
.93	.99627	1.01921	1.08555	1.00237	.93889
.94	.99729	1.01702	1.07285	1.00174	.94796
.95	.99814	1.01463	1.06030	1.00121	.95692
.96	.99883	1.01205	1.04792	1.00077	.96576
.97	.99935	1.00929	1.03570	1.00043	.97449
.98	.99972	1.00636	1.02364	1.00019	.98311
.99	.99993	1.00326	1.01174	1.00004	.99161
1.00	1.00000	1.00000	1.00000	1.00000	1.00000
1.01	.99993	.99659	.98841	1.00004	1.00828
1.02	.99973	.99304	.97697	1.00019	1.01644
1.03	.99940	.98936	.96569	1.00043	1.02450
1.04	.99895	.98553	.95456	1.00077	1.03246
1.05	.99838	.98161	.94358	1.00121	1.04030
1.06	.99769	.97755	.93275	1.00175	1.04804
1.07	.99690	.97339	.92206	1.00238	1.05567
1.08	.99600	.96913	.91152	1.00311	1.06320
1.09	.99501	.96477	.90112	1.00394	1.07062
1.10	.99392	.96031	.89086	1.00486	1.07795
1.11	.99274	.95577	.88075	1.00588	1.08518
1.12	.99148	.95115	.87078	1.00699	1.09230
1.13	.99013	.94646	.86094	1.00820	1.09933
1.14	.98871	.94169	.85123	1.00951	1.10626

TABLE D.4 (Continued)

M	T_0/T_0^*	T/T^*	p/p^*	p_0/p_0^*	V/V^*
1.15	.98721	.93685	.84166	1.01092	1.1131
1.16	.98564	.93195	.83222	1.01243	1.1198
1.17	.98400	.92700	.82292	1.01403	1.1264
1.18	.98230	.92200	.81374	1.01572	1.1330
1.19	.98054	.91695	.80468	1.01752	1.1395
1.20	.97872	.91185	.79576	1.01941	1.1459
1.21	.97685	.90671	.78695	1.02140	1.1522
1.22	.97492	.90153	.77827	1.02348	1.1584
1.23	.97294	.89632	.76971	1.02566	1.1645
1.24	.97092	.89108	.76127	1.02794	1.1705
1.25	.96886	.88581	.75294	1.03032	1.1764
1.26	.96675	.88052	.74473	1.03280	1.1823
1.27	.96461	.87521	.73663	1.03536	1.1881
1.28	.96243	.86988	.72865	1.03803	1.1938
1.29	.96022	.86453	.72078	1.04080	1.1994
1.30	.95798	.85917	.71301	1.04365	1.2050
1.31	.95571	.85380	.70535	1.04661	1.2105
1.32	.95341	.84843	.69780	1.04967	1.2159
1.33	.95108	.84305	.69035	1.05283	1.2212
1.34	.94873	.83766	.68301	1.05608	1.2264
1.35	.94636	.83227	.67577	1.05943	1.2316
1.36	.94397	.82698	.66863	1.06288	1.2367
1.37	.94157	.82151	.66159	1.06642	1.2417
1.38	.93915	.81613	.65464	1.07006	1.2467
1.39	.93671	.81076	.64778	1.07380	1.2516
1.40	.93425	.80540	.64102	1.07765	1.2564
1.41	.93178	.80004	.63436	1.08159	1.2612
1.42	.92931	.79469	.62779	1.08563	1.2659
1.43	.92683	.78936	.62131	1.08977	1.2705
1.44	.92434	.78405	.61491	1.09400	1.2751
1.45	.92184	.77875	.60860	1.0983	1.2796
1.46	.91933	.77346	.60237	1.1028	1.2840
1.47	.91682	.76819	.59623	1.1073	1.2884
1.48	.91431	.76294	.59018	1.1120	1.2927
1.49	.91179	.75771	.58421	1.1167	1.2970
1.50	.90928	.75250	.57831	1.1215	1.3012
1.51	.90676	.74731	.57250	1.1264	1.3054
1.52	.90424	.74215	.56677	1.1315	1.3095
1.53	.90172	.73701	.56111	1.1367	1.3135
1.54	.89920	.73189	.55553	1.1420	1.3175
1.55	.89669	.72680	.55002	1.1473	1.3214
1.56	.89418	.72173	.54458	1.1527	1.3253
1.57	.89167	.71669	.53922	1.1582	1.3291
1.58	.88917	.71168	.53393	1.1639	1.3329
1.59	.88668	.70669	.52871	1.1697	1.3366

TABLE D.4 (Continued)

M	T_0/T_0^*	T/T^*	p/p^*	p_0/p_0^*	V/V^*
1.60	.88419	.70173	.52356	1.1756	1.3403
1.61	.88170	.69680	.51848	1.1816	1.3439
1.62	.87922	.69190	.51346	1.1877	1.3475
1.63	.87675	.68703	.50851	1.1939	1.3511
1.64	.87429	.68219	.50363	1.2002	1.3546
1.65	.87184	.67738	.49881	1.2066	1.3580
1.66	.86940	.67259	.49405	1.2131	1.3614
1.67	.86696	.66784	.48935	1.2197	1.3648
1.68	.86453	.66312	.48471	1.2264	1.3681
1.69	.86211	.65843	.48014	1.2332	1.3713
1.70	.85970	.65377	.47563	1.2402	1.3745
1.71	.85731	.64914	.47117	1.2473	1.3777
1.72	.85493	.64455	.46677	1.2545	1.3809
1.73	.85256	.63999	.46242	1.2618	1.3840
1.74	.85020	.63546	.45813	1.2692	1.3871
1.75	.84785	.63096	.45390	1.2767	1.3901
1.76	.84551	.62649	.44972	1.2843	1.3931
1.77	.84318	.62205	.44559	1.2920	1.3960
1.78	.84087	.61765	.44152	1.2998	1.3989
1.79	.83857	.61328	.43750	1.3078	1.4018
1.80	.83628	.60894	.43353	1.3159	1.4046
1.81	.83400	.60463	.42960	1.3241	1.4074
1.82	.83174	.60036	.42573	1.3324	1.4102
1.83	.82949	.59612	.42191	1.3408	1.4129
1.84	.82726	.59191	.41813	1.3494	1.4156
1.85	.82504	.58773	.41440	1.3581	1.4183
1.86	.82283	.58359	.41072	1.3669	1.4209
1.87	.82064	.57948	.40708	1.3758	1.4235
1.88	.81846	.57540	.40349	1.3848	1.4261
1.89	.81629	.57135	.39994	1.3940	1.4286
1.90	.81414	.56734	.39643	1.4033	1.4311
1.91	.81200	.56336	.39297	1.4127	1.4336
1.92	.80987	.55941	.38955	1.4222	1.4360
1.93	.80776	.55549	.38617	1.4319	1.4384
1.94	.80567	.55160	.38283	1.4417	1.4408
1.95	.80359	.54774	.37954	1.4516	1.4432
1.96	.80152	.54391	.37628	1.4616	1.4455
1.97	.79946	.54012	.37306	1.4718	1.4478
1.98	.79742	.53636	.36988	1.4821	1.4501
1.99	.79540	.53263	.36674	1.4925	1.4523
2.00	.79339	.52893	.36364	1.5031	1.4545
2.01	.79139	.52526	.36057	1.5138	1.4567
2.02	.78941	.52161	.35754	1.5246	1.4589
2.03	.78744	.51800	.35454	1.5356	1.4610
2.04	.78549	.51442	.35158	1.5467	1.4631

TABLE D.4 (Continued)

M	T_0/T_0^*	T/T^*	p/p^*	p_0/p_0^*	V/V^*
2.05	.78355	.51087	.34866	1.5579	1.4652
2.06	.78162	.50735	.34577	1.5693	1.4673
2.07	.77971	.50386	.34291	1.5808	1.4694
2.08	.77781	.50040	.34009	1.5924	1.4714
2.09	.77593	.49697	.33730	1.6042	1.4734
2.10	.77406	.49356	.33454	1.6161	1.4753
2.11	.77221	.49018	.33181	1.6282	1.4773
2.12	.77037	.48683	.32912	1.6404	1.4792
2.13	.76854	.48351	.32646	1.6528	1.4811
2.14	.76673	.48022	.32383	1.6653	1.4830
2.15	.76493	.47696	.32122	1.6780	1.4849
2.16	.76314	.47373	.31864	1.6908	1.4867
2.17	.76137	.47052	.31610	1.7037	1.4885
2.18	.75961	.46734	.31359	1.7168	1.4903
2.19	.75787	.46419	.31110	1.7300	1.4921
2.20	.75614	.46106	.30864	1.7434	1.4939
2.21	.75442	.45796	.30621	1.7570	1.4956
2.22	.75271	.45489	.30381	1.7707	1.4973
2.23	.75102	.45184	.30143	1.7846	1.4990
2.24	.74934	.44882	.29908	1.7986	1.5007
2.25	.74767	.44582	.29675	1.8128	1.5024
2.26	.74602	.44285	.29445	1.8271	1.5040
2.27	.74438	.43990	.29218	1.8416	1.5056
2.28	.74275	.43698	.28993	1.8562	1.5072
2.29	.74114	.43409	.28771	1.8710	1.5088
2.30	.73954	.43122	.28551	1.8860	1.5104
2.31	.73795	.42837	.28333	1.9012	1.5119
2.32	.73638	.42555	.28118	1.9165	1.5134
2.33	.73482	.42276	.27905	1.9320	1.5150
2.34	.73327	.41999	.27695	1.9476	1.5165
2.35	.73173	.41724	.27487	1.9634	1.5180
2.36	.73020	.41451	.27281	1.9794	1.5195
2.37	.72868	.41181	.27077	1.9955	1.5209
2.38	.72718	.40913	.26875	2.0118	1.5223
2.39	.72569	.40647	.26675	2.0283	1.5237
2.40	.72421	.40383	.26478	2.0450	1.5252
2.41	.72274	.40122	.26283	2.0619	1.5266
2.42	.72129	.39863	.26090	2.0789	1.5279
2.43	.71985	.39606	.25899	2.0961	1.5293
2.44	.71842	.39352	.25710	2.1135	1.5306
2.45	.71700	.39100	.25523	2.1311	1.5320
2.46	.71559	.38850	.25337	2.1489	1.5333
2.47	.71419	.38602	.25153	2.1669	1.5346
2.48	.71280	.38356	.24972	2.1850	1.5359
2.49	.71142	.38112	.24793	2.2033	1.5372

TABLE D.4 (Continued)

M	T_0/T_0^*	T/T^*	p/p^*	p_0/p_0^*	V/V^*
2.50	.71005	.37870	.24616	2.2218	1.5385
2.51	.70870	.37630	.24440	2.2405	1.5398
2.52	.70736	.37392	.24266	2.2594	1.5410
2.53	.70603	.37157	.24094	2.2785	1.5422
2.54	.70471	.36923	.23923	2.2978	1.5434
2.55	.70340	.36691	.23754	2.3173	1.5446
2.56	.70210	.36461	.23587	2.3370	1.5458
2.57	.70081	.36233	.23422	2.3569	1.5470
2.58	.69953	.36007	.23258	2.3770	1.5482
2.59	.69825	.35783	.23096	2.3972	1.5494
2.60	.69699	.35561	.22936	2.4177	1.5505
2.61	.69574	.35341	.22777	2.4384	1.5516
2.62	.69450	.35123	.22620	2.4593	1.5527
2.63	.69327	.34906	.22464	2.4804	1.5538
2.64	.69205	.34691	.22310	2.5017	1.5549
2.65	.69084	.34478	.22158	2.5233	1.5560
2.66	.68964	.34267	.22007	2.5451	1.5571
2.67	.68845	.34057	.21857	2.5671	1.5582
2.68	.68727	.33849	.21709	2.5892	1.5593
2.69	.68610	.33643	.21562	2.6116	1.5603
2.70	.68494	.33439	.21417	2.6342	1.5613
2.71	.68378	.33236	.21273	2.6571	1.5623
2.72	.68263	.33035	.21131	2.6802	1.5633
2.73	.68150	.32836	.20990	2.7035	1.5644
2.74	.68038	.32638	.20850	2.7270	1.5654
2.75	.67926	.32442	.20712	2.7508	1.5663
2.76	.67815	.32248	.20575	2.7748	1.5673
2.77	.67704	.32055	.20439	2.7990	1.5683
2.78	.67595	.31864	.20305	2.8235	1.5692
2.79	.67487	.31674	.20172	2.8482	1.5702
2.80	.67380	.31486	.20040	2.8731	1.5711
2.81	.67273	.31299	.19909	2.8982	1.5721
2.82	.67167	.31114	.19780	2.9236	1.5730
2.83	.67062	.30931	.19652	2.9493	1.5739
2.84	.66958	.30749	.19525	2.9752	1.5748
2.85	.66855	.30568	.19399	3.0013	1.5757
2.86	.66752	.30389	.19274	3.0277	1.5766
2.87	.66650	.30211	.19151	3.0544	1.5775
2.88	.66549	.30035	.19029	3.0813	1.5784
2.89	.66449	.29860	.18908	3.1084	1.5792
2.90	.66350	.29687	.18788	3.1358	1.5801
2.91	.66252	.29515	.18669	3.1635	1.5809
2.92	.66154	.29344	.18551	3.1914	1.5818
2.93	.66057	.29175	.18435	3.2196	1.5826
2.94	.65961	.29007	.18320	3.2481	1.5834

TABLE D.4 (Continued)

M	T_0/T_0^*	T/T^*	p/p^*	p_0/p_0^*	V/V^*
2.95	.65865	.28841	.18205	3.2768	1.5843
2.96	.65770	.28676	.18091	3.3058	1.5851
2.97	.65676	.28512	.17978	3.3351	1.5859
2.98	.65583	.28349	.17867	3.3646	1.5867
2.99	.65490	.28188	.17757	3.3944	1.5875
3.00	.65398	.28028	.17647	3.4244	1.5882
3.50	.61580	.21419	.13223	5.3280	1.6198
4.00	.58909	.16831	.10256	8.2268	1.6410
4.50	.56983	.13540	.08177	12.502	1.6559
5.00	.55555	.11111	.06667	18.634	1.6667
6.00	.53633	.07849	.04669	38.946	1.6809
7.00	.52437	.05826	.03448	75.414	1.6896
8.00	.51646	.04491	.02649	136.62	1.6954
9.00	.51098	.03565	.02098	233.88	1.6993
10.00	.50702	.02897	.01702	381.62	1.7021
∞	.48980	0	0	∞	1.7143

TABLE D.5
Prandtl-Meyer Function

M	θ	μ	M	θ	μ
1.00	0	90.00	1.60	14.861	38.68
1.01	.04473	81.93	1.61	15.156	38.40
1.02	.1257	78.64	1.62	15.452	38.12
1.03	.2294	76.14	1.63	15.747	37.84
1.04	.3510	74.06	1.64	16.043	37.57
1.05	.4874	72.25	1.65	16.338	37.31
1.06	.6367	70.63	1.66	16.633	37.04
1.07	.7973	69.16	1.67	16.928	36.78
1.08	.9680	67.81	1.68	17.222	36.53
1.09	1.148	66.55	1.69	17.516	36.28
1.10	1.336	65.38	1.70	17.810	36.03
1.11	1.532	64.28	1.71	18.103	35.79
1.12	1.735	63.23	1.72	18.397	35.55
1.13	1.944	62.25	1.73	18.689	35.31
1.14	2.160	61.31	1.74	18.981	35.08
1.15	2.381	60.41	1.75	19.273	34.85
1.16	2.607	59.55	1.76	19.565	34.62
1.17	2.839	58.73	1.77	19.855	34.40
1.18	3.074	57.94	1.78	20.146	34.18
1.19	3.314	57.18	1.79	20.436	33.96
1.20	3.558	56.44	1.80	20.725	33.75
1.21	3.806	55.74	1.81	21.014	33.54
1.22	4.057	55.05	1.82	21.302	33.33
1.23	4.312	54.39	1.83	21.590	33.12
1.24	4.569	53.75	1.84	21.877	32.92
1.25	4.830	53.13	1.85	22.163	32.72
1.26	5.093	52.53	1.86	22.449	32.52
1.27	5.359	51.94	1.87	22.735	32.33
1.28	5.627	51.38	1.88	23.019	32.13
1.29	5.898	50.82	1.89	23.303	31.94
1.30	6.170	50.28	1.90	23.586	31.76
1.31	6.445	49.76	1.91	23.869	31.57
1.32	6.721	49.25	1.92	24.151	31.39
1.33	7.000	48.75	1.93	24.432	31.21
1.34	7.280	48.27	1.94	24.712	31.03
1.35	7.561	47.79	1.95	24.992	30.85
1.36	7.844	47.33	1.96	25.271	30.68
1.37	8.128	46.88	1.97	25.549	30.51
1.38	8.413	46.44	1.98	25.827	30.33
1.39	8.699	46.01	1.99	26.104	30.17
1.40	8.987	45.58	2.00	26.380	30.00
1.41	9.276	45.17	2.01	26.655	29.84
1.42	9.565	44.77	2.02	26.929	29.67
1.43	9.855	44.37	2.03	27.203	29.51
1.44	10.146	43.98	2.04	27.476	29.35
1.45	10.438	43.60	2.05	27.748	29.20
1.46	10.731	43.23	2.06	28.020	29.04
1.47	11.023	42.86	2.07	28.290	28.89
1.48	11.317	42.51	2.08	28.560	28.74
1.49	11.611	42.16	2.09	28.829	28.59
1.50	11.905	41.81	2.10	29.097	28.44
1.51	12.200	41.47	2.11	29.364	28.29
1.52	12.495	41.14	2.12	29.631	28.14
1.53	12.790	40.81	2.13	29.897	28.00
1.54	13.086	40.49	2.14	30.161	27.86
1.55	13.381	40.18	2.15	30.425	27.72
1.56	13.677	39.87	2.16	30.689	27.58
1.57	13.973	39.56	2.17	30.951	27.44
1.58	14.269	39.27	2.18	31.212	27.30
1.59	14.564	38.97	2.19	31.473	27.17

TABLE D.5 (Continued)

M	θ	μ	M	θ	μ
2.20	31.732	27.04	2.85	46.778	20.54
2.21	31.991	26.90	2.86	46.982	20.47
2.22	32.250	26.77	2.87	47.185	20.39
2.23	32.507	26.64	2.88	47.388	20.32
2.24	32.763	26.51	2.89	47.589	20.24
2.25	33.018	26.39	2.90	47.790	20.17
2.26	33.273	26.26	2.91	47.990	20.10
2.27	33.527	26.14	2.92	48.190	20.03
2.28	33.780	26.01	2.93	48.388	19.96
2.29	34.032	25.89	2.94	48.586	19.89
2.30	34.283	25.77	2.95	48.783	19.81
2.31	34.533	25.65	2.96	48.980	19.75
2.32	34.783	25.53	2.97	49.175	19.68
2.33	35.031	25.42	2.98	49.370	19.61
2.34	35.279	25.30	2.99	49.564	19.54
2.35	35.526	25.18	3.00	49.757	19.47
2.36	35.771	25.07	3.01	49.950	19.40
2.37	36.017	24.96	3.02	50.142	19.34
2.38	36.261	24.85	3.03	50.333	19.27
2.39	36.504	24.73	3.04	50.523	19.20
2.40	36.746	24.62	3.05	50.713	19.14
2.41	36.988	24.52	3.06	50.902	19.07
2.42	37.229	24.41	3.07	51.090	19.01
2.43	37.469	24.30	3.08	51.277	18.95
2.44	37.708	24.19	3.09	51.464	18.88
2.45	37.946	24.09	3.10	51.650	18.82
2.46	38.183	23.99	3.11	51.835	18.76
2.47	38.420	23.88	3.12	52.020	18.69
2.48	38.655	23.78	3.13	52.203	18.63
2.49	38.890	23.68	3.14	52.386	18.57
2.50	39.124	23.58	3.15	52.569	18.51
2.51	39.357	23.48	3.16	52.751	18.45
2.52	39.589	23.38	3.17	52.931	18.39
2.53	39.820	23.28	3.18	53.112	18.33
2.54	40.050	23.18	3.19	53.292	18.27
2.55	40.280	23.09	3.20	53.470	18.21
2.56	40.509	22.99	3.21	53.648	18.15
2.57	40.736	22.91	3.22	53.826	18.09
2.58	40.963	22.81	3.23	54.003	18.03
2.59	41.189	22.71	3.24	54.179	17.98
2.60	41.415	22.62	3.25	54.355	17.92
2.61	41.639	22.53	3.26	54.529	17.86
2.62	41.863	22.44	3.27	54.703	17.81
2.63	42.086	22.35	3.28	54.877	17.75
2.64	42.307	22.26	3.29	55.050	17.70
2.65	42.529	22.17	3.30	55.222	17.64
2.66	42.749	22.08	3.31	55.393	17.58
2.67	42.968	22.00	3.32	55.564	17.53
2.68	43.187	21.91	3.33	55.734	17.48
2.69	43.405	21.82	3.34	55.904	17.42
2.70	43.621	21.74	3.35	56.073	17.37
2.71	43.838	21.65	3.36	56.241	17.31
2.72	44.053	21.57	3.37	56.409	17.26
2.73	44.267	21.49	3.38	56.576	17.21
2.74	44.481	21.41	3.39	56.742	17.16
2.75	44.694	21.32	3.40	56.907	17.10
2.76	44.906	21.24	3.41	57.073	17.05
2.77	45.117	21.16	3.42	57.237	17.00
2.78	45.327	21.08	3.43	57.401	16.95
2.79	45.537	21.00	3.44	57.564	16.90
2.80	45.746	20.92	3.45	57.726	16.85
2.81	45.954	20.85	3.46	57.888	16.80
2.82	46.161	20.77	3.47	58.050	16.75
2.83	46.368	20.69	3.48	58.210	16.70
2.84	46.573	20.62	3.49	58.370	16.65

TABLE D.5 (Continued)

M	θ	μ	M	θ	μ
3.50	58.530	16.60	4.15	67.713	13.94
3.51	58.689	16.55	4.16	67.838	13.91
3.52	58.847	16.51	4.17	67.963	13.88
3.53	59.004	16.46	4.18	68.087	13.84
3.54	59.162	16.41	4.19	68.210	13.81
3.55	59.318	16.36	4.20	68.333	13.77
3.56	59.474	16.31	4.21	68.456	13.74
3.57	59.629	16.27	4.22	68.578	13.71
3.58	59.784	16.22	4.23	68.700	13.67
3.59	59.938	16.17	4.24	68.821	13.64
3.60	60.091	16.13	4.25	68.942	13.61
3.61	60.244	16.08	4.26	69.063	13.58
3.62	60.397	16.04	4.27	69.183	13.54
3.63	60.549	15.99	4.28	69.302	13.51
3.64	60.700	15.95	4.29	69.422	13.48
3.65	60.851	15.90	4.30	69.541	13.45
3.66	61.000	15.86	4.31	69.659	13.42
3.67	61.150	15.81	4.32	69.777	13.38
3.68	61.299	15.77	4.33	69.895	13.35
3.69	61.447	15.72	4.34	70.012	13.32
3.70	61.595	15.68	4.35	70.128	13.29
3.71	61.743	15.64	4.36	70.245	13.26
3.72	61.889	15.59	4.37	70.361	13.23
3.73	62.036	15.55	4.38	70.476	13.20
3.74	62.181	15.51	4.39	70.591	13.17
3.75	62.326	15.47	4.40	70.706	13.14
3.76	62.471	15.42	4.41	70.820	13.11
3.77	62.615	15.38	4.42	70.934	13.08
3.78	62.758	15.34	4.43	71.048	13.05
3.79	62.901	15.30	4.44	71.161	13.02
3.80	63.044	15.26	4.45	71.274	12.99
3.81	63.186	15.22	4.46	71.386	12.96
3.82	63.327	15.18	4.47	71.498	12.93
3.83	63.468	15.14	4.48	71.610	12.90
3.84	63.608	15.10	4.49	71.721	12.87
3.85	63.748	15.06	4.50	71.832	12.84
3.86	63.887	15.02	4.51	71.942	12.81
3.87	64.026	14.98	4.52	72.052	12.78
3.88	64.164	14.94	4.53	72.162	12.75
3.89	64.302	14.90	4.54	72.271	12.73
3.90	64.440	14.86	4.55	72.380	12.70
3.91	64.576	14.82	4.56	72.489	12.67
3.92	64.713	14.78	4.57	72.597	12.64
3.93	64.848	14.74	4.58	72.705	12.61
3.94	64.983	14.70	4.59	72.812	12.58
3.95	65.118	14.67	4.60	72.919	12.56
3.96	65.253	14.63	4.61	73.026	12.53
3.97	65.386	14.59	4.62	73.132	12.50
3.98	65.520	14.55	4.63	73.238	12.47
3.99	65.652	14.52	4.64	73.344	12.45
4.00	65.785	14.48	4.65	73.449	12.42
4.01	65.917	14.44	4.66	73.554	12.39
4.02	66.048	14.40	4.67	73.659	12.37
4.03	66.179	14.37	4.68	73.763	12.34
4.04	66.309	14.33	4.69	73.867	12.31
4.05	66.439	14.30	4.70	73.970	12.28
4.06	66.569	14.26	4.71	74.073	12.26
4.07	66.698	14.22	4.72	74.176	12.23
4.08	66.826	14.19	4.73	74.279	12.21
4.09	66.954	14.15	4.74	74.381	12.18
4.10	67.082	14.12	4.75	74.483	12.15
4.11	67.209	14.08	4.76	74.584	12.13
4.12	67.336	14.05	4.77	74.685	12.10
4.13	67.462	14.01	4.78	74.786	12.08
4.14	67.588	13.98	4.79	74.886	12.05

TABLE D.5 (Continued)

M	θ	μ	M	θ	μ
4.80	74.986	12.03	4.90	75.969	11.78
4.81	75.086	12.00	4.91	76.066	11.75
4.82	75.186	11.97	4.92	76.162	11.73
4.83	75.285	11.95	4.93	76.258	11.70
4.84	75.383	11.92	4.94	76.353	11.68
4.85	75.482	11.90	4.95	76.449	11.66
4.86	75.580	11.87	4.96	76.544	11.63
4.87	75.678	11.85	4.97	76.638	11.61
4.88	75.775	11.83	4.98	76.732	11.58
4.89	75.872	11.80	4.99	76.826	11.56
			5.00	76.920	11.54

Answers to Selected Even-Numbered Problems

CHAPTER 1

- **1.2b** 0.0517 lb
- **1.6** $40/\sqrt{10}$
- **1.8a** $\frac{1}{2}e^{2x} + y^2 z + C$
- **b** 6
- **1.10a** $8\hat{j}$
- **b** $16\hat{i}$
- **1.16a** 0.748
- **b** 0
- **1.18** $15\pi \cos \frac{\pi}{2} t$
- **1.20** 0.084 ft
- **1.22** -12.2 psi; 339 in. of H_2O
- **1.24** 0.7 ft
- **1.26** 4.4 psia
- **1.28** 18 ft
- **1.32** 6.74 ft
- **1.36** 16.75 psi
- **1.38a** 393 lb
- **b** 162 rpm
- **1.40** 339 ft; No
- **1.42** $0.000915/x$
- **1.44** 1.54 in.
- **1.48a** 3.41 ft
- **b** 0.019 ft
- **1.50** 4.99 sec
- **1.52** Yes

CHAPTER 2

- **2.6** 11,700 slug/sec
- **2.8** $-\rho \frac{\pi}{4}\left[(d_2^2 - d^2)\dot{h}_2 + \left(d_1 + \frac{2}{\sqrt{3}} h_1\right)^2 \dot{h}_1\right]$
- **2.10** $0.177 d_1^2 V_1 / Rw$
- **2.12** 3.33 in.
- **2.14a** 1.83 cfs
- **b** 3.58 cfs
- **2.16** 13.1 ft
- **2.18** $0.423 \dot{m}_1 V_1$
- **2.20** 30.1 fps
- **2.22** 30.7 hp; 48.2°

2.24	1.88 fps
2.26	1240 lb
2.28	169 lb
2.30	$M\dot{V}_c = 0.169(200 - V_c)^2$
2.32	0.575 in.
2.34	62.8 rad/sec
2.36	14.2 cfs
2.38	8.14 ft
2.40a	2200 psf
b	8740 psf
2.42	1.99 hp added
2.44	180 hp
2.46	2.0
2.48	52.5 lb
2.50	21.6 hp

CHAPTER 3

3.6	y^2
3.8	$C/(1 - e^{-C_2 x})$

CHAPTER 4

4.4a	400 fps
b	500 fps
c	20 lb
4.6	48.8 gal/min
4.8	25
4.10	4.44 rad/sec
4.12	56 knots; 600 knots
4.14a	1000 cfs; 100 rpm
b	1/5
4.18	$\dfrac{P}{\rho \omega^3 D^5} = f\left(\dfrac{\omega D^2 \rho}{\mu}\right)$
4.20	$\dfrac{\Delta p}{\rho V^2} =$ $f\left(\dfrac{V\rho d}{\mu}, \dfrac{\omega d}{V}, \dfrac{A}{d^2}, \dfrac{D}{d}\right)$
4.22	No dimensionless numbers
4.24	$\phi =$ $f\left(\dfrac{\omega d^2}{\nu}, \dfrac{D}{d}, \dfrac{h}{d}, \dfrac{w}{d}, \dfrac{q}{\omega d^3}, \dfrac{R}{d}\right)$

CHAPTER 5

5.4	0.002 fps
5.16	254 fps

CHAPTER 6

6.2	1610 psf
6.4	−0.0023
6.6	136 psf
6.8	$u = \dfrac{\lambda}{4}(r^2 - r_0^2) + U$
6.10	0.000485 psf
6.12	$\tau_0 = 6\mu V/h$
6.14	0.48 fps
6.16	20.8 psf
6.18	$u =$ $\dfrac{1}{2\mu}\dfrac{dp}{dx}(y^2 - hy) + \dfrac{U}{h}y$
6.20	0.0377 ft; 0.408 psf
6.22	0.000873 cfs
6.24	$u =$ $\dfrac{U}{1 - e^{k_2(T_2 - T_1)}}[1 - e^{-k_2(T_2 - T_1)(\frac{y}{h} - 1)}]$
6.26	3.92 ft-lb
6.28	$\mu = T\delta/2\pi R^3 \Omega H$
6.34	$u = e^{t/100} e^{10y}$
6.36	126 ft^{-1}

CHAPTER 7

7.6	$u = -\dfrac{c_1}{y} + c_2$
7.12	0.0414 fps; 174 lb
7.14	5.4 hp
7.18	1.0
7.20	7 in.
7.22	0.0865 cfs

CHAPTER 8

8.8	11.17 psi; 9.15 psi
8.14	$-A\theta$; $\mu \cos \theta / r$
8.18	$\mu \cos \theta / r$
8.20	1600 lb
8.24	(0, −3.94 ft)
8.28	$-\dfrac{q}{4\pi} \cos \theta$
8.30	$u =$ $\dfrac{q}{4\pi}(x^2 + y^2 + z^2)^{-3/2} x$
8.32	30 fps
8.34	2290 lb

CHAPTER 9

- **9.2** $\dfrac{1600}{40-3x}$; $-\dfrac{15{,}200}{(40-3x)^3}$
- **9.4** 0.875 in.
- **9.6** 0.375δ
- **9.8** $-\int_0^y \dfrac{\partial u}{\partial x}\,dy$
- **9.12** 14.4 lb
- **9.14** 3; 2.5
- **9.16** $1.83\sqrt{\dfrac{\mu x}{\rho U_\infty}}$; $0.73\sqrt{\dfrac{\mu x}{\rho U_\infty}}$
- **9.18** 0; 0.00127 fps; 0.00275 fps; 0.00275 fps
- **9.22a** 5260 ft
  - **b** 0.000987 ft
  - **c** 399 ft
- **9.24a** 8.95×10^{-5} ft
  - **b** 44.7 fps
  - **c** 0.0155 ft
  - **d** 0.103 ft
- **9.26** 0.904 hp

CHAPTER 10

- **10.4** Yes
- **10.6** 928 lb; up to 3.2 fps
- **10.8** 30,000 lb
- **10.10** 1.125
- **10.12** 39 mph
- **10.14a** 220 mph **b** 88 mph
- **10.16** 36%
- **10.18a** 4370 lb **b** 80.4 fps
- **10.20** 20%

CHAPTER 11

- **11.2** 654 fps
- **11.4** 0.885; 0.908; 0.908
- **11.6a** 0.322 fps **b** 0.061°
- **11.8** 37.4 psia; 0.00404 slug/ft^3; 833 fps
- **11.12** 2870 fps; 805 fps; 95.6 psia; 1120°R; 0.00717 slug/ft^3; 2.7×10^6 ft-lb/slug
- **11.14** 27.8 psia; 0.0151 slug/sec
- **11.16** 620 fps; 0.00243 slug/sec
- **11.18** 5.88 in.
- **11.20a** 4.37 psia
  - **b** 1030 fps; 7.77 psia
  - **c** 2010 fps; 0.441 psia
  - **d** 528 fps; 4.37 psia
- **11.22** 21,600 lb
- **11.24** 145 lb
- **11.26** 2.09 ft
- **11.28a** 9.65 ft **b** 1.66
- **11.30a** 1.27 **b** 0.170
- **11.32** 1.81 Btu/sec
- **11.34** 5.6 in.
- **11.36a** 1.34; 16.9 psia; 50.6 psia
  - **b** 0.670; 32.4 psia; 43.7 psia
- **11.38** 43°; 1.54
- **11.40** 0.799 psia; 236°R; 3250 fps; 26.1°
- **11.42** 0.0041

Index

Abbott, I. H., 473
Abernathy, F. H., 443
Acceleration, 29
  along a streamline, 50
  noninertial reference frame, 32
  substantial derivative, 30
Acosta, A. J., 165, 204, 399
Acoustics, 11, 348
Added mass, 470
  coefficient, 471
Adiabatic flow
  choking in, 509, 510
  with friction, 507
Adiabatic stagnation temperature, 485
Aerodynamic lift, 454
Air, properties of, 6, 8, 540
Angular momentum, 108
Average velocity
  between parallel plates, 284
  in a pipe, 278
Avogadro, A., 58
Axisymmetric flow, 386

Barbin, A. R., 332
Batchelor, G. K., 68, 204, 269, 305, 399
Bernoulli, D., 49, 175, 362, 392
Bernoulli constant, 50
  in rotational flow, 175
Bernoulli equation, 49, 174, 358
  in a gravitational field, 49
  in rotational flow, 175
  in an incompressible gas flow, 56
  in an inviscid flow, 50
  valid along a streamline, 175, 358
  valid throughout a region, 50, 362
Bird, R. B., 305
Blasius, H., 411, 413, 429
Blasius' formula, 429
Blasius solutions of boundary-layer equations, 411, 413
Body force, 95
Boiling, 13
Bolz, R. E., 68

Boundary conditions, 194
  free surface, 195
  inlet, 209
  inviscid flow, 195
  no-slip, 194
  porous surface, 195
  temperature, 195
Boundary layer, 245, 401
  actual dimensions, 406
  dimensionless, 408
  flow outside of, 246, 401
  laminar, 411
  laminar sublayer, 423
  skin friction, 414
  thickness, 401
    for a laminar flow, 414
    influence of pressure gradient, 437
    for a turbulent flow, 427, 430
  turbulent, 404
Boundary-layer equations, 409
  Blasius solutions of, 413
    friction coefficient from, 414
Boundary-value problem, 206
Boundary velocity, 77
Boussinesq, J. M., 316
Bradshaw, P., 350
Brighton, J. A., 69, 165
British thermal unit, 121
Bubble, pressure inside, 48
Buffer zone, 423
Bulk modulus of elasticity, 12

Capillary tube, 48
Carrier, G., 306
Cartesian coordinate system, 16
Cauchy, A. L., 364
Cauchy-Riemann equations, 364
Cavitation, 13, 185, 265, 449
  fixed, 265
  number, 266, 469
  supercavitation, 265
  traveling, 265
  vibratory, 265
  vortex, 265

INDEX

Center of pressure, 40
Characteristic quantity, 207
Choking
  in adiabatic flow, 509, 510
  in isothermal flow, 511
  in nozzle flow, 493
Circulation, 19, 250, 455
  in the boundary layer, 258
  time rate of change of, 255
  vortex, 371
Clauser, F. H., 425, 426, 444
Cockrell, D. J., 444
Coles, D. C., 444
Compressible flow, 220, 263, 474
  in a constant-area duct, 505
  in a nozzle, 490
  through a normal shock, 481
  through an oblique shock, 518
Compression waves, 523
Concentric cylinders, flow between, 242
Conduit flows, 294
Confined flow, 207
Conservation of mass
  differential equation, 168
  integral equation, 88
Conservative body force, 174
Conservative vector field, 20
Constitutive equations, 12, 185
Contact angle, 48
Continuity equation
  differential, 168
    cartesian coordinates, 169, 196
    cylindrical coordinates, 196
    spherical coordinates, 196
  fluctuating, 310
  integral, 88
  time-averaged, 310
Continuum, 5
Contraction loss coefficient, 341, 343
Control surface, 76, 80
Control volume, 72
  for a boundary layer, 415
  deformable, 78
  nondeformable, 74, 84
  steady flow, 84
Convective rate of change, 30
Converging-diverging nozzle, 264, 492
Coordinate transformations, 18
Corioles, G. D., 253, 260
Coriolis acceleration, 32
Correlation coefficient, 312, 318
Correlation functions, 320
Corrisin, S., 350
Couette, M., 288
Couette flow, 288
Creeping flow, 238
Critical area, 496

Critical pressure, 496
Critical velocity, 497
Cross product, 17
Curl, 17
Cylinder, 377
  with circulation, 381
Cylindrical coordinates, 196

Dailey, J. W., 68, 204, 236, 350, 444, 461
D'Alembert, J., 400
D'Alembert's paradox, 400
Deflection angle in oblique shocks, 519
Deformable control volume, 78
De Laval nozzle, 494
Density, 6
Derivative
  material, 29
  substantial, 30
  total, 30
Detached shocks, 522
Development length, 331
Diffuser, 355, 443
  stall regimes, 356
Diffusion of vorticity, 253
Diffusive process, 253
Dimensional analysis, 226
Dimensional constant, 14
Dimensionless integral, 213
Dimensionless variables, 207, 223
Dimensions, 14, 540
Displacement thickness, 414
  from Blasius solution, 415
  use of, 441
Dissipation in turbulent flow, 321
Dissipation function, 192
Divergence, 17
Dividing stream surface, 450
Dot product, 16
Doublet, 373
Drag coefficient, 454
  of airfoil, 466, 469
  of cylinder, 460
  of sphere, 458
  for various objects, 462
Drag force
  effect of added mass, 471
  on a body, 446
  due to a boundary layer, 414
  effect of cavitation, 468
  effect of compressibility, 468
  equations for, 454
  form drag, 457
  in a potential flow, 379
  pressure drag, 457
  viscous drag, 457
Droplet, pressure inside a, 48
Dryden, H. L., 444

INDEX

Eddy viscosity, 316
 related to mixing length, 318
Energy, internal, 8
 chemical, 9
 electrical, 9
 kinetic, 9
 magnetic, 9
 potential, 9
 specific, 112
 thermal, 9
Energy equation
 differential, 177, 192
  cartesian coordinates, 198
  cylindrical coordinates, 198
  spherical coordinates, 198
 integral, 112
Ensemble average, 309
Enthalpy, 8, 475
Entrance length
 in a channel, 284
 in a pipe, 275, 332
Entrance loss, 341
Entropy, 476
 rise across shock, 484
Equation of state, 8, 192
Equations of motion, 188
 for compressible flow, 220
 in cylindrical coordinates, 197
 in spherical coordinates, 197
Eskinazi, S., 399
Euler, L., 173, 353
Euler's equations, 173
 along a streamline, 356
 normal to a streamline, 356
Eulerian description of flow, 26, 29
Expansion waves, 523
Extensive properties, 7, 72

Fanno line, 510
Feliss, N. A., 296, 299
Fellows, J. R., 351
First law of thermodynamics, 8
Flat plate, laminar oscillations above, 289
Flow, types of. *See* Types of flow
Flow net, 380
Flow rate, 89
Flow work, 117
Fluid, definition of, 4
Fluid horsepower, 127
Fluidic resistor, 295
Fluid statics, 36
 accelerating reference frame, 43
Flux term, 76, 81
Force
 on a plane surface, 39
 on a curved surface, 41
Form drag, 457

Foss, J. F., 162
Fourier-Biot material, 195
Fox, R. W., 164
Free-stream flow, 191
Free-stream fluctuations, 191
Free-stream turbulence, 438
Free-stream velocity, 402
Friction coefficient, flat plate, 414, 426
Friction factor, 277
 flow between parallel plates, 284
 laminar flow in a pipe, 278
 turbulent flow in a pipe, 339
Friction loss
 in enlargements, 342
 in pipeline components, 340
Friction velocity, 330
Frictionless flow with heat transfer, 512
Froude, W., 215
Froude number, 215
 densiometric, 215
Fully developed flow, 271
 in a channel, 284
 in a pipe, 275, 331

g_c (dimensional constant), 14
Gas, definition of, 4
Gas constant, 8
Gas tables, 547
Gauss, C. F., 19, 168, 177
Gauss' theorem, 19, 177
Geometric similarity, 209
Gibbs' phase rule, 7
Giles, R. V., 68
Gradient, 17
Gravity vector, 189

Half-body, 376
Hamilton, J. B., 351
Hawkins, G. A., 68, 113
Hansen, A. G., 68, 164, 536
Harleman, D. R. F., 68, 204, 236, 350, 444, 461
Hauptman, E. G., 165, 204, 399
Head loss, 119, 277
Head-loss coefficient, 120, 341
Heat, 8, 113
 rate of heat flow, 112
Heat addition in a duct, 512
Hele-Shaw flow, 392
Heyer, D. E., 162
Hinze, J. O., 328, 333, 350, 444
Hirst, E. A., 444
Hoffman, R. L., 68
Homogeneous fluids, 186
Horner, S. F., 473
Hot-wire anemometer, 13
Housner, G. W., 68

Hudson, D. E., 68
Hughes, W. F., 67, 165
Hugoniot, H., 529
Hydraulic jump, 147
Hydraulic diameter, 207, 342
Hydraulically smooth surfaces, 330

Ideal fluid, 353
Images, 384
Incompressible fluid, 169
Incompressible air flow, 56, 498
Inertial body force, 95
Infinitesimal fluid element, 36
Instability, 241, 295
  between parallel plates, 296
  in a pipe, 296
Intensive properties, 7
Internal energy, 8, 113, 475
Inviscid flow, 173, 352
  boundary conditions, 195
  rotational, 390
Irrotational flow, 184, 353, 360
Irrotational vortex, 371
Isentropic process for a perfect gas, 11, 475
Isentropic stagnation pressure, 486
Isobaric process, 475
Isothermal flow, 511
  critical Mach number in, 511
Isothermal process, 475
Isotropic fluid, 185
Isotropic turbulence, 327

John, E. A. J., 536, 545
Jones, J. B., 113
Jones, J. R., 332
Joukowsky, N. E., 382

Kármán, integral equation, 417
  its use in calculations, 418
Kármán vortex street, 447
Karnitz, M. A., 296
Kayes, F. G., 68, 545
Keenan, J. H., 68, 545
Kinematic viscosity, 13, 189
Kinematics of flow, 25
Kinetic energy, 9, 113
  in turbulent flow, 321
Kinetic-energy correction factor, 137, 157, 342
Kinetic pressure, 208
Kleis, S. J., 346
Kline, S. J., 68, 236, 444
Kovasznay, L. S. G., 444
Kratz, A. P., 351
Kuethe, A. M., 473
Kutta, W., 382, 383, 455

Kutta condition, 383, 455
Kutta-Joukowsky theorem, 382

Lagrangian description of flow, 26, 29
Laminar flow, 190, 239
  in a boundary layer, 411
  in a channel, 282
  in a conduit, 294
  in a pipe, 274
  between parallel plates, 282
  between rotating cylinders, 286
Laminar sublayer, 423
Laplace, P. S., 18, 361, 367
Laplace equation, 18
  in cartesian coordinates, 361
Laplacian, 18
Laufer, J., 329, 336
Law of the wall, 422
Lay, J. E., 68
Leibniz, G. W., 76
Leibniz' rule, 144
Lift coefficient, 454
  of airfoil, 466
Lift, 446
  equations for, 454
Lightfoot, E. N., 305
Lighthill, M. J., 68, 269
Liquid, definition of, 4
Local rate of change, 30
Local skin friction coefficient, 414, 426
  Blasius' formula, 429
  power law, 430
Logarithmic velocity profile, 423
Loss cofficient, 120
Lumley, J. L., 68, 204, 350

Mach, E., 221, 479
Mach angle, 480
Mach cone, 480
Mach number, 58, 221, 477
  limit for incompressible flow, 58
Mach wave, 525
Manometry, 37
Mass, 15
Mass flux, 89
Material derivative, 29
Maximum velocity
  in a channel, 284
  in a pipe, 278
McCuskey, S. W., 68
McDonald, A. T., 164
Mean free path, 6
Mechanical properties, 7
Mechtly, E. A., 68
Meyer, Th., 526, 570
Milne-Thompson, L. M., 399
Minor losses, 340

INDEX

Mixing length, 317
Model, 206
Mollo-Christenson, E. C., 305, 350
Moment of inertia, 40
Moment of momentum, 107, 176
Momentum flux correction factor, 105, 136
Momentum, linear, 94
  differential momentum equation, 171, 192
  integral momentum equation, 94
  referred to noninertial coordinates, 95
  table for cartesian, cylindrical, spherical coordinates, 196
  time-averaged, 311
Momentum thickness, 414
  from Blasius solution, 415
  for turbulent flow, 429
Moody, L. F., 338
Moody diagram, 339
Morkovin, M. V., 444

Navier, M., 186, 240
Navier-Stokes equations, 188
  in cylindrical coordinates, 197
  in dimensionless form, 208
  in spherical coordinates, 197
  in turbulent flow, 311
  in vector form, 189
Newton, I., 14, 94
Newtonian fluid, 5
Newton's second law, 14
No-slip condition, 194
Nondeformable control volume, 74
Non-dimensionalizing, process of, 207, 406
Nonhomogeneous flow, 169
Noninertial reference frame, 31, 95
Normal stress, 115
Normalization, 207
  in boundary layer, 406
Nozzle flow, 490
  choked flow, 493
  converging-diverging, 492
  De Laval, 494
  mass rate of flow, 494
  overexpanded, 535
  underexpanded, 535

Oblique shocks, 482, 518
  deflection angle in, 519
  strong, 521
  weak, 521
One-seventh power law, 430
  comparisons with data, 434
Order-of-magnitude analysis, 407

Oscillatory flow, 448
Outer region, 424
Overexpanded nozzle, 535
Owczarek, J. A., 204, 269, 350, 443, 536

Pathline, 27
Perfect-gas law, 8
Persistence of irrotationality, 390
Piezometer tube, 50
Pipe flow
  completely turbulent regime, 340
  critical zone, 338
  friction factor, 277
  intermediate zone, 338
  laminar flow, 276, 338
  shear distribution, 332
  turbulent flow, 334
Pitot tube, 50
  in supersonic flow, 503
Poiseuille flow
  in a channel, 283
  in a pipe, 276
Porous surface, 195
Potenital energy, 9, 113
Potential flow, 353, 360
  axisymmetric flows, 386
  cylinder, 377
  cylinder with circulation, 381
  doublet, 373
  irrotational vortex, 371
  half-body, 376
  source near a wall, 384
  sources and sinks, 369
  sphere, 387
  stagnation flow, 368
  streamlined body, 382
  uinform flow, 368
  vortex, 371
Potential function, 20, 361
  solution for, 363
Prandtl, L., 221, 246, 317, 400, 526, 570
Prandtl boundary-layer equations, 409
Prandtl-Meyer function, 526, 573
Prandtl number, 221
Pressure, 8
  average, 186
  gage, 38
  kinetic, 208
  stagnation, 50
  static, 50
  thermodynamic, 8, 201
  total, 50
Pressure distribution
  on cylinder, 461
  on sphere, 458
Pressure drag, 457

Pressure gradient, 276
  adverse, 402, 452
  effects of, 436
  favorable, 436
Principal planes, 185
Production, turbulence energy, 322
Properties, thermodynamic, 7
Prototype, 206
Pump horsepower, 130

Radius of curvature, 48, 356
Rankine, W. J. M., 529
Rankine-Hugoniot relationship, 529
Ratio of specific heats, 475
Rayleigh, J. W. S., 504
Rayleigh pitot-tube formula, 504
Rectangular coordinate system, 16
Reducible curve, 250
Reference frames
  inertial, 31
  noninertial, 31
  rotating, 32
Relaminarization, 248
Relative roughness, 338
Relative velocity, 77
Resonance, 447
Reynolds, O., 205, 241, 282, 311, 403, 452, 460
Reynolds number, 208
  in boundary layers, 402
  critical, 298, 403
  high-Reynolds-number flows, 218
Reynolds stresses, 311
Reynolds transport theorem, 76, 80
Richardson, L. F., 326
Riemann, G. F. B., 364
Robertson, J. M., 399
Rocket, air-water, 159
Rosenhead, L., 68, 194, 269
Roshko, A., 448
Rotating cylinders, 286
Rotational flow, 390
Rough pipes, 338
Roughness, 338
  in pipes, 338
  relative, 338
  root-mean-square, 338

Sabersky, R. H., 165, 204, 399
Scalar potential function, 20, 361
Schetzer, J. D., 473
Schlichting, H., 204, 305, 319, 350, 444
Schultz-Grunow, F., 426
Secondary flow, 358
Second law of thermodynamics, 9
Self-similar flow, 422
Separated flow, 249

Separation, 248, 450
  condition for, 452
  control, 465, 467
Separation point, 249, 402
  influence of geometry, 450
  influence of pressure gradient, 436
  effect of turbulence, 453
Separation streamline, 249, 402
Separation of variables, 290
Shaft work, in energy equation, 85
Shapiro, A. H., 256, 269, 399, 473, 510, 536
Shear stress, 115
  in a laminar boundary layer, 414
  in a pipe, 276
  on rotating cylinders, 287
Shepard, D. G., 236
Shock waves, 263, 480, 481
  detached, 522
  entropy increase across, 484
  oblique, 482
  pressure ratio across, 484
  strong oblique, 521
  temperature ratio across, 484
  weak oblique, 521
Similarity solution, 413
Similarity transformations, 410, 412
Similitude, 209
  compressible flow, 220
  confined flow, 208
  gravity-influenced flow, 215
  high-Reynolds-number flow, 218
  oscillatory flow, 217
  parameters, 214
  surface-tension effects, 216
  turbulent flow, 314
Sink in potential flow, 369
Skin friction coefficient, 414, 426
  power law, 430
Sources in potential flow, 369
Sound, speed of, 12, 58, 221, 479
  for perfect gas, 12
Sovran, G., 444
Specific heats, 8
  ratio of, 8
Specific property, 7
Speed of sound, 12, 58, 221, 479
  for perfect gas, 12
Sphere
  drag coefficient of, 458
  potential flow about, 387
Spherical coordinates, 196
Stability, 241, 295
  theory, 296
  neutral stability curve, 298
Stagnation flow, 369
Stagnation pressure, 50, 486

Stagnation streamline, 53
Stagnation temperature, 458
Stall in a diffuser, 356
Starting vortex, 457
State
　equation of, 8, 192
　thermodynamic, 7
Static pressure, 50
　from pitot tube, 503
Steady flow, 27, 30, 76, 294
Stewart, R. W., 307, 350
Stewart, W. E., 305
Stivers, L. S. Jr., 473
Stokes, G. G., 186, 238, 240
Stokes' flow, 238
Stokes' hypothesis, 186
Stokes' theorem, 19
Strain rate
　normal, 181
　tensor, 181
Streakline, 27
Streamline, 28
　acceleration along, 50
　normal to potential line, 366
　validity of Bernoulli equation along, 175, 358
Streamfunction, 362
　for a boundary layer, 410
　cylinder, 378
　cylinder with circulation, 381
　doublet, 373
　irrotational vortex, 371
　half-body, 376
　stagnation flow, 368
　uniform flow, 368
　used to determine flow rate, 364
Streamwise coordinates, 357
Stress
　components, 186
　constitutive equations, 186
　in cylindrical coordinates, 199
　normal, 115
　shear, 115
　symmetric stress tensor, 176
　tensor, 173
　vector, 115
Strouhal, V., 217, 450
Strouhal number, 217, 461
Sublayer, laminar, 423
Subsonic flow, 491
Substantial derivative, 30
Sudden contraction, 343
Supersonic flow
　through a converging-diverging nozzle, 493
　through a finite angle, 518, 524
　over a body, 522

Superposition, principle of, 293, 375
Surface tension, 13, 47
　contact angle, 48
Surface force, 95
Surroundings, 7
System, 7, 72

Taylor, B., 30
Taylor, G. I., 305, 317, 325
Taylor hypothesis, 325
Taylor series, 18
Temperature, 8
　stagnation, 486
Tennekes, H., 350
Thermodynamic pressure, 8, 201
Thermodynamic properties, 7
　extensive, 7
　intensive, 7
　specific, 7
Thermodynamic system, 7
Thermodynamics, 7
　first law of, 9
　second law of, 10
Throat, of nozzle, 494
Time-averaged velocity, 244, 309
Time-rate-of-change term, 76, 81
Torque, 116
Total derivative, 30
Total pressure, 50
　from pitot tube, 50
Total temperature, 486
Transformation of coordinates, 18
Transformation of volume integral, 19, 168
Transition, 242
　in boundary layers, 402, 415
　in a channel, 282
　length, 247
　in a pipe, 274
　to a secondary flow, 242
　to turbulence, 242
Trefethen, L. M., 68
Turbine blades, 103
　horsepower, 125
Turbulence
　characteristics of, 307
　free stream, 191
Turbulent boundary layer, 422
　buffer zone, 423
　calculations, 427
　outer region, 424
　power-law form, 428
　self-similar, 422
　turbulent zone, 423
　velocity-defect law, 424
　viscous wall layer, 423
Turbulent bursts, 403

Turbulent flow, 190
  homogeneous, 327
  isotropic, 327
  kinetic energy in, 321
  mixing, 402
  phenomenological theories, 316
  in pipes, 331
  time-averaged quantities, 244
  velocity components in, 308
    average portion of, 308
    in boundary layers, 422
    fluctuating portion of, 308
Turbulent fluctuations, 244
Turbulent stresses, 311
Turbulent zone
  in a boundary layer, 423
Turning angle
  in supersonic flow, 526
Tuve, G. L., 68
Types of flow
  boundary layer, 245, 401
  compressible, 220, 263
  creeping, 238
  free-stream, 191
  fully developed, 271
  high-Reynolds-number, 218
  inviscid, 173
  isentropic nozzle flow, 491
  laminar, 190, 239
  oscillatory, 217
  secondary flow, 358
  separated, 249
  turbulent, 190

Underexpanded nozzle, 535
Uniform flow, 85, 368
Units, 14, 540
U-tube manometer, 38

Vallentine, H. R., 399
Van Dyke, M., 350, 444
Van Wylen, G. J., 68
Variables
  independent, 26
  dependent, 26
Vapor pressure, 13
Vector algebra, 16
Vector calculus, 17
Vector functions, 543
Vector identities, 544
Vector product
  dot product, 16
  cross product, 17
Velocity-defect law, 424
Velocity distribution
  in an annulus, 281

  in a boundary layer
    laminar, 406
    turbulent, 423
  logarithmic, 334
  between parallel plates, 283
  in pipes, 276, 334
  power-law profile, 337
Vena contracta, 343, 394
Vincenti, W. G., 350, 444
Viscosity, 12, 185, 539
  eddy, 316
  effect of temperature on, 539
  kinematic, 13
  second viscosity coefficient, 185
Viscosity pump, 304
Viscous drag, 457
Viscous superlayer, 435
Viscous wall layer, 423
von Doenhoft, A. E., 473
von Kármán, T., 417, 444, 447
von Kármán integral equation, 417
  its use in calculations, 418
von Kármán vortex street, 447
Vortex, 371
  strength of, 372
  stretching, 254, 307
  tip vortex, 468
Vortex lines, 184, 391
Vortex shedding, 447
Vortex tube, 184, 250
Vorticity, 183, 250
  diffusion, 253
  tensor, 181
  related to viscous effects, 367
Vorticity transport equation, 252
  for a plane flow, 254
  for an inviscid flow, 390

Wake, 450
Water, properties of, 539
Wave angle, of oblique shocks, 519
Weber, W. E., 216
Weber number, 216
Wedge angle, 520
Wehausen, J. V., 350, 444
Wetting angle, 48
Whitaker, S., 204
Work, 9
Work rate, 112
  flow work, 117
  inertial, 119
  shaft, 118
  shear, 119

Yih, C. S., 305
Yuan, Y. S., 68, 204, 305, 350, 521, 536

Zone of silence, 480